Plant Taxonomy

Plant Taxonomy

Edited by **Austin Balfour**

SYRAWOOD
PUBLISHING HOUSE
New York

Published by Syrawood Publishing House,
750 Third Avenue, 9th Floor,
New York, NY 10017, USA
www.syrawoodpublishinghouse.com

Plant Taxonomy
Edited by Austin Balfour

International Standard Book Number: 978-1-68286-270-4 (Hardback)

Printed in China.

Contents

Preface

The world is advancing at a fast pace like never before. Therefore, the need is to keep up with the latest developments. This book was an idea that came to fruition when the specialists in the area realized the need to coordinate together and document essential themes in the subject. That's when I was requested to be the editor. Editing this book has been an honour as it brings together diverse authors researching on different streams of the field. The book collates essential materials contributed by veterans in the area which can be utilized by students and researchers alike.

The study of botany and associated branches of biology, involves an advanced knowledge of plant taxonomy. The classification of plants into various categories aids not only the botanists, but also conservationists, agriculture scientists, genetic engineers, etc. This book is specifically designed to familiarize the reader with the key techniques of plant identification, classification and description, and concepts like alpha taxonomy, beta taxonomy, etc. While understanding the long-term perspectives of the topics covered within the book, it also makes an effort in highlighting their impacts as modern tools for the progress of botany as a science.

Each chapter is a sole-standing publication that reflects each author´s interpretation. Thus, the book displays a multi-facetted picture of our current understanding of applications and diverse aspects of the field. I would like to thank the contributors of this book and my family for their endless support.

Editor

1

Ampelomyces hyperparasites – occurrence and effect on the development of ascomata of *Erysiphales* species under conditions of anthropopressure

Ewa Sucharzewska*, Maria Dynowska, Dariusz Kubiak, Elżbieta Ejdys, Anna Biedunkiewicz
Department of Mycology, University Warmia and Mazury in Olsztyn, Oczapowskiego 1a, 10-719 Olsztyn-Kortowo, Poland

Abstract

Fungi of the genus *Ampelomyces* are the major antagonists of *Erysiphales* fungi being a significant group of phytopathogens. The hyperparasites attack various developmental stages of powdery mildews. As a result the infested ascomata do not reach the stage of maturity, do not form appendages nor ascospores, which is linked with a reduction of the source of primary infections. Studies conducted so far have mainly been devoted to *Ampelomyces* fungi application in the biological control of powdery mildews on crops, whilst a few only have been focused on the ecology of these mycoparasites on *Erysiphales* fungi infecting plants, especially in the urban environment. The present study addresses the natural occurrence and effect of *Ampelomyces* fungi on the development of ascomata of powdery mildews species. The material was collected in 2005–2009 in several cities of the northeastern Poland. First time ever *Ampelomyces* spores are reported in mature ascomata of *Erysiphales* with fully developed appendages. This phenomenon has been observed in the case of two species, viz.: *Erysiphe flexuosa* on *Aesculus* spp. and *E. vanbruntiana* var. *sambuci-racemosae* on *Sambucus racemosa* and is presumably linked with improved implementation, propagation and probably better survival during winter months.

Keywords: *Ampelomyces*, hyperparasites, *Erysiphales*, powdery mildews, ascomata, urban environment

Introduction

Species of the genus *Ampelomyces* Ces. ex Schltdl. are wide-spread hyperparasites found on more than 65 *Erysiphales* species [1]. Their hyphae penetrate the vegetative mycelium as well as conidiophores and conidia of the host fungus, thus forming inside them the spore-forming structures, i.e. pycnidia [2]. Hyperparasites may also colonize ascogonia and antheridia as well as young, immature chasmothecia of powdery mildews, transforming them into their own reproductive structures. Infested ascomata do not reach the stage of maturity, do not form appendages nor ascospores, which is linked with the reduction of the source of primary infections [3–7].

The capability of hyperparasites to infest different *Erysiphales* species, to suppress their sporulation as well as to inhibit the formation and development of chasmothecia has contributed to the advance of research addressing mainly the aspect of the application of *Ampelomyces* fungi in the biological control of powdery mildews [1,8–11]. From the economic point of view, *Erysiphales* species constitute a significant group of phytopathogens of multiple fruit plants and crops. Of high interest to researchers world-wide is also the problem of the genetic diversity of fungi of the genus *Ampelomyces*, which is still little known and requires in-depth examinations [12–16].

World and Polish literature provides a lot of information about presence *Ampelomyces* fungi on different *Erysiphales* species in natural condition ([11,17–24]. Relatively few investigations were focused on the ecology of these hyperparasites and their natural occurrence on various *Erysiphales* species under urban condition [25–27]. Therefore, the Department of Mycology of University of Warmia and Mazury in Olsztyn began studies on the occurrence of fungi from the genus *Ampelomyces* and their impact on the development *Erysiphales* in urban environment [28–30]. Data is also lacking on the effect of *Ampelomyces* fungi on the development of fruiting bodies of different species of powdery mildews infecting non-farming plants [29]. Works addressing this issue are devoted almost exclusively to a dangerous pathogen of grapevine – *Erysiphe necator* (*Uncinula necator*) – [3,5,6,31].

An analysis of the colonization of *Erysiphales* ascomata by *Ampelomyces* as well as the effect of mycoparasites on their development seems to be crucial in understanding the role of such antagonists in the natural dynamics of a powdery mildew population in its natural environment.

The objective of the study was to evaluate the occurrence of fungi of the genus *Ampelomyces* and their effect on the development of ascomata of species of powdery mildews in the urban environment.

* Corresponding author. Email: ewko@uwm.edu.pl

Material and methods

The mycological research was conducted in 2005–2009 in several cities of the northeastern Poland: Olsztyn, Frombork, Giżycko, Mrągowo and Kętrzyn. The experimental material were leaves of plants infested with *Erysiphales* fungi. The material were collected from different places of cities (parks, gardens, street and community lawns) always in the second half of the vegetative season (August-October), when mature fruiting bodies were visible on white mycelium. Ten leaves with symptoms of infestation by powdery mildews, randomly collected from each host plant, served as one sample.

Laboratory analyses were conducted to verify the presence of *Ampelomyces* fungi on the mycelium of the *Erysiphales* species examined. Stereomicroscope (Olympus SZX9) and light microscope (Olympus BX51) used to observe the mycelium, chasmothecia of *Erysiphales* species and hyperparasites from the genus *Ampelomyces*. Camera XC 50 and analytical software Analysis were used for graphic processing (documentation, measurement).

Tab. 1 The list of *Erysiphales* species and host plants.

No.	*Erysiphales* species	Host plants	Samples with *Erysiphales* species	Samples with *Ampelomyces* species
1	*Erysiphe adunca* (Wallr.) Fr.	*Salix cinerea* L.	3*	0
2	*Erysiphe alphitoides* (Griffon et Maubl.) Braun & Takam.	*Quercus robur* L.	26 (2*)	1
3	*Erysiphe artemisiae* Grev.	*Artemisia vulgaris* L.	10	8 (80%)
4	*Erysiphe asperifoliorum* Grev.	*Symphytum officinale* L.	4	2
5	*Erysiphe berberidis* (DC.)	*Mahonia aquifolium* (Pursh) Nutt.	2*	2
		Berberis vulgaris L.	31*	23 (69%)
		Berberis thunbergii DC.	3*	0
6	*Erysiphe cichoracearum* DC.	*Sonchus arvensis* L.	6	5
		Helianthus tuberosus L.	14	8 (65%)
7	*Erysiphe convolvuli* DC.	*Convolvulus arvensis* L.	9	2
8	*Erysiphe depressa* (Wallr.) Link	*Arctium lappa* L.	2	1
9	*Erysiphe flexuosa* Peck. Braun & Takam.	*Aesculus* spp.	27 (24*)	22 (81%)
10	*Erysiphe galeopsidis* DC.	*Lamium album* L.	13	2
11	*Erysiphe hedwigii* (Lév.) Braun & Takam.	*Viburnum lantana* L.	3	0
12	*Erysiphe heraclei* DC.	*Pastinaca sativa* L. s.str.	3	0
		Heracleum sphondylium L.. s.str.	6	6 (83%)
13	*Erysiphe hypophylla* (Nevod.) Braun & Cunningt.	*Quercus robur*	23 (20*)	7
14	*Erysiphe magnicellulata* Braun	*Phlox* sp.	6	5
15	*Erysiphe palczewskii* Braun & Takam.	*Caragana arborescens* Lam.	25 (20*)	16 (68%)
16	*Erysiphe polygoni* DC.	*Polygonum aviculare* L.	17	6
17	*Erysiphe syringae* Schwein.	*Syringa vulgaris* L.	24 (19*)	4
18	*Erysiphe tortilis* (Wallr.) Link	*Cornus sanguinea* L.	2	0
19	*Erysiphe trifolii* Grev.	*Caragana arborescens*	1	0
		Astragalus glycyphyllos L.	2	0
		Trifolium campestre Schreb.	21	7
20	*Erysiphe urticae* (Waller.) Blumer	*Urtica dioica* L.	5	1
21	*Erysiphe vanbruntiana* var. *sambuci-racemosae* (Braun) Braun & Takam	*Sambucus racemosa* L.	31 (26*)	28 (90%)
22	*Phyllactinia fraxini* (DC.) Homma	*Fraxinus excelsior* L.	12 (9*)	6
23	*Golovinomyces sordidus* (L. Junell) Helluta	*Plantago major* L. s.str.	14	12 (85%)
24	*Phyllactinia guttata* (Wallr.) Lév.	*Betula pendula* Roth.	23 (21*)	3
		Corylus avellana L.	7*	0
		Fagus sylvatica L.	2*	0
25	*Podosphaera fusca* (Fr.) Braun & Shishkoff	*Taraxacum* spp.	21	16 (76%)
26	*Sawadaea bicornis* (Wallr.) Homma	*Acer campestre* L.	2*	0
		Acer ginnala Maxim.	2 (1*)	0
		Acer negundo L.	5*	0
27	*Sawadaea tulasnei* (Fuckel) Homma	*Acer ginnala*	11 (7*)	1
		Acer platanoides L.	29 (26*)	15
	The total number of samples		447	209 (47%)

The number of samples with powdery mildew infected and uninfected by fungi from the genus *Ampelomyces* in the years 2005–2009. Nos. 3, 5, 6, 9, 12, 15, 21, 23, 25 denote species of *Erysiphales* with the particiaption of *Ampelomyces* fungi >50% in samples. * Data from 2005–2006 years [30].

Each infested leaf in a sample was observed under a stereomicroscope. Once hyperparasites were detected (symptoms of the presence of *Ampelomyces* fungi: change of mycelium color into brown, the presence of pycnidia on the mycelium), microscopic analyses were conducted for 10 young (yellow and orange pigmentation, lack of appendages) and 10 mature ascomata (brown pigmentation, branched appendages) collected at random from the surface of mycelium non-infested and infested by the hyperparasites. These analyses involved determinations of: (*i*) morphological features of mature ascomata (diameter, wall peridium pigmentation, the number and length of appendages); (*ii*) the development of asci with ascospores of mature ascomata (the presence of asci with ascospores; asci without ascospores; a lack of asci and ascospores); (*iii*) the presence of spores of *Ampelomyces* fungi in young and mature ascomata.

The fungi were identified by means of Braun's key [32]. The taxonomy and nomenclature were adopted after Braun and Takamatsu [33] and Braun et al. [34]. The host plants were identified using keys by Rutkowski [35]. The nomenclature was adopted after Mirek et al. [36].

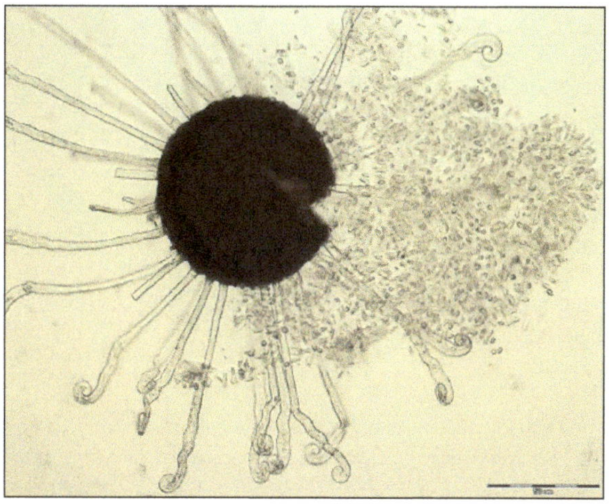

Fig. 1 *Erysiphe flexuosa* – spores of *Ampelomyces* fungi in mature ascomata. Scale bar: 50 µm.

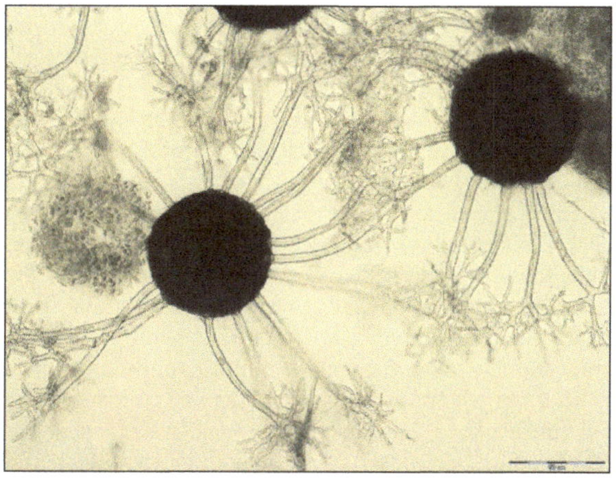

Fig. 2 *E. vanbruntiana* var. *sambuci-racemosae* – spores of *Ampelomyces* fungi in mature ascomata. Scale bar: 50 µm.

Results

In total 447 samples were collected, including 27 of *Erysiphales* species. The presence of *Ampelomyces* fungi was detected on 23 (47%) analyzed species of powdery mildews. In the case of 10 species, the contribution of hyperparasites in the samples exceeded 50%. The greatest percentage of hyperparasites was observed on four host species: *Erysiphe vanbruntiana* var. *sambuci-racemosae* (90% samples with *Ampelomyces* fungi), *Erysiphe sordida* (85%), *Erysiphe flexuosa* (81%), and *Erysiphe heraclei* (83%; Tab. 1).

In the case of all powdery mildews investigated, the analysis of the presence of *Ampelomyces* species in ascomata demonstrated the hyperparasites to infest young fruiting bodies without developed appendages. In 25 analyzed species of *Erysiphales*, fungi from the genus *Ampelomyces* infested only young ascomata. In turn, in two species: *Erysiphe flexuosa* and *E. vanbruntiana* var. *sambuci-racemosae* the hyperparasites were observed in young and mature chasmothecia with fully developed appendages (Fig. 1, Fig. 2). In the case of *Erysiphe flexuosa*, *Ampelomyces* fungi infested mature ascomata in 50% of the samples, whereas in the case of *E. vanbruntiana* var. *sambuci-racemosae* in 79%.

In both analyzed species some differences were noted in the morphological characteristics of mature ascomata collected from the surface of mycelium non-infested and infested by fungi from the genus *Ampelomyces*. The diameter of mature ascomata is smaller, wall peridium pigmentation is paler and the number of appendages is smaller in the infested ascomata than the un-infested ones (Tab. 2).

On both surfaces, analyses demonstrated the prevalence of chasmothecia with fully-developed appendages (>80%). In contrast, differences were observed in the development of asci and ascospores in the ascomata. In the case of *E. flexuosa*, analyses showed 90% of ascomata with developed asci and ascospores on the surface of mycelium free of hyperparasites and 72% of these on the surface of mycelium colonized by *Ampelomyces* fungi. The latter mycelium was also characterized by 7% contribution of mature ascomata infested by the hyperparasites and 16% contribution of empty ascomata, without any asci nor ascospores (Tab. 3). In the case of *E. vanbruntiana* var. *sambuci-racemosae*, the difference in the development of asci and ascospores turned out to be alike, i.e. 87% of the ascocarps containing asci and ascospores on the surface of mycelium without *Ampelomyces* fungi and as little as 60% of these on the surface of infested mycelium. In the latter, analyses showed additionally a very high percentage (38%) of *Ampelomyces* species in mature ascomata as well as 1% of empty chasmothecia (Tab. 3), which results in nearly 40% reduction of ascomata.

Discussion

Fungi of the genus *Ampelomyces* are considered to be major antagonists of *Erysiphales* species [11]. The range of their occurrence corresponds with the range of powdery mildews, which points to the ecological specialization of these hyperparasites. This has been confirmed in our study, where fungi of the genus *Ampelomyces* were noted on all investigated species, and with a high contribution in the samples (>50%).

Tab. 2 Morphological description of mature ascomata non-infested and infested by *Ampelomyces* fungi.

Morphological Features	EV		EF	
	A(−)	A(+)	A(−)	A(+)
Diameter (μm)	119–139 (−160)	(75–) 109 –119 (−139)	116–143	(68–) 99–116
Wall	brown	fawn brown	brown	fawn brown
Pigmentation	dark brown	brown	dark brown	brown
The number of appendages	18–27	(11–) 15–22	31–44	(7–) 13–31
Length of branched appendages (times as long as ascomata diameter)	1–1.5	1–1.5	0.5–1.5	0.5–1.5

A(−) – mature ascomata un-infested by *Ampelomyces* fungi; A(+) – mature ascomata infested by *Ampelomyces* fungi; EF – *Erysiphe flexuosa*; EV – *Erysiphe vanbruntina* var. *sambuci-racemosae.*

While infesting the host fungus, these hyperparasites colonize the mycelium, the conidial stages and young, yet immature fruiting bodies [3,5,17,25,32]. Ample investigations [5,31,37] have shown that *Ampelomyces* species are capable of infecting chasmothecia only at the early stage of their development, before the formation of appendages and development of walls. Young fruiting bodies are, obviously, susceptible to infections, whereas the mature ones are seemingly resistant. However, results of these investigations are only in part consistent with our findings. In the powdery mildews analyzed, the hyperparasites were very often detected only on the perfect stage and infested mainly young ascomata in various developmental stages – from white to orange ones with undeveloped appendages or at the early stages of development. They were transforming them into their own reproductive structures, yielding abundant production of conidial spores. In contrast, in the case of two species, *Erysiphe flexuosa* on *Aesculus* spp. and *E. vanbruntiana* var. *sambuci-racemosae* occurring on *Sambucus racemosa*, the presence of *Ampelomyces* fungi was additionally detected in mature ascomata with fully developed appendages, with an especially high percentage reaching 38% in *E. vanbruntiana* var. *sambuci-racemosae*. This is the first documented case of this phenomenon. The worldwide literature does not provide any information on the colonization of fully mature fruiting bodies by the hyperparasites. The presence of *Ampelomyces* fungi on apparently mature fruiting bodies of *Erysiphe necator* (*Uncinula necator*) was mentioned by Füzi [11]. This author described infested fruiting bodies as apparently mature due to the observed residual appendages and speculated on the terminated process of peridium formation before the infection with the hyperparasites.

Tab. 3 Percentage participation of mature ascomata with different degree of development of asci and ascospores from surface of mycelium non-infested and infested by *Ampelomyces* fungi.

The presences of asci, ascospores and spores of *Ampelomyces* fungi	EV		EF	
	S(−)	S(+)	S(−)	S(+)
ac(+) as(+) A(−)	87	60	90	72
ac(+) as(−) A(−)	11	1	7	5
ac(−) as(−) A(−)	2	1	3	16
ac(−) as(−) A(+)	0	38	0	7
	100%	100%	100%	100%

A – spores of *Ampelomyces* fungi; ac – asci; as – ascospores; S(−) – surface of mycelium non-infested by *Ampelomyces* fungi; S(+) – surface of mycelium infested by *Ampelomyces* fungi.

Infested fruiting bodies may be recognized by their fawn, milk-brown color and smaller sizes [5,6]. Results of our study confirm these changes of the colonized mature ascomata both in *E. flexuosa* as well as in *E. vanbruntiana* var. *sambuci-racemosae*.

However the infested ascomata possessed well developed appendages (>80%), whose development was very often synchronized with the development of asci and ascospores. In our study, apart from conidia of the hyperparasites, no other residues of those structures were observed in the ascomata. Therefore, it remains unclear when the chasmothecia are being infected. Being endoparasites, *Ampelomyces* fungi colonize structures of the host-fungi from the inside. The infection of fruiting bodies proceeds through thin hyaline hyphae of the mycelium that penetrate into the male sexual cell and ascogonium as well as into young chasmothecia, thus inhibiting their development [5]. The presence of fully developed appendages probably indicates the infection of ascomata at the final stage of their development.

The colonization of mature fruiting bodies is, probably, a phenomenon pointing to the adaptation of *Ampelomyces* species to a better overwintering and dissemination. Falk et al. [4] report that the survivability of *Ampelomyces* fungi in young fruiting bodies reaches as little as 1–3%. Thus, it may, be speculated on a low effectiveness of these hyperparasites in their natural environments, which has been proved by annual epiphytoses induced by powdery mildews. The low survivability of *Ampelomyces* fungi is certainly due to the immature wall of ascomata which at this developmental stage are not capable of surviving low temperatures. This has been confirmed by Mmbaga [38], who reported that fruiting bodies with just partly developed walls are not capable of surviving the winter. Therefore, the stage of colonizing mature chasmothecia – that completed the process of peridium formation and possess well developed appendages – by the hyperparasites seems to be understandable. Probably it provides suitable conditions for the hyperparasites for a better overwintering and transmission in their environments. The improvement of the adaptation strategy of *Ampelomyces* fungi may lead to changes in the dynamics of an *Erysiphales* population.

Thus, further researches of the occurrence of these hyperparasites of powdery mildews in their natural environment are advisable. They will enable recording the direction of those changes in a complicated system, viz. plant - parasite - hyperparasite.

References

1. Kiss L, Russell JC, Szentiványi O, Xu X, Jeffries P. Biology and biocontrol potential of *Ampelomyces mycoparasites*, natural antagonists of powdery mildew fungi. Biocontrol Sci Techn. 2004;14(7):635–651.
2. Sutton BC. The *Coelomycetes* Fungi Imperfecti with pycnidia acervuli and stromata. Kew: Commonwealth Mycological Institute; 1980.
3. Angeli D, Pellegrini E, Pertot I. Occurrence of *Erysiphe necator* chasmothecia and their natural parasitism by *Ampelomyces quisqualis*. Phytopathology. 2009;99(6):704–710.
4. Falk SP, Gadoury DM, Pearson RC. Survival of *Ampelomyces quisqualis* in parasited cleistothecia of *Uncinula necator*. Phytopathology. 1992;82:1119.
5. Falk SP. Parasitism of *Uncinula necator* cleistothecia by the mycoparasite *Ampelomyces quisqualis*. Phytopathology. 1995;85(7):794.
6. Füzi I. Natural parasitism of *Uncinula necator* cleistothecia by *Ampelomyces hyperparasites* in the south-western vineyards of Hungary. Acta Phytopathol Entomol Hung. 2003;38(1):53–60.
7. Gadoury DM, Pearson RC, Seem RC. Reduction of the incidence and severity of grape powdery mildew by *Ampelomyces quisqualis*. Phytopathology. 1991;81:122.
8. Elad Y, Kirshner B, Yehuda N, Sztejnberg A. Management of powdery mildew and gray mold of cucumber by *Trichoderma harzianum* T39 and *Ampelomyces quisqualis* AQ10. BioControl. 1998;43(2):241–251.
9. Falk SP, Gadoury DM. Partial control of grape powdery mildew by the mycoparasite *Ampelomyces quisqualis*. Plant Dis. 1995;79(5):483–490.
10. Sztejnberg A, Galper S, Lisker N. Conditions for pycnidial production and spore formation by *Ampelomyces quisqualis*. Ann Phytopathol Soc Japan. 1998;64:458–461.
11. Kiss L. A review of fungal antagonists of powdery mildews and their potential as biocontrol agents. Pest Manag Sci. 2003;59(4):475–483.
12. Szentiványi O, Kiss L, Russell JC, Kovács GM, Varga K, Jankovics T, et al. *Ampelomyces mycoparasites* from apple powdery mildew identified as a distinct group based on single-stranded conformation polymorphism analysis of the rDNA ITS region. Mycol Res. 2005;109:429–438.
13. Kiss L. Genetic diversity in Ampelomyces isolates, hyperparasites of powdery mildew fungi, inferred from RFLP analysis of the rDNA ITS region. Mycol Res. 1997;101(9):1073–1080.
14. Kiss L, Nakasone KK. Ribosomal DNA internal transcribed spacer sequences do not support the species status of *Ampelomyces quisqualis*, a hyperparasite of powdery mildew fungi. Curr Genet. 1998;33(5):362–367.
15. Kiss L, Vajna L. New approaches in the study of the genus *Ampelomyces quisqualis*, hyperparasite of powdery mildew fungi. In: Manka M, editor. Environmental biotic factors in integrated plant disease control: proceedings of the 3rd Conference FFPP. Poznań: Polish Phytopathological Society; 1995. p. 301–304.
16. Liang CH, Yang J, Kovács MG, Szentiványi O, Li B, Xu XM, et al. Genetic diversity of *Ampelomyces mycoparasites* isolated from different powdery mildew species in China inferred from analyses of rDNA ITS sequences. Fungal Divers. 2007;24:225–240.
17. Kiss L. Graminicolous powdery mildew fungi as new natural hosts of *Ampelomyces mycoparasites*. Can J Bot. 1997;75(4):680–683.
18. Kiss L. Natural occurrence of ampelomyces intracellular mycoparasites in mycelia of powdery mildew fungi. New Phytol. 1998;140(4):709–714.
19. Ranković B. Hyperparasites of the genus shape *Ampelomyces* on powdery mildew fungi in Serbia. Mycopathologia. 1997;139(3):157–164.
20. Schmidt A, Scholler M. Studies in *Erysiphales* anamorphs (II): Colutea arborescens, a new host for *Erysiphe palczewskii*. Feddes Repert. 2002;113(1–2):107–111.
21. Czerniawska B, Madej T, Adamska I, Błaszkowski J, Tadych M. Erysiphales and their hyperparasite, *Ampelomyces quisqualis*, of the Drawsko Landscape Park, Poland. Acta Mycol. 2000;35(1):79–84.
22. Adamska I, Czerniawska B. Nadpasożyty mączniaków prawdziwych roślin uprawnych i dziko rosnących Pojezierza Bobolickiego. Prog Plant Prot/Post Ochr Roślin. 2010;50(2):869–873.
23. Mułenko W, Majewski T, Ruszkiewicz-Michalska A. A preliminary checklist of micromycetes in Poland. Kraków: W. Szafer Institute of Botany, Polish Academy of Sciences; 2008.
24. Ruszkiewicz-Michalska M, Michalski M. Phytopathogenic micromycetes in central Poland. I. Peronosporales and Erysiphales. Acta Mycol. 2005;40:223–250.
25. Madej T, Antoszczyszyn S. *Ampelomyces quisqualis* Ces. (*Cicinnobolus cesatii* de Bary) w Szczecinie. Biul Inst Ochr Rośl. 1965;30:65–73.
26. Ruszkiewicz-Michalska M. *Cladosporium epichloës*, a rare European fungus, with notes on other fungicolous species. Polish Bot J. 2010;55(2):359–371.
27. Czerniawska B, Adamska I, Kocanowski J. Choroby grzybowe żywopłotowych drzew i krzewów liściastych Szczecina. Prog Plant Prot/Post Ochr Roślin. 2011;51(3):1142–1145.
28. Sucharzewska E, Dynowska M. Life strategies of *Erysiphe palczewskii* in the conditions of diversified anthropopressure. Acta Mycol. 2005;40(1):103–112.
29. Sucharzewska E, Dynowska M, Ejdys E, Biedunkiewicz A, Kubiak D. Hyperparasites of *Erysiphales* fungi in the urban environment. Polish J Natural Sci. In press.

30. Sucharzewska E, Dynowska M, Kempa AB. Occurrence of *Ampelomyces* – hyperparasites of powdery mildews (*Erysiphales*) infesting trees and bushes in the municipal environment. Acta Soc Bot Pol. 2011;80(2):169–174.
31. Gadoury DM. Initiation, development, dispersal, and survival of cleistothecia of *Uncinula necator* in New York vineyards. Phytopathology. 1988;78(11):1413–1421.
32. Braun U. A monograph of the *Erysiphales* (powdery mildews). Beih Nov Hedw. 1987;89:1–700.
33. Braun U, Takamatsu S. Phylogeny of *Erysiphe*, *Microsphaera*, *Uncinula* (*Erysipheae*) and *Cystotheca*, *Podosphaera*, *Sphaerotheca* (*Cystotheceae*) inferred from rDNA ITS sequences – some taxonomic consequences. Schlechtendalia. 2000;4:1–33.
34. Braun U, Cunnington JH, Brielmaier-Liebetanz U, Ale-Agha N, Heluta V. Miscellaneous notes on some powdery mildew fungi. Schlechtendalia. 2003;10:91–95.
35. Rutkowski L. Klucz do oznaczania roślin naczyniowych Polski niżowej. 2nd ed. Warszawa: Polish Scientific Publishers PWN; 2004.
36. Mirek Z, Piękoś-Mirkowa H, Zając A, Zając M. Flowering plants and pteridophytes of Poland – a checklist. Kraków: W. Szafer Institute of Botany, Polish Academy of Sciences; 2002. (Biodiversity of Poland; vol 1).
37. Falk SP, Gadoury DM. Reduced production of cleistothecia by *Uncinula necator* due to the mycoparasite *Ampelomyces quisqualis*. Phytopathology. 1994;84:543.
38. Mmbaga MT. Winter survival and source of primary inoculum of powdery mildew of dogwood in Tennessee. Plant Dis. 2000;84(5):574–579.

2

Traditionally used wild edible greens in the Aegean Region of Turkey

Yunus Dogan*

Buca Faculty of Education, Dokuz Eylul University, 35150 Buca-Izmir, Turkey

Abstract

Turkey has the largest coastal area in the Mediterranean, possesses an extraordinarily rich flora, and a great traditional knowledge. This diversity of plants naturally affects the traditional use of plants and is reflected in the rich Turkish cuisine. Consequently, the Mediterranean Diet (whose typical components are wild greens) constitutes one of the important elements of Turkish cuisine. For this reason, the aim of this study was to determine the consumption of wild edible green plants for the Aegean Region of Turkey and to establish the similarities to or differences from consumption in other regions and other cuisine in the Mediterranean Basin. This study compiles and evaluates the ethnobotanical data currently available. There were 111 taxa that were identified as wild edible greens in the study area belonging to 26 different families. Asteraceae (21 taxa) were the most commonly consumed as food. It was followed by Boraginaceae with 19 taxa, Apiaceae with 15 taxa and Lamiaceae with 7 taxa, respectively. *Rumex* and *Erodium* were the most represented genera with 4 species. *Tamus communis* and *Asparagus acutifolius*, Mediterranean elements and distributed in all of the Mediterranean Basin, are among the most widely consumed wild plants in the area. Wild edible plants are consumed in a variety of ways. The most common type of consumption (79 taxa) was in salads. The fact that the majority of the plants used in the area are consumed in salads shows the close relationship between the local diet and the concept of the Mediterranean Diet. As a result, very promisingly, there is a renewed or increasing interest in consuming wild food plants as part of this diet.

Keywords: wild edible greens, ethnobotany, traditional knowledge, consumption, Mediterranean Diet, Aegean, Turkey

Introduction

The type of food consumed reflects the identity of people and the use of wild food plants is a very good indicator of traditional local knowledge. The main characteristics of the Mediterranean Diet has been described as mainly composed of vegetables, salads, fruits and spices, whole-grain cereals, pasta, nuts, legumes, olive oil, seafood, a moderate consumption of wine with meals, poultry consumed in low-to-moderate amounts, and a relatively low consumption of red meat [1–3]. By supplying micronutrients, vitamins, and minerals, wild plants play an important role in complementing staple foods [4]. Many of these wild foods are common and productive, as well as being highly nutritious, palatable, and easily harvested [5]. They are also important as a source of income for poor communities, as well as being a source of food and considered a healthy diet by many. Diets consumed by Mediterranean peoples have been a subject of interest since antiquity, with more recent investigations focused on their evident health benefits.

However, the Mediterranean Basin contains many different cultures, religious beliefs, and ecologic backgrounds. This has resulted in many diets, sharing a multitude of elements. But at the same time, these diets differ in distinct local or regional traditions. Food and medicinal uses have always been the most relevant reasons for plant management in the folk traditions of the Mediterranean Region, and they still continue to be so, even in cultures that are progressively losing their close relationship with nature [2].

The studies show that Mediterranean Diet contains important dietary components that may contribute to a lower risk of cancer. Several micro-components with antioxidant potential are underlying factors that define the health benefits of this type of diet. The relative longevity of the Mediterranean people is attributed to the type of diet they consume [6]. The Mediterranean Diet is an example of sustainable food production. It is a dietary pattern that can combine taste and health, environmental protection, biodiversity protection, and consumption of local and seasonal products [7].

Turkey, which has the largest coastal area in the Mediterranean, possesses an extraordinarily rich flora and a great traditional knowledge. Due to its climate and geographical position, Turkey, with its 10000 taxa, is one of the richest countries in Europe and the Middle East, in terms of flora [8]. A total of ⅓ of the plant taxa of Turkey is endemic [9]. Anatolian people have been profiting from this variety and using plants as food and medicine since the Paleolithic [10]. This

* Email: yunus.dogan@deu.edu.tr

diversity of plants naturally affects the traditional use of plants and is reflected in the rich Turkish cousine. Consequently, the Mediterranean Diet constitutes one of the important elements of Turkish cuisine.

Despite all the great qualities of the Mediterranean Diet, in many Mediterranean regions, including Turkey, the traditions regarding the diet are under the risk of disappearing, and therefore there is an urgent need to study them [11]. In this modern era where the fast-food culture is spreading worldwide and the transfer of ethnobotanical knowledge from old generations is diminishing, this study will be helpful for both recording the information regarding the consumption of wild edible plants as a food source and providing people with new ideas. For this reason, the aim of present study was to review the consumption of wild edible green plants and establish the similarities to or differences from consumption in other regions and other cuisine in the Mediterranean Basin.

Several previous studies have described the traditional knowledge about the plants in the research area and the uses and different needs for them such as everyday household items, traditional crafts, etc. [8,12–17].

General characteristics of the study area

CLIMATE. The Mediterranean climate is dominant in the study area, particularly along the shores of the Mediterranean, Aegean, and Marmara regions. In the Mediterranean climate, summers are hot and dry; winters are mild and rainy. The maximum precipitation falls in the winter; the minimum falls in summer. A long and arid period predominates from mid-May till mid-October. In April and May, clear and still weather conditions alternate with rain showers and cold. The difference between summer and winter precipitation is very high. The average annual precipitation is 600 to 1000 mm. The annual average temperature is 18 to 20°C. Mountains in the Aegean Region lie perpendicular to the shore and this allows the Mediterranean climate to reach inland [18–20].

VEGETATION. The study area includes the Aegean Region of Turkey (Fig. 1). It is geographically situated in the western part of Turkey. Markgraf [21] has divided the Mediterranean Region into two subdivisions: the west and the east provinces. According to this division, the study area lies in the East Mediterranean Province. In the Mediterranean Basin there is huge topographic, climatic, and geographic variability giving rise to an astounding array of species and habitat diversity [22].

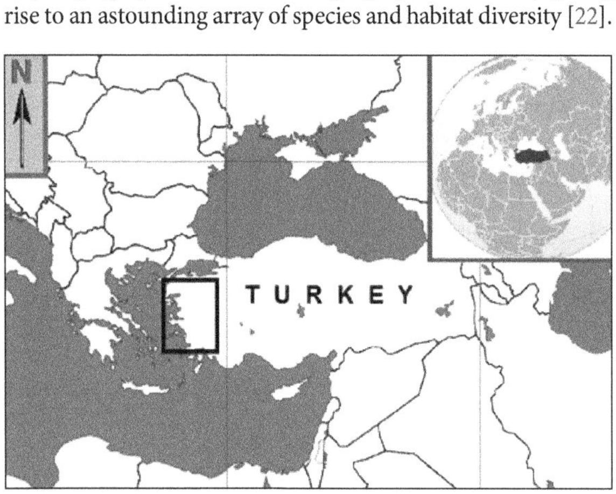

Fig. 1 The map of the study area.

The vegetation in general consists of sclerophyllous shrubs and trees. The typical Mediterranean species are *Styrax officinalis* L., *Rhamnus palaestinus* Boiss., *Rhamnus punctatus* Boiss., *Arbutus andrachne* L., *Cistus creticus* L., *Satureja thymbra* L., *Salvia triloba* L., and *Sarcopoterium spinosum* (L.) Spach. Some of the representative species found in both the east and west provinces are *Quercus ilex* L., *Quercus coccifera* L., and *Pistacia terebinthus* L. The East Mediterranean Province exhibits a strong influence from the Irano-Turanian region. The typical dominant species in the Mediterranean Region is *Quercus ilex*. This species is replaced in the East Mediterranean Province by *Quercus coccifera* and among the tree species are *Ceratonia siliqua* L., *Olea europaea* L. subsp. *oleaster* (Hoffmanns. & Link) Negodi, and *Pistacia lentiscus* L. The maquis was formed as a result of the destruction of tree formations. They develop on the slopes which lack soil cover deep enough to support forest vegetation and consist of the sclerophyllous species of shrubs, together with some climbers like *Lonicera*, *Clematis*, and *Asparagus*.

The borders of the Mediterranean Region in Turkey were drawn by Zohary [23]. They correspond with the southern border of the distribution of *Pinus brutia* Ten. The latter is an indicator of the true Mediterranean conditions. In some places this border is drawn by the *Quercus ithaburensis* Decne. subsp. *macrolepis* (Kotschy) Hedge, but never by plants like *Quercus cerris*, *Pinus nigra* Arn., and *Juniperus* spp. Black pine is a typical indicator of the oro- and sub-Mediterranean vegetation. This vegetation is phytogeographically closer to euxinian vegetation than the Mediterranean one. The region close to Central Anatolia serves as a transition zone between the Mediterranean, the semi-desert belt, and the steppes. The climate and vegetation in general show the same characteristics as those of the Mediterranean [24].

Material and methods

This study compiles and evaluates the ethnobotanical data currently available. More than seventy literature sources were reviewed [25–101] including six from the study area [25–30]. Plant names are given according to Davis [31]. A literature review was carried out firstly for the Aegean Region, which is target study area, and then for all of Turkey and the wild edible green plants used in these areas were identified. According to Turner et al. [5], edible wild plants are root vegetables, edible greens, fleshy fruits, grains, seeds, and nuts. This study focused on the wild edible greens. It includes leaves, stems, and shoots. In addition, based on field observations, the consumption of some plant taxa is reported here for the first time. The plants' scientific names, families, local and English names, parts consumed, type of consumption and related sources are also given in Tab. 1.

Results and discussion

Altogether, 111 taxa of wild edible greens have been used in the study area (Tab. 1). *Rumex* and *Erodium* are the most represented genera, with four species. *Amaranthus*, *Tragopogon*, and *Papaver* were represented by three species. *Allium*, *Anchusa*, *Chenopodium*, *Chrysanthemum*, *Eryngium*, *Lamium*, *Lepidium*, *Scorzonera*, *Silene*, *Sinapis*, *Sisymbrium*, *Smyrnium*, *Sonchus*, and *Urtica* were represented by two species (Fig. 2).

Tab. 1 Traditionally used wild edible greens of the Aegean Region of Turkey.

Scientific name	Family	Local name	English name	Edible green part	Recipe	Use in the study area	Citation out of study area
Allium ampeloprasum L.	Liliaceae	Karaköremen	wild leek	leaf	Consumed as pastry. Fried with egg. Cooked with olive oil, prepared as vegetable pie.	[25]	[46,48,49,68,80]
Allium subhirsutum L.	Liliaceae	Köremen	hairy garlic	leaf	Consumed as pastry. Fried with egg. Cooked with olive oil, prepared as vegetable pie.	[25]	[81]
Amaranthus chlorostachys Willd.	Amaranthaceae	sirken, tilkikuyruğu	slim amaranth	shoot and leaf	Chopped and added to pastries and salads, or fried in a skillet and consumed with eggs.	[29]	
Amaranthus retroflexus L.	Amaranthaceae	horoz ibiği	common amaranth	shoot and leaf	Prepared as stew, meal, pie. Consumed raw in salads, or added to salads after boiling.	[27]	[2,11,34,37,38,41,52,61, 71,77]
Amaranthus viridis L.	Amaranthaceae	deli sirken, tilkikuyruğu	slender amaranth	leaf	Olive oil and lemon is added after boiling and consumed as salad.	[25]	
Anagallis arvensis L.	Primulaceae	fare kulağı	scarlet pimpernel	leaf	Consumed as meal on its own.	[27]	[32,75]
Anchusa azurea Mill.	Boraginaceae	Sığırdili	wild bugloss	shoot and leaf	Consumed roasted.	[27,29]	[2,4,46,75]
Anchusa undulata L. subsp. *hybrida* (Ten.) Cout.	Boraginaceae	ballıkotu, kocaot	alkanet	leaf	Added to pastries.	[25,29]	[4,37]
Anethum graveolens L.	Apiaceae	dereotu	dill	aboveground	Added to salads, cooked with many vegetables.	[27]	[82]
Arum maculatum L.	Araceae	yılan yastığı	lords-and-ladies	leaf	Leaves are stuffed, consumed as meal on its own meal, or roasted.	[27]	[33,34]
Asparagus acutifolius L.	Liliaceae	kuşkonmaz	spiny asparagus	young shoot	Roasted like black bryony. Fried either alone or with black bryony and egg is added.	[25,26,30]	[2,4,11,42,44,48,51,66]
Bellis perennis L.	Asteraceae	koyungözü	daisy	leaf	Cooked with rice. In salads with other vegetables.	author's observations	[38,61,71,75,83]
Berula erecta (Huds.) Coville	Apiaceae	sukazayagı	cut leaf water parsnip	aboveground	Added to salads or eaten alone.	[25]	
Beta spp.	Chenopodiaceae	yabani pancar	beet	whole plant	Eaten raw, fried with egg.	[25]	[4,11,34,38,39,45,46,48, 49,52–54,56,58,61,63, 66,68,71,80,84,85]
Brassica nigra (L.) Koch	Brassicaceae	karahardal	black mustard	young shoot and leaf	Consumed as salad.	[25]	[2,49,59,71]
Capsella bursa-pastoris (L.) Medik	Brassicaceae	çobançantası	shepherd's purse	aboveground	Consumed in meal, roast, soup, or salads.	[25,27,29,30]	[2,4,32–34,37,38,41,45, 47,50,60,61,63,66,83]
Cardamine uliginosa M. Bieb.	Brassicaceae	sukerdimesi	bitter cress	shoot and leaf	Consumed raw in salads.	[25]	
Cardaria draba (L.) Desv.	Brassicaceae	kediotu	whitetop	shoot and leaf	Consumed raw in salads.	[27]	[34,71]
Centaurea solstitialis L.	Asteraceae	çakırdikeni	yellow star thistle	leaf	Boiled and seasoned with olive oil and lemon, and eaten as refreshing food.	[25]	
Chenopodium album L.	Chenopodiaceae	sirken	lamb's quarters	aboveground	Boiled, prepared as salad with lemon and olive oil. Fried with onion.	[26]	[2,4,11,36–39,41,45,49, 53,56,61,71,82,86–89]
Chenopodium polyspermum L.	Chenopodiaceae	sirken	all-seed	aboveground	Roasted. Fried with onion.	[27]	
Chondrilla juncea L.	Asteraceae	çengel sakızı	rush skeleton weed	aboveground	Added to pastries and salads.	[29]	[37,45,60,80]

Tab. 1 *(continued)*

Scientific name	Family	Local name	English name	Edible green part	Recipe	Use in the study area	Citation out of study area
Chrysanthemum coronarium L.	Asteraceae	dallama	crown daisy	leaf	Eaten raw.	[41]	[4,46]
Chrysanthemum segetum L.	Asteraceae	sarı papatya	corn marigold	leaf	Eaten raw.	[25]	
Cichorium intybus L.	Asteraceae	yabani hindiba	chicory	aboveground	Leaves can be eaten raw in salads, used in yogurt salad, cooked (sautéed with onion) seasoned with oil and lemon, consumed in soup, in stew or fried with garlic and oil.	[26,27]	[2,4,11,34,37,41,42,45–49, 52,53,61,63,66,68,71,72, 80,82,84,85,90,92]
Cirsium arvense (L.) Scop.	Asteraceae	köygöçerten	creeping thistle	leaf	Consumed as meal or stuffed.	[27]	[4,41,53,91]
Coriandrum sativum L.	Apiaceae	kişniş	coriander	aboveground	Added to salad as fresh plant. Fried with onion. Chopped into various soups.	author's observations	[4,38]
Crithmum maritimum L.	Apiaceae	denizmarulu, kayakoruğu	rock samphire	leaf	Leaves used freshly in salads.	[25]	[61]
Daucus carota L.	Apiaceae	daraklık, yabani havuç	wild carrot	shoot and leaf	Consumed as meal, and stew.	[25,27,30]	[2,4,50,53,54,61,71]
Echinophora tenuifolia L. subsp. *sibthorpiana* (Guss.) Tutin	Apiaceae	çördük, tarhana otu	tarhana herb	aboveground	Added to tarhana.	[25,27,29]	[78,79]
Echium italicum L.	Boraginaceae	ayıkulağı	Italian viper's bugloss	leaf	Consumed roasted.	[27]	[32,54,91]
Eremurus spectabilis Bieb.*	Liliaceae	çiriş	foxtail lily	leaf	Cooked with onion.	author's observations	[42]
Erodium cicutarium (L.) L'Hérit	Geraniaceae	iğnelik	redstem filaree	aboveground	Consumed fried. Added to stuffing of "sheet iron pastry" (sac böreği).	[25,27,28]	[32,37]
Erodium hoefftianum C. A. Meyer	Geraniaceae	dönbaba, iğnelik	stork's bill	aboveground	Consumed fried. Added to stuffing of sheet iron pastry.	[25]	
Erodium malacoides (L.) L'Herit.	Geraniaceae	dönbaba, iğnelik	Mediterranean stork's bill	aboveground	Consumed fried. Added to stuffing of sheet iron pastry.	[25]	
Erodium moschatum (L.) L'Herit.	Geraniaceae	dönbaba, iğnelik	musky stork's bill	aboveground	Consumed fried. Added to stuffing of sheet iron pastry.	[25,81]	[83]
Eruca sativa Mill.	Brassicaceae	yabani roka	rocket salad	leaf	Consumed raw in salads.	[29]	[2,45,47,50,52,59]
Eryngium campestre L.	Apiaceae	Boğa dikeni	field eryngo	shoot and leaf	Prepared as pie or stew. Eaten raw in salads with oil, lemon, and vinegar.	[25,27]	[4,34,40,45,47–49,52,83]
Eryngium creticum Lam.	Apiaceae	göz dikeni	Crete an eryngo	leaf	Eaten raw in salads. Leaves cooked with onion and olive oil.	[25]	[50,72]
Falcaria vulgaris Bernh.	Apiaceae	kazayağı	sickle weed	shoot and leaf	Cooked as meal.	[25]	[38,39,93]
Foeniculum vulgare Mill.	Apiaceae	arapçaçı	fennel	aboveground	Cooked with lamb meat. Fried with onion and egg. Added to pilaf. Chopped and added to salads. Cooked with olive oil and prepared as vegetable pie.	[25–27, 29,30]	[2,4,11,42,44–50,53,55, 56,58,60–63,65,66,68, 72,81–83,85,92]
Fumaria officinalis L.	Fumariaceae	şahtere	fumitory	leaf	Consumed as salad and pie.	[27]	[32,75]
Galega officinalis L.	Fabaceae	keçisakalı	goat's rue	aboveground	Consumed as stew.	[27]	[32,34]
Geranium dissectum L.	Geraniaceae	turnagagası	cutleaf geranium	leaf	Consumed as meal and pie.	[27]	

Tab. 1 *(continued)*

Scientific name	Family	Local name	English name	Edible green part	Recipe	Use in the study area	Citation out of study area
Gundelia tournefortii L.	Asteraceae	kengerotu	Tournefort's gundelia	young stem and leaf	Consumed as roasted, salad, and pickle.	[27]	[33,39,50,59,72,94–96]
Lactuca serriola L.	Asteraceae	acı marul, eşek marulu	prickly lettuce	leaf	Consumed as salad.	[27,29,30]	[4,11,37,45,48,49,53–55, 58,61,63,66,80,90]
Lamium amplexicaule L.	Lamiaceae	ballıbaba	henbit deadnettle	aboveground	Consumed as meal, added to pastry.	[27]	
Lamium moschatum Mill.	Lamiaceae	lünlün otu	musk deadnettle	aboveground	Consumed as meal, added to pastry.	[25]	
Lavatera cretica L	Malvaceae	develik, gömeç	cornish mallow	aboveground	Added to pastry.	[25]	
Lepidium sativum L.	Brassicaceae	tere	cress	leaf	Eaten fresh, added to salad.	[25]	[50]
Lepidium spinosum Ard.	Brassicaceae	kerdime	pepper cress	leaf	Eaten fresh, added to salad.	[25]	
Limonium sinuatum (L.) Mill.	Plumbaginaceae	sahil karanfili	wavyleaf sea lavender	young shoot and leaf	Put in olive oil and eaten raw as salad.	[25]	
Malva sylvestris L.	Malvaceae	ebegümeci	common mallow	aboveground	Cooked with minced meat or olive oil. Prepared as soup or put in pastry, fried with other herbs.	[25–28,30]	[2,4,32,34,37–42,45,47,49, 50,53,54,59,61,63,66,71, 72,78,83,89-91,97,98]
Mentha aquatica L.	Lamiaceae	Su nanesi	water mint	aboveground	Consumed as pie, soup, roasted, and salad.	[27]	[49,54]
Mercurialis annua L.	Euphorbiaceae	yer fesleğeni	annual mercury	leaf	Consumed as soup, and meal.	[27]	[34,52]
Nasturtium officinale R. Br.	Brassicaceae	su teresi	watercress	aboveground	Fried with onion. Eaten as salad.	[27,29,30]	[2,34,38,40,45,47,49,59, 61,66,72,83,88,89,91]
Oenanthe pimpinelloides L.	Apiaceae	kazayağı	corky-fruited water-dropwort	stem and leaf	Consumed as salad, and meal.	[25,30]	[82]
Onopordum tauricum Willd.	Asteraceae	eşek dikeni	Taurian thistle	stem	Consumed as salad.	[27]	
Opopanax hispidus (Friv.) Griseb.	Apiaceae	sarıot	opopanax	leaf	Boiled and then cooked with egg	[25]	[40]
Origanum onites L.	Lamiaceae	Izmir kekiği	Turkish oregano	leaf	Leaves are freshly eaten in breakfast, added to various salads.	author's observations	
Ornithogalum narbonense L.	Liliaceae	akbaldır	star of Bethlehem	aboveground	Consumed as meal.	[27]	
Oxalis pes-caprae L.	Oxalidaceae	ekşi yonca	Bermuda buttercup	young stem and leaf	Consumed mixed with green salad.	[25]	[2,4,54,71]
Papaver dubium L.	Papaveraceae	meşkük haşhaşı	long-head poppy	young shoot	Consumed mixed with green salad. Used as pastry stuffing. Cooked with olive oil, and prepared as vegetable pie. Consumed as meal, roasted, and soup.	[25]	[33,39]
Papaver guerlekense Stapf.	Papaveraceae	gelincik		young shoot	Consumed mixed with green salad. Used as pastry stuffing. Cooked with olive oil, and prepared as vegetable pie. Consumed as meal, roasted, and soup.	[25]	
Papaver rhoeas L.	Papaveraceae	gelincik	corn poppy	young shoot	Consumed mixed with green salad. Used as pastry stuffing. Cooked with olive oil, and prepared as vegetable pie. Consumed as meal, roasted, and soup.	[25–27,30]	[2,4,11,37,41,45–49,52–54, 56,58,59,61,63,66,68,80–83, 87,89,90]

Tab. 1 *(continued)*

Scientific name	Family	Local name	English name	Edible green part	Recipe	Use in the study area	Citation out of study area
Paracaryum aucheri (A. DC.) Boiss.	Boraginaceae	elköpürten	paracaryum	aboveground	Added to pastries and salads.	[29]	
Pistacia terebinthus L.	Anacardiaceae	menengic, çitlenbik	terebinth	young shoot	Fresh shoots are eaten alone. Added to pickle. Fried with onion and egg.	[29]	[2,59]
Plantago lanceolata L.	Plantaginaceae	sinirliot	ribwort plantain	leaf	Consumed as stuffed, pie, and salad.	[27]	[2,34,53,54,61,66,67,71,75]
Polygonum cognatum Meisn.*	Polygonaceae	madımak	knot-grass	aboveground	After chopping, fried with egg or made into soup with wheat grains and yogurt. Prepared as pastry.	[27,28]	[32,33,36–41]
Portulaca oleracea L.	Portulacaceae	semiz otu	purslane	aboveground	Consumed freshly as salad or with yogurt and garlic. Cooked with onion. Cooked as soup with other vegetables.	[25,27,28]	[2,4,32–34,37–39,45,47–50, 52,55,56,58–65,73,78, 80–82,87]
Ranunculus ficaria L.	Ranunculaceae	yağlıot,katırnalı	lesser celandine	leaf	Prepared as salad after boiling.	[25]	[47,53,61,75,83]
Raphanus raphanistrum L.	Brassicaceae	turpotu	wild radish	fresh shoot and leaf	Prepared as salad with olive oil, garlic, and lemon after boiling. Consumed as meal, roasted, stew, stuffed and soup.	[25,27,30]	[2,4,34,40,47,49,53,61,71, 82,83,88,90,92]
Rapistrum rugosum (L.) All.	Brassicaceae	küçük yabani turp	turnip weed	aboveground	Added to salads and pastries, boiled and seasoned with olive oil, garlic, and lemon.	[29]	[4,47,49]
Rheum ribes L.*	Polygonaceae	ışgın	rhubarb	fresh shoot and young petiole	Skin is removed and consumed raw or salad is made with olive oil and lemon. Fried with oil and egg is added.	author's observations	[33,36,39,40,42]
Rosmarinus officinalis L.	Lamiaceae	kuşdili	rosemary	leaf	Added to salads as fresh or to meat dishes or other dishes as spice after dried.	[27]	[2,4,45,48,55,58,59,87,92]
Rumex acetosella L.	Polygonaceae	kuzukulağı	sheep sorrel	young shoot and leaf	Due to sour taste, commonly consumed as salad or raw. Prepared as salad with yogurt. Consumed as stuffed, pie, soup, meal, and roasted.	[25,27,30]	[4,33,34,37,38,47,54–56, 61–63,66,89,91]
Rumex crispus L.	Polygonaceae	kıvırcık labada	curled dock	leaf	Consumed as stuffed, pie, soup, meal, salad, and roasted.	[27,28]	[4,34,37–41,48,49,53–56, 61,63,71,74,75,85,90]
Rumex obtusifolius L.	Polygonaceae	yabani labada	broad-leaf dock	leaf	Consumed as stuffed, pie, soup, meal, salad, and roasted.	[27]	[4,55,62]
Rumex patientia L.	Polygonaceae	labada, evelik	patience dock	leaf	Fried with onion and egg. Added to pastry. Consumed as stuffed, pie, soup, meal, salad, and roasted.	[27,30]	[34,50,71,82]
Salicornia europaea L.	Amaranthaceae	deniz börülcesi	glasswort	young shoot	Boiled and prepared as salad with olive oil, lemon and garlic.	[25]	[34,82]
Sanguisorba minor Scop.	Ranunculaceae	çayır düğmesi	salad burnet	young shoot and leaf	Eaten in salads.	[25]	[61,75]
Satureja thymbra L.	Lamiaceae	taş kekiği	thyme-leaved savory	leaf	Its leaves are eaten raw in the breakfast. Added to salads. It is a spice plant.	author's observations	
Scandix pecten-veneris L.	Apiaceae	kişkiş	shepherd's-needle	young shoot and leaf	Used in salads. Prepared as vegetable pie.	[25]	[4,63,81]
Scolymus hispanicus L.	Asteraceae	şevket-i bostan	blessed thistle	bark of the root and young basal leaf stalk	Cooked with lamb meat.	[27]	[2,4,42,45–49,59–61, 63,65,68,70,80,82]
Scorzonera cana (C. A. Meyer) Hoffm.	Asteraceae	yakıotu, iskorçina	dog's scorzonera	leaf	Young leaves sprinkled with salt are eaten. Consumed as meal and in salads.	[25,28]	[37–39]

Tab. 1 *(continued)*

Scientific name	Family	Local name	English name	Edible green part	Recipe	Use in the study area	Citation out of study area
Scorzonera elata Boiss.	Asteraceae	tekesakalı	viper's grass	leaf	Young leaves sprinkled with salt are eaten. Consumed as meal and in salads.	[25]	
Sedum rubens L.	Crassulaceae	kayaüzümü	red stonecrop	leaf	Consumed raw in salads.	[25]	
Senecio vulgaris L.	Asteraceae	kanaryaotu	groundsel	aboveground	Prepared as stew.	[27,30]	[37,75]
Silene italica (L.) Pers.	Caryophyllaceae	nakıl	Italian catchfly	aboveground	Prepared as stew. Consumed raw in salads. Cooked with olive oil, and prepared as vegetable pie.	[25]	
Silene vulgaris (Moench.) Garcke	Caryophyllaceae	gıvışgan otu	bladder campion	aboveground	Prepared as stew. Consumed raw in salads. Cooked with olive oil, and prepared as vegetable pie.	[25,27]	[4,32,34,44,46–49,53–55, 60–63,66,68,69,75,80,81, 85,87,90]
Sinapis alba L.	Brassicaceae	beyaz hardal	white mustard	aboveground	Roasted with egg, or boiled and made into salad.	[27,30]	[2,4,42,49,52,59,61,82,89]
Sinapis arvensis L.	Brassicaceae	hardal otu	wild mustard	aboveground	Roasted with egg, or boiled and made into salad. Also consumed as vegetable soup.	[25,27,29]	[11,34,37,40–42,45,49,50, 59–61,71,75]
Sisymbrium altissimum L.	Brassicaceae	bülbül otu	tumble mustard	aboveground	Added to pastries.	[29]	[37]
Sisymbrium officinale (L.) Scop.	Brassicaceae	akhardal	hedge mustard	shoot and leaf	Consumed raw in salads.	[27]	[4,34,45,49,53,61,86,90]
Smilax aspera L.	Liliaceae	gıcır, silcan	sarsaparilla	young shoot and leaf	After boiling, fried with onion and egg or cooked with rice. Consumed as pickled. Young shoots are eaten raw as snack.	[27]	[2,4,34,47,49,59,61]
Smyrnium connatum Boiss. et Kotschy	Apiaceae	sarıkırek	yellow Alexanders	leaf	Leaf stalks are eaten raw. Added to pickle.	[25]	
Smyrnium olusatrum L.	Apiaceae	yabani kereviz	Alexanders	leaf	Cooked. Leaf stalks are eaten raw. Added to pickle.	[25]	
Solanum nigrum L.	Solanaceae	istifno, köpek üzümü, giritotu	black nightshade	young shoot and leaf	Boiled with courgettes, prepared as salad with lemon, garlic, and olive oil.	[25,27,30]	[2,34]
Sonchus asper (L.) Hill subsp. *glaucescens* (Jordan) Ball	Asteraceae	sütlot, sütlen	spiny sow thistle	leaf	Fried with onion and egg. Boiled and prepared as salad. Used as pastry stuffing. Eaten in soup or raw in salad seasoned with oil and lemon.	[29,30]	[2,4,37,45,48,49,53,54,60,61, 65,80,83,90]
Sonchus oleraceus L.	Asteraceae	eşek marulu, yalancı marul	sow thistle	leaf	Fried with onion and egg. After boiled prepared as salad or consumed fresh as salad. Used as pastry stuffing. Cooked with olive oil.	[25,27]	[2,4,34,45–49,53,55,56,58,60, 61,63,71,75,80,81,85,92]
Stellaria media (L.) Vill.	Caryophyllaceae	kuşotu	chickweed	aboveground	Prepared freshly as salad. Used as pastry stuffing.	[25,30]	[34,45,47,63,71,75,83,88]
Tamus communis L.	Dioscoreaceae	sarmaşık	black bryony	young shoot	Fresh shoots fried with onion and egg is added.	[25,27]	[2,34,44–52,67,68]
Taraxacum spp.	Asteraceae	radika, karahindiba	dandelion	aboveground	Added to pastries and green salads. Boiled and prepared as salad with olive oil, garlic, and lemon. Consumed as stew, and meal.	[25,27–30]	[2,4,11,34,42,45–48,53–56, 58,60,63,66,68,75,92,99]
Thlaspi perfoliatum L.	Brassicaceae	çayır akça çiçeği	cotswold pennycress	aboveground	Consumed as meal, salad, and pie.	[27]	
Thymbra spicata L.	Lamiaceae	zahter	za'atar	leaf	Leaves are eaten in breakfast as fresh. Commonly consumed as salad. A spice plant.	author's observations	[35]
Tragopogon latifolius Boiss.	Asteraceae	yemlik	salsify	shoot and leaf	Added to pastries and salads.	[29]	[4,41]
Tragopogon longirostris Bisch. ex Sch. Bip.	Asteraceae	tekesakalı	long-beaked goat's beard	shoot and leaf	Added to pastries and salads.	[25]	[100]

Tab. 1 *(continued)*

Scientific name	Family	Local name	English name	Edible green part	Recipe	Use in the study area	Citation out of study area
Tragopogon longirostris Bisch. ex Sch. Bip.	Asteraceae	tekesakalı	long-beaked goat's beard	shoot and leaf	Added to pastries and salads.	[25]	[100]
Tragopogon porrifolius L.	Asteraceae	tekesakalı	salsify	shoot and leaf	Consumed as salad, or stewed with yogurt.	[27]	[47,48]
Tussilago farfara L.	Asteraceae	öksürükotu	coltsfoot	leaf	Consumed stuffed.	[27]	
Urtica dioica L.	Urticaceae	dalgan, ısırgan	stinging nettle	young shoot and leaf	Fried with onion and mallow. Added to pastry stuffing. Added to salad as fresh. Consumed as soup, and stew.	[25,28]	[2,4,33,34,37,38,40–42,45,47, 52–56,58,61–63,66,67,71,77, 78,83,87–89,91,92,97,99,101]
Urtica urens L.	Urticaceae	ısırgan	dwarf nettle	young shoot and leaf	Fried with onion and mallow. Added to pastry stuffing. Added to salad as fresh. Consumed as soup, and stew.	[27,30]	[4,32,34,39,49,54–56,63,101]

* Does not grow in the study area. Collected outside the study area, brought to the markets and sold.

There were difficulties in identifying the species reported as *Beta*, considering they are given at the genus level. While *Beta vulgaris* is given as wild in some studies, the use of *Beta maritima* is reported in others. However, *Beta vulgaris* has no wild localities in the Turkish flora. Another curiosity is related to *Taraxacum*. Although there is no *Taraxacum officinale* in Turkey [31], some authors, possibly due to difficulties in identification, have reported the consumption of this species as food. In addition, more than fifty *Taraxacum* taxa are naturally distributed in Turkey and most of them are consumed as food. Therefore, this genus is given as *Taraxacum* spp. in this study.

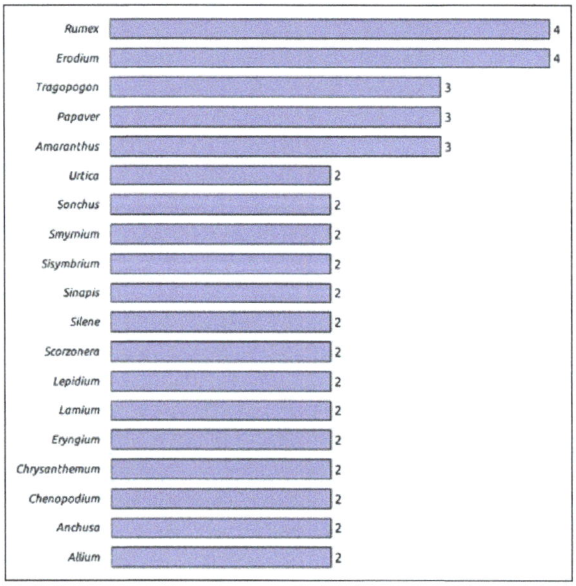

Fig. 2 Plant genera with the highest number of taxa in the study area.

Wild edible green vegetables identified in this study belong to 26 different families. Asteraceae are best represented (21 different taxa, 18% of the wild green vegetables). Other well represented families are Boraginaceae with 19 taxa, Apiaceae with 15 taxa, and Lamiaceae with 7 taxa, respectively (Fig. 3). The other 22 families have less representation, between one to six taxa each.

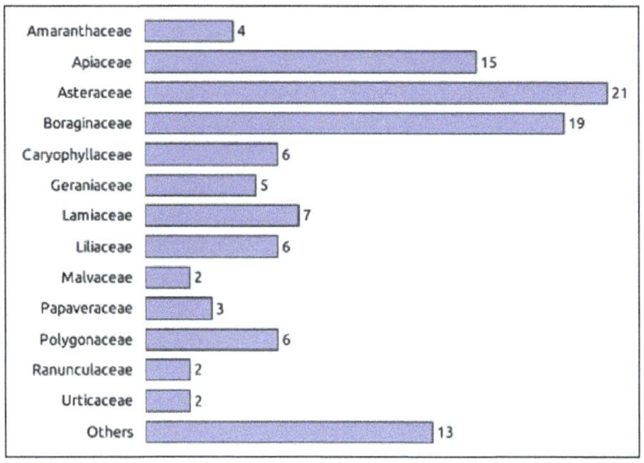

Fig. 3 Plant families with the highest number of taxa in the study area.

Wild edible plants are consumed in a variety of ways. The identified consumption types are divided into nine main categories: green salad (fresh plants and dressing), salad after boiling, salad with yogurt, raw/fresh, main dish (sautéed with oil and onion, also may have different vegatables, rice, meat etc. added), pastry/pie, soup, pickle, and stuffed. However, it must be noted that many species could be classified in more than one category. The most common type of consumption is as a salad (Fig. 4) with three sub-categories under this consumption type: green salad, salad after boiling, and salad with yogurt. Among them, only salad with yogurt is consumed without oil. All other salad types (76 taxa) are prepared with oil, as it is a typical Mediterranean region. Because olive oil is readily available in the region, it is preferred in all salads and most foods. In addition to olive oil, vinegar, lemon, and garlic are also added to salads, depending on the type of the salad. It has already been reported that salad and vegetable dishes prepared as traditional recipes in Turkish cuisine make use of many local wild plants [10,32]. Another category, following the consumption as salads, is the main dish category with 71 taxa. The pastry/pie category is represented by 42 taxa. The consumption habits of the Turkish people play an important role in the high number for this category. Due to the partiality for food made with dough, various tastes are obtained by mixing many plants in pastries. Twenty-two taxa are consumed as soup. Another category of consumption is stuffed foods. In this study area, stuffed rolls (so called "sarma" rolls) is a popular dish made especially with rice, but sometimes cracked wheat (bulgur) is preferred. Sarma rolls are primarily wrapped with the leaves of the grapevine (*Vitis vinifera* L.) and then those of cabbage (*Brassica oleracea* L.). However, nine other taxa are used to wrap sarma as well. Among these, *Arum maculatum*, which is a poisonous plant, so the leaves of this plant is consumed as sarma only after being boiled [27,33]. This type of consumption has not been reported outside the study area, while Redžić [34] reported consumption of ground parts of this plant as mush (from Bosnia-Herzegovina).

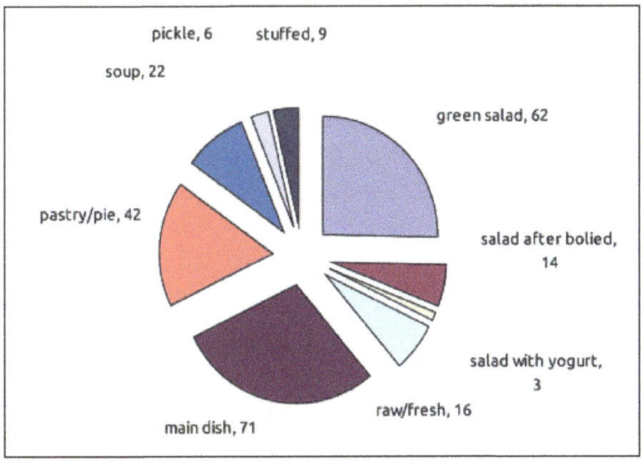

Fig. 4 Ways of consumption for the wild edible green plants in the study area.

Out of the listed wild greens some of them are both collected from the wild as well as cultivated. Plants such as *Daucus carota*, *Eruca sativa*, *Allium ampeloprasum*, *Beta vulgaris*, *Foeniculum vulgare*, *Coriandrum sativum*, *Lepidium sativum*, and *Portulaca oleracea* are mostly sold as crops. However, wild plants are either collected from nature and consumed or sold in the open markets, as well. Especially wild *Eruca sativa* and *Lepidium sativum* are commonly consumed as a salad or raw.

The consumption of plants in an area especially that of wild edible plants as food, is closely associated with the socio-cultural features of the population. Particularly, the increase in ethnic variety in the area is supportive of a variety of plant usage and occurrence of more recipes. In light of this, Izmir province in the study area, with a population of over 3 million and the 3rd most populated city in Turkey, in the past hosted immigration from many areas. This fact has naturally influenced the consumption of a variety of wild edible plants. For instance, Cretan immigrants have had a great impact in this regard. In this study, it was concluded that the information regarding the consumption and recipes of plants such as *Scolymus hispanicus*, *Solanum nigrum*, and *Taraxacum* spp. may have been transferred to the area by immigrants from Crete.

It must be noted that the species whose flowering tops or green parts are collected in small amounts for seasoning are not included in the study (e.g. *Thymbra spicata*, *Satureja thymbra*, *Origanum onites*, *Laurus nobilis* L., *Capparis ovata* Desf., and *Capparis spinosa* L.). *T. spicata* (za'atar), *O. onites* (oregano), and *S. thymbra* (thyme-leaved savory) are sold as wild in many open markets of Izmir. They are dried and used as a spice, while very rarely consumed fresh. Fresh leaves of *O. onites* are consumed raw even for breakfast in and around Izmir-Beydag. Similarly, *S. thymbra* and *T. spicata* are consumed raw or added to salads. "Za'atar salad", a type of salad mostly known and consumed in Eastern and Southeastern Anatolia, is not commonly consumed in the study area [35]. Green, unripe fruits of *Amygdalus communis* L. are consumed very commonly in the spring months in the study area. They were not included here due to the fact that technically they are not "green vegetables". Both cultivated and wild forms are collected and sold. However, the fruit dries and hardens in a very short time and its seeds are consumed as a snack or added to food or desserts for taste.

Rheum ribes and *Eremurus spectabilis* do not occur in the study area and *Polygonum cognatum* is very rare. However, they are common in Middle and Eastern Anatolia. *P. cognatum* is commonly consumed in various parts of Turkey [27,28,32,33,36–41], as well as *Rheum ribes* [33,36,39,40,42]. It was observed that these plants are only sold in Izmir (in the study area). There is only one reference to the consumption of *E. spectabilis* [42] in Turkey and no references from outside the country. This could be explained by the fact that these plants are Irano-Turanian elements and therefore are not distributed in the Mediterranean Basin. Market surveys showed, that sellers collect the plants in various parts of Middle and Eastern Anatolia and then transport and sell them in the markets of Izmir, but the indigenous population never consumes these plants. They are mostly consumed by immigrants from Middle and Eastern Anatolia or people who spent their childhood there. These people also bring their unique ways of using the plants and food culture. This is one of the factors that increase the variety of dishes in the region. Similarly, salad made with *Thymbra spicata*, an Irano-Turanian element, is commonly consumed in Eastern and Southeastern Anatolia. Also *Scorzonera elata* and *S. cana*, though are not commonly consumed in the study area, are sold in small amounts in some markets, beacuse they are eaten by immigrants from Middle and Eastern Anatolia. Although the two species are used in some parts of Turkey [25,28,37,39], there are no reports of their use in other parts of the Mediterreanean. Also, there is no report for the use of *Origanum onites* and *Satureja thymbra* outside of the study area. The usage of these plants is being reported for the first time based on our field observations.

Tamus communis and *Asparagus acutifolius*, Mediterranean elements distributed in all of the Mediterranean Basin, are among the most widely consumed wild plants in the area. An interesting fact is that one of the toxic plants identified in the study is black bryony (*T. communis*) [43,44]. The plant has a bitter taste because of its saponin content. Young shoots with leaves are the least toxic parts of the plant. They have been traditionally consumed after cooking, which destroys the toxic principles [44].

Although *T. communis* and *A. acutifolus* are among the most commonly consumed species throughout the study area, usually young shoots are fried with onion [25,27] and egg [25,26,42], they are not consumed in other, non-Mediterranean, parts of Turkey. On the other hand the two species are widely used in other parts of the Mediterranean, e.g. Spain, Italy, Slovenia, Croatia, Crete, Sardinia, and Sicily, where their tender leaves are eaten both cooked and raw, in soups, and omelets [4,11,44–66]. Pieroni et al. [45] and Menendez-Baceta et al. [67] have reported that *T. communis* is first boiled and then cooked with egg or in other ways. It is thought that the cooking method aims to eliminate the bitter taste coming from the saponin. However, despite its bitter taste, it is also consumed raw, but only in Spain [4,68]. The plant was probably used as food already in the times of Pedanius Dioscorides, physician, pharmacologist, and botanist (40–90 AD). According to many scholars Dioscorides mentiones this plant in his famous five-volume encyclopedia "De Materia Medica" [49] (though not everyone agrees, see [69]). This would show that some food habits have remained the same from antiquity. The plant formerly belonged to the genus *Dioscorea*, dedicated to Dioscorides. Similarly to *Tamus communis*, *Solanum nigrum* and *Arum maculatum* are also toxic. All the three plants are consumed after boiling. That way the bitter taste in the leaves is removed. Among the three, *S. nigrum* has the most toxic content. The greatest toxicity was detected in the unripe green berries. This species is important especially for Izmir in the study area. Among the locals, it is known with its Greek name "istifno". One of the most common salads made by Cretan immigrants is prepared with this plant. The salad is prepared by boiling fresh shoots and leaves and then by adding fresh courgette, olive oil, and garlic. The use of the plant declines towards the inner parts of the Aegean region, the reason being is that the Cretan immigrants did not settled there. Interestingly, in Cyprus, only the fruits of the plant are consumed and there is no record of consumption of the green parts [59]. The usage of *S. nigrum*'s fruits is common in many parts of the world. There are records of similar types of consumption for this study area [25,27]. Redžić [34] and Hadjichambis et al. [2] reported that only fresh shoots of the plant are consumed in Bosnia and Herzegovina, and Greece.

Another typical Crete food, served even in choice restaurants, is prepared with *Scolymus hispanicus* (the bark of the root and young basal leaf stalks of the plant with lamb). Theophrastus, known as the father of botany (371–287 BC), and Pliny the Elder (23–79 AD) mention the use of this plant as food in their works [49]. The consumption of blessed thistle, which is a Mediterranean element, is not very common especially around Izmir and outside of the places where the Cretan immigrants reside. It is a commonly consumed plant in the Mediterranean countries. In Cyprus, fresh stem and leaves are consumed raw, as well as fried or fried with eggs, after boiling [59]. In Spain, the inner parts of the prickly basal leaves are boiled and then usually lightly fried with a bit of garlic, cured ham, and scrambled eggs [70]. Neither this type of consumption nor the preparation with lamb is reported for the study area.

Daucus carota is a commonly consumed plant. Its roots are commonly used in salads and other foods. In various parts of the area, the leaves and roots of the wild form are used in the preparation of food. Shoots and leaves are consumed in the fall and winter months, while the roots are collected and consumed in the spring. Redžić [71] reported that, apart from the roots and leaves, its seeds are also used as food. Additionally, in some parts of Turkey, especially in the Mersin area, their roots are used in preparation of a sweet called "cezerye" and are widely consumed in all of Turkey.

An endemic plant, *Papaver guerlekense*, a species similar to *P. rhoeas*, is consumed mixed in green salads, used as pastry stuffing, as a main dish, roasted, in a pie, salad, and soup [25].

Pistacia terebinthus is not a common plant in the study area. It is generally collected from nature and its fresh shoots consumed raw. It is sold in the markets in small amounts. Due to its resin content, it is added to pickles for taste and also cooked with onion and egg. It is only in Cyprus that the plant is used in a similar way [59]. Although there are no literature data on the use of its seeds, a type of coffee is made of the, called "Kenger coffee", which is consumed widely instead of normal coffee in the Eastern Anatolian region (Y. D., unpublished data).

In this study, the Malvaceae family is represented by *Lavatera cretica* and *Malva sylvestris*. The former is not commonly consumed [25]. It is usually fried with other herbs and added to pastry. However, the other member of the family, *M. sylvestris*, is one of the most widely consumed plants in the study area, as well as in other parts of Turkey and all of the Mediterranean Basin. The plant is cooked with minced meat or olive oil, prepared as a soup, put in pastry, or fried with other herbs. In the Aegean part of Turkey it is usually eaten in pastry. Apart from this type of use, common mallow is consumed raw in salads in Northern Italy [66] and Lebanon [72]. This type of use is not encountered anywhere else.

In addition to *M. sylvestris*, which is widely distributed and commonly consumed in the study area, various *Malva* species such as *Malva moschata* L., *Malva cretica* Cav., *Malva nicaeensis* All., and *Malva parviflora* L. can be found, although not so commonly. Although these plants can also be consumed similar to *M. sylvestris*, there are no reports of their use. In other areas, *M. neglecta* Wallr. is consumed like *M. sylvestris* [32,33,36,39,41,73–77].

Plantago lanceolata is one of the plants that is consumed as stuffed, in a pie, and as a salad in the study area, but in other parts of Turkey its fresh leaves are consumed raw. Only Menendez-Baceta et al. [67] reports its fresh leaves being consumed raw in Spain. The closest to this type of use in this area is to chop it up and add to salads. Pieroni [53,54] reports its use as soup and Redžić [34] reports it as cooked vegetables.

Chrysanthemum coronarium (Asteraceae) is one of the rare plants whose leaves are consumed only raw [25]. Similar usage has only been reported for Palestine [50], whereas in Spain it is cooked [4]. There are no other reports for its use. However, consumption of *C. segetum* in the same manner as *C. coronarium* has been reported by Ertug [25] for Bodrum, which is included in the study area. *Rumex acetosella* is one of the plants that is consumed raw in the study area. It is eaten directly from the field and, in addition to its many other uses. Its green leaves are consumed widely in rural areas due to their sour taste. Plants that are sold in the markets are mostly consumed as a salad [27]. It is commonly consumed in Asian Minor and the whole of Europe.

Purslane (*Portulaca oleracea*) is a plant that is widely distributed and consumed as food from Central Asia to Central America, from Myanmar to Brazil and from the U. K. to Yemen [49]. It is a commonly consumed plant in the study area and in the whole Mediterranean Basin. Its aboveground parts are added to green salads [27], especially its fresh leaves which are mixed with yogurt and consumed as a salad widely in the study area. They are also commonly cooked. Both the wild and cultural forms exist and the wild forms are preferred in the rural areas, whereas the cultivated varieties are preferred in the urban areas. Purslane, a plant recommended for medicinal use by Dioscorides, has been used as a medicinal food all over the world since antiquity [49,56].

The leaves of *Sonchus oleraceus*, used in the study area and mentioned already by Dioscorides as food, are consumed as a salad and soup in many parts of the world [49]. In addition to this type of use, it is commonly consumed as a pastry and main dish, as well as boiled and prepared as a salad.

Salicornia europaea (glasswort), which is only consumed along the shores in the Western Anatolia, is a succulent and halophyte plant that grows in salt marshes and seashores. It is commonly consumed but only in areas close to the coast as a salad after boiling. Garlic and especially olive oil are added. It is also eaten in other parts of the Mediterranean [34].

One of the most interesting plants used in the study area is *Echinophora tenuifolia* subsp. *sibthorpiana*. It is a very well known and commonly used plant everywhere in Turkey. It is known as "tarhana herb" and is consumed as an addition to the tarhana soup, a soup unique to Anatolia. This taxon is added to tarhana when fresh to give a pleasant aroma and used as an aromatic food preserver in order to prevent pickles from frothing [25,27,29,78,79]. No other use of this plant has been reported outside of Anatolia.

In Western Anatolia, the most commonly eaten wild edible greens are *Asparagus acutifolius*, *Beta* spp., *Cichorium intybus*, *Foeniculum vulgare*, *Malva sylvestris*, *Papaver rhoeas*, *Portulaca oleracea*, *Silene vulgaris*, *Taraxacum* spp., and *Urtica dioica*. These taxa are also widely consumed in other parts of the Mediterranean Basin. Some species, such as *Tamus communis*, are very commonly consumed in the study area according to the author's unpublished observations, although this is not supported by earlier publications. An ethnobotanical research was carried out by Hadjichambis et al. [2] between 2003–2006 in seven countries from the Eastern Mediterranean (Cyprus, and Greece), the Western Mediterranean (Italy, and Spain), the Adriatic/Balkan Mediterranean (Albania) and the North-African Mediterranean (Egypt, and Morocco). It was a European Union-funded RUBIA Project and wild and semi-domesticated food plant consumption was investigated in these countries. As a result of the survey, 294 wild food plant taxa were documented. Asteraceae was the most represented family with 63 taxa. Of the identified plants 21% belonged to this family compared to 19% in this study. Similar results were obtained in another study carried out in a part of the study area [30]. The aforementioned wild edible plants are consumed as food in various ways; however, they include some invasive weeds and species under agricultural pest management. In many parts of Turkey, especially in Western Anatolia, some species such as *Anagalis arvensis*, *Cichorium intybus*, *Fumaria officinalis*, *Lactuca serriola*, *Lamium amplexicaule*, *Lamium moschatum*, *Malva sylvestris*, *Papaver rhoeas*, *Raphanus raphanistrum*, *Sinapis arvensis*, *Sisymbrium officinale*, *Stellaria media*, and *Tragopogon latifolius* are considered as invasive field pests and are controlled with agricultural chemicals. They cause a loss of yield in crops, especially in grains, and therefore various herbicides are applied, which decreases the availability of wild greens.

Conclusion

In total, 111 taxa that were identified as wild edible greens in the study area belong to 26 different families. Asteraceae is the most commonly consumed family and the results of this study show parallels with other Mediterranean areas.

The majority of the plants identified are species widely distributed in the Mediterranean Basin, due to the fact that the Aegean Region has a Mediterranean climate. In addition, the eastern borders of the region are adjacent to the Irano-Turanian Phytogeographical Region and therefore it contains species unique to that region. As a consequence, there exist species that are not consumed in other Mediterranean countries.

The fact that the majority of the identified plants in the area are consumed as a salad show the local diet has strong Mediterranean features. As a result, very promisingly, there is a renewed or increasing interest in consuming wild food plants as a part of the diet.

Acknowledgements

The author express his gratitude to Mesut Koyuncu (Tokat University, Tokat), Dr. Ilker Ugulu (Balikesir University, Balikesir), Dr. Anely Nedelcheva (Sofia University, Sofia), Robert J. Kelley (Ege University, Izmir), and Dr. Munir Ozturk (Ege University, Izmir) for their help, constructive comments, and support in the preparation of this manuscript. The author wishes to thank especially those who were interviewed for the key information used in this paper, for their time, and willingness to share their knowledge and experience.

References

1. Hu FB. The Mediterranean diet and mortality – olive oil and beyond. N Engl J Med. 2003;348(26):2595–2596.
2. Hadjichambis ACH, Paraskeva-Hadjichambi D, Della A, Giusti EM, De Pasquale C, Lenzarini C, et al. Wild and semi-domesticated food plant consumption in seven circum-Mediterranean areas. Int J Food Sci Nutr. 2008;59(5):383–414.
3. Petrillo PL. Biocultural diversity and the Mediterranean diet. In: Burlingame B, Dernini D, editors. Sustainable diets and biodiversity directions and solutions for policy, research and action. Rome: FAO; 2012. p. 224–229.
4. Tardío J, Pardo-De-Santayana M, Boyd R. Ethnobotanical review of wild edible plants in Spain. Bot J Linn Soc. 2006;152(1):27–71.
5. Turner NJ, Łuczaj ŁJ, Migliorini P, Pieroni A, Dreon AL, Sacchetti LE, et al. Edible and tended wild plants, traditional ecological knowledge and

agroecology. Crit Rev Plant Sci. 2011;30(1-2):198–225.

6. Pérez-López FR, Chedraui P, Haya J, Cuadros JL. Effects of the Mediterranean diet on longevity and age-related morbid conditions. Maturitas. 2009;64(2):67–79.
7. Iannetta M, Colucci F, Presenti O, Vitali F. Food and energy: a sustainable approach. In: Burlingame B, Dernini D, editors. Sustainable diets and biodiversity directions and solutions for policy, research and action. Rome: FAO; 2012. p. 274–279.
8. Dogan Y, Nedelcheva AM, Obratov-Petkovic D, Padure IM. Plants used in traditional handicrafts in several Balkan countries. Indian J Tradit Know. 2008;7:157–161.
9. Ugulu I, Baslar S, Yorek N, Dogan Y. The investigation and quantitative ethnobotanical evaluation of medicinal plants used around Izmir province, Turkey. J Med Plant Res. 2009;3(5):345–367.
10. Wetherilt H. Evaluation of Urtica species as potential sources of important nutrients. Dev Food Sci. 1992;29:15–26.
11. Pieroni A, Nebel S, Santoro RF, Heinrich M. Food for two seasons: culinary uses of non-cultivated local vegetables and mushrooms in a south Italian village. Int J Food Sci Nutr. 2005;56(4):245–272.
12. Dogan Y, Baslar S, Mert HH, Ay G. Plants used as natural dye sources in Turkey. Econ Bot. 2003;57(4):442–453.
13. Nedelcheva AM, Dogan Y, Guarrera P. Plants traditionally used to make brooms in several European countries. J Ethnobiol Ethnomed. 2007;3(1):20.
14. Dogan Y, Nedelcheva AM, Baslar S. Plant patterns of silk based needlework, a traditional handcraft in Turkey. Indian J Tradit Know. 20120;9(4):640–643.
15. Ugulu I, Baslar S. The determination and fidelity level of medicinal plants used to make traditional Turkish salves. J Altern Complement Med. 2010;16(3):313–322.
16. Nedelcheva AM, Dogan Y, Obratov-Petkovic D, Padure IM. The traditional use of plants for handicrafts in southeastern Europe. Hum Ecol. 2011;39(6):813–828.
17. Ugulu I. Fidelity level and knowledge of medicinal plants used to make therapeutic Turkish baths. Ethno Med. 2012;6(1):1–9.
18. Kocman A. Ege ovalarinin iklimi. Izmir: Ege Universitesi Edebiyat Fakultesi; 1993.
19. Ugulu I, Dogan Y, Kesercioglu T. The vascular plants of Buca Faculty of Education Campus (Izmir): contribution to educational practices. Eurasia J Biosci. 2012;6:11–23.
20. Yener H. Determination of horizontal and vertical distribution of tree species in Turkey via Shuttle Radar Topography Mission (SRTM) satellite data and geographic information system: the case of Crimean pine (*Pinus nigra*). Afr J Biotechnol. 2012;11(7):1659–1673.
21. Markgraf F. Genetische Beziehungen der Mittelmeerflora. Ber Dtsch Bot Ges. 1934;52:68–79.
22. Padilla M, Capone R, Palma G. Sustainability of the food chain from field to plate: the case of the Mediterranean diet. In: Burlingame B, Dernini D, editors. Sustainable diets and biodiversity directions and solutions for policy, research and action. Rome: FAO; 2012. p. 230–241.
23. Zohary M. Geobotanical foundations of the Middle East. Stuttgart: Gustav Fischer Verlag; 1973.
24. Ozturk M, Secmen O, Kondo K. Vegetation in Aegean region of Turkey. Mem Fac Integ Arts-Sci Hiroshima. 1983;8:53–62.
25. Ertug F. Wild edible plants of the Bodrum area (Mugla, Turkey). Turk J Bot. 2004;28:161–174.
26. Kaya I, Incekara N, Nemli Y. Ege Bölgesi'nde Sebze Olarak Tüketilen Yabani Kuskonmaz, Sirken, Yabani Hindiba, Rezene, Gelincik, Coban Degnegi ve Ebegümecinin Bazı Kimyasal Analizleri. Ingredients of some weeds consumed as food in Aegean region. Tarım Bilimleri Dergisi. 2004;14(1):1–6.
27. Dogan Y, Baslar S, Ay G, Mert HH. The use of wild edible plants in western and central Anatolia (Turkey). Econ Bot. 2004;58(4):684–690.
28. Kargioglu M, Cenkci S, Serteser A, Evliyaoglu N, Konuk M, Kok MS, et al. An ethnobotanical survey of inner-west Anatolia, Turkey. Hum Ecol. 2008;36(5):763–777.
29. Kargioglu M, Cenkci S, Serteser A, Konuk M, Vural G. Traditional uses of wild plants in the middle Aegean region of Turkey. Hum Ecol. 2010;38(3):429–450.
30. Ahiskali M, Ari C, Selvi S. Edible wild plants and their consumption during winter in a rural village on Kazdagi (Mount Ida). Bocconea. 2012;24:195–198.
31. Davis PH. Flora of Turkey: and the East Aegean Islands. Edinburgh: Edinburgh University Press; 1965–1988. (vol 1–10).
32. Ozbucak TB, Kutbay HG, Akcin OE. The contribution of wild edible plants to human nutrition in the Black Sea region of Turkey. Ethnobotanical Leaflets. 2006;10:98–103.
33. Cakilcioglu U, Khatun S. Nitrate, moisture and ash contents of edible wild plants. J Cell Plant Sci. 2011;2(1):1–5.
34. Redzic SJ. Wild edible plants and their traditional use in the human nutrition in Bosnia-Herzegovina. Ecol Food Nutr. 2006;45(3):189–232.
35. Kizil S. Determination of essential oil variations of *Thymbra spicata* var. *spicata* L. naturally growing in the wild flora of East Mediterranean and Southeastern Anatolia regions of Turkey. Ind Crops Prod. 2010;32(3):593–600.
36. Alan R, Padem H. A research on the composition of *Rheum ribes* L., *Tragopogon albinervis* L., *Polygonum cognatum* Meissn., *Chenopodium album* L. and *Malva neglecta* Wallr. used in human nutrition in Erzurum. Gida Dergisi. 1989;14(5):281–287.
37. Ertug F. An ethnobotanical study in Central Anatolia (Turkey). Econ Bot. 2000;54(2):155–182.
38. Tosun I, Karadeniz B, Yuksel S. The nitrate content of some edible wild plants consumed in Samsun region. Ekoloji. 2003;12(47):32–34.
39. Turan M, Kordali S, Zengin H, Dursun A, Sezen Y. Macro and micro mineral content of some wild edible leaves consumed in eastern Anatolia. Acta Agric Scand B. 2003;53(3):129–137.
40. Ufuk O, Yusuf K, Maksut C. Ethnobotancal studies in the villages of the district of Ilica (Province Erzurum), Turkey. Econ Bot. 2004;58(4):691–696.
41. Yücel E, Güney F, Yücel Şengün I. The wild plants consumed as a food in Mihaliccik district (Eskisehir/Turkey) and consumption forms of these plants. Bio Di Con. 2010;3(3):158–175.
42. Abak K, Duzenli A. Use of some wild plants as vegetables in Turkey. Acta Hortic. 1989;242:107–114.
43. Hadad Chi GHR, Moradi Z. The amounts and distribution of diosgenin and saponin and their carbohydrate moiety of *Tamus communis* L. J Agric Sci Natur Resour. 2005;12(2):55–66.
44. Morales P, Carvalho AM, Sánchez-Mata MC, Cámara M, Molina M, Ferreira ICFR. Tocopherol composition and antioxidant activity of Spanish wild vegetables. Genet Resour Crop Evol. 2011;59(5):851–863.
45. Pieroni A, Nebel S, Quave C, Münz H, Heinrich M. Ethnopharmacology of liakra: traditional weedy vegetables of the Arbëreshë of the Vulture area in southern Italy. J Ethnopharmacol. 2002;81(2):165–185.
46. Morales P, Ferreira I, Carvalho A, Sanchez-Mata M, Camara M, Tardio J. Fatty acids profiles of some Spanish wild vegetables. Food Sci Tech Int. 2012;18(3):281–290.

47. Guarrera PM. Food medicine and minor nourishment in the folk traditions of Central Italy (Marche, Abruzzo and Latium). Fitoterapia. 2003;74(6):515–544.
48. Tardío J, Pascual H, Morales R. Wild food plants traditionally used in the province of Madrid, Central Spain. Econ Bot. 2005;59(2):122–136.
49. Lentini F, Venza F. Wild food plants of popular use in Sicily. J Ethnobiol Ethnomed. 2007;3(1):15.
50. Ali-Shtayeh MS, Jamous RM, Al-Shafie JH, Elgharabah WA, Kherfan FA, Qarariah KH, et al. Traditional knowledge of wild edible plants used in Palestine (Northern West Bank): a comparative study. J Ethnobiol Ethnomed. 2008;4(1):13.
51. Martins D, Barros L, Carvalho AM, Ferreira ICFR. Nutritional and in vitro antioxidant properties of edible wild greens in Iberian Peninsula traditional diet. Food Chem. 2011;125(2):488–494.
52. Di Tizio A, Łuczaj ŁJ, Quave CL, Redžić S, Pieroni A. Traditional food and herbal uses of wild plants in the ancient South-Slavic diaspora of Mundimitar/Montemitro (Southern Italy). J Ethnobiol Ethnomed. 2012;8(1):21.
53. Pieroni A. Gathered wild food plants in the upper valley of the Serchio River (Garfagnana), Central Italy. Econ Bot. 1999;53(3):327–341.
54. Pieroni A. Medicinal plants and food medicines in the folk traditions of the upper Lucca Province, Italy. J Ethnopharmacol. 2000;70(3):235–273.
55. Bonet MA, Vallès J. Use of non-crop food vascular plants in Montseny biosphere reserve (Catalonia, Iberian Peninsula). Int J Food Sci Nutr. 2002;53(3):225–248.
56. Rivera D, Obon C, Inocencio C, Heinrich M, Verde A, Fajardo J, et al. The ethnobotanical study of local Mediterranean food plants as medicinal resources in Southern Spain. J Physiol Pharmacol. 2005;56 Suppl 1:97–114.
57. Salvatore S, Pellegrini N, Brenna OV, Del Rio D, Frasca G, Brighenti F, et al. Antioxidant characterization of some Sicilian edible wild greens. J Agric Food Chem. 2005;53(24):9465–9471.
58. Scherrer AM, Motti R, Weckerle CS. Traditional plant use in the areas of Monte Vesole and Ascea, Cilento National Park (Campania, Southern Italy). J Ethnopharmacol. 2005;97(1):129–143.
59. Della A, Paraskeva-Hadjichambi D, Hadjichambis AC. An ethnobotanical survey of wild edible plants of Paphos and Larnaca countryside of Cyprus. J Ethnobiol Ethnomed. 2006;2(1):34.
60. Nebel S, Pieroni A, Heinrich M. Ta chòrta: wild edible greens used in the Graecanic area in Calabria, Southern Italy. Appetite. 2006;47(3):333–342.
61. Ghirardini M, Carli M, Del Vecchio N, Rovati A, Cova O, Valigi F, et al. The importance of a taste. A comparative study on wild food plant consumption in twenty-one local communities in Italy. J Ethnobiol Ethnomed. 2007;3(1):22.
62. Pardo-de-Santayana M, Tardío J, Blanco E, Carvalho A, Lastra J, San Miguel E, et al. Traditional knowledge of wild edible plants used in the northwest of the Iberian Peninsula (Spain and Portugal): a comparative study. J Ethnobiol Ethnomed. 2007;3(1):27.
63. Rivera D, Obón C, Inocencio C, Heinrich M, Verde A, Fajardo J, et al. Gathered food plants in the mountains of Castilla–La Mancha (Spain): ethnobotany and multivariate analysis. Econ Bot. 2007;61(3):269–289.
64. Aberoumand A, Deokule SS. Comparison of phenolic compounds of some edible plants of Iran and India. Pakistan J Nutr. 2008;7(4):582–585.
65. Nebel S, Heinrich M. Ta chòrta: a comparative ethnobotanical-linguistic study of wild food plants in a Graecanic area in Calabria, Southern Italy. Econ Bot. 2009;63(1):78–92.
66. Mattalia G, Quave CL, Pieroni A. Traditional uses of wild food and medicinal plants among Brigasc, Kyé, and Provençal communities on the Western Italian Alps. Genet Resour Crop Evol. 2012.
67. Menendez-Baceta G, Aceituno-Mata L, Tardío J, Reyes-García V, Pardo-de-Santayana M. Wild edible plants traditionally gathered in Gorbeialdea (Biscay, Basque Country). Genet Resour Crop Evol. 2012;59(7):1329–1347.
68. Sánchez-Mata MC, Cabrera Loera RD, Morales P, Fernández-Ruiz V, Cámara M, Díez Marqués C, et al. Wild vegetables of the Mediterranean area as valuable sources of bioactive compounds. Genet Resour Crop Evol. 2012;59(3):431–443.
69. Renner SS, Scarborough J, Schaefer H, Paris HS, Janick J. Dioscorides's *Bruonia melaina* is *Bryonia alba*, not *Tamus communis*, and an illustration labeled *Bruonia melaina* in the Codex Vindobonensis is *Humulus lupulus* not *Bryonia dioica*. In: Pitrat M, editor. Cucurbitaceae 2008: proceedings of the IXth EUCARPIA meeting on genetics and breeding of Cucurbitaceae. Avignon: INRA; 2008. p. 273–280.
70. Polo S, Tardío J, Vélez-del-Burgo A, Molina M, Pardo-de-Santayana M. Knowledge, use and ecology of golden thistle (*Scolymus hispanicus* L.) in Central Spain. J Ethnobiol Ethnomed. 2009;5(1):42.
71. Redzić S. Use of wild and semi-wild edible plants in nutrition and survival of people in 1430 days of siege of Sarajevo during the war in Bosnia and Herzegovina (1992–1995). Coll Antropol. 2010;34(2):551–570.
72. Batal M, Hunter E. Traditional Lebanese recipes based on wild plants: an answer to diet simplification? Food Nutr Bull. 2007;28(2 suppl):303–311.
73. Gurses O, Artik N. A research on the composition of *Beta vulgaris* var. *cicla* L., *Malva neglecta* Wallr., *Portulaca oleracea* L. and *Spinacia oleracea* L. Gida Dergisi. 1984;9(2):83–87.
74. Alan R, Padem H. A research on the composition of *Rumex crispus* L., *Atriplex petula* L., *Polygonum aviculare* L., *Prangos uechtritzii* Boiss. and *Tragopogon reticulatus* Boiss., used in human nutrition in Erzurum. Turk J Agric For. 1990;14:48–57.
75. Ayan I, Acar Z, Mut H, Basaran U, Asci O. Morphological, chemical and nutritional properties of forage plants in a natural rangeland in Turkey. Bangladesh J Bot. 2006;35(2):133–142.
76. Akcin OE, Ozbucak TB. Morphological, anatomical and ecological studies on medicinal and edible plant *Malva neglecta* Wallr. (Malvaceae). Pak J Biol Sci. 2006;9(14):2716–2719.
77. Sekeroglu N, Ozkutlu F, Deveci M, Dede O, Yilmaz N. Evaluation of some wild plants aspect of their nutritional values used as vegetable in eastern Black Sea region of Turkey. Asian J Plant Sci. 2006;5(2):185–189.
78. Sivga HO, Secmen O. Ethnobotanic survey of Isikli (Carpin), Dagdancik and Tokdemir in Gaziantep, Turkey. IUFS J Biol. 2009;68(1):19–26.
79. Chalchat JC, Ozcan MM, Figueredo G, Chalard P. The effect of harvest years on chemical composition of essential oil of pickling herb (*Echinophora tenuifolia* subsp. *sibthorpiana*) leaves used as medicinal plant. Acta Bot Hung. 2011;53(1):73–77.
80. Leonti M, Nebel S, Rivera D, Heinrich M. Wild gathered food plants in the European Mediterranean: a comparative analysis. Econ Bot. 2006;60(2):130–142.
81. Simopoulos AP. Omega-3 fatty acids and antioxidants in edible wild plants. Biol Res. 2004;37(2):263–277.
82. Uysal I, Onan S, Karabacak E, Celik S. Ethnobotanical aspects of Kapidag Peninsula (Turkey). Bio Di Con. 2010;3(3):15–22.

83. Kocyigit M, Ozhatay N. The wild edible and miscellaneous useful plants in Yalova province (northwest Turkey). J Fac Pharm Istanbul. 2009;40:19–29.
84. Lemordant D. Plantes utiles et toxiques de Tunisie. Fitoterapia. 1977;48:191–214.
85. Vanzani P, Rossetto M, De Marco V, Sacchetti L, Paoletti MG, Rigo A. Wild mediterranean plants as traditional food: a valuable source of antioxidants. J Food Sci. 2011;76(1):C46–C51.
86. Yildirim E, Dursun A, Turan M. Determination of the nutrition contents of the wild plants used as vegetables in Upper Coruh Valley. Turk J Bot. 2001;25:367–371.
87. Rigat M, Bonet MA, Garcia S, Garnatje T, Vallès J. Ethnobotany of food plants in the high river Ter Valley (Pyrenees, Catalonia, Iberian Peninsula): non-crop food vascular plants and crop food plants with medicinal properties. Ecol Food Nutr. 2009;48(4):303–326.
88. Dogru Koca A, Yildirimli s. Ethnobotanical properties of Akçakoca District in Düzce (Turkey). Hacettepe J Biol Chem. 2010;38(1):63–69.
89. Akgunlu SB. Mineral content and microbiological analysis of some wild edible vegetables consumed in Kilis and Gaziantep provinces [Master thesis]. Kilis: Kilis 7 Aralık University; 2012.
90. Pieroni A. Evaluation of the cultural significance of wild food botanicals traditionally consumed in Northwestren Tuscany, Italy. J Ethnobiol. 2001;21:89–104.
91. Pieroni A, Giusti M. Alpine ethnobotany in Italy: traditional knowledge of gastronomic and medicinal plants among the Occitans of the upper Varaita valley, Piedmont. J Ethnobiol Ethnomed. 2009;5(1):32.
92. Motti R, Antignani V, Idolo M. Traditional plant use in the Phlegraean Fields Regional Park (Campania, Southern Italy). Hum Ecol. 2009;37(6):775–782.
93. Shafaghat A. Volatile oil constituents and antibacterial activity of different parts of *Falcaria vulgaris* Bernh. growing wild in two localities from Iran. Nat Prod Res. 2011;25(4):368–373.
94. Lev-Yadun S, Abbo S. Traditional use of A'kub (*Gundelia tournefortii*, Asteraceae), in Israel and the Palestinian Authority area. Econ Bot. 1999;53(2):217–219.
95. Halabi S, Battah AA, Aburjai T, Hudaib M. Phytochemical and antiplatelet investigation of *Gundelia tournifortii*. Pharm Biol. 2005;43(6):496–500.
96. Coruh N, Sagdicoglu Celep AG, Ozgokce F, Iscan M. Antioxidant capacities of *Gundelia tournefortii* L. extracts and inhibition on glutathione-S-transferase activity. Food Chem. 2007;100(3):1249–1253.
97. Tardio J, Pardo-de-Santayana M. Cultural importance indices: a comparative analysis based on the useful wild plants of Southern Cantabria (Northern Spain). Econ Bot. 2008;62(1):24–39.
98. Razavi SM, Zarrini G, Molavi G, Ghader G. Bioactivity of *Malva sylvestris* L., a medicinal plant from Iran. Iran J Basic Med Sci. 2011;14(6):574–579.
99. Pardo-De-Santayana M, Tardio J, Morales R. The gathering and consumption of wild edible plants in the Campoo (Cantabria, Spain). Int J Food Sci Nutr. 2005;56(7):529–542.
100. Gencay A. Ethnobotanical aspects of Cizre (Sirnak) [Master thesis]. Van: Yuzuncu Yil Universitesi; 2007.
101. Ayan AK, Caliskan O, Cirak C. Economical importance of stinging nettle (*Urtica* spp.) and its cultivation. J Fac Agric OMU. 2006;21(3):357–363.

3

Edible wild plant use in the Faroe Islands and Iceland

Ingvar Svanberg[1]*, Sigurður Ægisson[2]

[1] Uppsala Centre for Russian and Eurasian Studies, Uppsala University, Box 514, 751 20 Uppsala, Sweden

[2] Hvanneyrarbraut 45, 580 Siglufjörður, Iceland

Abstract

This paper reviews the use of wild edible plants in the Faroe Islands and Iceland from the times of the first settlement of Norse people in the Viking age until today, with a special emphasis on the 18th, 19th and 20th centuries. Animal products have been an important source of nutrients for the islanders of northern Atlantic. Cultivation of cereals on the other hand has played a minor role, and had already been abandoned by late medieval times in Iceland and by the early 20th century on the Faroes. Crops such as potatoes, turnips and other roots were only grown in the small patches of cultivated soil. Wild plants have therefore been of some importance for the Faroese people and the Icelanders; in the last centuries especially for the rural poor and during times of recessions. The native *Angelica archangelica* L. was gathered in the wild and also cultivated in gardens for centuries. A few species have been part of the regular food staple. Some plants are still gathered and made into food products by small companies, especially in Iceland. In the Faroes, the economic aspect of edible wild plant taxa is mostly of historical interest, although a few products of *A. archangelica* are sometimes available. Two taxa have been exploited as regular food exclusively in Iceland: *Cetraria islandica* (L.) Arch. and *Elymus arenarius* L. Icelanders have used *C. islandica* from the early settlement days and continue to do so today, *E. arenarius* became obsolete as a food plant a century ago.

Keywords: wild food plants, algae, lichens

Introduction

Tapping little known traditions of local plant knowledge in rural areas is an important issue for contemporary ethnobiology. The traditional use of wild "plants" (vascular plants, fungi, algae, lichens) for food is a popular biocultural domain to research [1]. However, despite many field works and increasing interest in archival sources many areas of Eurasia still remain blank spots on the ethnobiological map. Marginal regions in the North belong to these little researched areas [2].

The populations of the Faroe Islands and Iceland share a common cultural heritage as descendants of the Norse Vikings that settled between the 9th and 10th centuries in these far-away places. Faroese people and Icelanders used the gifts of nature from the beginning, both for food and as remedies against illnesses, although most of what can be said about the earliest period despite a few archaeobotanical studies, is guesswork. From early medieval times we have information in the Eddas, the Icelandic sagas and the law literature [3].

* Corresponding author. Email: ingvar.svanberg@ucrs.uu.se

Background and methods

Iceland has a total area of 103000 km^2 and a population of about 320000 (2012), which makes it the most sparsely populated country in Europe. From 1262 to 1918 Iceland was part of the Norwegian and later the Danish monarchies. Iceland became independent in 1918 and a republic was declared in 1944. Icelandic has been codified as a literature language since medieval times. Traditional food culture was based on marine mammals, seabirds (meat, eggs), fish, seashells, cattle, sheep, horses and dairy products. While animal husbandry and hunting have been of importance, agriculture was very limited. Some plants were gathered in the landscape, otherwise vegetables have been of very little importance [4–6].

The Faroes, located in the northeastern North Atlantic Ocean, constitutes a group of 17 populated (out of 18) islands with a total area of 1400 km^2 and a population of nearly 50000 people (2012). The Faroe Islands are still under the sovereignty of Denmark, but have their own government. Faroese has been used as a colloquial language beside Danish until the last century, when it became a written language. Animal products did and still do dominate the traditional food culture. As in the case of Iceland, it is very much based on fish, seabirds, whale (especially *Globicephala melas* Traill), sheep, and to some extent domestic geese. Various bivalves, e.g. *Mytilus edulis* (L.), *Modiolus modiolus* (L.), and *Ensis ensis* (L.), have been eaten in some villages, especially during times of food shortage. Cereals and vegetables have played a minor role in the traditional cuisine, of which potatoes and other root vegetables have been the most common [7–10].

Landscape and flora

According to a 12th century source Iceland was "forested from mountain to sea shore". However, it was deforested due to overexploitation (fuel, construction material) and grazing. Animal husbandry was a major livelihood. Wide areas were used for grazing and scattered meadows were used for haymaking. Very small remnants of the former tree flora still exist. The landscape is part of the Arctic region and the vegetation consists mainly of grassland. Large areas are non-vegetated or only sparsely vegetated. There are 490 vascular plant taxa, 460 mosses, 700 lichens and 2100 fungi species growing wild in Iceland [11]. Approximately 30 wild taxa are reported as having been utilised as food. Faroe Islands are part of the Atlantic region and the Faroese people have access to over 400 vascular plant taxa, 350 mosses, 470 lichens and 600 species of fungi [12]. Only around 20 have been used as food.

Until the nineteenth century, the settlers mostly subsisted on shepherding, hunting and fishing. Wild plants have always played a small but nevertheless important role in the economy of the populations of Iceland and the Faroes. Seaweed has been harvested as fodder and manure, grass has been used to thatch the houses, and turf has played an important role as fuel. Some plants and especially lichens have been used for dying wool, while the grass and hay production was crucial for keeping sheep, horses and some cattle. Cultivation of crops has been of minor importance; mostly barley in small fields and more recently potatoes. The soil is used for the production of grass and to some extent potatoes, the latter an innovation adopted in the islands in the middle of the 19th century. The meadows with grass and many flower plants are harvested in the early fall for production of fodder. In the towns and villages, there are also gardens with some vegetables and ornamental flowers [4,5,7,10,13].

Sources for this review

This review is based on archaeological, botanical, ethnographical and historical research, travelogues, ethnobotanical studies available, and on field research on the Faroes and Iceland [14–17]. The native scholars Eggert Ólafsson (1726–1768), Bjarni Pálsson (1719–1779) and Jens Christian Svabo (1746–1824) travelled, inspired by Carl Linnaeus travelogues, on behalf of the Danish authorities Iceland and Faroe Islands respectively, and gathered the best material on traditional plant use already in the 18th century [18,19]. A review of local plants some of which had been used until very recently was published by the Faroese botanist Rasmus Rasmussen (1871–1962) in 1946 [20]. The reverend Björn Halldórsson (1724–1794) in Sauðlauksdalur compiled already in 1783 a commentary on Icelandic useful plants [21]. Icelandic ethnologist Lúðvík Kristjánsson (1911–2000) has put together a monumental work on gathering activities along the shores and in the sea of Icelandic coasts, including the exploitation of plants [22].

Results

Very few taxa have been used as food plants. Some of the plants eaten are the same as the peasantry in Norway and other parts of Scandinavia utilised for the same purpose. However, a few plants are unique for these regions [23]. The wild plants that have been used for food are presented according to the categories given by Turner et al. [24].

Root vegetables

Very few underground plant parts have been used in the Faroes and Iceland. The rhizomes of *Angelica sylvestris* L. were eaten in the Faroes in the 18th century and still in end of the 19th century [15,20]. Roots of *Angelica archangelica* L. were considered to be the most important part of the plant in Iceland, and "going mountain rooting" was the term applied to the gathering of this valuable source of nutrition [3]. From Iceland there are data available on the use of *Bistorta vivipara* (L.) S. F. Gray eaten raw or made into bread [18,21].

A couple of species were used both in the Faroes and in Iceland. The use of the nutritious roots of *Potentilla anserina* L., eaten boiled or roasted, is the most interesting [15,18,21]. It has been used as food also in other Norse-influenced areas of Shetland Isles, Scotland and Norway [25]. The small, rich in starch, nodules attached to rhizomes of *Equisetum arvense* L. have been eaten in the Faroes and Iceland. Since they are sweet-tasting the habit has survived among children until rather recently. The nodules were gathered in spring-time while tilling the soil [26,27].

Edible greens

Ever since the first settlement of Iceland and the Faroese, wild growing *Angelica archangelica* L. has been used for human consumption and has also been important as medicinal herb. It was also eaten fresh, picked in the wild on the mountainsides. In Iceland all parts of the plant were used – the seeds, leaves, stalks and roots and were eaten fresh, fried or boiled (often in milk); or prepared in some other way [28].

The most commonly used green plant both on the Faroes and in Iceland was probably *Rumex acetosa* L. It was usually served with meat, fish or as a flavour to bird soup. It is still recommended in a cookbook as tasting well with boiled puffin [15,29]. Generally speaking, the common sorrel (*Rumex acetosa*) is the most popular wild food plant also in Iceland. It was used almost in the same way as in the Faroes [18,21].

Lady's smock, *Cardamine pratensis* L., was used in soup in the Faroes. The use of *Taraxacum* sp. tender leaves as a spring vegetable is mentioned from the Faroes in 1800. The Faroese children still suck the sweet nectar from the flowers of *Vicia cracca* [15].

An interesting green plant is *Cochlearia officinalis* L. that was eaten against scurvy and to promote good health among herdsmen and fowlers. Its use is also recorded in historical sources from Shetland, the Faroe Islands and Iceland as well as northern Norway [15,20,30]. This use of the plant was already known in the Faroe Islands in the late 17th century and persisted until this century. Lucas Debes (1673) writes that God has blessed the Faroe Islands with this plant species, as well as with some other herbs like cress, veronica and sorrel; also remedies against scurvy. Scurvy grass was considered good for one's health and also gave strength and stamina thus preventing exhaustion, caused by a shortage of food [31]. In Iceland, it is probably first mentioned in the saga of Grettir the Strong (early 14th century); but as a precious food its story seems to begin in the 17th century, in the island Grímsey off the northern coast. It was usually gathered from late August till the end of September if it was to be used as food for humans (also during winter if the ground was not covered by ice and snow). Scurvy grass was widely used to make soup and porridge and to eke out flour in bread making, as well as being eaten fresh as salad. The root was eaten raw or boiled. Some of these methods, mostly the last ones mentioned, are still in practice, here and there.

The use of *Oxyria digyna* (L.) Hill. in the Faroes is only historical. Jens Christian Svabo describes in the 1780s how people flavoured flour gruel with alpine mountain sorrel to make the dish sour [19]. The same could almost be said about Iceland, where it is common all over. Nowadays people mostly consume the plant fresh to quench their thirst while on tour walking in the wilderness. It is also used in salads at home, or the leaves are boiled in water and the mixture drunk cold as refreshment. Others use it in various soups and porridges and in recipes instead of lemons. The seeds are sometimes used in bread-baking, although only on a small scale. In Iceland *Honckenya peploides* (L.) Ehrh., is fermented and in that state used as food, like sauerkraut. *Oxalis acetosella* L., *Rumex longifolius* DC., *Stellaria media* L., *Chamerion latifolium* (L.) Holub. and *Epilobium angustifolium* L. have also been used in soup or stews or consumed fresh as salad. Also *Silene acaulis* L. was gathered, boiled and eaten with butter [18,21,22,32].

Berries

Due to sheep grazing, wild berries were (and are) scarce in the Faroese mountains and mostly eaten on the spot when found (*Vaccinium myrtillus* L.). The only one yielding some harvest is the *Empetrum hermaphroditum* Hagerup, picked and usually eaten on the spot. Berries of *Cornus suecica* L. were eaten by children and youth in the Faroes [15,19,20].

The Icelandic people have doubtless used wild berries, including the crowberry, *Empetrum nigrum* L. and *Empetrum hermaphroditum* Hagerup, in food since the time of settlement although other berries, such as bog bilberry, *Vaccinium uliginosum* L., and bilberry, *Vaccinium myrtillus* L., were generally considered better. According to the law of the Icelandic Commonwealth, it was permissible to eat berries on other people's land, but those attempting to gather them and take them away could be penalised. Berry fields were recorded as fringe benefits in many old farm registers. Crowberry wine was made at Skálholt in early centuries, possibly because of a lack of access to foreign communion wine [4,18].

In more recent times the most usual way to eat fresh crowberries has become an accompaniment to the Icelandic skyr (curdle milk), soups, puddings or the mixture of skyr and porridge known as "hræringur". A good way to preserve vitamin C in a land where this was sorely lacking was to place the berries between layers of skyr and store in this way until the dark days of winter. Berries were also used to flavour whey and add to its nutritional value. In present times, there are more ways on offer for the storage of berries; they can be frozen, kept as preserves or boiled to produce cordial, juice or jelly. They are also delicious in sauces.

Grains and seeds

Leymus arenarius (L.) Hochst. has been commonly gathered as food along the southern coast of Iceland until the beginning of last century. It was used as a bread additive and has probably been gathered for human consumption since the Viking era [17].

Lichens and algae

Iceland is famous for using lichens, especially Iceland moss, *Cetraria islandica* (L). Arch., as regular food. We will deal with that separately below. The lichen *Umbillicaria cylandrica* (L.) Delise ex Duby is mentioned as famine food in the 18th century [18].

Algae have been of importance as a source for food in many maritime communities worldwide [33]. This is also the case with the Faroes and Iceland. Some species of seaweed – especially *Palmaria palmata* (L.) Weber & Mohr, *Saccharina latissima* (L.) C. E. Lane, C. Mayes, Druehl, & G. W. Saunders, *Ulva lactuca* L., and *Alaria esculenta* (L.) Greville – were gathered along the shores on some Faroese islands. Seaweed was still gathered as food in the 1930s [10,15,20].

In Iceland various species of seaweed such as dulse, *Pamaria palmata*, carrageen moss, *Chondrus crispus* Stackhouse, winged kelp (*Alaria esculenta*) and others were gathered. *Mastocarpus stellatus* (Stackhouse) Guiry, *Laminaria digitata* (Hudson) J. V. Lamouroux, and *Laminaria hyperborea* (Gunnerus) Foslie were gathered in times of famine. Especially *P. palmata* has long been used as food and a source of essential nutrition in times of famine. The oldest law texts deal with the right to collect dulse, as this was an important economic crop, collected, transported to inland farms and sold in exchange for other goods. Dulse was, and still is, widely distributed along the coast, although it is more abundant on the west and southwest coasts than elsewhere. A certain amount was gathered in the spring, but mostly the harvesting took place during spring-time at the end of June and again at the end of August. Dulse was mostly eaten with dried fish along with butter or some other kind of fat, or added to stews, porridges and blood sausages; later occasionally, with the addition of potatoes or turnips. It was used also to eke out rye flour in making bread. Nowadays dulse is gaining popularity in Iceland again, this time as health food. It is often used as a snack to chew on, consumed dry and in whole pieces, or as seasoning for soups and bread, as well as in pasta, fish and meat dishes [4,22].

Carrageen moss, *Chondrus crispus*, was in earlier times gathered in Iceland in much the same manner as *P. palmata*, and only came second in popularity to that seaweed. It is found on shores in the south and west. The species was mostly used in a kind of porridge or prepared into "fjörugrasahlaup" (Carrageen moss jelly) which was put into sour milk. Nowadays carrageen is mostly used in sashimi garnishes, as a soup ingredient and (less commonly) in salads or sold dried and ground as health seasoning [22,27].

Spices and flavourings

Ligusticum scoticum L. was mostly considered as a medicinal plant in the Faroes, but the aromatic young plant could have been used as a spice instead of parsley according to one source. *Oxyria digyna* L. was mixed in the gruel in order to make the dish sour [15,20].

Pinguicula vulgaris L. was used instead of garlic in Iceland. Berries of *Juniperus communis* L. and the leaves of *Thymus praecox* Opiz have been used to flavour sour milk, *Carum carvi* L. was also used as a spice in Iceland [18].

Beverages

Dryas octopetala L. has been used as a substitute for tea in Iceland [18]. Ómar Gunnarsson, a native of Húsavík, is the only winemaker in Iceland, and the only one in the world to produce a wine made entirely from organic crowberries. This wine is made up of 80 per cent crowberries and 19 per cent rhubarb, the rest being wild blueberries and Icelandic herbs. This makes his wine highly unique due to its powerful antioxidant/anti-aging ingredients (Fig. 1). In the Faroes, a tea was brewed from the flowers of *Thymus praecox* Opiz. [20].

Specific West Norse food plants

There is of course a relationship between the use and availability of native plants [34]. Some taxa are known to have been used since the arrival of the Norse settlers, and are still specific for the Faroe Islands and Iceland. We will discuss two examples that have had particular cultural significance there.

Angelica gardens

Angelica has been used in Iceland, Greenland, in the Faroe Islands, and in northern Scandinavia. When the Vikings began trading expeditions in the 9th century, angelica was an important commodity. Growing angelica in primitive gardens has a long tradition in the Nordic countries, and was earlier common in the Faroes. Just before World War II, an observer wrote that angelica-gardens were disappearing [15,28]. However, well-kept angelica-gardens enclosed with stones have, until recently, been a source of pride for households in the Faroe Islands (Fig. 2). The angelica was harvested in the summer, especially for the Midsummer Festival and it could be gathered until the Festival of St. Olaf (29 July). Angelica stems were traditionally served together with thick, creamy sour milk, this being offered as a treat. It was also eaten fresh, picked either from the garden or in the wild on the mountainsides. Popular cookbooks still have recipes including this plant, for instance angelica mixed into rhubarb soup or as jelly. Since World War II it has been used mostly as an addition to rhubarb jam. Nowadays few people actually use it at home, but it is included in dishes served by catering companies offering Faroese local food on their menus. Rhubarb *Rheum raponticum* L., which was promoted as a food plant for home gardens in the first decade of the 20th century, has very much replaced the garden angelica [15,29,31,35,36].

In Iceland, there are many swathes of angelica and special angelica gardens were known to exist in olden days. Right from the time of settlement angelica has been used for human consumption. Many toponyms still remind us of places for harvesting angelica. Grágás, the law book of the Icelandic Commonwealth (930–1262/ 1264), refers to penalties incurred for the theft of angelica, some as serious as being outlawed or fined heavily if the plant was stolen from someone else's garden. Angelica was an important way of eking out the household larder. All parts of the plant could be used – the seeds, leaves, stalks and roots – and were eaten fresh, fried (even grilled) or boiled (often in milk); or prepared in some other way. It depended on the region, however, how the plant was used and it appears to have been appreciated more in the eastern and southern parts of Iceland. Public knowledge of angelica and its cultivation has gradually disappeared during the past years, but interest in this plant has awakened once more with the health revolution of recent years and changing attitudes and awareness of people regarding food and nutrition supplements. We must not forget the Angelica Root Spirit, also exported to other countries. Angelica is also used in the kitchen as a fresh herb on meat or fish. The newest use for this versatile plant is as principal fodder for lambs to bring out a better spicy taste and fragrance in the meat on our table – right from the inside. During recent years, The University of Iceland has conducted studies into the garden angelica and rumours of its medicinal powers have gathered strength. It is now marketed as the Ginseng of the North [18,28,33,37].

Iceland moss as a food plant

Sources indicate that Iceland moss, *Cetraria islandica* (L.) Arch., has always played a major role in Icelandic culinary tradition possibly from the time of the earliest settlement. It is mentioned in the law collection Grágás. This lichen was in many ways a lifesaver for the people of those long ago days. The possibility of lichen gathering was, therefore, an advantage attached to some farms, seen as added benefit; especially if the lichen-covered areas were not too difficult to reach. It could be chopped up or just eaten whole. Those farms that were fortunate enough to own a piece of land with Iceland moss as a fringe benefit sent a group of people each year to collect enough supplies to last through the coming winter. The participants were mostly women and young people under the direction of one adult man. The group travelled on horseback and was equipped with tents, food and skin bags in which to put the gathered lichens. It was considered that two barrels of moss contained the nutritional equivalent of one barrel of flour. Many Icelandic idioms have their origins in these lichen-gathering trips. The longer excursions had for the most part been discontinued by the beginning of the 20th century [18,21,38,39].

Fig. 1 Crowberry wine "Evening Sun" produced in Iceland (Photograph by Sigurður Ægisson).

Fig. 2 Old angelica garden at Gjógv, Faroe Islands (Photograph by Ingvar Svanberg).

Examples of the traditional type of recipe which included Iceland moss are:

(*i*) water gruel (the moss is steeped in water and cut up using a special curved knife; hot water is poured over the moss; flour or rice is added. The gruel is then boiled and served with milk),

(*ii*) moss milk (whole pieces of lichen are boiled in milk and the mixture eaten as a soup). Iceland moss is used to eke out the flour in bread and in the making of the Icelandic blood pudding, especially in times of great hardship,

(*iii*) Iceland moss tea (the lichen is boiled in water, then strained and the stock is used for drinking and was considered very wholesome). This beverage was also used for medicinal purposes, especially with regard to respiratory illnesses such as colds, pneumonia, asthma and tuberculosis. Some think it best to boil the lichens for only 5–15 minutes, as it is believed to become bitter if boiled for the whole hour; however, after 2–3 hours it takes on a sweet taste.

The use of Iceland moss has increased in the wake of the health revolution, with its desire to return to natural ways. In all the better bakeries in 21st century Iceland, e.g. in Akureyri and Reykjavík, it is possible to buy bread with Iceland moss as one of its ingredients and this is thought to be extremely good (Fig. 3). Recent research has proved that the old belief in Iceland moss as a very useful product is not just a whim to be smiled at, and there is now a wide variety of health products available containing this ingredient [39].

Fig. 3 Icelandic bread made with *Cetraria islandica*-flour (Photograph by Sigurður Ægisson).

During the Enlightenment the idea of using Iceland moss as a surrogate food was transferred from Iceland to Scandinavia and diffused into learned circles also elsewhere. In the late 18th and through the 19th century Iceland moss was promoted as a famine food in Norway, Finland, Sweden, Germany, Russia and even France. Inspired by the Icelandic traditional use, many botanists and economists, especially in time of famine and wants, published small booklets with recipes on lichen bread, lichen gruel and lichen pancakes, based on *Cetraria islandica* (L.) Arch. [40].

Discussion

Native, potentially edible plants in Iceland and the Faroese are few. Local plants play hardly any economic role as food in the contemporary Faroe Islands, although there is a degree of interest in reviving the use of garden angelica, and some people still prepare food with angelica as an ingredient. *Rumex acetosa* L. is sometimes used in homemade food.

Some plants are still gathered and made into food products by small companies, especially in Iceland. There is also an increasing interest in the use of local plants [41,42]. Two taxa have been exploited as food exclusively in Iceland: *Cetraria islandica* (L.) Arch. and *Elymus arenarius* L. Although propaganda efforts by authorities and enlightened persons in the 18th and 19th century in Scandinavia and other countries tried to promote Iceland moss as a surrogate food in times of food crises, it never became accepted by the peasantry. However, the Icelanders have used it since the early settlement days and continue to do still today. Wild grasses have been harvested for food elsewhere (e.g. [43]), but the use of lyme-grass seems to be restricted to Iceland.

References

1. Svanberg I, Łuczaj Ł, Pardo-de-Santayana M, Pieroni A. History and current trends of ethnobiological research in Europe. In: Anderson EN, Pearsall DM, Hunn ES, Turner NJ, editors. Ethnobiology. Hoboken NJ: Wiley-Blackwell; 2011. p. 191–214.
2. Łuczaj Ł, Szymański WM. Wild vascular plants gathered for consumption in the Polish countryside: a review. J Ethnobiol Ethnomed. 2007;3(1):17.
3. Guðjónsson SV. Folkekost og sundhedsforhold i gamle dage: belyst igennem den oldnordiske litteratur. København: Arnold Busck; 1941.
4. Jonsson G. Changes in food consumption in Iceland, 1770–1940. Scand Econ Hist Rev. 1998;46(1):24–41.
5. Zutter C. Wood and plant-use in 17th–19th century Iceland: archaeobotanical analysis of Reykholt, Western Iceland. Environmental Archaeology. 2000;5:73–82.
6. Tómasson T. Iceland: country report to the FAO international conference on plant genetic resources. Leipzig: FAO; 1996.
7. Guttesen R. Food production, climate and population in the Faeroe Islands 1584–1652. Geogr Tidsskr. 2004;104:35–46.
8. Guttesen R. Plant production on a Faeroese farm 1813–1892, related to climatic fluctuations. Geogr Tidsskr. 2001;101:67–76.
9. Rasmussen H. Die Nahrungsforschung auf den Färöer. Ethnol Eur. 1971;5:49–52.
10. Joensen JP. Fiskafólk. Ein lýsing av føroyska húshaldinum í slupptíðini. Tórshavn: Føroya Sparikassi; 1982.
11. Kristinsson H. A guide to the flowering plants and ferns of Iceland. Reykjavik: Örn og Örlygur; 2010.
12. Jóhansen J. Studies in the vegetational history of the Faroe and Shetland Islands. Tórshavn: Føroya Fróðskaparfelag; 1985. (Annales Societatis Scientiarum Færoensis; vol 11).
13. Harshberger JW. The gardens of the Faeroes, Iceland, and Greenland. Geogr Rev. 1924;14(3):404.
14. Jóhansen J. Medicinal and other useful plants in the Faroe Islands before ad 1800. Bot J Scotl. 1994;46(4):611–616.
15. Svanberg I. The use of wild plants in the Faroe Islands 1590–1990: a contribution to Scandinavian ethnobotany. Svenska Linnésällskapets Årsskrift. 1998;1996–1997:81–130.

16. Svanberg I. Plant knowledge as indicator of historical cultural contacts: tanning in the Atlantic fringe. In: Pieroni A, Vandebroek I, editors. Traveling cultures and plants: the ethnobiology and ethnopharmacy of migrations. Oxford: Berghahn Books; 2007. p. 227–244.
17. Guðmundsson G. Gathering and processing of lyme-grass (*Elymus arenarius* L.) in Iceland: an ethnohistorical account. Veg Hist Archaeobot. 1996;5(1-2):13–23.
18. Olafsen E. Vice-Lavmand Eggert Olafsens og Land-Physici Biarne Povelsen Reise igiennem Island, franstaltet af Videnskabernes Sælskab i Kiøbenhavn. Sorø: Videnskabernes Sælskab; 1772. (vol 1–2).
19. Svabo JC. Indberetninger fra en Rejse i Færøe 1782 og 1783. København: Selskabet til Udgivelse af færøske Kildeskrifter og Studier; 1959.
20. Rasmussen R. Gróðranýtsla fyrr í tíðini: til matna, til lækningar, til litagerðar og til ídnaðar. Tórshavn: Landnám; 1946.
21. Halldórsson B. Grasnytjar. København: August Friedrich Stein; 1783.
22. Kristjánsson L. Íslenzkir sjávarhættir. Reykjavík: Menningarsjóður; 1980. (vol 1–5).
23. Høeg OA. Planter og tradisjon Floraen i levende tale og tradisjon i Norge 1925–1973. Oslo: Universitetsforlaget; 1974.
24. Turner NJ, Łuczaj ŁJ, Migliorini P, Pieroni A, Dreon AL, Sacchetti LE, et al. Edible and tended wild plants, traditional ecological knowledge and agroecology. Cr Rev Plant Sci. 2011;30(1-2):198–225.
25. Holmboe J. Mura (*Potentilla anserina* L.) og dens bruk som matnyttig plante i ældre og nyere tid. Svensk Botanisk Tidskrift. 1928;22:160–168.
26. Svanberg I. Field horsetail (*Equisetum arvensis*) as food. Fróðskaparrit. 1997;45:45–55.
27. Mohr N. Forsøg til en islandsk naturhistorie, med adskillige oekonomiske samt andre anmærkninger. København: Christian Friderik Holm; 1786.
28. Svanberg I. Kvannen (*Angelica archangelica*) på Färöarna och Island. Gardar: Årsbok för Samfundet Sverige-Island i Lund-Malmö och Samfundet Sverige-Färöarna. 2008;38:22–28.
29. Skaale Ö, Johannesen M. Matur og matgerð. 5th ed. Tórshavn: Bókaforlagið Grønalið; 1990.
30. Nordal A. Über einige norwegische volksmedizinische Skorbut-Pflanzen, und ihren Vitamin-C-Gehalt. Nytt Mag Naturv. 1939;79:193–231.
31. Debes LJ. Færoæ et Færoa Reserata, det er: Færøernis oc færøeske Inbyggeris beskrifvelse. København: Matthias Jørgensøn; 1673.
32. Bjarnason AH. Íslensk flóra með litmyndum. Reykjavík: Iðunn; 1983.
33. Aaronson S. Algae. In: Kiple KF, Ornelas KC, editors. The Cambridge world history of food. Cambridge: Cambridge University Press; 2000. p. 231–249.
34. Albuquerque UP, Hanazaki N. Five problems in current ethnobotanical research - and some suggestions for strengthening them. Hum Ecol. 2009;37(5):653–661.
35. Joensen HD. Um hvonnina í Føroyum, serliga um nýtsluna til matna og um hvannsárið. Varðin. 1951;29:129–154.
36. Simonsen M. Konurák. Tórshavn: Mentunargrunnur Studentafelagsins; 1989.
37. Fosså O. Angelica: from Norwegian montains to the English trifle. In: Hosking R, editor. Wild food: proceedings of the Oxford Symposium on Food and Cookery, 2004. Totnes: Prospect books; 2006. p. 131–142.
38. Mehler M. From self-sufficiency to external supply and famine: foodstuffs, their preparation and storage in Iceland. In: Klápste J, Sommer P, editors. Processing, storage, distribution of food food in the Medieval rural environment. Turnhout: Brepols; 2011. p. 173–186.
39. Gísladóttir H. Substitutes for corn in Iceland. In: Lysaght P, editor. Food from nature. Uppsala: The Royal Gustavus Adolphus Academy for Swedish Folk Culture; 2000. p. 148–154. (Acta Academiae Regiae Gustavi Adolphi LXXI;).
40. Nelson MC, Svanberg I. Lichens as food: historical perspectives on food propaganda. Svenska Linnésällskapets Årsskrift. 1987;1986–1987:7–51.
41. Whitney CW. A survey of wild collection and cultivation of indigenous species in Iceland and the Faroe Islands [Master thesis]. Witzenhausen: University of Kassel; 2011.
42. Whitney CW, Gebauer J, Anderson M. A survey of wild collection and cultivation of indigenous species in Iceland. Hum Ecol. 2012;40(5):781–787.
43. Łuczaj ŁJ, Dumanowski J, Köhler P, Mueller-Bieniek A. The use and economic value of manna grass (*Glyceria*) in Poland from the middle ages to the twentieth century. Hum Ecol. 2012;40(5):721–733.

4

Historical ethnobotanical review of wild edible plants of Estonia (1770s–1960s)

Raivo Kalle[1], Renata Sõukand[2]*

[1] Institute of Veterinary Medicine and Animal Sciences, Estonian University of Life Sciences, Kreutzwaldi 62, 51014 Tartu, Estonia

[2] Estonian Literary Museum, Vanemuise 42, 51003 Tartu, Estonia

Abstract

This paper is a historical ethnobotanical review of wild plants used by the residents of present day Estonia during the 1770s–1960s. Twenty two sources addressing historical ethnographical accounts of the use of wild food plants were analysed. The use of 149 taxa of vascular plants (over 6% of Estonian vascular flora) and two lichens has been recorded. Although the data does not allow for reliable determination of the frequency of use of specific taxa among the population, general conclusions on the preferences for specific dishes made of wild food plants can be made. While the category of snacks covers the largest proportion of species used, a substantial addition to food rations was provided by bread ingredients (used predominantly in famine times), green vegetables used for making soup, and later jams and other dishes of wild berries. Also beverages (tea and coffee substitutes), beer and beer-like drinks were widely made, and the saps of several tree species were consumed in fresh and fermented form. The most important species, according to the criterion of diversity of use, were *Carum carvi*, *Urtica dioica*, and the wild berries *Vaccinium vitis-idaea* and *Vaccinium myrtillus*.

Keywords: wild food plants, historical ethnobotany, wild snacks, food culture, ethnographic and folkloristic records, wild edible plants

Introduction

In the quite recent past, during times of war, crop failure or other cases of food shortage, wild food plants have made a substantial addition to the human diet [1,2]. Nowadays, for the vast majority of the population of modern urbanized Europe, wild food plants seem to be of secondary importance as a resource for human nutrition; their gathering requires much more human labour than cultivated crops and vegetables do, their habitats are far from human settlements, and due to the loss of contact with nature, people simply do not know (or cannot recognize) wild plants. Nevertheless, even in Europe there are still many rural locations, where wild food plants are traditionally used on a daily basis. Such locations in southern and Western Europe are much appreciated by researchers in the field of ethnobotany [3–13]. In Northern Europe the research seems to be completed and it seems that not much can be added to it, as in countries with high industrialization and safe social settings the need for wild food is close to zero and the existing ethnographic resources are already thoroughly researched (for a few examples see [14–16]).

A different situation exists in Eastern Europe within the post-socialist countries. The relatively recent open access to western goods still enables us to remember the use of wild plants on a national level, and research in these areas often gives fruitful results [17,18]. Moreover, some of these countries, like Estonia and its neighbours, have extensive ethnographic collections. While in some of the countries, such as Poland [19–21], such resources have been thoroughly analysed, the content of the others is still unknown to the rest of the world, as they are either not used after collection, or their analyses are published in the native languages only. This makes cross-cultural and geographical analysis difficult or even impossible [21].

To contribute a small stroke to the pattern of the use of wild plants of Europe from the Estonian side, we need first to review the historical use of wild food plants. The written legacy is rather thin: the few publications in German touching on the use of local species for food in the territory of present-day Estonia were written by Baltic German botanophiles [22–29]. Moreover, they often described Estonia and Livonia [22,23] together, some even including Courland [27,28], and they cover only the period until the mid-19th century. Subsequent times are covered by the memory of respondents to folkloristic and ethnographic surveys, and later discussed by a few enthusiasts in Estonian only – dominantly through popular science publications. Hence, the aim of the study is to critically review all the publications concerning wild edibles covering the period 1770s–1960s. The data collected after 1970 has not yet been reflected in any reliable written source, except for a few of the authors' publications.

* Corresponding author. Email: renata@folklore.ee

Material and Methods

Estonia – ecological and historical background

Estonia belongs to the boreo-nemoral vegetation zone, its vegetation is very diverse. Forests, mires and grasslands alternate with cultivated land; meadows and grasslands constitute up to one fifth of it, whereas almost half of the territory is covered with forests, including one third of the area covered by peaty soils, with mixed forest dominating [30–33]. The number of known native and naturalized vascular plant species (with micro-species) of Estonia is estimated at 2200 [34,35].

Present-day Estonia covers an area of 45 thousand km^2 with a population of 1.3 million inhabitants. Within the researched period, until 1918, Estonia was divided between two provinces of Russia (Estonia and Livonia); part of the latter now belongs to Latvia. Also a small area of the territory of present Estonia (Narva and its surroundings) belonged to the St. Petersburg province [36].

Definition of wild food plants

In the context of this article, the term "wild food plants" covers the popular category of wild plants in Estonia, referring mainly to plants growing without intended cultivation. It includes predominantly native and naturalized species, but also cultivars, if the sources suggest parts that are not usually eaten (like *Secale creale* L. and *Linum usitatissimum* L.), or plants, which are cultivated for non-food purposes (like *Aesculus hippocastanum* L.). Some of the species grow in the wild as well as in cultivated settings (like *Armoracia rusticana* Gaertn. et al. and *Ribes nigrum* L.).

The sources of the review

The documentation of traditions of plant use in Estonia began along with the start of documentation of Estonian flora [22], published by Baltic German Estophile, Pastor August Wilhelm Hupel (1737–1819). Although the list of plants was most probably compiled by the student of Carl Linneaus, Jakob Benjamn Fisher (1731–1793) [34], additions on the use and local names were made by Hupel. Hupel's later work also contained some relevant data [23].

Another important figure in Estonian ethnobotany was a German doctor residing on Island Ösel, Johann Wilhelm Ludvig von Luce (1756–1842). He not only wrote an ethnographic overview of the use of plants growing on his home island [24], but also educated peasants on how to survive the times of famine by relying on wild plants [25]. In the same period there was one more attempt to popularise the use of the plants, by another Baltic German Estophile Pastor Johann Heinrich Rosenplänter (1782–1846), but this book is still a manuscript [29]. For more details on the three aforementioned figures and their impact on Estonian ethnobotany follow Kalle and Sõukand [37].

Although there were a few more pieces of information here and there [26–28,38], gathered until the beginning of the 20th century, they rather repeat previous authors and do not differentiate suggetions and ethnographic use reports.

The later publications could be divided into two categories: a critical review of previous authors, and folklore and ethnographical data, including published collections of folklore texts.

The oldest ethnographical data on the use of wild food plants originate from the last decades of the 19th century. However, the systematic collection of relevant information started in the 1930s, when the Estonian National Museum and Estonian Folklore Archives published several questionnaires (e.g. [39–42]), addressing, among other subjects, the use of wild food plants. As the purpose of those collections was to save "antiquity", the data was left untouched until "better times".

The first researcher to analyse responses to the questionnaires from 1930 was ethnographer Aliise Moora (1900–1992). She published her short overviews in the then only popular natural science journal, Eesti Loodus [43–46]; also her monograph on the food of the peasants [47] contains some bits on the use of wild edible plants. Some pieces of this archival data, unpublished elsewhere, were also used in the authors' recent historical overview article in Estonian [48].

Nevertheless, the greatest ever collector and writer of the legacy of wild food plants was a doctor of botany, and for most of his life just a simple schoolteacher, Gustav Vilbaste (1885–1967). He collected his data on his own, mostly in 1920s–1930s, but some also later, until the 1960s. His profound analyses of the use of local plants contain not only his own data, but also critical reviews of earlier publications [49,50]. Still his most important work on wild edible plants is only a manuscript [51]. Another work, a monograph on the local plant names [52], was published almost 30 years after his death. Kalle and Sõukand [53] provide more information on Vilbaste's ethnobotanical impact.

We also use two sources publishing plain, uncommented folklore and ethnographic texts. The first of them, providing a few excerpts on edible wild plants within the framework of farm-life, was published before WW II [54]. The latter publication is an electronic database, containing predominantly data on the medicinal use of the plants, with a few pieces on wild edibles [55]; from it, only records collected before 1970 were evaluated. There are also two publications based on this database [56,57], but only the one published by Sõukand and Kalle [57] has been included in the present review, as the other deals, exceptionally, with medicinal plants. Moreover, as the article on "teetaimed" [57] is based on HERBA database [55], then if the plant was used only for making infusions and no additional data on other uses was available in the database, the published article became the only source in Tab. 1.

Altogether, 22 publications were used for this review. Nevertheless, as some of the publications rely on the same sources or repeat previous statements, and all the results are obtained using different methodologies, it is not reasonable to count the importance of the species based on the number of citations.

Methodology of plant identification

All accessible sources addressing historical ethnographical accounts of the use of wild food plants were analysed. Historical sources written by Baltic Germans and Gustav Vilbaste contained predominantly Latin names of the plants. In the few cases where Gustav Vilbaste had over-identified the species listed by Baltic Germans, the identification was critically evaluated and Vilbaste's identification was considered correct (for example *Urtica urens* L. was changed to *Urtica dioica* L., *Arctium lappa* L. to *Artium tomentosum* Mill.). In the European context the last change may seem unreasonable, since *A. lappa* is a widely eaten species all over Europe. Nevertheless, in Estonia *A. lappa* could be found in only a few places on the coastline, while the uses originate from south Estonia, where *A. lappa* does not grow [58].

The written resources based on folklore and ethnographic collections [43–47,54,55] were critically evaluated and plant

Tab. 1 List of wild food plants used in Estonia in 1770s–1960s.

Family	Latin name	Local name	Parts used	Mode of use	Citing source
Aceraceae	*Acer platanoides* L.	Vaher	sap, flowers, flower buds	drink (fresh and rarely fermented), snacks, infusions, for making food instead of water	[22,24,55,57]
Alliaceae	*Allium oleraceum* L.	Metslook	leaves, shoots, bulbs	spices	[25]
	Allium schoenoprasum L.	Murulauk, leemlaugud	leaves, bulbs	snacks, spices, soup	[24,48,52,55]
	Allium spp.	Lauk	leaves	snacks, salad	[48]
	Allium ursinum L.	Karulauk	leaves	salad, snacks	[23,48]
Araceae	*Acorus calamus* L.	Kalmus	rhizomes*	unspecified food*, snacks	[23,55]
	Calla palustris L.	Soovõhk	rhizomes*	cooked for food*	[28,51]
Asteraceae	*Achillea millefolium* L.	Raudrohi, verihein	aerial part, flowers, stalks	infusions, additive to beer	[47,55,57]
	Antennaria dioica (L.) Gaertn.	Kassikäpp	aerial part	infusions	[57]
	Anthemis tinctoria L.	Rumaska	flowers	infusions	[57]
	Arctium tomentosum Mill.	Kobruleht, takjas	leaves, peeled young stalks*	soup	[22,23,45]
	Artemisia absinthium L.	Koirohi, pänül	aerial part	additive to beer, beer-like drinks	[27,47,48,55]
	Centaurium erythraea Rafn	Maasapp, põldhumalad	aerial part	additive to beer	[22,24,47, 50,52]
	Cichorium intubus L.	Sigur	roots	coffee substitute	[55]
	Cirsium arvense (L.) Scop. var. *mite* (Wimm. et Grab.) Lange	Ohakas, ohtja	young shoots	soup, bread ingredient	[22,23,45, 51,55]
	Matricaria chamomilla L.	Kummelid	flowers	infusions	[29]
	Matricaria spp.	Kummel	flowers, aerial part	infusions	[57]
	Scorzonera humilis L.	Mustjuur	brown dust found in flowers	snacks	[52]
	Sonchus spp.	Piimohakas, ohtja	aerial part	soup, snacks	[22,29,51,55]
	Tanacetum vulgare L.	Reinvars	aerial part	beer additive	[47]
	Taraxacum officinale Webb	Võilill, seaõitsed	leaves, flowers, roots	salad, snacks, coffee substitute, soup*	[25,27,45,55]
	Tragopogon pratensis L.	Kohv, piimhain	young stalks	snacks	[52,55]
	Tussilago farfara L.	Paiseleht	leaves	soup	[28,45]
Berberidaceae	*Berberis vulgaris* L.	Kukerpuu, paburits	fruits, leaves	snacks, infusions, jam, spices for fermented cucumbers, additive to deserts, wine	[22,28,48, 55,57]
Betulaceae	*Alnus* spp.	Lepp	cambium, wood, catkins	snacks, for smoking meat, bread ingredient	[45,47,48, 54,55]
	Betula spp.	Kask, kõiv	sap, leaves, catkins, cambium, bark, sawdust	drink (fresh and fermented), snacks, infusions, bread ingredient, salad, vinegar	[22,24,44,45, 47,54,55,57]
Boraginaceae	*Anchusa officinalis* L.	Mesilill, imikas	nectar, young shoots*	snacks, soup*, salad*	[28,29,52]
	Pulmonaria officinalis L.	Imikas, söögilill	nectar	snacks	[52,55]
Brassicaceae	*Alliaria petiolata* (M. Bieb.) Cavara & Grande	Salukõdrik	aerial part*	unspecified food*	[22]
	Armoracia rusticana P. Gaertn., B. Mey. & Scherb.	Mädarõigas, mäerõigas	roots, leaves	spices for fermented cucumber, additive to food, meat dishes	[48,55]
	Bunias orientalis L.	Tõlk, rakvere raibe	young leaves, (peeled) stalks	snacks, soup	[51]
	Capsella bursa-pastoris (L.) Medik.	Hiirekõrv	aerial part	infusions	[57]
	Cardamine amara L.	Salat, kressid	leaves	salad	[49,52,55]
	Cardamine pratensis L.	Salat, kressid	leaves	salad	[49]
	Crambe maritima L.	Merekapstad	leaves*	soup*	[25]
	Sinapis arvensis L.	Telg, tõlk	seeds, shoots	mustard, soup	[23,24,45,55]
	Thlaspi arvense L.	Litterhein	seeds	spices for soup and blood sausage	[55]

Tab. 1 (continued)

Family	Latin name	Local name	Parts used	Mode of use	Citing source
Cannabaceae	*Humulus lupulus* L.	Humal	fruits, young shoots	additive to beer, beer-like drinks, bread ingredient, soup*, salad*	[22,24,27,45,47,48]
Caprifoliaceae	*Lonicera xylosteum* L.	Kukepuu, ukekuusmaa, sadakordne	fruits	snack	[22,24]
Chenopodiaceae	*Atriplex* spp.	Malts	young shoots	soup	[25,55]
	Chenopodium album L.	Malts, hanemalts	young shoots, seeds	soup, bread ingredient	[29,45,51,52,55]
	Viburnum opulus L.	Lodjapuu, koera õidpuu	fruits	snacks, bread ingredient, jam	[24,48,55]
Cladoniaceae	*Cladonia* spp.	Valge sammal, sammal (?)	thallus	bread ingredient	[38,45,47]
Convallariaceae	*Maianthemum bifolium* (L.) F. W. Schmidt	Viinamari, orava marjad, metsviinamari	fruits	snacks	[29,52,55]
Cornaceae	*Cornus sanguinea* L.	Komppuu, kontpuu	fruits	snacks	[55]
Corylaceae	*Corylus avellana* L.	Sarapuu, pähklipuu	nuts, catkins, buds	snacks, desserts ingredient, bread ingredient, seasonal beverage	[24,27,28,45,47,48,51,55]
Crassulaceae	*Jovibarba sobolifera* (Sims) Opiz (?)	Maarjasõnajalg	young leaves	soup	[55]
	Sedum maximum (L.) Suter	Tuhkkartul	radial tuber, leaves*, young shoots*	cooked for snacks, soup*, salad*	[28,52]
Cupressaceae	*Juniperus communis* L.	Kadakas	cones, twigs, wood	snacks, infusions, jam, drink, spices for food, beer, beer-like drinks, smoking meat, additive to fermented birch sap	[22,24,28,43,47,48,54,55,57,62]
Cyperaceae	*Schoenoplectus tabernaemontani* (C. C. Gmel.) Palla	Merekõrkjas	radial tuber	snacks	[54]
	Unidentified Cyperaceae	Villpea, sootutid, soohein	stalks	snacks	[45]
Dryopteridaceae	Dryopteridaceae (?)	Sõnajalg	leaves, roots	bread ingredient	[47,48,55]
Equisetaceae	*Equisetum arvense* L.	Kesatilgad, kitsenisad, piibusk, seatilk, savipähklid,	spring shoots, radial tuber	soup, snacks (also cooked on fire), bread ingredient	[45,51,55]
Ericaceae	*Arctostaphylos uva-ursi* (L.) Spreng.	Leesikad, siapohl	leaves, fruits	infusions, snacks	[55,57]
	Calluna vulgaris (L.) Hull.	Kanarbik	flowers, aerial part, seeds	infusions, bread ingredient	[45,55,57]
	Empetrum nigrum L.	Kukesilmad, varesemarja	fruits	snack, wine	[28,55]
	Pyrola rotundifolia L.	Talihaljak	leaves	infusions	[57]
	Rhododendron tomentosum Harmaja	Sookail, sookikas, sookaer	aerial part	additive to beer, beer-like drinks	[22,47,48,55]
	Vaccinium myrtillus L.	Mustikas	fruits, leaves, flowers, aerial part	snacks, jam, soup, additive to desserts, infusions	[28,46,51,54,55]
	Vaccinium oxycoccos L.	Kuremari, karbala	fruits	snacks, additive to dessert and sauerkraut	[46,48,51,54,55]
	Vaccinium uliginosum L.	Sinikas	fruits	snacks, wine	[29,51,54,55]
	Vaccinium vitis-idaea L.	Pohl, palohka, poolgas	fruits, flowers, aerial part	snacks, jam, salad, additives to other desserts, stored under water, infusions	[28,29,46,48,51,54,55,57]
Fabaceae	*Lathyrus tuberosus* L.	Seapähkel	radial tuber*	food, unspecified*	[23]
	Lotus corniculatus L.	Virapool	aerial part	infusions	[57]
	Trifolium spp.	Ristik, ristikhein, must härjapea, valge härjapea	flowers	infusions	[55,57]
	Vicia cracca L.	Kurehernes, hiirehernes	seeds	snacks	[29,49]
	Vicia faba L.	Uba	stalks	additive to beer, for colouring	[47]
Fagaceae	*Quercus robur* L.	Tamm	acorn, bark, leaves	coffee substitute, spices for lacto-fermented cucumbers, bread ingredient, spices for beer-like drinks	[24,27,28,45,55,57]
Grossulariaceae	*Ribes alpinum* L.	Mage sõstar, imalad, mammuspuu, naestõmari	fruits	snacks	[52]

Tab. 1 *(continued)*

Family	Latin name	Local name	Parts used	Mode of use	Citing source
	Ribes nigrum L.	Sitikas, must sõstar	fruits, leaves, twigs	snacks, spices for lacto-fermented cucumber, birch sap, infusions, bread ingredient, species for beer-like drinks	[45,47,48,55, 57,62]
	Ribes uva-crispa L.	Kikerberi	leaves	infusions	[57]
Hippocastanaceae	*Aesculus hippocastanum* L.	Kastan	seeds	coffee substitute, jam	[28,55]
Hypericaceae	*Hypericum* spp.	Naistepuna, viinalilled, viinapunad	aerial part, flowers	infusions, spices for blood sausages, white pudding and vodka	[29,52,55,57]
Hypolepidaceae	*Pteridium aquilinum* (L.) Kuhn	Sõnajalg	rhizomes	bread ingredient	[45,48,55]
Lamiaceae	*Galeopsis tetrahit* L.	Mesililled, imikad	nectar	snacks	[29,52]
	Lamium album L.	Iminõges, imikas, imik, emanõges, naistenõges, miiksmaasikad	nectar, flowers, young plants	snacks, infusions, soup	[28,45,52,55]
	Mentha spp.	Münt, vehverments, piparmünt, põldmünt, vesimünt	aerial part, leaves, flowers	infusions, additive to food e.g. blood sausages and white pudding, species for vodka	[50,55,57]
	Origanum vulgare L.	Pune, vorstirohi	aerial part, leaves, flowers	infusions, spices for a variety of food, blood sausages and white pudding, additive to beer, soup	[27,28,47,50, 55,57]
	Prunella vulgaris L.	Metspiparmünt	aerial part	infusions	[57]
	Thymus serpyllum L.	Liivatee, viinapuna	aerial part, flowers, leaves	infusions, spices for food, additive to vodka	[28,50,55,57]
Linaceae	*Linum usitatissimum* L.	Lina	capsule	bread ingredient	[45]
Malvaceae	*Malva pusilla* Sm.	Kassirattad, kassinaerid	aerial part	soup*, snacks	[25,55]
Menyanthaceae	*Menyanthes trifoliata* L.	Ubaleht	aerial part	additive to beer	[24,28,48]
Myricaceae	*Myrica gale* L.	Porss	leaves, twigs	additive to beer	[24,28,47,48]
Nymphaeaceae	*Nymphaea candida* C. Presl	Vesiroos	rhizomes*	milled in food*	[28,51]
Oleaceae	*Syringa vulgaris* L.	Sirinäs	flowers	snacks for luck	[55]
Onagraceae	*Epilobium angustifolium* L.	Kaporuski	young stalks*, leaves	soup*, stew*, infusions	[27,28,57]
Orchidaceae	*Orchis morio* L.	Jänese munad	tubers*	bread ingredient*, porridge*	[25]
	Platanthera bifolia (L.) L. C. Rich.	Jänese munad	tubers*	bread ingredient*, porridge*	[25]
Oxalidaceae	*Oxalis acetosella* L.	Jänese hapuoblikas, kikikapsas, saksamaa oblikad, jänese kapstad	leaves, flowers	snacks, soup	[24,25,27,45, 51,54,55]
Papaveraceae	*Fumaria officinalis* L.	Kolmekõrraline hain	aerial part	infusions	[57]
	Papaver somniferum L.	Moonid	seeds	snacks, additive on home-backed roll	[55]
Parmeliaceae	*Cetraria islandica* (L.) Ach.	Nõmmesammal, põdrasammal, islandi sammal, liivasamblad, sammal (?)	thallus	bread ingredient, infusions	[25,38,45,47, 55,57]
Pinaceae	*Picea abies* (L.) H. Karst.	Kuusk	shoots (young), needles, cambium, cones, resin	snacks, smoking of meat	[45,48,54,55]
	Pinus sylvestris L.	Pedaja, mänd, pettäi	shoots (young), cambium, cones, bark	snacks , bread ingredient, smoking of meat	[45,47,48,55]
Plantaginaceae	*Plantago major* L.	Teeleht	leaves	infusions, soup, salad*	[22,27–29,48, 50,57]
Poaceae	*Bromus secalinus* L.	Lusted, kaer	seeds	distilled vodka, beer ingredient	[55]
	Phragmites australis (Cav.) Trin. ex Steud.	Roog	roots	snacks	[48,54]
	Poaceae	Aganad, kõrred	soft part of stalk, chaff	snacks, bread ingredient	[45,47,48]
	Secale cereale L.	Rukis	young crops	infusions	[57]
Polygonaceae	*Polygonum amphibium* f. *terrestre* (Leers) S. F. Blake	Seavinnal	roots	snacks	[55]

Tab. 1 *(continued)*

Family	Latin name	Local name	Parts used	Mode of use	Citing source
	Rumex acetosa L.	Oblikas, hapuoblikas	leaves	snacks, soup, pie	[23,25,29,45, 51,54,55]
	Rumex acetosella L.	Hapuoblikas	leaves, young stalks	snacks	[24,52]
Polypodiaceae	*Polypodium vulgare* L.	Magusjuur, suhkrupuu, lagrits	rhizomes	snacks	[55]
Primulaceae	*Lysimachia nummularia* L.	Trudamorda	aerial part	infusions	[57]
	Lysimachia vulgaris L.	Metsvits	medulla*	snacks*	[27,50]
	Primula veris L.	Nurmenukk, käekaatsed	flowers, leaves, steams, pistil, aerial part, nectar, flower stalks	snacks, infusions, snacks for luck, wine*, beer-like drinks*	[24,27,49,54, 55,57]
Ranunculaceae	*Caltha palustris* L.	Varsakabi, konnakapsad	aerial part, buds*	bread ingredient, substitute for capers*	[24,45,47,49]
	Ranunculus ficaria L.	Kanakoole	young shoots, tubers	salad (tubers were used only after soaking in salted water)	[23,50,51]
	Thalictrum aquilegiifolium L.	Ängelhein, kirnputk, koerkuseputk, koerputk	flowers	infusions	[57]
Rosaceae	*Alchemilla vulgaris* auct. (coll.)	Krooklehed, kortsleht	leaves	infusions	[55,57]
	Crataegus spp.	Leivamari, viirpuu, leivapuu	fruits	snacks, bread ingredient	[48,55]
	Filipendula ulmaria (L.) Maxim.	Angervaks, viinarohi	flowers, aerial part	infusions, spices for vodka	[52,55,57]
	Filipendula vulgaris Moench	Angerpist, viinarohi	flowers, aerial part, radial tuber*	infusions, bread ingredient*, spices for vodka	[28,49,55,57]
	Fragaria vesca L.	Maasikas, metsmaasikas	fruits, leaves, flowers, aerial part	snacks, jam, infusions, additive to dessert	[24,27,29,46, 51,54,55,57]
	Fragaria viridis A. Duch.	Muulukas, mullikmaasikad	fruits	snacks, jam, additive to dessert	[49,51,55]
	Geum rivale L.	Mesikupp, karukellad, ärjamürakad, härjapea, surnu sukapael	flowers, nectar, roots*	snacks, infusions, spices for food*	[24,52,55,57]
	Geum urbanum L.	Maamõõl	young shoots*, roots	food*, spices*	[24,27]
	Malus domestica Borkh.	Õunapuu	fruits, leaves, wood	snacks, wine, vinegar, infusions, additive to sauerkraut, smoking of meat	[48,54,55,57]
	Malus sylvestris Mill.	Metsõunapuu	fruits	snacks, wine	[24,29,48,51]
	Potentilla erecta (L.) Raeusch	Tedre madar, tedremaran	rhizomes	cooked for food, additive to vodka	[29,45,48]
	Prunus cerasus L.	Kirsipuu	fruits, leaves, flowers	snacks, infusions, spices for fermenting cucumbers	[55,57]
	Prunus domestica L. subsp. *insititia*	Kreegipuu	leaves	infusions	[57]
	Prunus domestica L.	Ploomipuu	flowers	infusions	[57]
	Prunus padus L.	Toomingas	fruits, leaves, flowers	snacks, spices for fermenting cucumbers, snacks for luck	[24,46,51, 54,55]
	Rosa spp.	Kibuvits, orjavits, roos	fruits, flowers	snacks, infusions, jam, kissel, bread ingredient, coffee, dessert	[24,48,54, 55,57]
	Rubus caesius L.	Põldmurakas, põldmari	fruits	snacks, wine, jam	[28,49,52, 54,55]
	Rubus chamaemorus L.	Murakas, käbalad	fruits, leaves, flowers	snacks, infusions, jam	[28,29,51,54, 55,57]
	Rubus idaeus L.	Vaarikas, vabarn	fruits, leaves, twigs	snacks, infusions, additive to fermented cucumbers, jam, coffee substitute, wine	[24,27,51,54, 55,57]
	Rubus saxatilis L.	Lillakas	fruits	snacks, wine	[51,55]
	Sorbus aucuparia L.	Pihlakas	fruits, flowers, sap	snacks, jam, wine, infusions, bread ingredient, kvass	[27,44,46–48, 51,54,55]
	Sorbus intermedia (Ehrh.) Pers.	Leivamari, pooppuu	fruits	snacks, bread ingredient	[48,55]
	Sorbus rupicola Hedl.	Kirsi-pooppuu	fruits	bread ingredient	[55]

Tab. 1 *(continued)*

Family	Latin name	Local name	Parts used	Mode of use	Citing source
Salicaceae	*Populus tremula* L.	Haab	wood, sawdust	smoking of meat, bread ingredient, vinegar	[47,48,55]
	Salix spp.	Paju	leaves	bread ingredient	[45]
Scrophulariaceae	*Pinguicula vulgaris* L.	Võipätakas	aerial part	additive for souring the milk	[27,28,50]
	Rhinanthus spp.	Robirohi, luutsihain	nectar	snacks	[50,52]
	Verbascum spp.	Üheksavägine	flowers	infusions	[57]
Solanaceae	*Solanum tuberosum* L.	Kartul	aerial part	soup	[54,55]
Tiliaceae	*Tilia* spp.	Pärn, niin, pähnapuu	flowers, buds, sap, twigs and flower buds	infusions, snacks	[44,55,57]
Typhaceae	*Typha* spp.	Hunditõlv, hundipurik, hundikurk	rhizomes, young shoots*	snacks, salad*, stew*	[24,28,45,48]
Umbelliferae	*Aegopodium podagraria* L.	Naat, varesnaat	young leaves, shoots and stalks	soup, bread ingredient in spring time	[22,24,25,45, 51,54]
	Angelica sylvestris L.	Pütski, heinputk, söögiputk	young shoots, peeled stalks, aerial part	soup, snacks, infusions	[45,49,51,54, 55,57]
	Anthriscus sylvestris (L.) Hoffm.	Pütski, penipütsk	young shoots, peeled stalks	soup, snacks	[45,49,51]
	Carum carvi L.	Köömen, köömlid	seeds, leaves, stems	spices for variety of food, bread, blood sausages and white pudding, vodka, sauerkraut, fermented cucumbers, birch sap and other preserves, snacks, soup, infusions, additive to beer-like drinks and cottage cheese	[22,23,25,29, 47,48,51,55, 57,62]
	Heracleum sphondylium L. (?)	Karuputk, natid	young shoots, peeled stalks	soup, snacks	[22–24,27,48]
	Peucedanum palustre (L.) Moench	Soo-ingver, ingver	roots	spices for home-made Baltic sprat	[49]
	Pimpinella anisum L.	Ingver, maaingver	roots	spices for food	[55]
Urticaceae	*Urtica dioica* L.	Nõges, kõrvenõges, raudnõges	leaves, aerial part (young), stalks, shoots, roots	soup, infusions, bread ingredient, for smoking meat	[22,23,25,26, 45,48,51,54, 55,57]
Valerianaceae	*Valeriana officinalis* L.	Palderjan	aerial part, flowers, roots	infusions	[57]
Violaceae	*Viola odorata* L.	Kannike	flowers	dessert, spices for syrup and vinegar*	[28,55]

(?) – name did not have enough details for specific identification. * Data of uncertain ethnic origin.

names identified through using local names [52]. Identification changes were made where necessary, relying on the work of Sõukand and Kalle [55]. For example identifications followed by Moora [45]: *Dryopteris* spp. to *Pteridium aquilinum* (L.) Kuhn. The previously unidentified species *Sinapis arvensis* L. was also added to the list based on its local name ("tõlk"). Based on the same principles two plants identified in sources relying on folklore were excluded: *Angelica archangelica* L. and *Conium maculatum* L. The latter was named kaljaputk, indicating its use for making kvass. Regardless of the fact that some nations do use deadly poisonous plants for food after certain preparations (e.g. [59]), the text claiming the use of kaljaputk did not contain any other reliable information, nor a description of its preparation allowing for its identification as *C. maculatum*.

If the local plant name listed in the sources did not allow for identification on the species level, a taxon was identified as a genus. The list of all species that could potentially be eaten within the particular genus is provided in Tab. 2. Nevertheless, two taxa listed in folklore and written sources did not allow for identification below the family (Cyperaceae and Poaceae), and in both families the softer part of the stalks of several species were eaten as snacks.

The texts in which plant identification credibility (discussed by Łuczaj [60]) was very low or the taxon was impossible to detect, were left aside.

The reliability and completeness of the presented data

Compared to Estonian serfs, the food selection of Baltic German landlords was much more diverse and contained many wild spices, used to make salads and soups; as well as numerous wild berries used to make juice, wine, salads and jams. Therefore, Baltic Germans were trying to educate local peasants through books and personal advice [22,25].

Memories of the earliest folklore texts reach back to the mid-19th century. Therefore they do not contain the specific species found in Hupel's and Luce's publications or later interpretations in Vilbaste's study [51] (see uses marked with "*" in Tab. 1). The original publications are written in a manner that does not enable us to understand where, or, more particularly, who used to make salad from the leaves of *Taraxacum officinale*

Tab. 2 The species potentially named in texts where the taxa could be identified to the genus only. Species are presented in the order of assumed use frequency.

Genera	Species potentially used
Allium spp.	*A. vineale* L., *A. schoenoprasum* L., *A scorodoprasum* L., *A oleraceum* L., *A. ursinum* L.
Alnus spp.	*A. glutinosa* (L.) Gaertn., *A. incana* (L.) Moench
Atriplex spp.	*A. hortensis* L., *A. patula* L.
Betula spp.	*B. pendula* Roth, *B. pubescens* Ehrh.
Cladonia spp.	*C. stellaris* (Opiz) Pouzar & Vězda, *C. stygia* (Fr.) Ruoss, *C. mitis* Sandst., *C. arbuscula* (Wallr.) Flot.
Crataegus spp.	*C. palmstruchii* Lindm., *C. rhipidophylla* Gand., *C. sanguinea* Pall.
Hypericum spp.	*H. perforatum* L., *H. maculatum* Crantz
Matricaria spp.	*M. chamomilla* L., *M. recutita* L.
Mentha spp.	*M. aquatica* L., *M.* × *piperita*, *M. crispa* L., *M. arvensis* L.
Primula spp.	*P. veris* L., *P. farinosa* L., *P. elatior* (L.) Hill
Rhinanthus spp.	*R. angustifolius* C. C. Gmel. s.l., *R. minor* L.
Rosa spp.	*R. rugosa* Thunb, *R. majalis* Herrm., *R. vosagiaca* N. H. F. Desp., *R. subcanina* (H. Christ) Dalla Torre et Sarnth., *R. mollis* Sm.
Salix spp.	*S. myrsinifolia* Salisb., *S. aurita* L., *S. cinerea* L., *S. phylicifolia* L., *S. triandra* L.
Sonchus spp.	*S. arvensis* L., *S. oleraceus* L.
Tilia spp.	*T. cordata* Mill., *T. platyphyllos* Scop.
Trifolium spp.	*T. repens* L., *T. pratense* L., *T. montanum* L., *T. aureum* Pollich, *T. spadiceum* L.
Typha spp.	*T. angustifolia* L., *T. latifolia* L.
Verbascum spp.	*V. nigrum* L., *V. thapsus* L.

Webb, soup of leaves of *Crambe maritima* L. or peeled stems of *Arctium tomentosum* Mill.; who used to make various food from the rhizomes of *Nymphaea candida* C. Presl, *Calla palustris* L., *Acorus calamus* L., the tubers of Orchidaceae and *Lathyrus tuberosus* L. or the aerial parts of *Alliaria petiolata* (M. Bieb.) Cavara et Grande; or who used the buds of *Caltha palustris* L. as a substitute for capers, etc. The actual use of all those species by Estonian peasants cannot be confirmed, but cannot be disproved either. These species could have been used by local Baltic Germans, Swedes, or Russians, or on the territory of Livonia.

Although the majority of the cited authors use folkloristic and ethnographic sources, the whole existing archival data on the use of wild food plants in Estonia has never been digitized nor thoroughly systematized. Until this is done, it is impossible to state if any more taxa were historically used.

Results and discussion

During the research period, altogether 149 vascular plant taxa (approx. 6.6 % of native and naturalized flora of Estonia) and 2 lichens were consumed (Tab. 1). If we exclude from the calculation those 19 species used for making infusions (recreational teas) only, there will be 130 taxa left, covering almost 5.8% of the Estonian vascular flora. All the taxa belong to 123 genera from 56 families [largest Rosaceae (22 taxa) and Asteraceae (16 taxa)]. The list includes 18 tree taxa, 15 shrubs, 11 subshrubs and one vine. Among the herbaceous plants 76 are perennials, 6 biennials and 5 annuals.

Most of the taxa were used in a variety of ways, utilizing various parts of the plant, only a few plants were used for one specific purpose (usually infusions or snacks only). The species that were used for infusions only had popularly recognized medicinal properties, and snacks were occasionally eaten raw. Among the most diversely consumed plants we can single out *Carum carvi* L., *Urtica dioica* L., and a variety of wild berries (*Vaccinium vitis-idaea* L., *Vaccinium myrtillus* L., etc) and fruits (*Sorbus aucuparia* L., *Rosa* spp., etc).

Snacks

Seasonal snacks form the largest group of wild plants used, covering almost half of all the species (72). Most of the snacks (fruits, nectar, leaves, buds, cambium, stalks, etc.) were eaten in outdoor conditions without prior processing, while some were cooked [as tubers of *Sedum maximum* (L.) Suter and *Equisetum arvense* L.], frozen (such as fruits of *Viburnum opulus* L., *Malus sylvestris* Mill. and *Sorbus aucuparia* L.), or dried (*Corylus avellana* L. and *Vaccinium myrtillus* L.). While in earlier ethnobotanical research snacks were associated predominantly with children [60,61], in 19th century Estonia snacks were eaten by both children and adults, although more recently snacks tended to be consumed by children. Still, specific plants have been used as children's snacks only, such as the rhizomes of *Polypodium vulgare* L (also used as medicine instead of *Glycyrrhiza glabra* L.), the dark "coffee" dust in the flowers of *Scorzonera humilis* L., a species used for sucking nectar, and a few others. Sweet snacks with sugar added were rather uncommon among peasants until the first decades of the 20th century due to the lack of sugar, but after that many berries were eaten with sugar and milk (later cream) added. Some other sweets were also made of flowers as candy, for example of *Viola odorata* L. and *Rosa* spp.

Beverages

Beverages made of wild plants are the second largest group of wild food consumed. As many as 54 vascular plants and one lichen were used to make infusions [57]. In the time of scarcity, the infusion made of well-tasting plants was especially appreciated as it gave better taste to the food. Twelve taxa were used for making or flavouring beer and eight for beer-like beverages. Several parts (acorns, roots, twigs) of six species were used to make coffee. Also saps of different trees were used (fresh or fermented), the most widespread being saps of *Betula* spp. and

Acer platanoides L. (see also [62]). Fruits of only eight taxa were used for making wine; this practice started only at the beginning of the 20th century, when sugar became more accessible.

Bread ingredients

In the 18th and the first half of the 19th century the main foods for peasants were bread and other products made of flour [47]. Thus every crop failure caused by bad weather brought along large-scale famine [63]. This explains why a large group of vascular plants (30) and the thallus of two lichens were used as bread ingredients. A vast variety of plant parts was used for filling bread, including (dried) bulbs, leaves, shoots, cambium and bark. Some plants, especially powdered roots and young shoots, were added to the bread to "prolong" the flour; laced bread was called "näljaleib" (hunger-bread) in times of famine, and "hädaleib" (need-bread) before the new crop. Some of the ingredients that had pleasant taste were used also for flavouring bread in better times, for example *Sorbus* spp. fruits, also *Carum carvi* fruits were used, along with providing better taste to ensure longer preservation of the bread.

During the second half of the 19th century, when potato became widespread, the nutritional situation of peasants became better. The need for wild plants as bread ingredients gradually diminished, until it almost disappeared by the 1960s, partly also because, due to socio-economic changes, bread was then made at home by only a very few people. Still, few wild ingredients were later also used in making home-made rolls.

Soups and gruel

Greens (*U. dioica*, *Rumex* spp. and *Cirsium* spp., being most mentioned in folklore among them) that popped up after the snowmelt, were used for making a variety of dishes such as soup and porridge during food shortages. This happened almost every year to a greater or lesser extent, and the quantity of the added greens depended on the stocks of flour and grain available, but also on regional and household food habits. Altogether, 26 species were used in the form of soup, which argues against the statement of Baltic German authors that the monotonous diet of Estonians contained a very limited amount of green vegetables [22,25]. Moreover, some sources from the 18th century complain that local peasants eat too many greens in the spring famines and die of it [63].

Spices

Altogether 18 taxa have been used for seasoning different foods, the majority of them very commonly. Eight taxa were used as ingredients for lacto-fermented cucumbers (mostly leaves, but also fruits, twigs and roots); three taxa were used for seasoning sauerkraut and another three for flavouring soured birch sap. Some of the herbs used for infusions were also used to season vodka (six taxa) and blood sausages (four taxa). Eight taxa were used as additives to beer. Seven taxa were used to smoke meat or fish; mostly wood, but also leaves and cones. Along with the seasoning purposes, spices were intended to improve the storing properties of the foods they were used in, for example a handful of *Carum carvi* in bread batter, a few segments of *Armoracia rusticana* root in cucumber preserves, a few inflorescences of *Rhododendron tomentosum* Harmaja. for a bucket of beer, etc.

Fruits and seeds

Of the fleshy fruits, 14 were fruits of wild shrubs and semi-shrubs and 11 were fruits of wild trees, all of them were consumed as snacks, but some were also used for making food, or as spices, and/or preserved for the winter in different ways. Until the end of 19th century the preservation of wild fruits was rather rare and mostly carried out in the form of drying or freezing, also under-water storage, and the fruits were used predominantly for healing purposes. Since the beginning of the 20th century, the wider making of jams (for food) began, reaching its peak in the1970s–1980s.

Although the majority of the seeds and dry fruits (7 species) were used as snacks only, there were also a few used for making food (nuts of *Corylus avellana* L. and acorns of *Quercus robur* L.). Also, the caryopses (grains) of *Bromus secalinus* L. were used to make vodka and beer; and seeds of *Sinapis arvensis* L. were used as the main component of mustard.

Salads

For many species, the green parts of the vegetable such as leaves and stalks were consumed. However, the use of wild green vegetables for making salad was rather limited, only 11 species were consumed. The use of salad seems to be a clear imitation of the landlords' food-habits. For example two local species *Cardamine amara* L. and *Cardamine pratensis* L. were named after *Lepidium* spp., cultivated and consumed in manors [49]. The use of the other greens for salad is of quite late origin (beginning of 20th century).

Underground parts

Underground parts of the plants were consumed rather rarely, being predominantly either famine food or snacks; altogether 24 taxa were used. Of them, 7 species with underground parts used were of uncertain ethnic origin (marked with "*" in Tab. 1; see "The reliability and completeness of the presented data" section for details). Underground parts were mostly used as bread ingredients, but also as snacks [*Equisetum arvense* L., *Phragmites australis* (Cav.) Trin. ex Steud., *Polypodium vulgare* L.] and cooked food [*Potentilla erecta* (L.) Raeusch, *Sagittaria sagittifolia* L.]. Roots of some taxa were also used as spices (*Armoracia rusticana* Gaertn. et al.), or for making infusions (*Valeriana officinalis* L.) or coffee (*Taraxacum officinale*).

Alien plants, literature and education

While the local habitants had a long time to acquire knowledge of the native plants, the quick introduction of alien plants into the diet of Estonians came about mostly through the popularizing literature. At the end of the 19th century, general literacy in Estonia was the highest in all of the Russian Empire, being almost 90% at the end of the 19th century [64]. The first alien taxon of cultivated ornamental plants introduced for food was *Caragana arborescens* Lam. [65], and although there are no records of its use during the period of the research, probably due to its very early publication date, its seeds and flowers were eaten later (unpublished fieldwork results). Nevertheless, the use of the conkers of *Aesculus hippocastanum* L. is clearly of literature origin, as its use for making coffee was widely propagated at the beginning of the 20th century (for example see [66,67]).

The use of native plants was also acquired through the literature, like the use of the thallus of *Cetraria islandica* (L.) Ach. as a bread ingredient from [25] and the teachings of local Baltic Germans. Estonian pioneer doctor and the author of national epic Friedrich Reinhold Kreutzwald (1803–1882) reports that flour made of soaked and dried thallus saved town citizens of Võru from famine in the 1830s, as the town's mayor supported its use for making bread [38]. Another lichen taxon, *Cladonia* spp., was also used instead of *C. islandica*, as people often did not differentiate them on the name level [68].

Nevertheless, most of the teachings communicated through the literature were not accepted. For example the very popular first herbal written in Estonian by Baltic German pastor Otto August Jannau (1800–1865) teaches the use of *Glyceria* spp. for making manna porridge [69], but despite the fact that this plant was widely eaten in parts of central Europe [70], Estonian peasants did not adopt its use, probably because they already had potatoes and severe famines were only history at the time the book was published.

The practice of making of jam from wild berries most probably first came from the book by local schoolteacher and pomologist Jaan Spuhl-Rotalia (1859–1916), who taught the use of almost all wild berries growing in Estonia [66]. Also, the making of jams was taught on numerous cooking courses conducted in the first half of the 20th century all over Estonia.

Future discussion

It is difficult to compare the use of wild food plants in Estonia with that of neighbouring regions (Latvia, Finland, Russia), as they lack detailed ethnographic reviews. Still, large proportions of Rosaceae and Asteraceae make Estonia similar to other eastern European countries, e.g. Poland and Slovakia [21,61]. The most important detail needing attention is the low use of salads made of fresh leafy wild vegetables. That may be caused by their short-term availability in Estonia, but also by the need for hot food due to climatic conditions and the need for easier access to nutrients. In general, the large proportion of use as snacks among wild edible plants indicates a relatively good knowledge of the plants and their edibility among the population of Estonia in the past. Notable consumption of wild berries up to near-modern times reflects the fact that they are better appreciated than green vegetables.

Acknowledgments

The authors acknowledge the Governmental Research and Development programme "Estonian language and cultural memory" (EKKM09-84) for supporting the digitalisation of Estonian herbal lore. The research has been supported by ESF grant ETF9419. Many thanks to Heldur Sander for valuable references, to Tõnu Ploompuu for help with identifying species listed in ethnographic sources [*Pteridium aquilinum* (L.) Kuhn and *Cetraria islandica* (L.) Ach.], to Sarah Luczaj for language editing, to the guest editor Łukasz Łuczaj for support and inspiration and managing editor Piotr Otręba for willingness to help in improving the final version of the manuscript.

References

1. Turner NJ, Łuczaj ŁJ, Migliorini P, Pieroni A, Dreon AL, Sacchetti LE, et al. Edible and tended wild plants, traditional ecological knowledge and agroecology. Cr Rev Plant Sci. 2011;30(1-2):198–225.
2. Redžić S. Use of wild and semi-wild edible plants in nutrition and survival of people in 1430 days of siege of Sarajevo during the war in Bosnia and Herzegovina (1992–1995). Coll Antropol. 2010;34(2):551–570.
3. Pieroni A, Nebel S, Quave C, Münz H, Heinrich M. Ethnopharmacology of liakra: traditional weedy vegetables of the Arbëreshë of the Vulture area in southern Italy. J Ethnopharmacol. 2002;81(2):165–185.
4. Guarrera PM. Food medicine and minor nourishment in the folk traditions of Central Italy (Marche, Abruzzo and Latium). Fitoterapia. 2003;74(6):515–544.
5. Tardío J, Pascual H, Morales R. Wild food plants traditionally used in the province of Madrid, central Spain. Econ Bot. 2005;59(2):122–136.
6. Pieroni A, Nebel S, Santoro RF, Heinrich M. Food for two seasons: culinary uses of non-cultivated local vegetables and mushrooms in a south Italian village. Int J Food Sci Nutr. 2005;56(4):245–272.
7. Rivera D, Obon C, Inocencio C, Heinrich M, Verde A, Fajardo J, et al. The ethnobotanical study of local Mediterranean food plants as medicinal resources in Southern Spain. J Physiol Pharmacol. 2005;56(1 suppl):97–114.
8. Tardio J, Pardo-De-Santayana M, Morales R. Ethnobotanical review of wild edible plants in Spain. Bot J Linn Soc. 2006;152(1):27–71.
9. Guarrera P, Salerno G, Caneva G. Food, flavouring and feed plant traditions in the Tyrrhenian sector of Basilicata, Italy. J Ethnobiol Ethnomed. 2006;2(1):37.
10. Pardo-de-Santayana M, Tardío J, Blanco E, Carvalho A, Lastra J, San Miguel E, et al. Traditional knowledge of wild edible plants used in the northwest of the Iberian Peninsula (Spain and Portugal): a comparative study. J Ethnobiol Ethnomed. 2007;3(1):27.
11. Leonti M, Nebel S, Rivera D, Heinrich M. Wild gathered food plants in the European Mediterranean: a comparative analysis. Econ Bot. 2006;60(2):130–142.
12. Parada M, Carrió E, Vallès J. Ethnobotany of food plants in the Alt Emporda region (Catalonia, Iberian Peninsula). J Appl Bot Food Qual. 2011;84(1):11–25.
13. Menendez-Baceta G, Aceituno-Mata L, Tardío J, Reyes-García V, Pardo-de-Santayana M. Wild edible plants traditionally gathered in Gorbeialdea (Biscay, Basque Country). Genet Resour Crop Evol. 2011;59(7):1329–1347.
14. Svanberg I, Nelson MC. Bone meal porridge, lichen soup, or mushroom bread: acceptance or rejection of food propaganda 1867–1868. In: Häkkinen A, editor. Just a sack of potatoes? Crisis experiences in European societies, past and present. Helsinki: Societas Historica Finlandiae; 1992. p. 119–147.
15. Svanberg I. The use of wild plants in the Faroe Islands 1590–1990: a contribution to Scandinavian ethnobotany. Swedish Linnean Society. 1998;1996–1997:81–130.
16. Svanberg I. Folklig botanik. Stockholm: Dialogos; 2011.
17. Redzic S. Wild edible plants and their traditional use in the human nutrition in Bosnia-Herzegovina. Ecol Food Nutr. 2006;45(3):189–232.
18. Pieroni A. Local plant resources in the ethnobotany of Theth, a village in the Northern Albanian Alps. Genet Resour Crop Evol. 2008;55(8):1197–1214.
19. Łuczaj Ł, Kujawska M. Botanists and their childhood memories: an underutilized expert source in ethnobotanical research. Bot J Linn Soc. 2012;168(3):334–343.
20. Łuczaj Ł. Changes in the utilization of wild green vegetables in Poland since the 19th century: a comparison of four ethnobotanical surveys. J Ethnopharmacol. 2010;128(2):395–404.
21. Łuczaj Ł, Szymański WM. Wild vascular plants gathered for consumption in the Polish countryside: a review. J Ethnobiol Ethnomed. 2007;3(1):17.
22. Hupel AW. Topographische Nachrichten von Lief- und Ehstland: Nebst

vollständigen Register über alle drey Bände. Riga: Hartknoch; 1777. (vol 2).
23. Hupel AW. Die gegenwärtige Verfassung der Rigischen und der Revalschen Statthalterschaft: Zur Ergänzung der topographischen Nachrichten von Lief- und Ehstland. Riga: Hartknoch; 1789.
24. von Luce JWL. Topographische nachrichten von der insel Oesel, in medicinischer und ökonomischer hinsicht. Riga: W. F. Häcker; 1823.
25. von Luce JWL. Nou ja abbi, kui waesus ja nälg käe on. Riga: J. H. Gressel; 1818.
26. Petri JC. Ehstland und die Ehsten, oder historisch-geographisch-statistisches Gemälde von Ehstland: Ein Seitenstück zu Merkel über die Letten. Gotha: Ettingerschen Buchhandlung; 1802. (vol 1).
27. Friebe WC. Oekonomisch-technische flora für Liefland, Ehstland und Kurland. Riga: Hartmann; 1805.
28. Wiedemann FJ, Weber E. Beschreibung der phanerogamischen Gewäschse Esth-, Liv- und Curlands mit möglichst genauer Angabe der Fundorte und der geographischen Verbreitung nebst Andeutung über den Gebrauch in medicinischer, technischer und öconomischer Beziehung. Reval: Franz Kluge; 1852.
29. Rosenplänter JH. Õppetuse katse ma Rohtudest ja pudest, mis J. H. Rosenplänter kirjutanud [Manuscript]. Tartu: Estonian Literary Museum; 1831.
30. Valk U. Eesti Sood. Tallinn: Valgus; 1988.
31. Paal J. Rare and threatened plant communitiesof Estonia. Biodivers Conserv. 1998;7(8):1027–1049.
32. Peterson K. Nature conservation in Estonia. Tallinn: Huma; 1994.
33. Laasimer L. Eesti NSV taimkate. Tallinn: Valgus; 1965.
34. Kukk T. Eesti taimestik. Tallinn: Estonian Academy Publishers; 1999.
35. Ööpik M, Kukk T, Kull K, Kull T. The importance of human mediation in species establishment: analysis of the alien flora of Estonia. Boreal Env Res. 2008;13:53–67.
36. Wikipedia, the free encyclopedia [Internet]. Estonia. 2012 [cited 2012 Aug 21];
37. Kalle R, Sõukand R. Collectors of Estonian folk botanical knowledge. BJES. 2011;1:213–229.
38. Kreutzwald FR. Leivast ja leiva-jätkust. In: Tarto- ja Wõrroma kalender ehk Täht-ramat 1842 ajastaja päle, perran meie Issanda Jesusse Kristusse sündimist. Tartu: Schünmann; 1841. p. 42–51.
39. Manninen I, Linnus F. Söögid, joogid, maitsetaimed. Söögikombed. Rahvateaduslik küsimuskava. Tartu: Estonian National Musem; 1937. (Küsimusleht; vol 10).
40. Tampere H. Mahl. Tartu: Estonian National Musem; 1934. (Küsimusleht; vol 5).
41. Sion V. Seene-, marja-, pähkli- ja muu taimetoidu korjamisest. Tartu: Estonian National Musem; 1947. (Küsimusleht; vol 43).
42. Viidalepp R. Juhiseid rahvaluulekogujatele: mida ja kuidas koguda. Tartu: Estonian National Musem; 1936. (ERA Küsimuskava; vol 3).
43. Moora A. Kuidas vanasti kadakamarju kasutati. Eesti Loodus. 1984;6:378–380.
44. Moora A. Mida vanasti kasemahlast tehti. Eesti Loodus. 1982;5:298–300.
45. Moora A. Mida vanasti loodusest leivakõrvaseks korjati. Eesti Loodus. 1981;8:489–497.
46. Moora A. Marjad rahvatoidus. Eesti Loodus. 1980;8:588–592.
47. Moora A. Eesti talurahva vanem toit. 2nd ed. Tartu: Greif; 2007.
48. Kalle R, Sõukand R. Ajalooline ülevaade eestlaste looduslikest toidu- ja ravimtaimedest. In: Kusmin T, Meikar T, editors. Metsa kõrvalkasutus Eestis. Tartu: Estonian Society of Foresters; 2011. p. 29–44. (Akadeemilise Metsaseltsi Toimetised; vol 25).
49. Vilberg G. Meie kodumaa taimi rahva käsitluses. Tartu: Loodusevaatleja; 1934. (vol 1).
50. Vilberg G. Meie kodumaa taimi rahva käsitluses. Tartu: Loodusevaatleja; 1935. (vol 2).
51. Vilbaste G. Söödavaid taimi [Manuscript]. Tartu: Estonian Literary Museum; 1950.
52. Vilbaste G. Eesti taimenimetused. Tallinn: Estonian Academy of Sciences; 1993. (Emakeele Seltsi Toimetised; vol 20).
53. Kalle R, Sõukand R. Gustav Vilbaste (1885–1967) and ethnobotany in Estonia. Pioneers in European ethnobiology. In: Svanberg I, Łuczaj Ł, editors. A schoolteacher with a mission: preserving disappearing plant names. Uppsala: Royal Gustavus Adolphus Academy. In press.
54. Loorits O. Lugemispalu metsaelust ja jahindusest. 2nd ed. Tartu: Estonian Literary Museum; 2004. (Endis-Eesti Elu-Olu; vol 2).
55. Sõukand R, Kalle R. HERBA: historistlik Eesti rahvameditsiini botaaniline andmebaas [Internet]. 2008 [cited 2012 Aug 21]; Available from:
56. Sõukand R, Kalle R. Change in medical plant use in Estonian ethnomedicine: a historical comparison between 1888 and 1994. J Ethnopharmacol. 2011;135(2):251–260.
57. Sõukand R, Kalle R. The use of teetaimed in Estonia, 1880s–1990s. Appetite. 2012;59(2):523–530.
58. Kukk T, Kull T, editors. Eesti taimede levikuatlas. Tartu: Estonian University of Life Sciences; 2005.
59. Kang Y, Łuczaj ŁJ, Ye S. The highly toxic *Aconitum carmichaelii* Debeaux as a root vegetable in the Qinling Mountains (Shaanxi, China). Genet Resour Crop Evol. 2012;59(7):1569–1575.
60. Łuczaj ŁJ. Plant identification credibility in ethnobotany: a closer look at Polish ethnographic studies. J Ethnobiol Ethnomed. 2010;6(1):36.
61. Łuczaj Ł. Ethnobotanical review of wild edible plants of Slovakia. Acta Soc Bot Pol. 2012;81(4):245–255.
62. Svanberg I, Sõukand R, Łuczaj Ł, Kalle R, Zyryanova O, Dénes A, et al. Uses of tree saps in northern and eastern parts of Europe. Acta Soc Bot Pol. 2012;81(4):343–357.
63. Seppel M. Näljaabi Liivi- ja Eestimaal 17. sajandist 19. sajandi alguseni. Tartu: University of Tartu Press; 2008. (Dissertationes Historiae Universitatis Tartuensis; vol 15).
64. Vahtre L. Äärmuslikkus ja äärmusetus Eesti ajaloos. In: Rohumets I, editor. Lehed ja tähed: looduse ja teaduse aastaraamat. Tallinn: Loodusajakiri; 2004. p. 99–105. (vol 2).
65. Sander H, Elliku J, Roht U. Eesti parkide ja kollektsioonide levinumate võõrlehtpuude ja -põõsaste introduktsioonist ja metsistumisest. In: Magnus R, editor. Uurimusi Eesti loodusteaduste ajaloost. Tartu: Estonian Naturalists' Society; 2008. p. 78–102. (Eesti Looduseuurijate Seltsi aastaraamat; vol 85).
66. Spuhl-Rotalia J. Kodumaa marjad. Viljandi: A. Peet; 1898.
67. Lehmann K. Kohvi-aseainete ja söögisinepi valmistamine vähemal ja suuremal mõedul. 2nd ed. Tallinn: J. Offril & Mõte; 1909.
68. Vilbaste G. Sammalde, samblike ja vetikate nimesid. Gustav Vilbaste rahvaluulekogu (1907–1966) [Manuscript]. Tartu: Estonian Literary Museum; 1953 . (vol T3).
69. Jannau OA. Ma-rahva Koddo-Arst, ehk, lühhikenne juhhataja, kuida iggaüks mõistlik innimenne ommas maias ja perres, kui kegi haigeks saab, agga arsti ep olle sada, võib aidata. Tartu: H. Laakmann; 1857.
70. Łuczaj ŁJ, Dumanowski J, Köhler P, Mueller-Bienick A. The use and economic value of manna grass (*Glyceria*) in Poland from the middle ages to the twentieth century. Hum Ecol. 2012;40:721–733.

5

Geometric parameters of the apical meristem and the quality of phyllotactic patterns in *Magnolia* flowers

Danuta Wiss, Beata Zagórska-Marek*
Institute of Experimental Biology, Wrocław University, Kanonia 6/8, 50-328 Wrocław, Poland

Abstract

The ratio of primordium size to the meristem size (*P/M* ratio) is regarded by some geometrical models of phyllotaxis as the parameter, which determines the quality of spiral and whorled patterns of lateral organ arrangement. This assumption was tested on floral meristems in four genets representing four *Magnolia* taxa: *M.* × *salicifolia*, *M. stellata*, *M. denudata* and *M. acuminata*. In successive zones of *Magnolia* flower, lateral organs are initiated in specific phyllotactic patterns and at specific values of the meristem and primordia sizes. The elements of perianth, usually positioned in three trimerous whorls, are initiated as large primordia on relatively small meristem. The switch in the identity of primordia, from tepals to stamens is accompanied by an abrupt increase in the size of the meristem and decrease in the primordia size. Small values of *P/M* ratio and frequent occurrence of qualitative transformations of phyllotaxis contribute to the exceptionally rich spectrum of spiral patterns in androecium zone. New spiral patterns emerge when bigger primordia of carpels are initiated on the meristem, which at the same time starts diminishing in size either abruptly (*M.* × *salicifolia*, *M. stellata*, *M. acuminata*) or slowly (*M. denudata*). Spiral patterns identified in gynoecia have lower numbers of parastichies than the patterns of androecia and occur in frequencies specific for the genet. Although noted ranges of the meristem and primordia sizes, justify the occurrence of phyllotactic patterns observed in successive zones of *Magnolia* flower, they do not explain genet-specific frequencies of the patterns observed in gynoecium zone. The lack of straightforward relationship between frequency of the patterns and *P/M* ratio in gynoecium suggests that more complex geometrical factors or factors of non-geometrical nature are engaged in determination of *Magnolia* floral phyllotaxis.

Keywords: floral phyllotaxis, *Magnolia*, flower, organography, apical meristem, patterning, plant development

Introduction

Architecture of the aerial parts of plant is determined in the shoot apical meristem generating primordia of leaves, flowers and floral organs in regular phyllotactic patterns. In higher plants shoot apical meristem is organised into central zone with stem cells contributing to the meristem self-renewal and providing cells, which give rise to the internal tissues of the stem in the meristematic rib zone and cells incorporated into primordia in the peripheral zone. While self-perpetuating ability of the meristem allows preserving its general size and shape, initiation of primordia leads to cyclical changes of the meristem size during plastochrone – from minimal at the primordium inception to maximal just prior to the emergence of the next primordium. The size and shape of primordia are defined by the number of founder cells sequestered into primordia and subsequent polarization and rate of their growth and proliferation. Geometrical features of the meristem and primordia are regarded by some theoretical models as the most important determinants of phyllotactic pattern quality [1–5] [see also Iterson (1907; cited in [5])]. However, some data suggest that the order, in which primordia are initiated on the meristem, is determined by yet undefined signal emanating from differentiated tissues in the stem [6–9].

Several numerical and geometrical rules can be applied to the identification and description of phyllotactic patterns, what allows discerning their two major categories – whorled and spiral patterns. Whorled patterns result from simultaneous initiation of two or more lateral organs at the same level of the meristem in such a way, that elements of successive whorls can be connected along vertical, parallel lines called ortostichies. The number of ortostichies is always double in relation to the merosity, which express the number of organs initiated per whorl. In spiral phyllotactic patterns lateral organs are connected by spiral lines, i.e. parastichies winding either clockwise (chirality S) or counterclockwise (chirality Z). Among several discernable sets of parastichies two, interesting each other at the angle closest to 90° are the most conspicuous. Their numbers, often denoted as $\boldsymbol{m}$:$\boldsymbol{n}$, constitute so called expression of the pattern and correspond to two successive terms of mathematical phyllotactic sequence, in which every term can be derived from the sum of the previous two. The

* Corresponding author. Email: beata@biol.uni.wroc.pl

most often encountered expressions belong to the Fibonacci sequence 1,2,3,5,8,13,... but may also belong to Lucas sequence 1,3,4,7,11,... or any other. Various phyllotactic sequences representing numbers of parastichies in sets of alternating S and Z chiralities, start with different terms what guarantees that any expression can be assigned only to one, unique pattern. Hence, every spiral pattern can occur in many of various possible expressions, which adopt one of two alternative chiral configurations. The chirality of the pattern is determined by the chirality of the first term in the phyllotactic sequence, which is related to imaginary line known as ontogenetic spiral. The number of ontogenetic spirals is the basis of further sub-categorization of spiral patterns into monojugies and multijugies. While in monojugate spiral patterns all primordia can be connected by single ontogenetic spiral, two or more are necessary in multijugate patterns and respectively two or more primordia are initiated at the same level of the meristem. For example, main Fibonacci bijugy with parastichies numbers belonging to the series 2,4,6,10,16,... represents multijugy with two ontogenetic spirals and two primordia initiated at consecutive levels of the meristem.

Phyllotactic patterns of vegetative and generative structures play important role in plant classification and tend to be more constant for low rank taxonomic groups. Additionally, in the vast majority of higher plants we observe Fibonacci numbers of spirals in vegetative development, or Fibonacci numbers of organs in floral whorls – pentamerous in dicotyledons and trimerous in monocotyledons. Nonetheless, variability of phyllotaxis can be sometimes noted even in taxa representing low levels of taxonomic hierarchy, what is exemplified by intraspecific variation of patterns in vegetative shoots of *Abies balsamea* [10], in inflorescences of *Carlina acaulis* [11] or in flowers of *Magnolia acuminata* [12]. Diversity of spiral patterns can be especially spectacular in ancient flowers of magnolids with high number of floral organs [13], e.g. *Michelia fuscata* [14] and *Drimys winteri* [15]. Among such plants *Magnolia* stands out, exhibiting considerable species- or even genet-specific diversity of phyllotaxis in gynoecia [12].

Variability of patterns stems not only from their different spectra in different taxa in comparable stages of development but also from the fact, that phyllotactic patterns can quantitatively or qualitatively transform during ontogeny of a single plant [16,17]. Quantitative transformations occur through the increase or decrease in the number of parastichies within the same phyllotactic sequence and are quite common. Qualitative transformations result from changes in parastichies numbers bringing up the transition from one phyllotactic pattern to the other. In plant ontogeny qualitative transformations occur often together with the change in the phase of development, but only rarely within organs of the same identity [10,12,18–20].

Substantially much is already known about molecular mechanism governing primordia initiation on the meristem. It was demonstrated in series of experiments that auxin transport, dependent on subcellular localization of auxin efflux and influx carriers within meristematic L1 layer, plays the central role in this process [21–23]. Nonetheless we still do not know why some plants initiate organs in uniform and stable phyllotaxis while others exhibit more or less diverse patterns of organ arrangement. Many theoretical models of phyllotaxis try to address that problem by use of computer simulations reproducing creation of patterns during plant development [24–31]. All these models are consistent with Snow and Snow concept [32–34], which implicate that organ arrangement is a consequence of spatial influence which older primordia exert on newly initiated ones. Considering that primordia of definite size simply form in the first available space on the meristem surface, this concept suggests that the quality of phyllotactic pattern depends on the ratio of the primordia size to the size of the meristem (*P/M* ratio). In whorled patterns primordia share equal space on the meristem circumference and the *P/M* ratio is determined directly by the number of elements initiated

Fig. 1 *M. stellata* – an example of a typical structure of *Magnolia* flower. Only inner two of the three trimerous whorls of the perianth elements can be seen as complete. The large, narrow, white tepals, numbered increasingly in ontogenetic sequence, are followed here by one tepaloid stamen and then by small numerous stamens in androecium zone. The uppermost zone – the green gynoecium of cone-like appearance, is composed of many apocarpic carpels. In this young genet growing in Przecławice garden (not included into this study) tepalody is not yet fully expressed. C – carpels; S – stamens; T – tepals; TS – tepaloid stamen.

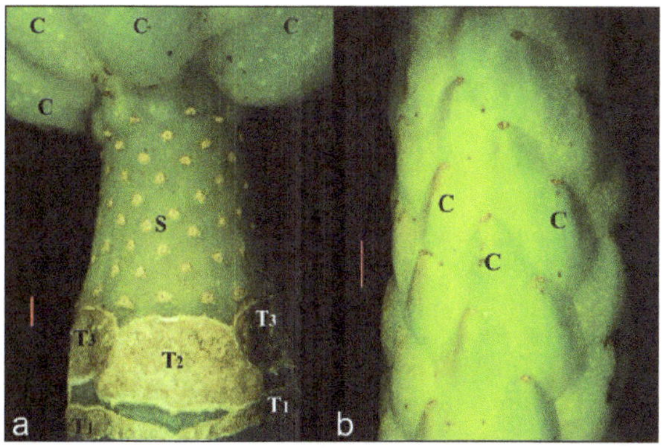

Fig. 2 *M. acuminata* – the axis of maturing generative shoot is clearly divided into distinct floral zones, as shown by the pattern of scars left by the floral parts, which fall off after blooming period. **a** Three regular trimerous whorls of tepal scars in the perianth zone and the spiral pattern of scars left by stamens on conical surface of androecium. **b** Ripening apocarpic ovaries arranged on cylindrical gynoecium in clear helical pattern, the lines connecting them (parastichies) winding up clockwise (S) or counterclockwise (Z) are easy to discern. C – ovaries; S – stamen scars; T – tepal scars. Scale bars (red): **a** 1 mm; **b** 3 mm.

in the whorl. The relationship between the P/M ratio and the quality of spiral phyllotaxis was first elucidated for idealised system of densely packed, circular primordia, distributed along parastichies intersecting at the right angle [1]. However, in the system with defined numbers of parastichies the value of P/M ratio varies depending on the angle of parastichies intersection, which is determined by the shape of primordia or the way they pack on the meristem surface [2–4]. Consequently, every expression of any spiral pattern is characterised by a range of the P/M ratio values rather than a single specific value of that parameter.

Geometrical models of phyllotaxis [1–5] [see also Iterson (1907; cited in [5])] suggest that spectra of patterns and their ontogenetic stability in the population of plants or lateral branches are under direct control of genetic mechanisms determining the size of the meristem and primordia. Molecular studies, focusing on *Arabidopsis* as a model plant, have revealed complexity of regulatory networks influencing geometrical parameters of the meristem. The *WUS-CLV* signalling loop was identified as the main genetic mechanism, which determines the meristem size through the control over the stem cell number in the central zone [35–37]. The size of the primordia founder cell poll is defined by gene *ARGOS* [38], but sequestration of these cells is accompanied also by repression of the genes *KNOX*, which specify meristematic state of the cells in the peripheral zone [39,40]. Data concerning subsequent proliferation and expansion of the primordia founder cells are still incomplete, although some components of the mechanisms underlying these processes have been characterized [41,42]. In addition, numerous studies have identified genetic factors which affect the meristem size specifically in generative phase of development [43–46] or influence both traits – the size of the meristem and primordia [47–49]. While wild type *Arabidopsis* plants have rather uniform, stable phyllotaxis in vegetative and generative development, mutations in most of the genes involved in the specification of the meristem and/or primordia size cause disorganisation of the pattern. Thus *Arabidopsis*, so well genetically defined, seems to be not suitable enough system for inquiries into the origins of phyllotactic diversity in plants.

In this paper we describe how geometric parameters of the meristem influence the quality of phyllotactic patterns in *Magnolia* flowers. *Magnolia*, exhibiting interspecific variability of phyllotaxis in gynoecia [12] and having different sizes of primordia of different identity [12,50,51], is the excellent model for this type of study. We focused on four different taxa – *M.* × *salicifolia*, *M. stellata*, *M. denudata* and *M. acuminata*, taking into account all three zones of the flower – perianth, androecium and gynoecium (Fig. 1, Fig. 2), to have a better view of developmentally changing floral phyllotaxis.

Material and methods

Collection of the material

Four taxa of *Magnolia*, each represented by one, individual tree (genet) were selected for this study: *M.* × *salicifolia* (hybrid of *M. salicifolia* and *M. kobus*), *M. stellata* (both growing in Botanical Garden of the University of Wrocław), *M. denudata* and *M. acuminata* (growing in Wrocław's public parks).

After short flowering period, which occurs in spring or early summer, depending on species and environmental conditions, floral buds emerge on lateral branches of *Magnolia*. Each newly formed bud contains apical meristem, which quickly initiates all primordia of floral organs, successively: tepals, stamens and carpels. During autumn and winter all initiated primordia and floral axis slightly elongate and differentiate into embryonic, generative shoot, which will fully develop in the next flowering season (Fig. 2). Both types of the material – meristems initiating organ primordia of different identity and embryonic generative shoots were collected in relevant seasons and examined.

Identification of patterns

Randomly collected floral buds containing meristems at the stage of initiation of either tepals, stamens or carpels were fixed in FAA (70% alcohol, formaldehyde and acetic acid in ratio 9:0.5:0.5), dehydrated in ethanol series and embedded in paraffin. Buds were then sectioned transversally at 7–9 μm thickness and some of them also longitudinally (Fig. 3) at 5–7 μm thickness using Leica microtome. Sections were stained in safranine and fast green [52] and then photographed using Olympus BX50 light microscope connected with digital camera and "Silicon Graphics" workstation INDY. Sequence of the photomicrographs of the transverse sections from consecutive levels of a single floral bud enabled reconstruction of its content. Each reconstruction was represented by drawing of all initiated primordia mapped around central point of the meristem in plain, polar projection.

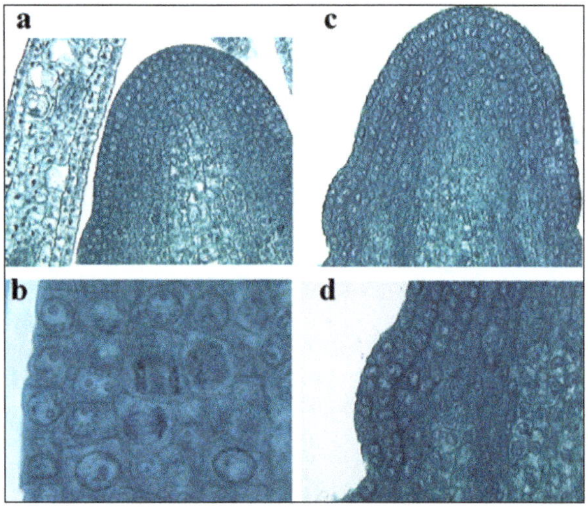

Fig. 3 a–d Longitudinal sections of two floral meristems of *M. acuminata* being at different stages of the youngest primordium initiation in gynoecium zone. **b** In magnified view of the lateral, organogenic zone of the apex (shown in **a**) the mitoses of the first periclinal divisions can be seen; they mark the onset of a new primordium, the formation of which surprisingly takes place well below the surface, covered by still flat two layers of tunica. **d** Swelling up primordium of the more advanced apex (shown in **c**) is already elevated, but still clearly covered by L1 and L2 layers.

Reconstructions were used for identification of phyllotactic patterns in the successive zones of the flower. The numbers ***m*:*n*** of contact parastichies in two opposite sets of spirals, intersecting each other at the angle closest to 90°, were counted separately for each zone. The chirality of parastichies was also denoted, depending on direction of their winding from the base to the top of the zone, by index S (clockwise direction) or Z (anticlockwise direction). Alternate extrapolation of index S or Z from identified numbers of parastichies ***m*:*n*** to the first number in corresponding phyllotactic sequence, allowed determining chiral configuration of the pattern. Method adapted for identification of patterns on a meristem, although precise, limited the examination to small sample of 19–34 buds per tree. Thus the study of phyllotaxis was extended over material extracted from autumn and winter buds, where the early processes of growth and determination do not affect the arrangement of carpels. Identification of phyllotactic patterns in gynoecium zone of embryonic shoots was done under binocular, by counting the numbers of contact parastichies.

Floral meristem measurements and calculation of parameter *A*

Meristem circumference (*M*) and its arc occupied by primordium (*P*) were measured on photomicrographs taken at the level of the meristem where the last initiated primordium had the widest transversal dimension. Measurements were performed using MapInfo Professional 6.5SCP. In this way the range of most probable values of parameter *P* and *M* was identified for successive flower zones, taking into account mean values and standard deviation. The variation of parameters *P* and *M* was expressed as coefficients of variation (*V*). Knowing the ratio of parameter *P* and *M* we could calculate angular value of the meristem arc occupied by primordium (parameter *A*). According to Church [1] the value of parameter *A* for any pattern expression ***m*:*n***, at right angle β of parastichies intersection can be determined using equation:

$$A = 360°\sqrt{m^2 + n^2} \quad (1)$$

To calculate the value of parameter *A* in the systems where angle β is different than 90° Church's equation was modified to the following form:

$$A = 360°\sqrt{m^2 + n^2 +/- 2mn\cos\beta} \quad (2)$$

In real systems with known phyllotactic patterns angles β were determined by comparison of estimated values *A* with theoretical values of that parameter, obtained on the basis of Equation (1) or (2) given above. The range of estimated values of parameter *A* and values of angle β, compared with theoretical values of *A*, allowed us to predict the probability of occurrence of different patterns in the successive zones of the particular flower.

Results

Perianth

PHYLLOTACTIC PATTERNS

In all examined trees the first initiated elements of flower were typically three trimerous whorls of tepals (Fig. 1, Fig. 2, Fig. 4, Tab. 1). Unequal sizes and slightly asymmetric distribution of the elements of the whorl along meristem circumference suggested, that they were not always initiated simultaneously but sometimes in a very rapid spiral sequence.

Some exceptions from trimerous pattern of tepals were noted. In one bud of *M.* × *salicifolia* first two whorls were tetramerous, with four elements evenly distributed along the meristem circumference. Another rare modification was one element missing in the first whorl (noted in one *M. stellata* bud) and occurrence of two elements instead of one in the last whorl (noted in four *M.* × *salicifolia* buds). Microscopic

Tab. 1 Phyllotactic patterns of analysed magnolia genets.

Genet	Perianth	*N*	Androecium	*N*	Gynoecium	*N*
Magnolia ×	3:3	8	-	-	-	-
salicifolia	3:3*	1	-	-	-	-
	3:3	20	**7s:12z**	1	-	-
			7z:13s	1		
			8z:9s	1		
			9z:11s	1		
			9z:13s	1		
			7z:11s	1	**4s:7z**	4
			7s:11z ▸ 7s:9z	1		
			8z:11s → 7z:11s	2		
			7s:11z	1	**4z:7s**	1
			9z:10s → 8z:10s ▸ 7s:8z → 6s:7z	1	**5s:7z**	1
			8z:11s ▸ 8z:9s	1	**5s:8z**	5
			8z:13s	1		
			9z:12s ▸ 8z:11s	1		
			9z:12s → 8z:12s	1		
			9z:13s → 9z:12s ▸ 8z:9s	1		
			7z:9s ▸ 5z:8s	1	**5z:8s**	4
			7z:9s ▸ 5z:9s → 5z:8s	1		
			8s:13z → 8s:12z	2		
	3:3*	4	10z:11s → 9z:11s → 8z:11s	1	**5s:8z**	3
			10z:12s → 9z:12s → 9z:11s ▸ 8z:9s	1		
			9z:13s → 8z:13s	1		
			8s:12z ▸ 8s:10z	1	**5z:8s**	1
	4:4 → 3:3	1	10z:13s ▸ 10z:11s	1	**6z:10s**	1
Magnolia	3:3	2	-	-	-	-
stellata	3:3 → 3s:5z	2	-	-	-	-
	3:3s:5z → •3	1	-	-	-	-
	3:3 → 4z:5s	2	-	-	-	-
	3:3 → 4s:5z	3	**6z:8s**	1	-	-
			8s:14z	1	-	-
			8z:10s → **8z:9s**	1	-	-
	3:3 → 4z:5s	2	**8z:12s**	1	-	-
			6z:11s → **6z:10s**	1	-	-
	3:3 → 5:5	1	**8s:9z**	1	-	-
	3:3 → 3z:4s	2	9z:10s ▸ 7z:9s	1	6z:9s → **5z:8s**	1
			9z:13s → 8z:13s	1	**5s:8z**	1
	3:3 → 3z:4s → 3z:6s	1	9s:9z ▸ 7s:9z ▸ 5s:8z	1	**5s:8z**	1
	3:3 → 3z:5s	1	8z:11s ▸ 8z:9s ▸ 8z:6s	1	**5s:8z**	1
	3:3 → 4z:5s	3	6z:10s	1	**6z:10s**	1
			7s:10z → 7s:9z	1	**6s:9z**	1
			9s:12z → 9s:11z	1	**5z:8s**	1
	3:3 → 4s:5z	3	8s:13z	1	**5z:8s**	2
			7z:11s	1		
			9z:13s ▸ 8s:11z	1	**5s:8z**	1
	3:3 → 5:5	4	8s:13z ▸ 8s:10z	1	**5z:8s**	3
			8s:13z → 8s:12z ▸ 8s:5z	1		
			8s:13z ▸ 8s:9z	1		
			9z:12s → 9z:11s ▸ 8z:10s	1	**5s:8z**	1
	3:3 → 5s:7z	1	7s:9z	1	**6s:9z**	1

Tab. 1 *(continued)*

Genet	Perianth	*N*	Androecium	*N*	Gynoecium	*N*
Magnolia	3:3	6	-	-	-	-
denudata	3:3	13	**8z:12s**	1	-	-
			8s:13z	2	-	-
			8z:13s	3	-	-
			10z:14s	1	-	-
			8s:12z → 8s:13z	1	**5z:8s**	3
			8s:13z	2		
			10s:11z ► 9z:10s → 10z:10s	1	**6z:9s**	2
			10s:15z ► 9s:13z	1		
			10z:15s ► 10z:9s	1	**6s:10z**	1
Magnolia	3:3	10	-	-	-	-
acuminata	3:3	15	**9z:14s**	1	-	-
			11s:12z	1	-	-
			11s:15z	2	-	-
			11s:14z → **11s:13z**	1	-	-
			12z:15s ► **11z:12s**	1	-	-
			8s:13z	3	**5z:8s**	3
			9z:15s ► 9z:13s ► 9z:11s	1	**4z:6s**	1
			11s:12z ► 10s:11z	1	**6z:9s**	1
			11s:12z ► 9z:11s	1	4s:7z → **4s:6z**	1
			11z:13s ► 9z:10s	2	**5s:8z**	1
					5z:7s	1
			13z:14s ► 9z:13s	1	**4z:6s**	1

Patterns of successive zones of single meristem are arranged horizontally with final pattern underlined in bold type. Changes in phyllotactic patterns occurring as numerous transformations at the same level of the meristem, but in different sectors of its circumference are denoted by "►", typical transformations by "→". Asterisks mark the occurrence of two elements in place of one in the last whorl ("*") or lack of one element in the first whorl ("•"). *N* – total number of particular patterns or transformations identified in successive zones of a flower.

analysis of *M.* × *salicifolia* meristem at the early stage of the initiation of such atypical whorl suggested, that the additional element could be created through unequal cleavage of primordium, which was initially much bigger than two other elements of the whorl.

In contrast to *M.* × *salicifolia*, *M. denudata* and *M. acuminata*, where initiation of perianth elements was always restricted to three whorls, in *M. stellata* tepalody occurred – the additional 5 to 16 tepals were initiated in places topologically corresponding to stamens (Fig. 1). These additional elements (tepaloid stamens) were arranged in different spiral patterns (Tab. 1).

SIZE OF THE MERISTEM AND PRIMORDIA

Small numbers of meristems initiating elements of perianth were found during collection of buds from each individual tree. The significant differences between minimal and maximal values of the meristem circumference (*M*) and its arc occupied by primordium (*P*) were noted between successive whorls and between examined genets (Fig. 5, Tab. 2). Those differences, underlined by high variation coefficients, could be only partially ascribed to natural variation. Meristems initiating large primordia, like in the perianth zone of *Magnolia*, are considerably smaller in the minimal phase of plastochrone – just after the initiation than in the maximal phase – just prior to the initiation of the next primordium. The collection of the meristems in different phases of plastochrone and their small number undoubtedly influenced noted differences of the meristem and primordia sizes. Thus, the real extent to which the meristem and primordia changed their sizes between whorls, and in the case of *M. stellata* also after the switch to initiation of tepaloid stamens, could not have been properly estimated.

Androecium

PHYLLOTACTIC PATTERNS

The greatest diversity of phyllotactic patterns was noted in androecium zone of examined magnolias; 20 different patterns were identified in *M.* × *salicifolia* androecia (*N* = 25), 16 in *M. stellata* (*N* = 21), 8 in *M. denudata* (*N* = 13) and 13 in *M. acuminata* (*N* = 15). Besides well known patterns, like Fibonacci pattern (8:13) and its bijugy (6:10) or Lucas pattern (7:11), also various rare monojugies and multijugies were found in this zone (Fig. 4, Tab. 1). Numbers of parastichies expressing phyllotactic patterns were generally lower in *M.* × *salicifolia* and *M. stellata* than in *M. denudata* and *M. acuminata* (Fig. 4).

Whereas in the case of *M.* × *salicifolia*, *M. stellata* and *M. acuminata* the richness of phyllotaxis was so considerable that none of the patterns was recognised as dominating, in *M. denudata* the Fibonacci pattern was identified in the seven of thirteen examined buds. The great variability of phyllotaxis in androecia was clearly increased by the common occurence of qualitative transformations. Some of them were caused by dissapearence of single parastichy while others, what was very specific for androecium zone, by disappearence of multiple parastichies at the same level but at different points of the meristem circumference. In some cases, when transformations occured at two different levels of the meristem, even three phyllotactic patterns coexisted in one androecium (Tab. 1).

SIZE OF THE MERISTEM AND PRIMORDIA

The change in the identity of initiated primordia from tepals to stamens was accompanied by increase in the size of the meristem and decrease in the size of primordia. Similar numbers in opposite sets of parastichies in androecia of *M.* × *salicifolia* and *M. stellata* corresponded with similarity of geometrical parameters of the meristem. The mean values of parameters *M* and *P* in *M.* × *salicifolia* were equal to 939.18 μm and 65.05 μm respectively, compared to 864.99 μm and 65.69 μm in *M. stellata*. Meristems of *M. denudata* and *M. acuminata*, which had higher numbers of parastichies had also apparently larger meristems (mean value of *M* = 1149.94 μm for *M. denudata* and 1262.61 μm for *M. acuminata*) and slightly larger primordia (mean value of *P* = 75.38 μm for *M. denudata* and 70.32 for *M. acuminata*; Fig. 5, Tab. 2).

The broad range of low values of parameter *A*: 20.35°–30.23° in *M.* × *salicifolia*, 21.96°–34.35° in *M. stellata*, 20.4°–27.1° in *M. denudata* and 16.77°–23.77° in *M. acuminata* (Tab. 3), together with high values of variation coeficients (Tab. 2), theoretically guaranteed wide spectra of phyllotactic patterns in androecia. Prevalence of Fibonacci pattern in *M. denudata* androecia was probably caused by the stability of angles β, which in every examined bud were equal or very close to 90°. Similarly, greater variability of phyllotaxis determined for *M.* × *salicifolia* than for *M. stellata*, was probably caused by a wider

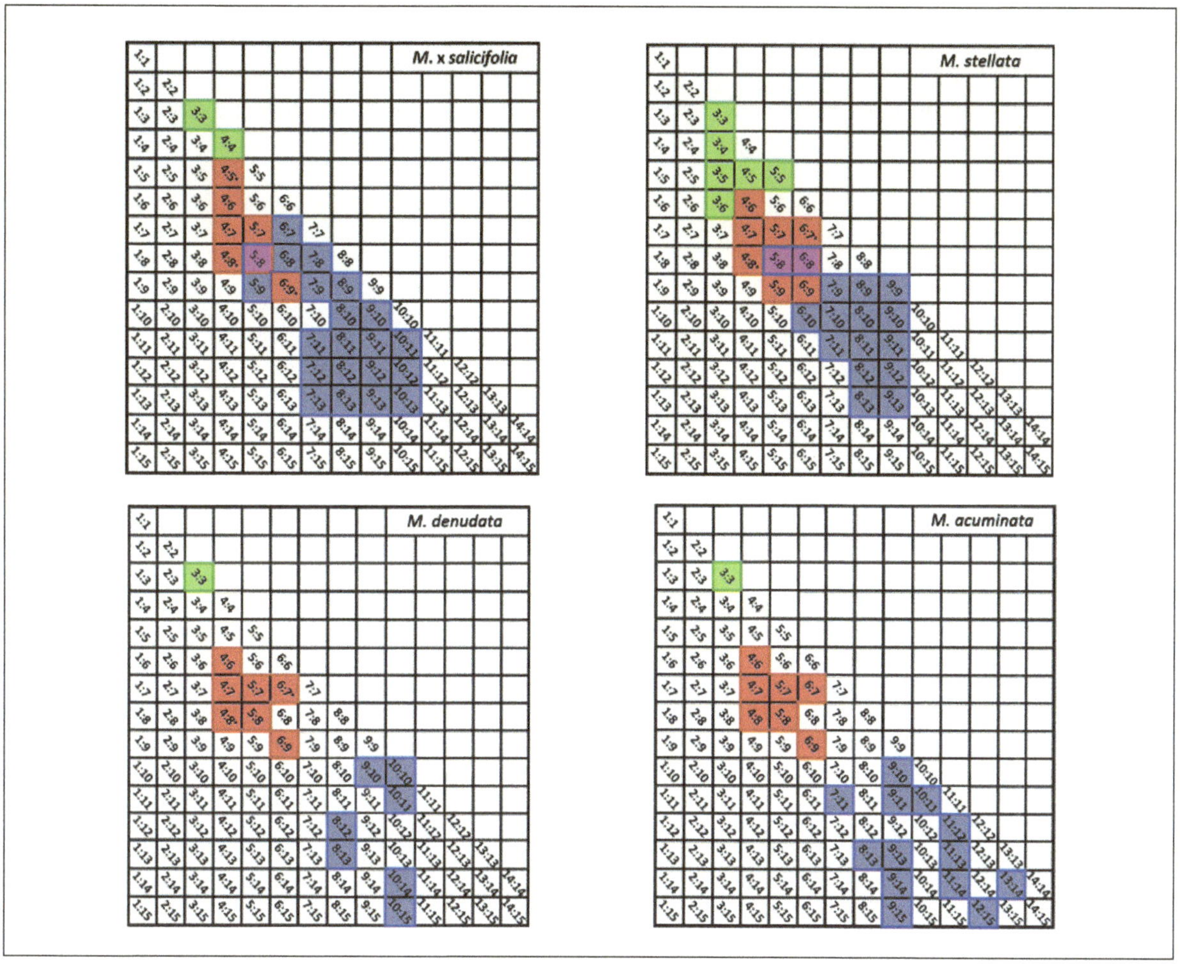

Fig. 4 Universal phyllotactic grids [12] presenting all ***m:n*** phyllotactic patterns of analysed magnolias. Patterns of perianth, androecium and gynoecium are marked with green, blue and red colour respectively. Purple colour denotes patterns identified in gynoecia as well as androecia. Asterisks mark the patterns identified exclusively as involved in rare qualitative transformations in gynoecia, occurring either before or after the change in parastichy number.

range of estimated angles β in the stamen zone of the later specimen (60°–105° in the case of *M.* × *salicifolia* compared to 75°–90° in *M. stellata*; Tab. 3).

Gynoecium

PHYLLOTACTIC PATTERNS

Analysis of phyllotaxis in *Magnolia* gynoecia had a very detailed character – in addition to the meristems also a large number of the embryonic, generative shoots was collected from each tree during autumn and winter. Shoots of *M.* × *salicifolia*, *M. stellata* and *M. denudata* were collected twice, during two successive seasons to check, whether the spectra of their patterns in gynoecium change seasonally. In *M.* × *salicifolia* shoots, only three pure patterns were identified: noticeably dominating Fibonacci pattern (83.54% of 407 shoots and 83.52% of 261 shoots in two successive seasons), Lucas pattern as the second in frequency and Fibonacci bijugy as the most rare (Tab. 4). Fibonacci phyllotaxis dominated also in gynoecia of *M. stellata* (90.09% of 111 shoots and 78.49% of 172 shoots in two successive seasons) and *M. denudata* (88.5% of 87 and 85.53% of 159 shoots in two successive seasons). In *M. stellata* among 6 other patterns Fibonacci bijugy (6:10) was the second in frequency (Tab. 4), while in *M. denudata* all 4 additional patterns occurred in comparable frequency (Tab. 4). In the case of *M. acuminata*, the embryonic shoots were collected only during one season. Fibonacci pattern prevailed but occurred in slightly less than a half of the analysed shoots (48.05% of 77 shoots in 2004), whereas among 6 other patterns some occurred in relatively high frequency – 5:7 and 4:6 (12.99% each), 6:9 and 4:7 (11.69% each; Tab. 4).

No major preference in the direction of the ontogenetic spiral was noted – numbers of generative shoots with configurations S and Z were similar within each phyllotactic pattern in all specimens (Tab. 4). Examination of the meristems (Tab. 1) revealed, that the patterns most frequent in gynoecium zone were initiated usually (5:8 and 4:7 in *M.* × *salicifolia*, 5:8 in *M. stellata* and *M. acuminata*) or exclusively (5:8 in *M. denudata*, and 4:6 in *M. stellata*) in continuity with the same pattern in androecium zone, whereas the rare patterns in gynoecium were preceded by qualitatively different but also uncommon androecium patterns.

In contrast to androecium zone qualitative transformations in gynoecium zone were not common – 13 transformations were identified in *M.* × *salicifolia* embryonic shoots, 9 in *M. stellata* and 3 in *M. denudata* (Tab. 4). No qualitative transformations were found during inspection of *M. acuminata*

Tab. 2 Mean, minimal and maximal values of the meristem circumference and its arc occupied by primordium, standard deviation and variation coefficient in successive zones of examined flowers.

Genet	The zone of flower	*M*, *P*	*N*	X_s (µm)	X_{min} (µm)	X_{max} (µm)	*S* (µm)	*V* (%)
M.* × *salicifolia	Perianth 1	*M*	2	1140.09	1055.06	1225.13	120.25	10.54
		P	6	263.67	210.14	305.12	40.31	15.29
	Perianth 2	*M*	3	455.4	400.12	513.85	56.93	12.5
		P	9	61.91	65.06	98.59	38.87	62.78
	Perianth 3	*M*	4	578.27	482.14	674.37	86.73	14.99
		P	12	93.78	67.86	116.66	43.79	62.25
	Androecium	*M*	5	939.18	834.61	1015.38	68.2	7.26
		P	5	65.05	54.69	73.24	8.09	12.44
	Gynoecium	*M*	20	742.49	663.36	845.26	51.57	6.94
		P	20	87.1	73.58	100.17	7.53	8.64
M. stellata	Perianth 2	*M*	2	476.54	452.98	500.11	23.56	4.94
		P	6	64.71	58.99	73.89	5.39	8.33
	Perianth (TS)	*M*	5	797.14	571.25	934.78	177.15	22.22
		P	5	90.54	81.58	99.55	7.53	8.31
	Androecium	*M*	6	864.99	710.71	996.95	115.58	13.36
		P	6	65.69	57.79	71.87	5.86	8.91
	Gynoecium	*M*	15	765.85	668.39	812.96	39.71	5.18
		P	15	81.72	70.78	93.08	7.33	8.97
M. denudata	Perianth 2	*M*	4	897.87	850.09	933.93	44.85	4.99
		P	12	157.9	150.64	168.31	6.61	4.18
	Perianth 3	*M*	2	769.66	745.82	793.5	36.87	4.79
		P	6	114.22	101.54	130.05	14.39	12.6
	Androecium	*M*	7	1149.94	1068.96	1234.37	54.62	4.74
		P	7	75.38	65.51	87.5	7.1	9.42
	Gynoecium	*M*	6	1178.77	1089.33	1409.1	118.79	10.07
		P	6	109.1	97.22	140.9	16.09	14.75
M. acuminata	Perianth 2	*M*	3	974.12	897.79	1080.83	95.22	9.77
		P	9	162.65	128.75	205.83	66.66	40.98
	Perianth 3	*M*	7	523.88	432.91	633.34	64.25	9.77
		P	21	86.76	56.94	119.87	14.71	16.96
	Androecium	*M*	6	1262.61	1178.57	1396.15	81.57	6.46
		P	6	70.32	59.52	80.74	7.69	10.93
	Gynoecium	*M*	9	1033.11	713.76	1233.34	156.9	15.18
		P	9	114.94	92.86	129.16	13.37	11.63

Terms perianth 1, 2, 3 and (TS) refer to the first, second and third whorl of the perianth and tepaloid sepals respectively. *M* – meristem circumference; *P* – meristem arc occupied by primordium; *S* – standard deviation; TS – tepaloid sepals, *V* – variation coefficient; X_s, X_{max}, X_{min} – mean, maximal and minimal values of *M*, *P*, *S* and *V*, respectively.

embryonic shoots although on one of the meristems carpel arrangement changed from Lucas pattern into the main Fibonacci bijugy (Tab. 1). Transformations in the embryonic shoots occurred usually by disappearance of one parastichy, in few cases at two different levels of the system, e.g. in *M. stellata* (5s:8z → 5s:7z → 4s:7z) or *M. denudata* (6z:7s → 5z:7s → 4z:6s). Occasionally two parastichies were removed from the system at one point of the shoot circumference and in one case as many as three although not in the same set (transformation 5z:8s → 4z:6s in *M.* × *salicifolia*). Transformation 5s:7z → 5s:8z in *M. denudata* was exceptionally evoked by addition of the new parastichy to the existing set of parastichies (Tab. 4).

Some of identified transformations involved a very rare patterns. This enriched phyllotactic spectra of examined genets. Patterns 6:7 and 4:8 in *M. stellata* and *M. denudata* and also 4:5, 4:8, 5:7 and 6:9 in *M.* × *salicifolia* were identified exclusively as associated with transformations (Fig. 4).

SIZE OF THE MERISTEM AND PRIMORDIA

After the switch from initiation of stamens to initiation of carpels, the size of primordia increases while meristem becomes smaller and eventually terminates after production of the last carpel. The number of organs in gynoecium differed between examined genets: *M.* × *salicifolia* initiated on the average 67 carpels (estimated for *N* = 18 shoots), *M. stellata* 53 (*N* = 13), *M. denudata* 68 (*N* = 10) and *M. acuminata* 39 (*N* = 7).

M. × *salicifolia* meristems were analysed at different stages between initiation of 11th and 51st primordium (examined

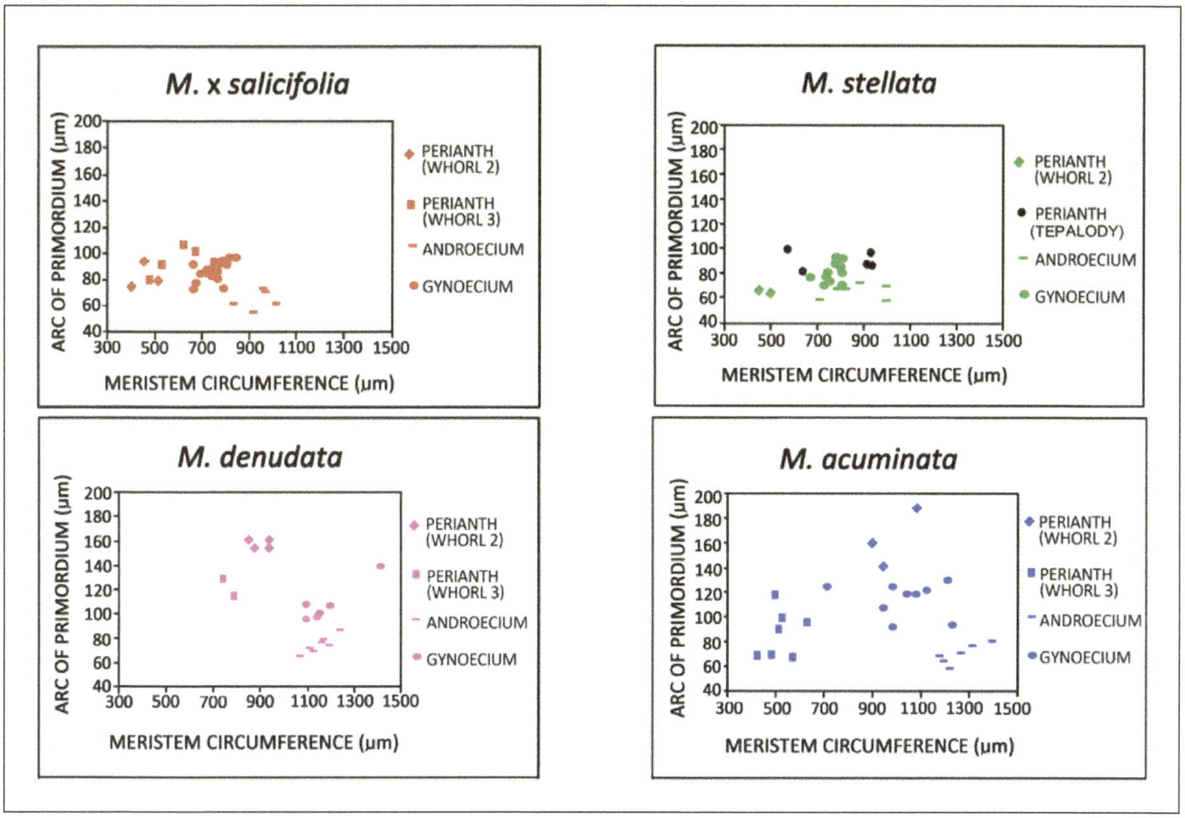

Fig. 5 The circumference of the meristem and its arc occupied by lateral organ primordium in the successive floral zones of analysed magnolias. Detailed description in the text.

sample contained $N = 20$ meristems). At mean values $M = 742.49$ μm and $P = 87.1$ μm (Fig. 5, Tab. 2) the most probable values of parameter A ranged between 36.25° and 49.31° (Tab. 3). Angles β were usually close to 105° ($N = 10$), in few cases close to 90° or 75° (in both cases $N = 4$) and the most rarely to 120° ($N = 2$; Tab. 3). Comparison of empirical data with theoretical values A accounted for the occurrence of patterns 5:8, 4:7 and 4:6 in gynoecia of this specimen but also indicated, that patterns 5:7 and 6:9, never noted as pure ones, should have been initiated as often as dominating Fibonacci pattern.

Meristems of *M. stellata*, which initiated from 14 to 34 primordia ($N = 15$), were slightly larger (mean $M = 765.85$ μm) while their primordia were smaller (mean value of $P = 81.72$ μm) compared to *M.* × *salicifolia* (Fig. 5, Tab. 2). The range of estimated values A – 33.24°–42.06° at the most common angle $\beta = 90°$ ($N = 6$), similarly like in the case of *M.* × *salicifolia*, suggested occurrence of patterns 5:8, 6:9 and 5:7 as the most common in *M. stellata* gynoecia. According to theoretical data main Fibonacci bijugy, noted as the second pattern in frequency should have occurred only at rare angle $\beta = 105$ ($N = 3$) and should not be more often than Lucas pattern, which was predicted to occur only when angle β is close to 75° ($N = 3$). Although all examined specimens had similar numbers of parastichies in gynoecia, *M. denudata* and *M. acuminata* had noticeably larger primordia and meristems than *M.* × *salicifolia* and *M. stellata*.

Only 6 meristems, at the stage of initiation from 10th to 22nd carpel primordium, were examined in *M. denudata*. Unlike the meristems of the other specimens, meristems of *M. denudata* initiating carpels were larger (mean value of $M = 1178.77$ μm) than the meristems initiating stamens but the carpel primordia were typically larger (mean value of $P = 109.1$ μm) than the stamen primordia (Fig. 5, Tab. 2). Wide spectrum of estimated values A – between 25.8° and 42.51°, the high variation of parameters P and M (Tab. 2) and variable values of angle β (two meristems with angle $\beta = 60°$, two with angle $\beta = 75°$ and single meristems with $\beta = 90°$ and 105° were found) theoretically guaranteed the occurrence of many different phyllotactic patterns. However, according to theoretical data main Fibonacci bijugy (4:6) and Fibonacci trijugy (6:9) should have occurred as the most frequent patterns in gynoecia of this specimen.

Meristems of *M. acuminata* were analysed at different stages between initiation of the 22nd to 35th carpel primordium ($N = 9$ meristems). The mean value of M was equal to 1033.11 μm and had the highest variation ($V = 15.18\%$) among examined specimens (Tab. 2). *M. acuminata* had also the largest carpel primordia with mean $P = 114.94$ μm (Fig. 5, Tab. 2). The wide range of estimated values A – from 30.72° to 52.71°, and variable angles β (4 meristems with angle $\beta = 90°$, 2 with $\beta = 75°$ and 105° and single with $\beta = 60°$) explained the wide spectrum of phyllotactic patterns. Nonetheless, like for the other specimens a comparison of empirical data with theoretical values A indicated that frequency of particular patterns should have been different than identified one – patterns 5:8, 5:7 and 4:7 should have occurred as the most common and in equal frequencies in *M. acuminata* gynoecia.

Some features, like the size of the meristem and primordia in gynoecium zone, were similar for *M.* × *salicifolia* and *M.*

Tab. 3 The ranges of parameter *P* and *M* values (calculated taking into account their mean values and standard deviations), estimated values of parameter *A* and angles *β* in androecia and gynoecia of examined flowers.

Genet	The zone of flower	*M*, *P*	Range of *M* and *P*	Parameter *A*	Angle *β*
M.* × *salicifolia	Androecium	*M*	870.97–1007.39	20.35°–30.23°	60°–105°
		P	56.95–73.14		
	Gynoecium	*M*	690.91–794.06	36.25°–49.31°	75°–120°
		P	79.97–94.64		
M. stellata	Androecium	*M*	749.41–980.57	21.96°–34.35°	75°–90°
		P	59.83–71.52		
	Gynoecium	*M*	762.14–805.56	33.24°–42.06°	75°–120°
		P	74.39–89.05		
M. denudata	Androecium	*M*	1095.32–1204.56	20.4°–27.1°	90°
		P	68.28–82.48		
	Gynoecium	*M*	1059.98 –1297.56	25.8°–42.51°	60°–105°
		P	93.01 –125.19		
M. acuminata	Androecium	*M*	1181.04–1344.18	16.77°–23.77°	75°–90°
		P	62.63–78.01		
	Gynoecium	*M*	876.21–1190.01	30.72°–52.71°	60°–105°
		P	101.57 –128.31		

stellata and distinguished them from *M. denudata* and *M. acuminata*. The last two genets, as opposed to *M.* × *salicifolia* and *M. stellata*, did not have common expressions of phyllotactic patterns in gynoecium and androecium zone (Fig. 4). This suggested that the value of parameter *A* was changing more drastically between androecium and gynoecium zone in *M. denudata* and *M. acuminata*. Examination of transverse sections of the meristems in all specimens, and also longitudinal sections of *M.* × *salicifolia* and *M. acuminata* showed, that in every case the change in the meristem size after the switch from initiating stamens to carpels occurred gradually. Thus, the drastic change in the parameter *A* value between androecium and gynoecium zone in *M. denudata* and *M. acuminata* was probably caused by abrupt change in the size of initiated primordia between those two zones.

Discussion

During ontogeny of *Magnolia* flower the changes in the initiated organ identity are accompanied by the changes in the meristem and primordia sizes and alterations in organ arrangement. Typically whorled pattern of perianth transforms into spiral pattern of androecium, which is subsequently replaced by a new spiral pattern of gynoecium. Our study revealed that the range of the meristem and primordia sizes in successive zones of the flower is characteristic for the selected individual trees. However, the significance of these geometrical parameters in determination of characteristic phyllotactic spectra differs in successive zones of the flower.

The most stable part of the analysed *Magnolia* flowers is perianth, where relatively large tepals are asymmetrically distributed around meristem circumference in three trimerous whorls. The distribution and unequal sizes of the elements may suggest that they in fact are initiated spirally in the sequence of very short plastochrones. Such mechanism of the perianth initiation was noted earlier in *M. denudata* [53] and in some other plants [54–58]. The rare discrepancies from the basic trimerous pattern of the perianth initiation include: (***i***) tetramerous whorls in *M.* × *salicifolia*; (***ii***) the element missing in the predicted position of the whorl in *M. stellata* or (***iii***) initiation of two elements instead of one in *M.* × *salicifolia*. Similar anomalies were identified in *M. acuminata* gynoecia, where lack or occurrence of additional elements was noted within spiral pattern, in the course of single parastichy [59]. Clearly spiral mode of tepals initiation was observed in *M. stellata* where perianth elements develop in the region topologically belonging to androecium on meristems which are larger than in the zone of trimerous pattern. Variations in magnolia perianth discussed recently by Zagórska-Marek [60] may provide an additional clues on genetic regulation of the identity and distribution of floral structures by homeotic genes.

In the development of *Magnolia* flower relatively large meristems initiating small primordia, yielding low values of *P*/*M* ratio, are specific for androecium. Our analysis revealed exceptional richness of spiral phyllotactic patterns with high numbers of parastichies in this zone. In contrast to our results, Erbar and Leins [61] identified usually eight and rarely seven or nine parastichies on meristems of *M. stellata* initiating stamens and carpels. However, their method of pattern identification, based on estimation of the number of parastichies in one chiral group only, was incorrect. Eight parastichies in one of two chiral groups can be ascribed to the Fibonacci pattern as well as to the pattern 8:11, similarly seven parastichies can be ascribed to Lucas pattern but also to the pattern 7:10. Although rare, and even considered as non-existing [62], patterns 8:11 and 7:10 were identified in the course of present analysis – in androecia of *M.* × *salicifolia* and *M. stellata*, and earlier in vegetative development of *Abies balsamea* [10], *Pinus mugo* [63] and *Rebutia cacti* [64]. Phyllotactic spectra in *Magnolia* androecia are significantly enriched by qualitative transformations, occurring very often even in single meristems. Two alternative mechanisms have been proposed as underlying phyllotactic transformations. Transition of one pattern into another can result from the local fluctuations of growth on the meristem circumference [16,17], or can be induced by gradual changes affecting the growth along the whole meristem circumference [26–28,31,65]. Although we cannot predict, which of those

Tab. 4 Phyllotactic patterns and their transformations identified in the gynoecia of embryonic shoots.

Genet	Season	Pattern	S	Z	Transformations	*N*	Z/S → S/Z
Magnolia salicifolia	2000–2001	5:8	166	174	5z:8s → 4z:6s	1	–
		4:7	26	28	4z:7s → 4z:6s	2	+
		4:6	4	9	5z:7s → 4z:6s	1	+
					5z:7s → 4z:7s	2	–
					4s:6z → 4s:5z	1	–
	2001–2002	5:8	128	90	6z:9s → 5z:8s	1	–
		4:7	14	15	5z:8s → 4z:7s → 5z:8s	2	–/+
		4:6	9	5	4z:7s → 4z:6s	1	+
					5z:7s → 4z:6s	1	+
					4s:8z → 4s:7z	1	–
Magnolia stellata	2001–2002	6:8	-	1	4z:7s → 4z:6s	2	+
		5:8	55	45	5s:7z → 4s:7z	1	–
		5:7	-	1	6z:9s → 5z:8s	1	–
		4:7	1	-	4z:6s → 4z:5s	1	–
		4:6	4	4			+
	2002–2003	6:9	1	3	5s:8z → 5s:7z → 4s:7z	1	+/–
		5:9	1	1	4z:6s → 3z:6s	1	+
		5:8	50	85	4z:8s → 4z:7s	1	–
		5:7	3	1	6z:7s → 4z:8s	1	+
		4:7	1	1			
		4:6	14	11			
Magnolia denudata	2001–2002	6:9	1	2	5s:8z → 4s:8z	1	+
		5:8	46	31			
		5:7	-	1			
		4:7	1	1			
		4:6	2	2			
	2002–2003	6:9	-	3	5s:7z → 5s:8z	1	+
		5:8	64	72	6z:7s → 5z:7s → 4z:6s	1	+/+
		5:7	6	3			
		4:7	2	4			
		4:6	4	1			
Magnolia acuminata	2004	6:9	4	5			
		6:7	-	1			
		5:8	19	18			
		5:7	5	5			
		4:8	-	1			
		4:7	5	4			
		4:6	3	7			

In three of four genets the data were collected for two consecutive seasons. The S and Z configurations of ontogenetic helix for each pattern were distinguished and the number of cases in each class counted. In case of transformations it was noted if they occurred with (+) or without (–) a change in the direction of ontogenetic helix.

mechanisms is the most adequate for explanation of changing phyllotaxis in *Magnolia* androecia, we can expect that fluctuations of growth affecting meristem and/or primordia size are considerable at this stage of floral ontogeny.

Geometrical models of phyllotaxis [1–5] [see also Iterson (1907; cited in [5])] assume, that the potential of meristems to initiate stable or variable patterns should be reflected in the degree of variation in the meristem and primordia sizes. Variation of the meristem size is decisive for the variability of whorled patterns in vegetative development of *Hippuris* [66], *Equisetum* [67] and whorled and spiral patterns in *Sedum maximum* [68]. In *abphyl1* mutants of maize alteration of phyllotaxis, from distichous to decussate pattern, is accompanied by changes in the size of vegetative meristem and initiated primordia of leaves [69–71]. High diversity of phyllotaxis in *Magnolia* androecia is similarly guaranteed by considerable variation of both parameters – meristem and primordia sizes. Additionally all examined specimens, except *M. denudata*, have the variable angles of parastichy intersection in this zone. Constancy of the angles of parastichy intersection, reflecting stable mode of primordia packing, may condition slight preference towards initiation of stamens in pure Fibonacci pattern in *M. denudata*.

The development of generative meristems producing flowers is determinate, what means that their activity decreases and eventually ceases with the initiation of the last primordium [72]. Genetic mechanism inducing determinate state of the meristem can be triggered at different stages of generative development [36,73]. In *M.* × *salicifolia*, *M. stellata* and *M. acuminata* flowers such developmental switch probably occurs together with the change in the identity of initiated primordia from stamens to carpels, when the mean value of the meristem size decreases comparing to that in preceding stage of organogenesis. Similar reduction in the meristem size, occurring concurrently with the beginning of carpels initiation, was noted in *Michelia fuscata* [54]. In *M. denudata* the early stages of gynoecium formation are concomitant with further increase in the mean value of the meristem size. This implies that the determinate mechanism of the meristem growth may be induced in the flowers of this specimen relatively late – after the initiation of certain defined number of carpels, similarly like in *Ranunculus* [56]. Because the measurements of analysed geometrical parameters were taken at various stages of androecium and gynoecium development mean values of the meristem and primordia sizes do not inform how exactly these parameters change during ontogeny of a single flower. The change in the *P/M* ratio between androecium and gynoecium probably always has quite drastic character in *M. denudata* and *M. acuminata* meristems which, as opposed to *M.* × *salicifolia* and *M. stellata*, never initiate patterns with common numbers of parastichies in these two zones. Our analysis suggests that drastic alterations in *P/M* ratio values are caused by abrupt increase in the primordia size at the beginning of carpel initiation while the changes in the size of the meristem occur gradually. Quite clear visual change in primordia size in a generative part of the floral axis of *Magnolia acuminata* at the time of transition from initiating stamens into carpels has already been documented on scanning electron microscopic images [12].

In gynoecia of all examined magnolias the main Fibonacci pattern dominates but its frequency, as well as the occurrence and frequency of additional patterns are genet-specific. Some of the additional patterns were noted as exclusively accompanying qualitative transformations, but they did not broaden phyllotactic spectra to the extent observed in the androecium zone, where transformations are particularly frequent. Examination of patterns in gynoecia of *M.* × *salicifolia* and *M. stellata* confirmed the results of earlier analysis performed for the same trees [12] – as the second in frequency *M.* × *salicifolia* initiated Lucas pattern while *M. stellata* the main Fibonacci bijugy. Among additional patterns identified in *M. denudata* and *M. acuminata* none was markedly more frequent than others. Previous analysis of gynoecia patterns, conducted in other genets of these species [12], indicated much greater diversity of phyllotaxis in *M. denudata* and significant intraspecific variation of phyllotactic spectra in *M. acuminata*. Depending on a tree, gynoecia of *M. acuminata* exhibited dominance of either Fibonacci pattern or others like main Fibonacci bijugy or even trijugy [12].

Four genets examined in this study differ in the meristem and primordia sizes, determining specific ranges of *P/M* ratio, and in the degree of variation of the angles of parastichy intersection in gynoecia. The values of these parameters suggest slightly different phyllotactic spectra than those identified and quite different frequencies of the patterns. This implies that the arrangement of primordia on the meristem surface, at least in *Magnolia* gynoecia, depends not only on the ratio of the organ size to the size of the meristem, but also on other geometric factors, or factors of different than purely geometrical nature.

It has been shown in computer simulations that geometrical parameters of the initial pattern in the perianth and the way in which the *P/M* ratio evolves during flower ontogeny, are important for the quality and stability of phyllotaxis in *Magnolia* gynoecium [31]. The importance of the initial pattern for the arrangement of newly initiated primordia have been demonstrated for many plant systems [10,74–76] or systems generated in computer simulations [27,28]. It is also supported by our observation, that patterns frequently noted in gynoecia are often initiated in continuity with the same pattern in androecium zone, while uncommon patterns of gynoecia are preceded by patterns of androecia regarded as rare in nature. On this basis we can explain why examined tree of *M. denudata* has less variable phyllotaxis of gynoecia, than suggested by the analysis of the ratio *P/M* or former analysis of patterns in gynoecia of the other specimen of this species [12]. It is probable that stamens, which in this specimen are often initiated in pure Fibonacci pattern, exert stabilising influence on the positioning of carpel primordia and restrict the number of their possible arrangements. This stabilising effect can be additionally enhanced by relatively late shift from indeterminate to determinate growth of the meristem, what excludes potential fluctuations of growth in the early stages of gynoecium pattern formation. In the recent theoretical study it has been shown that phyllotactic patterns differ in their developmental stability and that the Fibonacci pattern and its bijugy are the most stable [65]. This might explain the persistence of these prevailing patterns in a course of initiation of generative parts on the magnolia floral axis.

All these theoretical considerations are based on the assumption of the autonomy of apical meristem generating certain, defined organographic order. Numerous data implicate though, that meristems act as non-autonomous organogenic centres, in which initiation process is guided by the signals flowing acropetally from differentiated tissues of the stem. Primary vascular system is proposed as developmental blueprint for primordia positioning, based on interrelationship between lateral organ arrangement and vascular architecture demonstrated in many plants [6,7,14,77,78]. Although the exact mechanism of the interaction between these two systems and the nature of the signalling molecule mediating between them remain undefined, there are some new exciting data on *PIN1* mutants of *Arabidopsis thaliana* suggesting possible role of auxin transported acropetally in provascular strands [9]. In vegetative development of *Torreya* the degree of meristem autonomy most probably changes with the identity of primordia – bijugate pattern of the needles is determined by differentiating protoxylem, but bud scales are initiated in decussate pattern independently of any signal from differentiated tissues [8]. It is possible that the degree of meristem autonomy changes also in the development of *Magnolia* flower. Our analysis suggests that the arrangement of stamens is dictated by purely geometrical clues – the size of primordia relative to the meristem size and the angle of parastichies intersection. In gynoecia the ratio *P/M* determine only the general number of parastichies but exact spectra and frequencies of initiated patterns seem to be specified either by the way in which this parameters evolve from the beginning of the flower ontogeny or by the signals flowing from primary vascular system, as suggested by the first symptoms of primordium initiation (mitoses of the first periclinal divisions) taking place beneath L2 layer of the floral apex (Fig. 3). First scenario can be verified by use of computer simulation, the second through the analysis of the relationship between development of organ primordia and corresponding vascular traces. Both approaches were beyond the scope of our analysis and are awaiting further studies.

Acknowledgements

This research is a part of the PhD Thesis of the first author, presented to the Faculty of Biological Sciences at the University of Wrocław, Poland.

References

1. Church AH. On the relation of phyllotaxis to mechanical laws. London: Williams & Norgate; 1904.
2. Richards FJ. The geometry of phyllotaxis and its origin. Symp Soc Exper Biol. 1948;2:217–245.
3. Richards FJ. Phyllotaxis: its quantitative expression and relation to growth in the apex. Phil Trans R Soc B. 1951;235(629):509–564.
4. Richards FJ. Spatial and temporal correlations involved in leaf pattern production at the apex. In: Milthrope FL, editor. The growth of leaves. London: Butterworths; 1956. p. 66–76.
5. Erickson RO. The geometry in phyllotaxis. In: Dale JE, Milthorpe FL, editors. The growth and functioning of leaves: proceedings of a symposium held prior to the thirteenth International Botanical Congress at the University of Sydney, 18–20 August 1981. Cambridge: Cambridge University Press; 1983. p. 53–88.
6. Larson PR. Development and organization of the primary vascular system in *Populus deltoides* according to phyllotaxy. Am J Bot. 1975;62(10):1084–1099.
7. Larson PR. Phyllotactic transitions in the vascular system of *Populus deltoides* Bartr. as determined by ^{14}C labeling. Planta. 1977;134(3):241–249.
8. Banasiak A, Zagórska-Marek B. Signals flowing from mature tissues to SAM determine the phyllotactic continuity in successive annual increments of the conifer shoot. Acta Soc Bot Pol. 2006;75(2):113–121.
9. Banasiak A. Putative dual pathway of auxin transport in organogenesis of *Arabidopsis*. Planta. 2011;233(1):49–61.
10. Zagórska-Marek B. Phyllotactic patterns and transitions in *Abies balsamea*. Can J Bot. 1985;63(10):1844–1854.
11. Szymanowska-Pułka M. Phyllotactic patterns in capitula of *Carlina acaulis* L. Acta Soc Bot Pol. 1994;65(3–4):229–245.
12. Zagórska-Marek B. Phyllotaxic diversity in *Magnolia* flowers. Acta Soc Bot Pol. 1994;63(2):117–137.
13. Endress P. Floral phyllotaxis and floral evolution. Bot Jahrb Syst. 1987;108(2–3):417–438.
14. Tucker SC. Phyllotaxis and vascular organization of the carpels in *Michelia fuscata*. Am J Bot. 1961;48(1):60.
15. Doust AN. The developmental basis of floral variation in *Drimys winteri* (Winteraceae). Int J Plant Sci. 2001;164(4):697–717.
16. Zagórska-Marek B. Phyllotaxis triangular unit: phyllotactic transitions as the consequence of apical wedge disclinations in a crystal-like pattern of the units. Acta Soc Bot Pol. 1987;56(2):229–255.
17. Meicenheimer RD, Zagórska-Marek B. Consideration of the geometry of the phyllotaxic triangular unit and discontinuous phyllotactic transitions. J Theor Biol. 1989;139(3):359–368.
18. Gomez-Campo C. Phyllotactic patterns in *Bryophyllum tubiflorum* Harv. Bot Gaz. 1974;135(1):49–58.
19. Gola EM. Phyllotaxis diversity in *Lycopodium clavatum* L. and *Lycopodium annotinum* L. Acta Soc Bot Pol. 1996;65(3–4):235–247.
20. Kwiatkowska D. Intraspecific variation of phyllotaxis stability in *Anagallis arvensis*. Acta Soc Bot Pol. 1997;66(3–4):259–271.
21. Stieger PA, Reinhardt D, Kuhlemeier C. The auxin influx carrier is essential for correct leaf positioning. Plant J. 2002;32(4):509–517.
22. Reinhardt D, Pesce ER, Stieger P, Mandel T, Baltensperger K, Bennett M, et al. Regulation of phyllotaxis by polar auxin transport. Nature. 2003;426(6964):255–260.
23. Bainbridge K, Guyomarc'h S, Bayer E, Swarup R, Bennett M, Mandel T, et al. Auxin influx carriers stabilize phyllotactic patterning. Genes Dev. 2008;22(6):810–823.
24. Veen AH, Lindenmayer A. Diffusion mechanism for phyllotaxis: theoretical physico-chemical and computer study. Plant Physiol. 1977;60(1):127–139.
25. Williams RF, Brittain EG. A geometrical model of phyllotaxis. Aust J Bot. 1984;32(1):43.
26. Douady S, Couder Y. Phyllotaxis as a dynamical self organizing process part I: the spiral modes resulting from time-periodic iterations. J Theor Biol. 1996;178(3):255–273.
27. Douady S, Couder Y. Phyllotaxis as a dynamical self organizing process part II: the spontaneous formation of a periodicity and the coexistence of spiral and whorled patterns. J Theor Biol. 1996;178(3):275–294.
28. Douady S, Couder Y. Phyllotaxis as a dynamical self organizing process part III: the simulation of the transient regimes of ontogeny. J Theor Biol. 1996;178(3):295–312.
29. Jonsson H, Heisler MG, Shapiro BE, Meyerowitz EM, Mjolsness E. An auxin-driven polarized transport model for phyllotaxis. Proc Natl Acad Sci USA. 2006;103(5):1633–1638.
30. Smith RS, Guyomarc'h S, Mandel T, Reinhardt D, Kuhlemeier C, Prusinkiewicz P. A plausible model of phyllotaxis. Proc Natl Acad Sci USA. 2006;103(5):1301–1306.
31. Zagórska-Marek B, Szpak M. Virtual phyllotaxis and real plant model cases. Funct Plant Biol. 2008;35(10):1025–1033.
32. Snow M, Snow R. Auxin and leaf formation. New Phytol. 1937;36(1):1–18.
33. Snow M, Snow R. Minimum areas and leaf determination. Proc R Soc B. 1952;139(897):545–566.
34. Snow M, Snow R. Regulation of sizes of leaf primordia by growing-point of stem apex. Proc R Soc B.

1955;144(915):222-229.

35. Lenhard M, Laux T. Stem cell homeostasis in the *Arabidopsis* shoot meristem is regulated by intercellular movement of CLAVATA3 and its sequestration by CLAVATA1. Development. 2003;130:3163-3173.

36. Sharma VK, Carles CC, Fletcher JC. Maintenance of stem cell populations in plants. Proc Natl Acad Sci USA. 2003;100(90001):11823-11829.

37. Zhao Y, Medrano L, Ohashi K, Fletcher JC, Yu H, Sakai H, et al. HANABA TARANU is a GATA transcription factor that regulates shoot apical meristem and flower development in Arabidopsis. Plant Cell. 2004;16(10):2586-2600.

38. Hu Y, Xie Q, Chua NH. The *Arabidopsis* auxin-inducible gene *ARGOS* controls lateral organ size. Plant Cell. 2003;15(9):1951-1961.

39. Ori N, Eshed Y, Chuck G, Bowman JL, Hake S. Mechanisms that control knox gene expression in the *Arabidopsis* shoot. Development. 2000;127(24):5523-5532.

40. Semiarti E, Ueno Y, Tsukaya H, Iwakawa H, Machida C, Machida Y. The *ASYMMETRIC LEAVES2* gene of *Arabidopsis thaliana* regulates formation of a symmetric lamina, establishment of venation and repression of meristem-related homeobox genes in leaves. Development. 2001;128(10):1771-1783.

41. Autran D, Jonak C, Belcram K, Beemster GTS, Kronenberger J, Grandjean O, et al. Cell numbers and leaf development in *Arabidopsis*: a functional analysis of the *STRUWWELPETER* gene. EMBO J. 2002;21(22):6036-6049.

42. Clay NK, Nelson T. The recessive epigenetic swellmap mutation affects the expression of two step II splicing factors required for the transcription of the cell proliferation gene *STRUWWELPETER* and for the timing of cell cycle arrest in the *Arabidopsis* leaf. Plant Cell. 2005;17(7):1994-2008.

43. Running MP, Fletcher JC, Meyerowitz EM. The *WIGGUM* gene is required for proper regulation of floral meristem size in *Arabidopsis*. Development. 1998;125(14):2545-2553.

44. Song JY, Leung T, Ehler LK, Wang C, Liu Z. Regulation of meristem organization and cell division by *TSO1*, an *Arabidopsis* gene with cysteine-rich repeats. Development. 2000;127(10):2207-2217.

45. Hauser BA, He JQ, Park SO, Gasser CS. TSO1 is a novel protein that modulates cytokinesis and cell expansion in *Arabidopsis*. Development. 2000;127(10):2219-2226.

46. Dornelas MC, van Lammeren AA, Kreis M. *Arabidopsis thaliana* SHAGGY-related protein kinases (AtSK11 and 12) function in perianth and gynoecium development. Plant J. 2000;21(5):419-429.

47. Leyser HMO, Furner IJ. Characterisation of three shoot apical meristem mutants of *Arabidopsis thaliana*. Development. 1992;116(2):397-403.

48. Laufs P, Dockx J, Kronenberger J, Traas J. *MGOUN1* and *MGOUN2*: two genes required for primordium initiation at the shoot apical and floral meristems in *Arabidopsis thaliana*. Development. 1998;125(7):1253-1260.

49. Para A, Sundås-Larsson A. The pleiotropic mutation *dar1* affects plant architecture in *Arabidopsis thaliana*. Dev Biol. 2003;254(2):215-225.

50. Xu FX. Floral ontogeny of two species in *Magnolia* L. J Integr Plant Biol. 2006;48(10):1197-1203.

51. Xu F, Rudall PJ. Comparative floral anatomy and ontogeny in Magnoliaceae. Plant Syst Evol. 2006;258(1-2):1-15.

52. Johansen DA. Plant microtechnique. New York: McGraw-Hill Book Co.; 1940.

53. Leins P, Erbar C. Floral developmental studies: some old and new questions. Int J Plant Sci. 1997;158(6 suppl):3-12.

54. Tucker SC. Ontogeny of the floral apex of *Michelia fuscata*. Am J Bot. 1960;47(4):266.

55. Lyndon RF. Phyllotaxis and the initiation of primordia during flower development in *Silene*. Ann Bot. 1978;42:1349-1360.

56. Meicenheimer RD. Relationships between shoot growth and changing phyllotaxy of *Ranunculus*. Am J Bot. 1979;66(5):557.

57. Uhl NW, Moore HE. Androecial development in six polyandrous genera representing five major groups of palms. Ann Bot. 1980;45(1):57-75.

58. Kirchoff BK. Shape matters: Hofmeister's rule, primordium shape, and flower orientation. Int J Plant Sci. 2003;164(4):505-517.

59. Zagórska-Marek B, Wiss D. Dislocations in the repetitive unit patterns of biological systems. In: Nation JB, editor. Formal descriptions of developing systems. Manoa: Kluwer Academic Publishers; 2003. p. 99-117.

60. Zagórska-Marek B. *Magnolia* flower – the living crystal. Magnolia. 2011;89:11-21.

61. Erbar C, Leins P. Zur Spirale in Magnolien-Blüten. Beitr Biol Pflanzen. 1982;56:225-241.

62. Jean RV. Phyllotaxis: a systemic study in plant morphogenesis. Cambridge: Cambridge University Press; 1994.

63. Vakarelov I. Changes in phyllotactic pattern structure in *Pinus* L. due to changes in altitude. In: Jean RV, Barabé D, editors. Symmetry in plants. Singapore: World Scientific; 1998. p. 213-230.

64. Gola EM. Phyllotactic spectra in cacti: *Mammillaria* species and some genera from *Rebutia* group. Acta Soc Bot Pol. 1997;66(3-4):237-257.

65. Szpak M, Zagórska-Marek B. Phyllotaxis instability – exploring the depths of first available space. Acta Soc Bot Pol. 2011;80(4):279-284.

66. McCully ME, Dale HM. Variations in leaf number in *Hippuris*. A study of whorled phyllotaxis. Can J Bot. 1961;39(3):611-625.

67. Bierhorst DW. Symmetry in *Equisetum*. Am J Bot. 1959;46(3):170.

68. Kwiatkowska D, Florek-Marwitz J. Ontogenetic variation of phyllotaxis and apex geometry in vegetative shoots of *Sedum maximum* (L.) Hoffm. Acta Soc Bot Pol. 1999;68(2):85-95.

69. Greyson RI, Walden DB. The ABPHYL syndrome in *Zea mays*. I. Arrangement, number and size of leaves. Am J Bot. 1972;59(5):466-472.

70. Greyson RI, Walden DB, Hume JA, Erickson RO. The

ABPHYL syndrome in *Zea mays*. II. Patterns of leaf initiation and the shape of the shoot meristem. Can J Bot. 1978;56(13):1545–1550.
71. Jackson D, Hake S. Control of phyllotaxy in maize by the *abphyl1* gene. Development. 1999;126(2):315–323.
72. Lyndon RF. The shoot apical meristem: its growth and development. Cambridge: Cambridge University Press; 1998.
73. Carles CC, Fletcher JC. Shoot apical meristem maintenance: the art of a dynamic balance. Trends Plant Sci. 2003;8(8):394–401.
74. Hernandez LF, Palmer JH. Regeneration of the sunflower capitulum after cylindrical wounding of the receptacle. Am J Bot. 1988;75(9):1253.
75. Endress P. Chaotic floral phyllotaxis and reduced perianth in *Achlys* (Berberidaceae). Bot Acta. 1989;102:159–163.
76. Lehmann NL, Sattler R. Irregular floral development in *Calla palustris* (Araceae) and the concept of homeosis. Am J Bot. 1992;79(10):1145.
77. Kwiatkowska D. The relationships between the primary vascular system and phyllotactic patterns of *Anagallis arvensis* (Primulaceae). Am J Bot. 1992;79(8):904.
78. Zagórska-Marek B, Banasiak A. Related to phyllotaxis interlocked systems of vascular sympodia and cortical resin canals in *Abies* and *Picea* shoots. Acta Soc Bot Pol. 2000;69(3):165–172.

6

Food uses of ferns in China: a review

Yujing Liu[1], Wujisguleng Wujisguleng[1,2], Chunlin Long[1]*

[1] College of Life and Environmental Sciences, Minzu University of China, Beijing 100081, China

[2] Mongolian Medicine College, Inner Mongolia Medical University, Hohhot 010110, China

Abstract

Edible ferns are some of the most important wild vegetables in China. This paper reviews their food uses. The history of eating ferns in China may go back as far as 3000 years. An ethnobotanical inventory of edible ferns was created, with 52 species (including 4 varieties), which were traditionally used. The potential species number of edible ferns was estimated as 144 species (including 4 varieties). The cuisines, products and chemical components of ferns were also summarized. The most commonly eaten fern, *Pteridium aquilinum* var. *latiusculum*, was discussed in terms of its toxicity, massive productivity and development strategies. Suggestions and recommendations were proposed for the future development of edible ferns in China.

Keywords: edible ferns, food uses, China, ethnobotany, *Pteridium aquilinum* var. *latiusculum*, pterosin, detoxification

Introduction

Wild gathering, together with hunting, used to be the only food source for our ancestors. Although a total of 82 species commodities (or 103 species taxonomically) contribute 90% of national per capita supplies of food plants in the world [1], wild plant gathering is still a tradition that has endured in many local communities [2]. For example, the indigenous peoples of different countries, like Mexico [3], China [4–6], Ethiopia [7], India [8,9], Kenya [10], Palestine [11], and Thailand [12] collect various wild plants for supplementary diets or substitutes for staple food in famine years.

Food uses of wild plants are not only common in developing countries, but also in Japan, Europe and North America. For example many wild plants are still consumed in Spain, Portugal and Italy [13–15]. In some European countries, however, the tradition of wild food plant collection and consumption has been lost for generations. In Poland, for example, among over a hundred wild edible plant species recorded, only a few species are still commonly gathered, including two green vegetables, 15 folk species of fruits and seeds and four taxa used for seasoning, or as preservatives. The traditions of use of other species have either been forgotten or are very rarely used [16]. Both American Indians and white people have collected wild vegetables, fruits and mushrooms for consumption, in the United States and Canada [17–19]. In Japan, wild vegetables used to be important food resources in the cold northern parts. Local communities have developed preservation methods including pickling, salting and drying of edible wild plants to be used throughout all the seasons [20]. The number of edible wild plants occurring in Japan is estimated to be over 1000 species [21], or even around 2000 species [22]. However, only around 25 or less species were utilized on a regional scale [23].

Edible ferns are some of the most common wild food plants collected by people around the world. The fern stems, rhizomes, leaves, young fronds and shoots, and sometimes the whole plants are used for food. As early as the 1940s, *Athyrium esculentum*, *Ceratopteris siliquosa*, *C. pteridoides*, *Dryopteris prolifera*, *Pteretis nodulosa*, *P. esculenta*, and *Helminthostachys* sp. were reported as consumed by the indigenous Filipino [23]. In recent years, more and more research has reported the food uses of pteridophytes in different parts of the world. The fronds of *Ophioglossum polyphyllum* and those of *O. nudicaule* are cooked as a vegetable in Nepal. The fronds of *Helminthostachys zeylanica* are eaten as a salad in rural areas of Malaysia and the Philippines. Tender leaves of *Botrychium lanuginosum* are regarded as a delicious vegetable in Nepal [24].

Tree ferns have often been used as food and starch in Hawaii [24]. Both young fronds and underground stems of Asplenium ensiforme are used for food in the Himalayas. In Malaysia, the *Blechnum orientalis* rhizomes are eaten, and the fronds of *Ceratopteris thalictroides* and *Diplazium esculentum* are used as vegetables. The rhizome and young shoots of Nephrolepis biserrata are edible. *Ophioglossum reticulatum* is also eaten as a salad and as a vegetable [25]. In India, stems of *Angiopteris* sp. are eaten for starch. *Pteris ensiformis*, *Helminthostachys zeylanica*, *Phymatosorus longissimus* and *Microsorum alternifolium* fern fronds are also used for food. The Maori and Oceanic precedents collected the rhizomes of *Petridium esculentum* and made food in spring and early

* Corresponding author. Email: long@mail.kib.ac.cn

summer when the sweet potato or taro was not available [24,26]. The starchy paste of the sporocarps of *Marsilea drummondii* is made into cakes and is eaten by the natives of Australia [26].

There are about 10000–12000 species of pteridophytes in the world. In North America, 420 ferns and fern allies have been recorded. Some Asian countries are rich in pteridophyte diversity. For instance, 639 species of ferns occur in Japan, about 1000 in the Philippines, 550 in Malaysia, more than 700 in Thailand, and about 600 in India [27–29].

China is one of the richest countries in pteridophyte species. About 2600 species, in 63 families and 230 genera have been reported in China, which occupies more than 20% of the global pteridophyte flora [27]. The Chinese people have traditionally used ferns for various purposes for many centuries. The indigenous uses include medicines (*Cibotium barometz*, *Davallia mariesii* and many others) [27,30,31], ornamentals (*Neottopteris nidus*, *Nephrolepis auriculata* and others) [27,32], and food as well.

The food uses of ferns in China have been recorded in many ancient books. Some fern species such as *Pteridium aquilinum* var. *latiusculum*, *Callipteris esculenta*, *Matteuccia struthiopteris*, *Osmunda japonica* and *Lygodium japonicum* are very common in the country [27]. Local people, especially those from ethnic minorities in rural areas, have kept the tradition of collecting and eating ferns for many generations.

So far, however, we cannot find any ethnobotanical investigative reports on edible ferns in China. And there is no calculation of the number of fern species, which have traditionally been eaten by local people. This review will try to give an outline of edible ferns for traditional food uses in the ethnic communities of China. The species number of pteridophytes with edibility will be estimated. The future development of edible ferns will also be proposed in the present paper.

Results

Records of edible ferns in ancient Chinese literature

Edible ferns were recorded for the first time in "Shi-Jing" ("Book of Odes"), about 3000 years ago. Two species of ferns, *Pteridium aquilinum* var. *latiusculum* (or "juecai" in Chinese) and *Osmunda japonica* (or "weicai" in Chinese), appeared in the ode [33]. These ferns are still often collected and eaten in rural areas.

The edible parts (fronds), collection time and morphology of ferns were described in the ancient books issued in the Three Kingdoms Period (called "San-Guo", about 1800 years ago). The significance of edible ferns in livelihood had also been recorded in the literature of that period [33,34].

The time when eating ferns was most popular, was during the Tang Dynasty (AD 618–907). Many poets described, praised or admired edible ferns in their poems. The fronds of *Pteridium aquilinum* var. *latiusculum* and *Osmanda japonica* were the parts most commonly consumed by people in that period [33,35].

The first written record of extracting starch for food from fern rhizomes was during the Song Dynasty (AD 960–1276) [36]. In the Yuan Dynasty (AD 1271–1368), the fern starch from *Pteridium aquilinum* var. *latiusculum* became very important and sometimes it became the main substitute for staple food in the famine years [33].

How many species of ferns have traditionally been used for food in China?

Edible ferns or fern products can be seen in local markets or supermarkets in China. However, nobody can tell how many species of ferns have traditionally been used for food in China. In Guizhou Province only, 64 edible fern species were reported in recent years, covering both traditional and potential edible species [36,37].

Based on literature studies [27,36–40], and our field investigations over the past 20 years, we estimated that there are 52 species (including 4 varieties) of edible ferns traditionally consumed as food by various ethnic groups throughout the country (Tab. 1). Generally, ethnic minorities in the southwest region (especially Yunnan, Guangxi, Guizhou and Sichuan provinces) eat more fern species. Some species like *Neottopteris nidus* taste very bitter, but the local people in southern Yunnan still eat the young leaves as a precious vegetable.

The most common species used as food in traditional communities are *Pteridium aquilinum* var. *latiusculum* ("juecai"), *Callipteris esculenta*, *Osmunda japonica* ("weicai"), *Pteridium revolutum* and *Ceratopteris thalictroides*. Among them, *Pteridium aquilinum* var. *latiusculum* is widely distributed in China. It grows massively and produces a lot of fronds in spring, while the rhizomes, with rich starch, can be harvested in the fall. This species has become the most popular edible fern, consumed by a billion people in China every year. For example, in the Zhuzhou County of Hunan Province only, the dry fronds of *Pteridium aquilinum* var. *latiusculum* ("juecai") reached 1200 tons in 2011, with a value of 70 million RMB (ca. 1 million in US dollars) [41]. An estimated over 1000 companies in China produce dry "juecai" and its products. Their annual production values reach 300 million US dollars.

Potential edible ferns in China

The stems, rhizomes, fronds, young leaves and shoots, or the whole plants of many fern species can be used for food. The starch in the stems or rhizomes can be harvested, processed and cooked for food, or used to make liquor. The fronds, young leaves and shoots can be eaten as vegetables.

Because of the inaccessibility of habitats, low biomass, lack of consumers, or for cultural, religious and legal reasons, many edible fern species have never, or hardly ever, been collected for food, or have not been recorded in any literature sources. For example, *Metapolypodium memeiense* (Christ) Ching, an epiphytic fern species in southwestern China, has been used as a taste-enhancer for cooking with vegetables by the Jinuo ethnic group in southern Yunnan. Tree ferns are CITES-listed species [42], and also listed in the "Red data book of China" [43]. These species were not included in Tab. 1. Only *Alsophila spinulosa* in the tree fern group is an exception because of its larger population. Sometimes the local people collected the starch from stems which fell down in the forests.

We estimated the potential species number of edible ferns, based on literature analysis [25,26,36–39], field surveys, edibility investigation, and phylogenic or taxonomic information. The total number reaches at least 144 species, including 4 varieties (Tab. 1, Tab. 2).

Among these 144 potential species (varieties) of edible ferns, the biggest groups are *Allantodia* (Athyriaceae, 30 spp.), *Pteris* (Pteridaceae, 20 spp.), *Angiopteris* (Angiopteridaceae, 18 spp.), *Coniogramme* (Hemionitiaceae, 15 spp.), *Alsophila* (Cyatheaceae, 14 spp.), *Athyrium* (Athyriaceae, 12 spp.), *Asplenium* (Aspleniaceae, 10 spp.) and *Lygodium* (Lygodiaceae, 10 spp.), in terms of species numbers in each genus. However, the fern

Tab. 1 Ethnobotanical inventory of ferns used for food in China.

No.	Species Name	Family Name	Edible Parts	Preparation	Notes	References
1	*Allantodia dilatata* (Blume) Ching	Athyriaceae	Tender leaves	Stir-fry, soup	Fresh or dry use	[36]
2	*Allantodia gigantea* (Bak.) Ching	Athyriaceae	Tender leaves	Stir-fry, soup	Fresh or dry use	[36]
3	*Allantodia himalayensis* Ching	Athyriaceae	Tender leaves	Stir-fry, soup	Fresh or dry use	[36]
4	*Allantodia spectabilis* (Wall. et Mett.) Ching	Athyriaceae	Tender leaves	Stir-fry, soup	Fresh or dry use	[27,36]
5	*Allantodia viridissima* (H. Christ) Ching	Athyriaceae	Tender leaves	Cooked after processing with salt	Sour taste	[36]
6	*Alsophila spinulosa* (Hook.) Tryon	Cyatheaceae	Starch in stems	Made into cakes or mixed with other food	Substitute for staple food	[37,39,40]
7	*Ampelopteris prolifera* (Retz.) Cope.	Thelypteridaceae	Young leaves and shoots	Stir-fry, soup	Fresh or dry use	[36–39]
8	*Angiopteris esculenta* Hieron.	Angiopteridaceae	Starch in rhizomes	Made into cakes or mixed with other food	Substitute for staple food	[38–40]
9	*Angiopteris fokiensis* Hieron.	Angiopteridaceae	Starch in rhizomes	Made into cakes or mixed with other food	Substitute for staple food	[37–40]
10	Asplenium unilaterale Lam.	Aspleniaceae	Rhizomes	Snack, salad	Fresh use	[36]
11	*Athyrium brevifrons* Nakai ex Kitag.	Athyriaceae	Starch in rhizomes	Cakes, noodles	Starch content 40-50%	[38–40]
12	*Athyrium multidentatum* (Doll) Ching	Athyriaceae	Fronds	Stir-fry, cooked with other food, soup	Fragrant, tender	[38,39]
13	*Athyrium pachyphyllum* Ching	Athyriaceae	Fronds	Stir-fry, cooked with other food, soup	Fresh or dry use	[36,39]
14	*Athyrium yokoscense* (Franch. et Sav.) Chrsit	Athyriaceae	Fronds	Stir-fry, cooked with other food, soup	Fresh use or processing	[36,39]
15	*Athyriopsis japonica* (Thunb.) Ching var. *oshimensis* (Christ) Ching	Athyriaceae	Young leaves	Stir-fry, cooked with other food, soup	Fresh or dry use	[36,38]
16	*Blechnum orientale* L.	Blechnaceae	Fronds and young red leaves	Fried or stir-fried	Fresh or dry use	[36,39]
17	*Callipteris esculenta* (Retz.) J. Sm. ex Moore et Houlst	Athyriaceae	Fronds	Stir-fry, soup, salad	Fresh	[27,36,38]
18	*Callipteris esculenta* (Retz.) J. Sm. ex Moore et Houlst. var. *pubescens* (Link) Ching	Athyriaceae	Fronds	Stir-fry, soup, salad	Fresh	[36,39]
19	*Ceratopteris thalictroides* (L.) Brongn.	Parkeriaceae	Young leaves	Stir-fry	Fresh	[27,37–39]
20	*Cibotium barometz* (L.) J. Sm.	Dicksoniaceae	Starch in rhizomes	Cakes, liquor	Substitute for staple food	[37–40]
21	*Coniogramme emeiensis* Ching et Shing	Hemionitiaceae	Fronds, starch in rhizomes	Cooked leaves as vegetable; noodles		[36,39]
22	*Coniogramme intermedia* Hieron.	Hemionitiaceae	Fronds, starch in rhizomes	Cooked leaves as vegetable; noodles		[36,39]
23	*Coniogramme intermedia* Hieron. var. *glabra* Ching	Hemionitiaceae	Fronds, starch in rhizomes	Cooked leaves as vegetable; noodles		[36,39]
24	*Coniogramme japonica* (Thunb.) Diels	Hemionitiaceae	Fronds, starch in rhizomes	Cooked leaves as vegetable; noodles		[36,39]
25	*Coniogramme jingangshanensis* Ching et Shing	Hemionitiaceae	Fronds, starch in rhizomes	Cooked leaves as vegetable; noodles		[36,39]
26	*Coniogramme robusta* Christ	Hemionitiaceae	Fronds, starch in rhizomes	Cooked leaves as vegetable; noodles		[36,39]
27	*Coniogramme rosthornii* Hieorn.	Hemionitiaceae	Fronds, starch in rhizomes	Cooked leaves as vegetable; noodles		[36,39]
28	*Coniogramme simillima* Ching ex Shing	Hemionitiaceae	Fronds, starch in rhizomes	Cooked leaves as vegetable; noodles		[36,39]
29	*Coniogramme taipaishanensis* Ching et Y. T. Hsieh	Hemionitiaceae	Fronds, starch in rhizomes	Cooked leaves as vegetable; noodles		[36,39]
30	*Coniogramme wilsoni* Hieron.	Hemionitiaceae	Fronds, starch in rhizomes	Cooked leaves as vegetable; noodles		[36,39]
31	*Cornopteris decurrenti*-alata (Hook.) Nakai	Athyriaceae	Fronds	Stir-fry, soup	Fresh or dry use	[39]

Tab. 1 *(continued)*

No.	Species Name	Family Name	Edible Parts	Preparation	Notes	References
32	*Cyrtomium fortunei* J. Sm.	Dryopteridaceae	Fronds	Stir-fry, soup	Processed for storage	[36,39]
33	*Drynaria baronii* (Christ) Diels	Dryopteridaceae	Starch in rhizomes	Source of starch		[39,40]
34	*Drynaria fortunei* (Kunze) J. Sm.	Drynariaceae	Starch in rhizomes	Cakes, liquor		[36,39,40]
35	*Lunathyrium acrostichoides* Ching	Athyriaceae	Fronds	Stir-fry, soup		[39]
36	*Lunathyrium coreanum* (Christ) Ching	Athyriaceae	Fronds	Stir-fry, soup		[39]
37	*Lygodium japonicum* (Thunb.) Sw.	Lygodiaceae	Young leaves	Stir-fry, soup		[27,37,39]
38	*Marsilea quadrifolia* L.	Marsileaceae	Young leaves	Stir-fry, soup		[27,36–39]
39	*Matteuccia intermedia* C. Chr.	Onocleaceae	Fronds	Stir-fry, soup		[36,39]
40	*Matteuccia orientalis* (Hook.) Trev.	Onocleaceae	Fronds	Stir-fry, soup		[36,39]
41	*Matteuccia struthiopteris* (L.) Todaro	Onocleaceae	Fronds, starch in rhizomes	Stir-fry, soup; starch for noodle-making	Delicious vegetable; starch content in rhizomes reaches 40–50%	[27,38–40]
42	*Nephrolepis auriculata* (L.) Trimen.	Nephrolepidaceae	Young leaves, tubers	Cooked as vegetable, tuber eaten as snack	Young tubers with sweetner	[36,39,40]
43	*Neottopteris nidus* (L.) J. Sm.	Aspleniaceae	Young leaves	Stir-fry, soup	Bitter taste	[36]
44	*Ophioglossum polyphyllum* A. Braun	Ophioglossaceae	Fronds	Soup	Fresh or dry use	[25]
45	*Osmunda japonica* Thunb.	Osmundaceae	Fronds, starch in rhizomes	Stir-fried or cooked with other food, soup; starch for making noodles, liquor	Fresh or dry uses, or processing. Delicious taste. Leaves contain starch, too	[37–40]
46	*Osmundastrum cinnamomeum* (L.) Presl.	Osmundaceae	Fronds	Stir-fried or cooked with other food, soup	Delicious taste	[37–40]
47	*Polypodiodes niponica* (Mett.) Ching	Polypodiaceae	Tender leaves	Cooked as vegetable	Bitter taste	[39]
48	*Pteridium aquilinum* L. var. *latiusculum* (Desv.) Underw.	Pteridiaceae	Fronds, starch in rhizomes	Stir-fried or cooked with other food, or processed. Starch used to make noodles, liquor, cakes and other products	Commonly called juecai. Fronds should be cooked in boiled water and then washed for a few hours. Widely distributed species in many areas	[27,37–40]
49	*Pteridium revolutum* (Bl.) Nakai	Pteridiaceae	Fronds, starch in rhizomes	Stir-fried or cooked with other food, or processed. Starch used to make liquor or cakes	Makes similar products as juecai	[27,37–40]
50	*Pteris wallichiana* Agardh	Pteridaceae	Tender leaves	Stir-fry	Fresh use	[27,37]
51	*Woodwardia japonica* (L. f.) J. Sm.	Blechnaceae	Starch in rhizomes	Cakes, noodles, liquor		[36–40]
52	*Woodwardia unigemmata* (Makino) Nakai	Blechnaceae	Starch in rhizomes	Cakes, noodles, liquor		[36,39]

products from these groups are rarely sold in local markets or are less consumed by the local people.

Traditional fern cuisines

The Chinese people in different areas have developed various methods to make fern dishes and food varieties. The most common dish is made of stir-fried fresh or dry fern fronds of *Pteridium aquilinum* var. *latiusculum* ("juecai"). There are various stir-fried dishes created by adding other food to the fern fronds. For example, chili peppers, Chinese leeks (*Allium tuberosum*), ham, or chicken can be stir-fried with fern fronds. Other species, like *Allantodia gigantea*, *Athyrium yokoscense*, *Callipteris esculenta*, *Lygodium japonicum*, *Matteuccia struthiopteris*, *Osmunda japonica* and *Osmundastrum cinnamomeum* have also been cooked following a similar formula.

Another popular dish is made of fern slices made from starch, either fried or steamed. The starch was usually extracted from the rhizomes of *Pteridium aquilinum* var. *latiusculum*. The slices could be prepared like Chinese bread, and then cooked in pots. Fern cakes are also common, made from fern starch. In the famine years, people collected the rhizomes in the mountains and extracted the starch as a substitute for staple food.

Both fern fronds and starch are used to make porridge (congee), together with rice or millet. The fronds can be either fresh or dried, but need to be cut into small slices.

Fern food products in China

Both in local markets and supermarkets, there are a lot of fern food products in China. These products can be

Tab. 2 List of all edible fern genera in China.

No.	Genus Name	Family Name	Edible Parts	Species No. in China	Edible species (estimated number)
1	*Allantodia* R. Br. emend. Ching	Athyriaceae	Young leaves	73	30
2	*Alsophila* R. Br.	Cyatheaceae	Starch in stems	14 (incl. 2 varieties)	14
3	*Ampelopteris* Kunze	Thelypteridaceae	Young leaves	1	1
4	*Angiopteris* Hoffn.	Angiopteridaceae	Rhizomes	20	18
5	*Asplenium* L.	Aspleniaceae	Young leaves	110	10
6	*Athyrium* Roth	Athyriaceae	Young leaves	82	12
7	*Athyriopsis* Ching	Athyriaceae	Young leaves	10	3
8	*Blechnum* L.	Blechnaceae	Fronds	1	1
9	*Callipteris* Bory.	Athyriaceae	Young leaves	3 (incl. 1 variety)	2
10	*Ceratopteris* Brongn.	Parkeriaceae	Young leaves	2	2
11	*Cibotium* Kaulf.	Dicksoniaceae	Starch in rhizomes	2	1
12	*Coniogramme* Fee	Hemionitiaceae	Young leaves, starch in rhizomes	20	15
13	*Cornopteris* Nakai	Athyriaceae	Fronds	11	6
14	*Cyrtomium* Presl.	Dryopteridaceae	Fronds,	40	4
15	*Drynaria* (Bory) J. Sm.	Drynariaceae	Satrch in rhizomes	9	8
16	*Gymnosphaera* Bl.	Cyatheaceae	Starch in stems	9	9
17	*Lunathyrium* Ching	Athyriaceae	Fronds	48 (incl. 10 varieties)	5
18	*Lygodium* Sw.	Lygodiaceae	Young leaves	10	10
19	*Marsilea* L.	Marsileaceae	Young leaves	3	1
20	*Matteuccia* Todaro	Onocleaceae	Fronds	3	3
21	*Metapolypodium* Ching	Polypodiaceae	Leaves	1	1
22	*Nephrolepis* Schott	Nephrolepidaceae	Young leaves, tubers	6	2
23	*Neottopteris* Schott	Aspleniaceae	Young leaves	11	2
24	*Ophioglossum* L.	Ophioglossaceae	Fronds	5	1
25	*Osmunda* L.	Osmundaceae	Young leaves, starch in rhizomes	3	3
26	*Osmundastrum* Presl.	Osmundaceae	Young leaves, starch in rhizomes	2	2
27	*Polypodiodes* Ching	Polypodiaceae	Tender leaves	12	3
28	*Pteris* L.	Pteridaceae	Tender leaves	50	20
29	*Sphaeropteris* Bernh	Cyatheaceae	Starch in stems	2	2
30	*Woodwardia* Smith	Blechnaceae	Starch in rhizomes	5	5

stored or preserved for later. Common products include dried fronds, salted fronds, packaged fronds, fern starch, fern starch noodles, fern starch cakes, and fern leaf tea. When making fern starch noodles and cakes, wheat flour can be added and mixed.

The fronds are mostly from two species, *Pteridium aquilinum* var. *latiusculum* ("juecai") and *Osmunda japonica* ("weicai"). The former occupies more than 80% of the fern food market. Both species are exported to international markets. The starch for making food products comes mainly from the rhizomes of *Pteridium aquilinum* var. *latiusculum*. The season for harvesting rhizomes is from September to November, when the aerial parts of the ferns become withered.

People use the fern starch to produce other products, such as liquor and soft drinks. Before they make liquor, other materials like rice, sweet potato or corn flour can be added to the fern starch.

Nowadays, many varieties of fern products have been developed. The canned ferns, spicily seasoned ferns, dehydrated ferns, salted ferns and other products can easily brought in the markets in all seasons.

Chemical components of edible ferns and their bioactivities

Many researchers have reported the chemical constituents of different edible ferns. Nine compounds have been isolated from the fronds of *Pteridium aquilinum* var. *latiusculum*, the most common edible fern in China. They are daucosterol, trans-tiliroside, adenosine, wallichoside, inkosterone, rutin, pterosin A, ponasterone A and quercetin [44].

Matteuccia struthiopteris is also a commonly consumed edible fern species. The compounds found in this species include 1-O-β-D-glucopyranosyl-(2S, 3R, 4E, 8z)-2-N-(2'-hydroxydocosanoyl) eicosasphinga-4,8-dienine, 1-O-β-galactosyl-(6→1)-α-β-galactosyl-2,3-O-dihexadecanoyl-glycerol, succinic acid, D-mannitol, demethoxymatteucinol, matteucinol, pinosylvin, matteuorien, pinosylvin 3-O-β-D-glucopyranoside, matteuorienate A [45,46].

Many chemical compounds are reported in Pteris [47,48]. These are, for example, apigenin-7-O-β-D-glucopyranosyl-4'-O-α-L-rhamnopyranoside, luteolin-7-O-β-D-glucopyranoside, apigenin-7-O-β-D-glucopyranoside, apigenin, luteolin, naringenin-7-O-β-D-neohesperidoside, apigenin-7-O-β-D-neohespedoside, apigenin-4'-O-α-L-rhamnopyranoside, (2S,3S)-pterosin C, isovanillic acid, ferulic acid, 2β,16α-dihydroxy-ent-kaurane 2-O-β-D-glucoside, 2β,6β,15α-trihydroxy-ent-kaur-16-ene, 9-hydroxy-15-oxo-ent-kaur-16-en-19-oic acid 19β-D-glucoside, 2β,14β,15α,16α,17-pentahydroxy-ent-kaurane, and 9-hydroxy-ent-kaur-16-en-19-oic acid. Some extracts from *Pteris* showed antitumor, antifungal and antibacterial activity.

Some of the compounds had an inhibitory effect on platelet aggregation, as well as an anti-inflammatory effect [49].

Pterosins, terpenoids, sterides, flavones, glucosides, aromatic, and pyrone compounds were isolated from *Cibotium barometz* [50]. Their pharmacological effects, such as anti-haemorrhagic, bacteriostatic, antiosteoporotic, anticancer, and hepatoprotective, have been tested [31].

A representative species in Athyriaceae, Callipteris esculenta, has also been studied phytochemically. Such compounds as β-sitosterol, stigmast-4-ene-6β-ol-3-one, stigmast-4-ene-3,6-dione, benzeneacetic acid, glycerol-1,3-dihexadecanoate, 3β-hydroxy-5α,8α-epidioxyergosta-6,22-diene, stigmast-4-ene-3β,6β-diol, stigmast-5-ene-3β,7α-diol, stigmast-4-ene-6α-o1-3-one, and daucosterol have been isolated from the fronds of this species [49].

Chemicals from *Nephrolepls cordifolia* have also been isolated. They were identified as β-si-tosterol, fern-9(11)-ene, oleanolic acid, myristic acid octadecylester, hentriacontanoic acid and triacontanol [51].

Discussion and suggestions

Toxicity of *Pteridium aquilinum* var. *latiusculum*

Bracken fern, or *Pteridium aquilinum*, is the most controversial species. Its fronds and starch have been eaten by billions of people in the world. However, it is a problematic, dangerous and poisonous plant due to its special chemical composition, especially the content of ptaquiloside (PT), a highly carcinogenic compound. It has been found that PT is present in all parts of the plant, including fronds, leaves, rhizomes, and roots. The highest concentration of PT (average at 3800 μg/g) is in the frond growing season, while the PT contents of roots are the lowest (5–230 μg/g). In the rhizomes, the PT contents are less than 1200 μg/g in general [52].

Different types of human carcinomas (oesophageal, gastric cancers, etc.) are produced by ingestion of bracken fern or milk from bracken-fed cattle [53]. Even nowadays, because of its wide distribution and easy accessibility, bracken fern has been collected as part of the diet by many poor people, from south Pacific islands to northern, temperate prairies.

Botanically, the taxon massively collected by local people in China as food is a variety of *Pteridium aquilinum*. Unfortunately, this lower taxon (*Pteridium aquilinum* var. *latiusculum*, in Chinese it is also called "juecai", "longzhuacai" or "ruyicai" commonly) has also been proven to be poisonous because PT has been detected in this subspecies [54]. Thus, we should warn the public that "juecai" (including fronds, starch and their products) is poisonous and harmful to human health.

However, it is a strong part of traditional food culture to collect and eat ferns in the most rural regions of China. And "juecai" (*Pteridium aquilinum* var. *latiusculum*) is the most important food fern among all edible pteridophytes for both household consumption and commercial purposes. We cannot stop Chinese people eating juecai. The only solution is to remove toxins from "juecai" in an easy and cost-effective way.

Fortunately, PT is unstable in water. This compound will also decompose under acidic or alkaline conditions. This is the reason why very few people have suffered from eating "juecai", because people usually wash it in water for hours or even a few days to remove the bitter flavor. Therefore detoxification is very easy when processing "juecai" and its products, even in the countryside. It is strongly encouraged that the fresh fronds or starch from *Pteridium aquilinum* var. *latiusculum* are washed in water, or placed under acidic or alkaline conditions, for a few hours before eating or making products.

Further research

Among China's 196 edible fern species, including the potentially edible species (144 spp.) and those traditionally used as food (52 spp.), only 10% have been studied phytochemically. Chemical constituents of other edible fern species, in particular the possible toxins, should be examined carefully.

The contents of PT in the *Pteridium aquilinum* var. *latiusculum* ("juecai") plant may be different from one place to another. Studies carried out in New Zealand, Australia, Costa Rica and Venezuela revealed that the PT contents for bracken fern fronds (from 12 stands of four geographically distinct habitats) were found between 210 and 2150 μg/g, this variation was positively correlated with development phases and altitudes [55]. Therefore, it is necessary to study the PT contents in different populations or geographical origins of "juecai".

Most fern species grow slowly ("juecai" is an exception). It is necessary to grow edible ferns for commercial development artificially or semi-artificially. As ferns produce numerous spores, mass propagation and production of edible ferns from spores for commercial production should be studied. The micropropagation techniques of edible ferns (especially the rare species) can be encouraged, too.

Suggestions for production and toxicity examination of edible ferns

In China, there are thousands of companies producing edible fern products. The contents of PT and other poisonous compounds in edible ferns and fern products should carefully be examined before they appear in commercial markets. Those with higher PT content must not be allowed onto the market. Raw materials from *Pteridium aquilinum* var. *latiusculum* and allied species should be double-checked.

The governmental food and drug administrative authorities at different levels should issue criteria for quality/safety control of edible fern products. Every edible fern enterprise should follow the criteria, especially the content limitation of PT in the products.

It will be helpful to identify the origin of raw fern materials, because the contents of PT are very different based on their geographical locations. The practice of placing a certificate of origins with low PT contents on edible fern products should be encouraged.

Recommendations of a few fern species

Based on our investigation, we recommend the following edible fern species for future commercial development in China:

(*i*) *Athyrium multidentatum* is a common fern species in northeast China. It is called "houtui" in Chinese, which means monkey legs because of the yellow hairs on the petioles. The fronds of *Athyrium multidentatum* are very tasty and fragrant. Both fresh and processed fronds can be stir-fried or cooked for soup.

(*ii*) *Matteuccia struthiopteris* also occurs in northeast China. In Chinese it is called "Huanggua Xiang", which means it has a cucumber ("Huanggua") flavor. The young leaves can be cooked as a delicious dish or used to make dumplings.

(*iii*) *Osmunda japonica* is a widely distributed species. Its production, however, has decreased over the last two decades.

People overharvested it because the products made from its tasty fronds were exported to Japan as well as consumed in domestic markets. Artificial cultivation of this fern will hopefully meet the demands of international markets.

(*iv*) *Ophioglossum polyphyllum* is distributed in the tropical and subtropical areas. In Tibet it is eaten in summer as a vegetable but also dried and stored for further consumption in winter. Previous studies revealed that it contains high levels of essential amino-acids that can complement the local diet [25]. It can be developed as a cultivated species in southern China.

Acknowledgements

This study was supported by the National Natural Science Foundation of China (Nos. 31161140345 & 31070288), the Ministry of Education of China through its 111 and 985 projects (Nos. B08044, MUC985-9 & MUC98506-01000101), and the Ministry of Science and Technology of China (2012FY110300).

References

1. Prescott-Allen R, Prescott-Allen C. How many plants feed the world? Conserv Biol. 1990;4(4):365–374.
2. Turner NJ, Łuczaj ŁJ, Migliorini P, Pieroni A, Dreon AL, Sacchetti LE, et al. Edible and tended wild plants, traditional ecological knowledge and agroecology. Cr Rev Plant Sci. 2011;30(1-2):198–225.
3. Quelites BR. Ethnoecology of edible greens. Past, present and future. J Ethnobiol. 1981;1(1):109–123.
4. Wujisguleng W, Khasbagen K. An integrated assessment of wild vegetable resources in Inner Mongolian Autonomous Region, China. J Ethnobiol Ethnomed. 2010;6(1):34.
5. Ge XG, Ning W, Fan WL. The status and development of wild vegetable in China. Applied Engineering Technology in Rural Areas: Greenhouse Horticulture. 2005(7):14–16.
6. Wang JR, Long CL. Ethnobotanical study of traditional edible plants of Jinuo nationality. Acta Bot Yunnan. 1995;17(2):161–168.
7. Balemie K, Kebebew F. Ethnobotanical study of wild edible plants in Derashe and Kucha Districts, South Ethiopia. J Ethnobiol Ethnomed. 2006;2:53.
8. Misra S, Maikhuri RK, Kala CP, Rao KS, Saxena KG. Wild leafy vegetables: a study of their subsistence dietetic support to the inhabitants of Nanda Devi Biosphere Reserve, India. J Ethnobiol Ethnomed. 2008;4(1):15.
9. Jain A, Sundriyal M, Roshnibala S, Kotoky R, Kanjilal PB, Singh HB, et al. Dietary use and conservation concern of edible wetland plants at indo-burma hotspot: a case study from northeast India. J Ethnobiol Ethnomed. 2011;7(1):29.
10. Orech FO, Aagaard-Hansen J, Friis H. Ethnoecology of traditional leafy vegetables of the Luo people of Bondo district, western Kenya. Int J Food Sci Nutr. 2007;58(7):522–530.
11. Ali-Shtayeh MS, Jamous RM, Al-Shafie' JH, Elgharabah WA, Kherfan FA, Qarariah KH, et al. Traditional knowledge of wild edible plants used in Palestine (Northern West Bank): a comparative study. J Ethnobiol Ethnomed. 2008;4(1):13.
12. Cruz-Garcia GS, Price LL. Ethnobotanical investigation of "wild" food plants used by rice farmers in Kalasin, Northeast Thailand. J Ethnobiol Ethnomed. 2011;7(1):33.
13. Pardo-de-Santayana M, Tardío J, Blanco E, Carvalho A, Lastra J, San Miguel E, et al. Traditional knowledge of wild edible plants used in the northwest of the Iberian Peninsula (Spain and Portugal): a comparative study. J Ethnobiol Ethnomed. 2007;3(1):27.
14. Ghirardini M, Carli M, Del Vecchio N, Rovati A, Cova O, Valigi F, et al. The importance of a taste. a comparative study on wild food plant consumption in twenty-one local communities in Italy. J Ethnobiol Ethnomed. 2007;3(1):22.
15. Lentini F, Venza F. Wild food plants of popular use in Sicily. J Ethnobiol Ethnomed. 2007;3(1):15.
16. Łuczaj Ł, Szymański WM. Wild vascular plants gathered for consumption in the Polish countryside: a review. J Ethnobiol Ethnomed. 2007;3(1):17.
17. Cavender A. Folk medical uses of plant foods in southern Appalachia, United States. J Ethnopharmacol. 2006;108(1):74–84.
18. Turner NJ. Food plants of interior First Peoples. Victoria: Royal British Columbia Museum; 2006.
19. Kuhnlein HV, Turner NJ. Traditional plant foods of Canadian indigenous peoples: nutrition, botany, and use. In: Katz S, editor. Food and nutrition in history and anthropology. Philadelphia PA: Gordon and Breach Sicnece Publishers; 1991. (vol 8).
20. Cetinkaya G. Challenges for the maintenance of traditional knowledge in the Satoyama and Satoumi ecosystems, Noto Peninsula, Japan. Hum Ecol Rev. 2009;16(1):27–40.
21. Miyazawa B, Tanaka C. Illustrated handbook of useful wild plants. Tokyo: Yokendo Press; 1948.
22. Sugiura T. Present situation of use of edible wild plants in our country and its possibilities in the future. Tokusan Joho. 2007;28(9):28–31.
23. Copela EB. Edible ferns. Am Fern J. 1942;32(4):121–126.
24. Leach H. Fern consumption in Aotearoa and its oceanic precedents. J Polyn Soc. 2003;112(2):141–156.
25. Lognay G, Haubruge E, Delcarte E, Wathelet B, Mathieu F, Marlier M, et al. *Ophioglossum polyphyllum* A. Braun in Seub. (Ophioglossaceae, Pteridophyta), a rare potherb in south central Tibet (T. A. R., P. R. China). Geo Eco Trop. 2008;32:9–16.
26. Mannan M, Maridass M, Victor B. A review on the potential uses of ferns. Ethnobotanical Leaflets. 2008;12:281–285.
27. Lu SG. Pteridology. Beijing: Higher Education Press; 2007.
28. Tryon RM. Ferns and allied plants, with special reference to tropical America. New York NY: Springer; 1983.
29. Schneider H, Schuettpelz E, Pryer KM, Cranfill R, Magallón S, Lupia R. Ferns diversified in the shadow of angiosperms. Nature. 2004;428(6982):553–557.
30. Chang HC, Huang GJ, Agrawal DC, Kuo CL, Wu CR, Tsay HS. Antioxidant activities and polyphenol contents of six folk medicinal ferns used as "Gusuibu". Botanical Studies. 2007;48:397–406.
31. Xu JX, Wang YL, Wang JJ, Tang W, Chen LJ, Long CL. Chemical constituents of Cibotium barometz and their bioactivities: a review. Natural Product Research and Development. In press;
32. Shi L. Ornamental ferns. Beijing: China Forestry Press; 2002.
33. Zhang YM. A preliminary investigation of the history of pteridophyte eating in China. Journal of Southwest Agricultural University (Social Science Edition). 2007;5(4):103–106.
34. Wu LJ. Notes to plants and animals in the Book of Odes. Shanghai: Shanghai Ancient Bookstore; 1937.
35. Peng DQ, Yang ZN. Collected poems of Tang Dynasty. Beijing: Zhonghua Book Company; 1985.
36. Yun XL, Zhao NW, Pan LT, Zhao JH. Resource distribution and exploitation of edible ferns in Guizhou (II). Journal of Anhui Agricultural Science. 2009;37(33):16317–16319.
37. Yun XL, Zhao NW, Pan LT, Zhao JH. Resource distribution and

exploitation of edible ferns in Guizhou (I). Journal of Anhui Agricultural Science. 2009;37(32):15911–15912.
38. Cui GY. Edible fern resources and development in China. Chinese Cuisine Research. 1998;6(1):21–28.
39. Dai XL, Li XG, Wu SF. List of Chinese edible pteridophytes. Quarterly of Forest By-Product and Speciality in China. 2003;4:5–6.
40. Cao QM, Zhong HY, Li ZH, Sun CB. Exploitation and utilization study of fern starch in China. Food Research and Development. 2007;28(12):168–171.
41. Pteridium fern [Internet]. 2012 [cited 2012 Sep 10].
42. CITES appendices [Internet]. 2012 [cited 2012 Sep 10].
43. Li YY, editor. National protected wild plants in Yunnan Province, China. Kunming: Yunnan Science and Technology Press; 2005.
44. Zhang F, Luo SZ, Gao BC, Ding LS. Chemical constituents from the sprout of Pteridium aquilinum var. latiusculum. Natural Product Research and Development. 2004;16(2):121–123.
45. Yang L, Wang MY, Zhao YY, Tu YY. Chemical constituents of the rhizome of Matteuccia struthiopteris. Yao Xue Xue Bao. 2005;40(3):252–254.
46. Zhang D, Yang L, Fu MH, Tu YY. Chemical constituents of the rhizome of Matteuccia struthiopteri III. China Journal of Chinese Materia Medica. 2008;33(14):1703–1705.
47. Ouyang DW, Yang PM, Kong DY. Chemical constituents from Pteris multifida Poir. Chinese Journal of Pharmaceuticals. 2008;39(12):898–900.
48. Gong XL, Chen ZH, Liang NC. Advances in study on chemical constituents and pharmacological activities of plants of genus Pteris. Zhongguo Zhong Yao Za Zhi. 2007;32(14):1382–1387.
49. Wang ZJ, Zhao QS, Peng LY, Cheng X. Study on the chemical constituents of *Callipteris esculenta* (Athyriaceae). Natural Product Research and Development. 2009;21:960–962.
50. Wu Q, Yang XW, Yang SH, Zou L, Yan J. Chemical constituents of *Cibotium barometz*. Natural Product Research and Development. 2007;19:240–243.
51. Liang ZY, Yang XS, Zhu HY, Hao XJ. Chemical constituents of *Nephrolepis cordifolia*. Guihaia. 2008;28(3):420–421.
52. Rasmussen LH, Lauren DR, Smith BL, Hansen HCB. Variation in ptaquiloside content in bracken [*Pteridium esculentum* (Forst. f) Cockayne] in New Zealand. N Z Vet J. 2008;56(6):304–309.
53. Alonso-Amelot ME, Castillo U, Smith BL, Lauren DR. Bracken ptaquiloside in milk. Food and Chemical Toxicology. 1996;34(11–12):1187.
54. Ojika M, Wakamatsu K, Niwa H, Yamada K. Ptaquiloside, a potent carcinogen isolated from bracken fern var.: structure elucidation based on chemical and spectral evidence, and reactions with amino acids, nucleosides, and nucleotides. Tetrahedron. 1987;43(22):5261–5274.
55. Smith BL, Seawright AA. Bracken fern (*Pteridium* spp.) carcinogenicity and human health – a brief review. Nat Toxins. 1995;3(1):1–5.

7

The use of wild plants as food in pre-industrial Sweden

Ingvar Svanberg*

Uppsala Centre for Russian and Eurasian Studies, Uppsala University, Box 514, 751 20 Uppsala, Sweden

Abstract

This paper is a review of the actual gathering and use of wild edible plants in the 18th and 19th centuries, with a brief concluding discussion on the present day use of wild plants as food within Sweden. The peasants and the nomads in pre-industrial Sweden utilised very few wild plant taxa as food. Many even despised the wild fruits and green plants. Some plants and fruits were earlier mostly eaten fresh on the spot, or gathered for consumption in bread, gruel or soup. Other fruits were dried or preserved in other ways. In times of food shortages the amount of wild plants increased in the diet, but still the peasantry and nomads were often able to use fish and game to provide enough nutrients.

With access to cheap sugar in the early 20th century wild fruits (*Vaccinium myrtillus* L., *V. vitis-idaea* L., and *Rubus chamaemorus* L.) increased in importance, especially among urban-dwellers and within food industry. In the last few decades fungi have also become part of the urban diet. Fifty years ago working class people gathered only *Cantharellus cibarius* (Fr.) and occasionally *Boletus edulis* Bull. Nowadays more taxa are utilised within the Swedish households, and especially the easy to pick *Cantharellus tubaeformis* (Pers.) has become very popular recently. Harvesting fruits and mushrooms in the forests is a popular pastime for many urban people, but also a source of income for immigrants and especially foreign seasonal labour. The only traditional green wild food plant that is regularly eaten in contemporary Sweden is *Urtica dioica* L.

Keywords: ethnobiology, plant gathering, wild food plants, famine food

Introduction

Although several studies provide information on the historical and contemporary exploitation of wild plants and non-timber products in Sweden and neighbouring Nordic countries, very few ethnobiological analyses of the use of wild food plants exist. The ethnobotanical encyclopaedias that have been published in Denmark, Norway and Sweden of course provide lots of descriptive data on plant use, including edible plants [1–3]. An ethnobotanical study about the Faroe Islands reviews most available historical data on plant use, including traditional food plants [4]. Scattered information is also at hand from Finland and Iceland [5,6]. The impact of the propaganda efforts in Sweden to promote the utilisation of wild plants, lichens and mushrooms has been analysed in a few scientific studies [7,8]. Sweden's change from a mycophobious to a rather mycophilious nation in recent years has been discussed in a couple of reviews [9,10].

Cultural diversity in Sweden (and northern Scandinavia) includes the Saami minority in the north, scattered groups of Roma and other Travellers, Finnish-speakers in the centre of Sweden and neighbouring parts of Norway who are nowadays completely assimilated, as well as Finnish-speakers in the province Norrbotten (including Meänkieli-speakers). Analyses of the use of food plants among the Saami exist, including the pioneering study by ethnologist Phebe Fjellström on the use of *Angelica archangelica* L. [11–15]. The folk botany among the various Finnish-speakers in the Scandinavian Peninsula is very little known, although rich data exists in the archives. For the Roma there is a small study on food among the Kalderaša subgroup which also includes data on wild plants [16].

The interest in researching folk botany and use of plants in their cultural historical and social contexts has increased the last 15 to 20 years. Much data has been published recently in Sweden, almost all of them in Swedish. Especially the many ethnobiological monographs, written by the present author, have analysed empirical data given in historical sources, dialect wordlists, older travelogues and handwritten records in the folk-life archives, and discussed in detail the various uses and beliefs on plant taxa in pre-industrial Sweden (e.g. [17–19]). Very few monographic articles are available in English though [11,20–23], apart from a relatively superficial review from 1951 by the ethnologist Sigurd Erixon [24].

This review of Swedish plants traditionally used as food and snacks in earlier times will fill a gap in the international ethnobiological literature. The concept of famine food and the impact of propaganda efforts will be discussed. A brief account on the contemporary use of wild plants in food culture concludes the presentation.

* Email: ingvar.svanberg@ucrs.uu.se

Material and methods

Sweden – the cultural and ecological setting

Contemporary Sweden covers an area of around 450300 km^2 and it forms the eastern part of the Scandinavian Peninsula. It includes also the islands of Gotland and Öland in the Baltic Sea. The climate is temperate. Nearly 60 per cent of the country is covered with forests and 15 per cent is located north of the Arctic Circle. Natural vegetation varies considerably due to the various climate zones and ecological settings with mountains, forest regions, coastal areas and the agricultural landscape. Eight vegetation zones can be distinguished in Sweden and the boreal zone and its sub-zones cover the most of the country.

In the 18th century Sweden was still poor, and despite considerable efforts being made, the provision of sufficient foodstuffs for the population was far from having been reliably secured. In some parts of the country the peasantry experienced frequent crop failures, and food crises and famine lurked constantly around the corner. Agriculture remained primitive and wild plants were essential for the subsistence economy, although not for human nutrition. During the 19th century land reforms, ditching projects, modernisation of agriculture and cattle breeding, better communications through steamboats and railways, and better health care that resulted in the epidemiological transition improved the situation for the population. Sweden became industrialised rather late, and in the 1880s a large part of the population was still rural and poor. Almost 1.2 million people migrated to North America between 1851 and 1930 [25].

Although still considered to be rather linguistically homogenous, Sweden also has a cultural diversity of many groups speaking various minority languages. Traditional minorities include the indigenous Saami, divided in various dialects and groups which have a background in full or semi-nomadic life styles. Due to the long historical unification with Finland (until 1809) various Finnish-speaking groups still exist within the contemporary borders. In the year 1700 the population was only 1.4 million, and in 1900 it had reached 5.1 million. The country has remained sparsely populated, especially in the north. Contemporary Sweden has a population of almost 9.5 million inhabitants (2012) including a lot of immigrant groups that have arrived since World War II as refugees and labour. They are estimated at around 15 per cent (1.3 million foreign born) of the population and originate from most countries of the world [26].

Export of timber products and iron ore has always been an important part of the economy, while the agricultural sector has fallen to only 2 per cent in recent years. Most people are urban, while still in the 1870s around 70 per cent lived in the countryside [25].

Sources used for this review

Wild plants gathered for food is a fascinating primordial bio-cultural domain. An early evidence of plant gathering from Sweden is the Roman author Tacitus who described in his "De origine et situ Germanorum" (ca. 98 A.D.) how the people who are considered to be the ancestors of the Saami gathered a kind of grass that was consumed as food. This has been interpreted to mean *Rumex acetosa* L., which has until very recently played an important role in Saami food culture. Although occasional data on plant use can be found in various older records of wild-plant harvesting from the 16th and 17th century, such as the reports on the ethnography of the Saami provided by clergymen in the 1670s, the real documentation of local plant use in Sweden began with Carl Linnaeus (1707–1778). His tour to Lapland and northern Scandinavia in the summer of 1732 is sometimes regarded as the starting point of ethnobotanical investigations in Sweden [27]. Thanks to his careful observations his travelogue provides us with the necessary data on where, when and from whom the data were collected, and he himself published "Flora lapponica" (1737) in Latin, where he reports also economic aspects of the plants described [28]. This book had an immense impact and became a model for botanists all over the world, but especially the data on the economic use of various plants had been passed on into many later books, thus creating a problem for later researcher to identify data originating in Linnaeus's writings, and local data [29–31]. Linnaeus published further travelogues, and they are full of information on his observations of the use of plants for food, medicine, dyes and other technical purposes among the peasantry in mid-18th century Sweden (the ones to Lapland 1732, Dalecarlia 1734, and Öland and Gotland in 1742 are available in English translations). Some of Linnaeus's prominent pupils also travelled in Sweden and gathered observations from the peasantry about useful plants. Among the most important pupils are Pehr Kalm (1716–1779), Johan Otto Hagström (1716–1792), and Anders Tidström (1723–1779). Without Linnaeus's impact and importance as a role model for later scholars the amount of data on the folk botany in Sweden would have been very mediocre [32].

Some dialectologists, ethnographers and local historians provide interesting first-hand information gathered in the field in the 19th and early 20th century on the use of wild plants. Since plant knowledge has been rather high among educated Swedes due to the importance of botany in the secondary school curricula until early 1960s, the information given is usually reliable and often contains scientific names, besides local and normalised Swedish phytonyms [33].

Important collectors of folk botanical data with information on food plants are Georg Bergfors (1882–1975), Gustav Fridner (1891–1981), Gullik Gulliksson (1894–1983), and Erik Modin (1862–1953) [33–37]. Ethnological and folklore archives also provide useful information. Such archives are to be found in Stockholm (the Nordic Museum), Gothenburg, Lund, Umeå, Uppsala and Östersund. For instance, in 1929 the Folklife Archives (present Swedish Institute for Language and Folklore) in Uppsala distributed a questionnaire on famine food in Sweden, which contained interesting data on plants used as emergency food in the late 19th century [38]. Among unpublished data collections, Lisa Johansson (1894–1982) is especially interesting since her manuscript also includes voucher specimens of the plants used in the Vilhelmina parish in northern Sweden. Another important manuscript has been compiled from the Edsele parish in Ångermanland by Frans Bergvall (1903–1995) [39,40].

The heavy loss in traditional plant knowledge in Sweden does not make further fieldwork especially appealing. On the other hand, the archives are still full of data which has not been analysed. The use and importance of wild plants among recent immigrants in Sweden is however worth field studies, for instance among Anatolian Turks, Chinese and Thais.

The impact of propaganda efforts

Food propaganda in Sweden, published as pamphlets, cookbooks, instructions, educational efforts, laws and as newspaper articles, is a common source for the contemporary use of many plants (fruits, mushrooms, etc.). Ever since the Age of Enlightenment, men (and a few women) of science have sought possible natural products worth exploiting on a large scale in order to improve the nutritional status of the population and of course the country's economy. The efforts have had two main goals: to benefit the national economy (by using local products or substituting valuable resources with others, for instance bark of *Pinus sylvestris* L. with lichens and mushrooms), and to improve the general health of the public. Luxurious imported products could be substituted with native wild plants. For instance, in 1746 the authorities published a list of 45 plant taxa, mostly native species that could be used as a substitute for tea. Imported tea was consumed by the upper classes in the early 18th century [41].

Carl Linnaeus particularly emphasised the use of wild plants as food resources and flour substitutes. Not content with just conducting inventories of folk knowledge of useful plants, he and his contemporaries compiled lists of plants (such as his "Ceres noverca arctorum" 1733, "Plantae esculenta patriae" 1752, "De pane diaetetico" 1757, "Macellum olitorium" 1760 and a memorial in Swedish in 1757), known to be used elsewhere [42,43]. Another man dedicated to investigating wild plants that could be used for food was the botanist and priest Petrus Holmberger (1745–1807) [32]. The question of using wild plants as substitutes was very much in vogue among learned men in the late 18th century. Many booklets, including several dissertations, on the issue were published from the 18th century until the early 20th century [7,42,44].

Ever since the 18th century the authorities have propagandised in different ways for new food among the peasantry with the goal of improving agriculture and diet. This propaganda intensified in conjunction with famine years and during recessions. In the pre-industrial period very few new food items were accepted or only slowly became part of diet of the common people. The intensive, almost 150-year-long propaganda effort to introduce lichens [especially *Cetraria islandica* (L.) Ach.] failed completely, and mushrooms did not became generally accepted until the 1930s and 1940s. It was among the urban population that finally wild mushrooms as a food item were accepted. Many other food plants were discussed as well [7–9].

In the early 20th century, especially during the two World Wars and the period in between, many efforts were made to increase the interest for wild plants as food [45,46]. New media like radio probably had some impact. The increasing interest in fruits was due to the access to sugar, while modern recipes provided by cookbooks, weekly magazines, newspapers and school-instructions probably also had an impact. Cordial, jam, marmalade, gelatine desserts, biscuits and confectionary are modern products that could be made with wild berries. The interest in hips of *Rosa* sp. (rich in vitamin C), which was popular in the early 20th century, was a result of such educational efforts. *Rose* hip soup is still popular although the main ingredient is mostly bought dried in grocery stores. Also the fruits of *Sorbus aucuparia* L. became popular as foodstuff used for various products. It is still used as a jam served with game [47,48].

Books, newspapers, magazines, radio and TV-programs still show how various wild plants can be used in the household. The right of public access to the wilderness (allemansrätten) that allows everyone to roam freely also on private land and to pick berries, mushrooms and wildflowers, is crucial for understanding the general interest for using these resources. Also the availability for the urban people through bicycles and later cars has been of importance for the general use of wild fruits and mushrooms in the 20th century [47,49].

The impact of the propaganda was probably small before Sweden became industrialised and urbanised, although the population, including the peasantry, as in most Lutheran countries had, due to the 1686 church law, a high literacy rate (almost 100 per cent) already in the 18th century and read almanacs and newspapers. Clergymen also played an important role in instructing the common people in health care, agricultural improvements and gardening. However, when researching historical sources we must consider that the data can be influenced by these propaganda efforts rather than reflecting folk knowledge. Also the records in the folklore archives are sometimes influenced by these texts and careful source criticism is therefore necessary. For instance, local plant names seem to have been completely replaced by book-names since World War II. Ghost data must be dismissed and the impact of written data must be discussed in every ethnobotanical study using historical information. Contextual data on when, where and by whom a specific plant was used must be at hand. Icelandic lichen has for instance never been used as substitute food stuff in Sweden (or any European country outside Iceland), despite this often being said in many historical reviews [7,31,50].

Results

The traditional uses of more than 100 taxa as food, spices as well as for beverages have been recorded. They are presented (with a couple of exceptions) according to the categories given by Turner et al. [51]. Some further species were used as tobacco substitutes or for flavouring tobacco. Many edible plants were used only by poor people or in connection with poor harvests and famine years. There seems to be a difference in use between the 18th and 19th centuries. Notwithstanding a wide range of efforts, the provision of sufficient foodstuffs for the population was far from having been reliably secured in the 18th century Sweden. The peasants were poorer and relied much more on subsistence economy in that century. Various green plants, used in soup and stews by the peasantry, are mentioned from the 18th century. Food crises due to crop failures were still common. The last severe starvation took place in 1772–1773 with high mortality rates [44,52].

During the 19th century rural people usually despised fruits and they were only eaten locally and usually on the spot among people living in the forests or in the northern mountain areas, or by children. Fruits were not worth harvesting due to their low energy content. Only a few green plants are mentioned in the sources, mostly eaten by the rural poor. Some plants were eaten for magical purposes by the peasantry. For instance, to eat the first flowers of *Hepatica nobilis* Schreb. and *Anemosa nemorosa* L. in springtime was seen as bringing luck, protecting against snake bites or being prophylactic in general [53].

Social differences in using wild plants existed. In the 18th and early 19th centuries the upper classes, living in cities, castles and mansions, had activity fields and consumption patterns influenced by French and German culture. Therefore, there was a market for products that was not used among the peasantry. Linnaeus observed for instance young peasant women selling *Gyromitra esculenta* (Pers. ex. Pers.) Fr. in Småland in 1741. Still in the 19th century the peasantry gathered the species in the same area for selling to the cities. Sugar was expensive and the peasantry could only occasionally make sweet cakes and desserts before industrialisation. However, at the mansions it was possible, and therefore there was a small market for wild fruits such as *Fragaria vesca* L., *Prunus avium* L., *Rubus arctica* L. and *Rubus plicatus* Weihe & Nees [3].

In southern Sweden grain production was predominant among the peasants, and they were also more reluctant to use wild plants as food. In the north the peasantry and the Saami were more dependent on gathering activities and a combination of animal husbandry, hunting, fishing and to some extent gathering of plant products. Fruits were more used among these categories. The rural poor were of course more active in gathering activities than the farmers. In the 20th century city-dwellers were eager to accept new food items, including wild mushrooms. For the post-World War II era it also seems that the working class preferred wild fruits, while the middle class has been interested in mushrooms [50].

Root vegetables (roots, corms, tubers and rhizomes)

Underground parts (roots, rhizomes, bulbs) have been recorded for several taxa. In the northern parts of Sweden the rhizomes of *Calla palustris* L. were widely used for making bread in the 18th and early 19th centuries. Rhizomes of *Menyanthes trifoliata* L. and *Nuphar lutea* (L.) Sm. were occasionally used as famine food. As elsewhere in Europe the bittersweet rhizomes of *Polypodium vulgare* L. have been eaten, especially by children. They were also recommended for their medicinal qualities (purgative) and therefore gathered and sold to the pharmacies. Occasionally, children today still gather and eat these rhizomes. The Finns in the northern and central parts of Sweden used the rhizomes of *Bistorta vipipara* (L.) Gray ground into flour when baking unleavened bread in the 18th century. Rhizomes of *Elytrigia repens* (L.) Desv. ex Nevski have been used as famine food in some parts of Sweden [3,44].

The use of the edible nodules attached to rhizomes of *Equisetum arvense* L., known to be eaten in Norway, Iceland and the Faroe Islands, is only documented from Ångermanland and the Finnish-speaking areas of Värmland and was probably an old habit that survived among children [54]. Root tubers of *Filipendula vulgaris* Moench were readily eaten by free-roaming hogs, but in some parts of western Sweden poor people in the 19th century gathered them and used them to sweeten porridge [3].

Edible greens (leaves, stems, shoots)

Very few green vegetables harvested in the wild have been used by the peasantry in Sweden. The Saami utilised *Angelica archangelica* L., *Rumex acetosa* L., *Cicerbita alpina* (L.) Wallr., *Oxyria digyna* L. and occasionally also the leaves of *Epilobium angustifolium* L. and *Alchemilla vulgaris* L., mainly for mixing it with reindeer milk [11–13,17,55].

Also the Swedish settlers in northern Sweden used especially *Cicerbita alpina* for food. Spring shoots of *Urtica dioica* L. have been generally eaten in Sweden, known already in 18th century. It was very commonly eaten, according to Linnaeus. Still many people gather it for making soup. It is also available in some weekly markets. *Rumex acetosa* L., earlier widely gathered, was mainly used by children during the 20th century. However, it has recently become part of haute cuisine, although usually replaced by the cultivated *Rumex rugosus* Campd. Immigrants from Anatolia still gather wild *R. acetosa* L. for food. Strobils of *Equisetum arvense* L. were earlier eaten in some parts of Sweden [3,56].

Leaves, normally used as animal fodder, were, in times of food shortages, used as additives in bread. There is evidence of using leaves of *Salix caprea* L. and *Tilia cordata* L. Buds of *Tilia cordata* and *Corylus avellana* L. were also used in flour for bread according to several sources. Fresh shoots of *Pinus sylvestris* was sometimes also utilised in bread [56].

Many other plants were earlier gathered for stews and soup, and their importance increased in famine years, for instance *Aegopodium podagraria* L., *Allium oleraceum* L., *Allium schoenoprasum* L., *Allium scorodprasum* L., *Anchusa officinalis* L., *Anthriscus sylvestris* (L.) Hoffm., *Aster tripolium* L., *Barbarea vulgaris* R. Br. (widely cultivated in the 18th century), *Circium palustre* (L.) Scop., *Cirsium helenioides* Hill, *Crambe maritima* L., *Galeopsis bifida* (Boenn.) Fries, *Campanula latifolia* L., *Hypochoeris maculata* L., *Lamium album* L., *Lamium purpureum* L., *Ranunculus ficaria* L., *Ribes uva-crispa* L. (leaves) and *Sinapis arvensis* L. Most of these data were recorded in the 18th century. Some plants, usually considered as fodder, have been eaten in times of food crisis, such as *Epilobium angustifolium* L. among the settlers in the north [3,24,35–38,40,56].

The itinerant Kalderaša Roma gathered in the late 19th and early 20th century *Chenopodium album* and *Taraxacum* and used these plants in salads and soup [16]. The former taxon is also gathered by Greek immigrants in Sweden.

The only Pteridophyta eaten today, besides the occasional use of polypody rhizomes among children, are the young stems of ferns [probably *Pteridium aquilinum* (L) Kuhn] which are gathered by North and South Koreans living in Sweden.

Berries and other fleshy fruits

Fruits were not often eaten, although cowberries and cloudberries played a not unimportant role as food for the rural poor in the northern part of the country [47,56] It was also possible to preserve them without sugar. Wild apple (*Malus sylvestris* Mill.) and wild pears (*Pyrus communis* L.) were used by the peasantry in Gotland and southern Sweden in the 18th century. The apples were used in soup, and sometimes regarded as food for poor people. Other fleshy fruits gathered were *Sambucus nigra* L., *Prunus avium* L., *Prunus padus* L., *Cratageus* sp., *Ribes nigrum* L., *Ribes spicatum* E. Robson, and *Sorbus intermedia* (Ehrh.) Pers. Fruits were dried and used as additives in pancakes or in bread, or eaten salted [3,28,34,56]. The traditional use of the sour fruits of sea-buckthorn, *Hippophaë rhamnoides* L., was restricted to fishermen in the north part of the Bothnian Bay. They used them for sauce eaten together with fish [3,28].

Naturalised *Ribes uva-crispa* L. berries were harvested by young people and sold in the cities, but the peasants also produced jam of it as was observed in Värmland in the mid-19th century. Most people seem to have rejected the berries of *Vaccinium uliginosum* L., believing they were toxic, but in some areas in northern parts of Sweden they were actually eaten, not only by children but also adults. Juniper pseudo-fruits (*Juniperus communis* L.) were used by the peasantry for producing sweet syrup in western Sweden in the 18th and 19th century [3,56].

The Saami gathered berries of *Empetrum hermaphroditum* (Hagerup) as food. It has been rarely used among the peasantry in the northern part of the country [13,28]. *Rubus chamaemorus* L. and *Vaccinium vitis-idaeus* L. were traditionally gathered and could be kept preserved especially in the north. The Finns in the central part of Sweden utilised large quantities of *V. vitis-idaea* berries in gruel, called "hillo" [3,56]. Fresh bilberries (*Vaccinium myrtillus*), in some areas completely despised as food, were sometimes elsewhere eaten with milk or cream, otherwise they were mostly eaten on the spot by children. The same is true for *Rubus idaeus*. With industrialisation and greater access to sugar some of these fruits became popular and are nowadays considered a must in Swedish food culture [3,36,56].

Grains, seeds, and nuts

Hazelnuts (*Corylus avellana*) were widely gathered in the southern and middle part of Sweden. Servants at manors and large farms even had free days in the summer in order to gather nuts for their own consumption (especially for Christmas). Hazelnuts were also sold on the market, and even exported in large quantities [47,49]. From Gotland and Öland we have reports from the 18th and 19th century respectively that poor people harvested large amounts of high quality nuts from the winter nests of field mice, *Apodemus flavicollis* (Melchior) [57]. Also beechnuts *Fagus sylvatica* L. were gathered and used in flour, especially during times of crop failure. They were used as additives in bread [3,56].

Seeds of *Chenopodium album* L. were widely gathered, dried, grained and mixed in flour for baking bread during crop-failures in northern Sweden [37,56]. However, grains of *Glyceria fluitans* (L.) R. Br. seem to have been gathered only very locally in Skillinge, Scania, in the mid-18th century. Linnaeus observed in Dalecarlia in 1734 that the locals used the grains of *Avena fatua* L. for flour [3,56].

Bark, buds and other edible plants

Traces of bark of *Pinus sylvestris* L. have been found in prehistoric breads from archaeological sites [58]. The innermost layer of *P. sylvestris* bark was commonly used to make bread among the peasantry in the northern Sweden still in the 18th century, but in the 19th century it was used mostly as flour substitute at times of famine. Bread mixed with bark flour was abandoned after the famine years 1867–1868. The energy content of bark flour is 82 kcal per 100 g. [56,59]. Due to the increasing economic importance of timber, the authorities were eager to replace the bark with other substances, such as lichens and mushrooms. The propaganda efforts increased in connection with crop failures. However, lichens and mushrooms were seen as cattle fodder and therefore generally rejected by the peasantry. The Saami have also been harvesting bark from pine trees. The inner bark was prepared by wrapping it in birch bark and put in the heat of a fire. The bark prepared in that way was sweet and eagerly eaten by the Saami. This kind of harvesting pine bark has been documented from late medieval times until the 19th century in Sweden [60].

The bark of *Fagus sylvatica* L. and *Ulmus glabra* L. were also used in southern Sweden as famine food in the early 19th century. In Småland the peasantry also used bark from *Tilia cordata*. Some sources also mention that the bark of *Betula* sp., and *Picea abies* (L.) H. Karst. was used for baking bread in times of famine [56].

Spices, flavourings and preservatives

Spices were grown in the villages and gardens of the manors for medicine or using for seasoning food. However, several plants were also gathered in the wild, such as *Artemisia vulgaris* L., *Carum carvi* L., *Mentha arvensis* L., *Origanum vulgare* L. and *Thymus serpyllum* L. Spruce resin was commonly eaten as a refreshing chewing gum during the long services in churches or when walking in the forests [3,61]. Also resin from *Prunus cereus* L. was eaten by peasant boys in eastern Sweden in early 19th century. *Haploporus odorus* (Sommerf.) Bondartsev & Singer is a bracket mushroom found on goat willow trunks in the northern part of Sweden. The species was widely used in cabinets and chests for its pleasant scent. However, it was also used to flavour cheese and bread [3,36–38].

A special case in Swedish folk botany was the use of the leaves of *Pinguicula vulgaris* L. and *Drosera* sp. in order to curdle milk when making a kind of dairy product known as "thickened milk" "tätmjölk". Whether the enzymes on the slimy plant leaves really contributed to the process is a matter of some debate, but most probably it was just an old folk belief. Locally, in the provinces of Värmland and Lapland, and adjacent areas of Norway, black slugs (*Arion ater* L.) were used to curdle the same kind of thick milk [62].

Hard liquor (brännvin) has traditionally been flavoured with plants, including wild species, such as *Achillea millefolium* L., *Carum carvi* L., *Galium odoratum* L., *Myrica gale* L., *Gymnadenia nigra* (L.) Reichb., *Peucedanum palustre* (L.) Moench., *Potentilla erecta* (L.) Räusch., *Primula veris* L., *Tanacetum vulgare* L., *Taraxacum* sp., as well as various fruits, such as *Fragaria vesca* L., *Prunus padus* L., *Ribes nigrum* L. and *Sorbus aucuparia* L. Still very popular is to gather the buds of *Hypericum perforatum* L. and to make a red light bitter with them [3,56].

When making beer the peasantry also used wild plants. Most common seems to have been *Myrica gale*, but also *Achillea millifolium* L., *Filipendula ulmaria* (L.) Maxim., *Gentianella campestris* L. (Börner), *Hypericum perforatum* L., *Juniperus communis* L. (pseudo-fruits), *Primula veris* L., *Menyanthes trifoliata* L., *Rhinanthus serotinus* (Schönh.) Oborny, and *Rhododendron tomentosum* Harmaja are mentioned in the sources [3,36,63].

A special kind of spices is the fragrant plants used in pipe tobacco and snuff. Sometimes (especially for children and elderly people) the plants substituted the real tobacco. Additives and surrogates were *Achillea millefolium* L., *Arctostaphylos uva-ursi* L., *Arnica montana* L., *Galium odoratum* L., *Solanum dulcamara* L., *Centaurea cyaneus* L., *Hypericum perforatum* L., *Menyanthes trifoliata* L. and *Prunus cereus* L. Snuff could be flavoured with scraps of the algae *Trentepohlia iolithus* (L.) Wallroth or with the fragrant grass *Anthoxanthum odoratum* L. The roots of *Angelica archangelica* L. and *Peucedanum palustre* (L.) Moench. were chewed by the Saami as substitutes for tobacco [3,13,20,34,36]. The plants used as tobacco substitutes have been reviewed in Norway and the Faroes. It could be a nice subject for a review covering a larger part of Europe [2,4].

Beverages

Birch (*Betula pendula* Roth., *B. pubescens* Ehrh.) sap has been gathered all over Sweden and was usually seen as a refreshing drink, but has also been used for gruel (made of birch sap and barley meal), in coffee and in some areas (e.g. Dalecarlia) made into ale by mixing with malt and yeast [64].

In the 18th and 19th century it was a widespread custom to make a beverage of the pseudo-fruits from *Juniperus communis* L. Also poor people could make this drink. Juniper pseudo-fruits were sometimes made into beer and also syrup. Making cordials, wine and liqueurs of wild fruits was practiced in the upper classes already in the 18th century (and probably earlier as well). Also shoots of *Picea abies* (L.) H. Karst were brewed into a beverage [3,56,63].

Drinks made of berries of *Vaccinium vitis-idaea* L. were common among the peasantry in certain parts of Sweden (for instance the Finns in Central Sweden) already in the 18h and 19th century. It does not require any sugar and it was a way to save the berries for the winter. In some areas the peasants made a drink of *Vaccinium oxycoccos* L. berries [3,56].

Making liqueurs out of wild fruits (e.g. *Prunus spinosa* L.) seems to be a more recent habit and has been a kind of vogue. Others are commercially produced. On the other hand, seasoning hard liquor with wild fruits like *Sorbus aucuparia* L. or *Fragaria vesca* L. are old habits stemming from the upper classes. Recipes of wine made of for instance *Vaccinium myrtillus* exists, but has not become popular. There was a time in the 1970s when it was a kind of fashion to make wine of *Taraxacum* sp. flowers. It was actually illegal to make home-made wine of the flowers from 1978, but the law was changed in 1994. Nowadays very few people make dandelion wine [3].

Among the Saami a hot drink made of the bracket fungus *Piptoporus betulinus* (Bull. ex Fr.) P. Karst. seems to have been common before coffee was introduced in the 1860s. During World War II when there was a shortage of imported food stuff, there was a revival of using the birch bracket fungus for making a hot drink among the Saami. It is still made into a drink by Saami children in Norway. Herbal tea has been made using *Matricaria recutita* L. already in the 1830s (and it is still sometimes gathered in the fields for the same purpose). Also flowers of *Trifolium pratense* L. have been used to make herbal tea [3].

Coffee surrogates are known already in the early 19th century, although coffee was still restricted to the upper classes. It was not until the mid-19th century coffee became widely accepted [65]. Coffee substitutes are mentioned in cookbooks and recipe booklets from the late 19th century and World War I. However, ethnographical data also mentions dried rhizomes of *Taraxacum* sp. and seeds of *Iris pseudacoris* L. Coffee substitutes made of rye and/or chicory were available through the trade during the war-times [3].

Children's snacks

Children have always eaten green plants as a kind of treat and change in a rather monotonous diet of cereals and dried fish or meat. Especially children tending cattle in the forests had to complement the small amount of food they were given by their masters with *Oxalis acetosella* L., *Rumex acetosa* L., *Angelica sylvestris* L., *Cicerbita alpina* (L.) Wallr. and green shoots from Norway spruce, *Picea abies* (L.) H. Karst. These were not just snacks, but real food keeping them healthy and helping them to get enough energy [3,66].

Children were and to some extant are still gathering plants as snacks (Tab. 1). Popular sour plants were *Oxalis acetosella* L. and *Rumex acetosa* L., locally also *Rumex acetosella* L. Also the immature pea-like fruits of *Vicia cracca* L. were popular. The immature fruits of *Capsella bursa-pastoris* (L.) Medik. have been eaten in southern Sweden. Sucking nectar from the flowers of *Primula veris* L., *Trifolium pratense* L. and various Lamiaceae species (*Lamium*, *Galeopsis*) were common. Seeds of *Bistorta vivipara* (L.) Gray were widely eaten by children. Capsules of the hairy cap moss, *Polytrichum commune* Hedw., were gathered and eaten by children during springtime, appreciated for their sweet taste. Eating *Chrysomyxa woroninii* Tranz., found on spruce shoots, has been a common practice among children. However, in the 19th century the Saami still gathered and ate them during the wanderings in forest areas as refreshment. Fruits were used too, including *Vaccinium oxycoccos* L. Some fruits were only utilised on the spot by children, for instance *Cornus suecica* L. and *Rubus saxatilis* L. [3,34,36,37,39,40,66].

Tab. 1 Plants traditionally eaten by children in Sweden [3,40,56,66].

Taxa	Plant part used
Oxalis acetosella	leaves
Rumex acetosa	leaves
Rumex acetosella	leaves
Primula veris	nectar
Galeopsis speciosa	nectar
Lamium album	nectar
Trifolium pratense	nectar
Prunus padus	fruits
Cornus suecica	fruits
Ribes alpinum	fruits
Ribes spicatum	fruits
Ribes uva-crispi	fruits
Vaccinium myrtillus	fruits
Vaccinium oxycoccos	fruits
Vaccinium uliginosum	fruits
Fragaria vesca	fruits
Empetrum hemaphroditum	fruits
Rubus caesius	fruits
Rubus chamaemorus	fruits
Rubus idaeus	fruits
Rubus saxatilis	fruits
Vicia cracca	seeds
Bistorta vivipara	seeds
Capsella bursa-pastoris	seeds
Betula sp.	sap
Picea abies	resin
Prunus cereus	resin
Equisetum arvense	nodules
Polypodium vulgare	rhizomes
Polytrichum commune	capsules
Chrysomyxa woroninii	fungus on spruce shoots

Discussion

Malnutrition and famine food

During the famine years at the end of the 16th century the peasantry used the chaff of frozen grain, flax fruits, bark, hazel buds, nettles, leaves, hay, straw, nut shells and vetch. At the end of the 19th century they were still using bark, leaves and straw in order to alleviate hunger [44,56]. Although Sweden was hit by several famines also including the 19th century, harvest failures never caused starvation through the total elimination of food [52] (Fig. 1).

The authorities' efforts to change people's food patterns by accepting new food items such as lichens and mushrooms were never accepted by the populace. Cultural barriers, taste preferences and distrust of those in power are factors explaining why

Fig. 1 Famine in Sweden in 1867. The boy is chewing on a shoe while his father harvests bark from a pine tree (From "Fäderneslandet", 1867).

Tab. 2 Emergency bread additives reported in the 19th century [7,56].

Product	Species
Underground parts	*Bistorta vivipara, Elytrigia repens, Calla palustris, Menyanthes trifoliata, Solanum tuberosum*
Bark flour	*Pinus sylvestris, Picea abies, Tilia cordata, Ulmus glabra, Fagus sylvatica*
Seeds	*Bistorta vivipara, Chenopodium album*
Buds, young shoots	*Corylus avellana* (buds, catkins), *Picea abies* (shoots), *Pinus sylvestris* (shoots)
Flowers, leaves and grass:	*Salix caprea, Rumex acetosa, Trifolium pratense, Tilia cordata, Epilobium angustifolium, Calluna vulgaris*, Poaceae
Fruits	*Fagus sylvatica, Vaccinium vitis-idaea, Sorbus intermedia*
Other substances	Chaff, straw, ears of grain, mash, horse manure, sawdust, soil, fish roe, bone meal

the peasantry rejected these stuffs. They used food they were used to in other times as well, although they had to substitute flour with other plants. The inner bark of pine was still used during the food crisis in the 1860s. Straw, ears of grain and chaff were also popular, as were potatoes and root vegetables like swedes and turnips. Very few wild plants were actually used as emergency food (Tab. 2). If the peasants had been on the verge of starving to death, they most certainly would have eaten anything available (such as leather, rodents, dogs etc.) [67].

Comparison with other Scandinavian countries

Almost the same taxa as in Sweden were used in Norway. Some differences are due to the species available [2]. However, some plants, although common in Sweden, were not used. An interesting example is the West Nordic custom (Norway, Faroe Islands, Iceland and Shetland Islands) of eating the nutritious roots of *Potentilla anserina* L. which is not known from Sweden [68]. Although Sweden has a very long coastline, seaweed has not played any role as food or famine food in the same way as in Norway, Faroe Islands, Iceland and the British Isles. Seaweed has been used only as manure or for technical purposes [3]. It is only very recently that a small company in Grebbestad, Bohuslän, has started to make bread (crisp bread, baguette) of flour made of seaweed, *Laminaria digitata* (Hudson) J. W. Lamouroux.

New trends in the 20th century and onwards

Today especially wild fruits play an important role in Swedish food culture (Tab. 3). With increased access to cheap sugar in the early 20th century wild fruits (*Vaccinium myrtillus*, *V. vitis-idaea*, and *Rubus chamaemorus*) became very popular, especially among urban people. The fruits were used in the households for cordial, jam, gruel and desserts. *V. vitis-idaea* can be eaten with meat, fish, game, dumplings ("palt"), black pudding, porridge, ice-cream, cakes, biscuits, and desserts [69]. While writing this the present author had a

Tab. 3 Wild fruits and mushrooms picked for local consumption in Sweden 1990, 2000, 2005 [80].

Products	Total quantity harvested/collected (kilogram)		
	1990	2000	2005
Bilberry (*Vaccinium myrtillus* L.)	4754000	4800000	4300000
Cowberry (*Vaccinium vitis-idaea* L.)	5950000	6000000	5850000
Rasberry (*Rubus idaeaus* L.)	2250000	2200000	1700000
Cloudberry (*Rubus chamaemorus* L.)	1900000	1900000	1325000
Mushrooms	8640000	8500000	8910000

1 litre of fruits = 0.5 kg fruits; 1 litre mushrooms = 0.6 kg mushrooms.

cowberry-cheesecake with his coffee at his break. Waffles with whipped cream and cloudberry jam are an appreciated treat among ordinary people and are often sold at festivals, fairs and other events all over the country. The furniture and product company IKEA serves Swedish meatballs with cream sauce and cowberry jam, instant rose hip soup and elderflower cordial in their department store restaurants and food stalls all over the world. Bilberries make everything from traditional jam to modern grappa [70]. Still only a fraction of the about 1000 million litres of fruits and 3600 million litres of mushrooms that are produced in Swedish forests every year are actually harvested [71].

Mushrooms were despised by the peasantry. However, mushrooms became accepted as food first among the French-inspired aristocracy in the 18th century and by the urban bourgeoisie in the late 19th century. Later also urban industrial workers became interested in picking mushrooms as a free food resource in the forests, especially after World War I. Fifty years ago working class people gathered only *Cantharellus cibarius* (Fr.) and occasionally *Boletus edulis* Bull. Nowadays more species are utilised in Swedish households, and especially the easily picked *Cantharellus tubaeformis* (Pers.) Fr. has become very popular since the 1970s, but also *Cantharellus lutescens* Fr., *Craterellus cornucopioides* (L.) Pers., *Hydnum repandum* L., *Macrolepiota procera* (Scop.) Singer, *Suillus luteus* (L.) Gray and occasionally also *Lactarius deterrimus* Gröger, just to mention a few, find their way into the basket. Many elderly people still only gather and use *Cantharellus cibarius*, being uncertain about the edibility of other taxa. Whether it is safe to eat *Gyromita esculenta* (Pers. ex Pers.) Fr. is disputed, but *Morchella elata* Fr. is gathered instead. Chinese immigrants, occasionally also others, pick *Marasmius oreades* (Bolton) Fr., in the urban lawns. Some cases of mushroom poisoning are reported every year [72]. A brochure available in many languages describing the 24 most common poisonous mushrooms in Sweden is published by Swedish Poison Information Centre (http://www.gic.se). Harvesting fruits and mushrooms in the forests is a popular pastime for many urban people (58 per cent of the population was picking fruits and mushrooms in 1997), but also a source of income for immigrants and especially foreign seasonal labour [47,73] (Fig. 2).

The only traditional green wild food plant that is regularly eaten in contemporary Sweden is *Urtica dioica* L. Occasionally some people gather *Aegopodium podagraria* L., once naturalised but nowadays considered an invasive weed, for food. Some wild species have become part of modern regional food culture. *Allium scorodoprasum* L. was traditionally used in coastal areas as a spring vegetable, especially in stews. On the island of Gotland, it has been harvested for centuries, and used as a remedy against spring fatigue. Today *A. scorodoprasum* L. is an ingredient of a popular soup which is served as a local speciality all over Gotland [74]. Another recent example is jam made of *Rubus caesius* L. fruits, hardly eaten before but now a regional speciality also in Gotland. The jam is usually eaten with a local kind of pancake flavoured with saffron [75]. Products (jam, cordial) made from berries of *Vaccinium oxycoccos* L. and *Empetrum* sp. are also available. The real success, however, has been the fruits of *Hippophaë rhamnoides* L., which began to be very popular a couple of decades ago. Although most fruits are nowadays harvested from cultivated plants, some people actually gather them from wild plants along the Baltic coastline. Jam and juice of wild sea buckthorn fruits can actually be bought at Uppsala's weekly market [76].

Many herbs were used as spices (and medicine) in the peasant society, but during industrialisation most of them were forgotten by the working class people. Nowadays herbs are coming back into the households and Swedes use a lot of culinary herbs, although most of them are cultivated in gardens or bought in stores. Wild spice plants are not gathered anymore. General trends sometimes increase the interest for wild plants, such as making tea of *Tilia cordata* L. or homemade cordials from the flowers of *Sambucus nigra* L., *Filipendula ulmaria* (L.) Maxim. or *Centaurea cyanus* (L.) Hill, and making elderberry "capers" of unripe fruits.

Many wild food plants are today known through the many handbooks available in bookshops. Newspapers also have articles on the subject each summer. It is hard to determine how many people actually try using these recipes. Some consider wild plants as healthier than cultivated plants; others appreciate them for being a free resource. *Alliaria petoliata* (M. Bieb.)

Fig. 2 Thai women selling chanterelles, *Cantharellus cibarius* (Fr.), and bilberries, *Vaccinium myrtillus* L., on a Saturday market in Uppsala in August 2012 (photograph by Navarana Ingvarsdóttir).

Cavara & Grande, *Capsella bursa-pastoris* (L.) Medik. and *Lamium album* L. are examples of plants mentioned in these modern recipes [77,78]. There are of course also specialists using and promoting wild plants as food, including individuals and small circles inspired by New Age ideologies or for other ideological and pseudo-scientific reasons (homeopathy). However, generally speaking their sources are not from the old folk-knowledge but from old and modern propaganda books (and nowadays also through the Internet).

Conclusions

Apart from aromatic fruits and some mushrooms, other wild plants are very little used today, despite propaganda efforts to create an interest in them. Sweden in the past must be regarded as a herbophobous society, according to the ethnobotanist Łuczaj's categorization [79]. Wild green plants have been and still are of very limited interest for most people. However, lettuce and other cultivated greens are easily available in the grocery stores throughout the year, and so are herbs. Cultivated fruits and vegetables are always available. Modern Swedes now eat a lot of vegetables in their daily meals. It is part of contemporary food culture.

Wild fruits and mushrooms have increased in importance in recent years and are very much used, not only in households but also in restaurants and food industry [80] (Tab. 3). Cowberry *Vaccinium vitis-idaea* L. jam can for instance be used with almost all kinds of dishes. The right of public access in the countryside, which entitles people to pick fruits, to gather mushrooms and to pursue various outdoor activities, is important for our understanding of the landscape and its products. However, it is also a source of income for many people, previously for people in the countryside in the 20th century, nowadays for immigrants (especially Thai women) and foreign seasonal workers (Tab. 4). Now and then there is also an interest in green plants. Survival courses are given not only by the military, but today also by high schools, educational TV-programs and instructions given in evening classes, with information on how to use wild plants as food. People lost in the forests know how to use the available resources. Many people are used to outdoor life. An angler, who in the summer of 2012 was lost for almost two weeks in the forests of Jämtland, survived by eating spruce shoots and unripe fruits of *Rubus chamaemorus* L. before he was found again unharmed by his experiences.

Tab. 4 Food products of wild taxa available at the weekly market on Fyris torg, Uppsala, in 18 August 2012.

Species	Cordial	Jam	Frozen leaves
Bilberry (*Vaccinium myrtillus* L.) fruits	×	×	
Cowberry (*Vaccinium vitis-idaea* L.) fruits	×	×	
Wild raspberry (*Rubus idaeaus* L.) fruits		×	
Cloudberry (*Rubus chamaemorus* L.) fruits		×	
Wild strawberry (*Fragaria vesca* L.) fruits	×		
Blackthorn (*Prunus spinosa* L.) fruits	×		
Elder (*Sambucus nigra* L.) flowers	×		
Nettle (*Urtica dioica* L.) shoots/leaves			×

Acknowledgments

The author owes his sincere gratitude to Professor Łukasz Łuczaj (Poland), Professor Marie Clark Nelson (Sweden) and Mr Nicholas Redman (U. K.) for comments on the first draft of this article and for help of various kinds.

References

1. Brøndegaard VJ. Folk og flora: Dansk etnobotanik. København: Rosenkilde & Bagger; 1978. (vol 1–4).
2. Høeg OA. Planter og tradisjon: floraen i levende tale og tradisjon i Norge 1925–1973. Oslo: Universitetsforlaget; 1974.
3. Svanberg I. Folklig botanik. Stockholm: Dialogos; 2011.
4. Svanberg I. The use of wild plants in the Faroe Islands, 1590–1990: a contribution to Scandinavian ethnobotany. Svenska Linnésällskapets Årsskrift. 1998;1996–1997:81–130.
5. Manninen I. Überreste der Sammlerstufe und die Notnahrung aus dem Pflanzenreich bei den nordeurasischen, vorzugsweise den finnischen Völkern. Eurasia Septentrionalis Antiqua. 1931;6:30–48.
6. Guðmundsson G. Gathering and processing of lyme-grass (*Elymus arenarius* L.) in Iceland: an ethnohistorical account. Veget Hist Archaeobot. 1996;5:13–23.
7. Nelson MC, Svanberg I. Lichens as food: historical perspectives on food propaganda. Svenska Linnésällskapets Årsskrift. 1987;1986–1987:7–51.
8. Svanberg I, Nelson MC. Bone meal porridge, lichen soup, or mushroom bread: acceptance or rejection of food propaganda 1867–1868. In: Häkkinen AJ, editor. Just a sack of potatoes? Crisis experiences in past and present societies. Helsinki: Societas Historica Finlandiae; 1992. p. 119–147.
9. Egardt B. Svenskarna och svampmaten. Rig. 1954;37:33–44.
10. Klintberg B. Kuttrasju: folkloristiska och kulturhistoriska essäer. Stockholm: Norstedt; 1998.
11. Fjellström P. *Angelica archangelica* in the diet of the Lapps and the Nordic peoples. In: Furumark A, Lagercrantz S, Nesheim A, Widegren G, editors. Lapponica: essays presented to Israel Ruong 26.5.1963. Uppsala: Uppsala University; 1964. p. 99–115. (Studia Ethnographica Upsaliensia; vol 21).
12. Aronsson KÅ. Användning av ängssyra (*Rumex acetosa*) bland samerna. In: Svanberg I, Tunón H, editors. Samisk etnobiologi: människor, djur och växter i norr. Nora: Nya doxa; 2000. p. 253–258.
13. Svanberg I. Samisk etnobiologi. In: Svanberg I, Tunón H, editors. Samisk etnobiologi: människor, djur och växter i norr. Nora: Nya doxa; 2000. p. 13–58.
14. Svanberg I. Ethnobotany. In: Kulonen UM, Seurujärvi-Kari I, Pulkkinen R, editors. The Saami: a cultural encyclopedia. Helsinki: Suomalaisen Kirjallisuuden Seura; 2005. p. 104–105.
15. Svanberg I. The Sami use of *Lactuca alpina* as a food plant. Svenska Linnésällskapets Årsskrift. 2002;2000-2001:77–84.
16. Tillhagen CH. Food and drink among the Swedish Kalderaša Gypsies. Journal of the Gypsy Lore Society, 3rd Ser. 1957;34:25–52.
17. Svanberg I. Fjällsyra, *Oxyria digyna*. In: Tunón H, Pettersson B, Iwarsson M, editors. Människan och floran: etnobiologi i Sverige. Stockholm: Wahlström & Widstrand; 2005. p. 277. (vol 2).
18. Pettersson B, Svanberg I. Hassel, *Corylus avellana*. In: Tunón H, Pettersson B, Iwarsson M, editors. Människan och floran: etnobiologi i Sverige. Stockholm: Wahlström & Widstrand; 2005. p. 158–160. (vol 2).
19. Svanberg I. Svinmålla, *Chenopodium album*. In: Tunón H, Pettersson B, Iwarsson M, editors. Människan och floran: etnobiologi i Sverige. Stockholm: Wahlström & Widstrand; 2005. p. 197. (vol 2).
20. Svanberg I. "The Lapps chew this root a lot": milk parsley (*Peucedanum palustre*) in Sami plant knowledge. In: Fransson U, Svendin M, Bergerbrant S, Androschuck F, editors. Cultural interaction between east and west. Archaeology, artefacts and human contacts in northern Europe.

Stockholm: Stockholm University; 2007. p. 328–330.

21. Svanberg I. Garden angelica (*Angelica archangelica*). In: Cumo CM, editor. Encyclopedia of cultivated plants. Santa Barbara CA: ABC-CLIO. In press.
22. Svanberg I. Caraway (*Carum carvi*). In: Cumo CM, editor. Encyclopedia of cultivated plants. Santa Barbara CA: ABC-CLIO. In press.
23. Svanberg I. Tansy (*Tanacetum vulgare*). In: Cumo CM, editor. Encyclopedia of cultivated plants. Santa Barbara CA: ABC-CLIO. In press.
24. Erixon S. Surviving primitive gathering in the Nordic countries. Folk-Liv. 1951;14–15:95–102.
25. Sporrong U. The Swedish landscape: the regional identity of historical Sweden. In: Jones M, Olwig K, editors. Nordic landscapes: region and belonging on the Nordic edge of Europe. Minneapolis MN: University of Minnesota Press; 2008. p. 141–156.
26. Svanberg I. Cultural and language pluralism in contemporary Nordic countries. In: Fossum Grønn J, editor. Nordic voices: literature from the Nordic countries. Oslo: Nordbok; 2005. p. 8–17.
27. Balick MJ, Cox PA. Plants, people, and culture: the science of ethnobotany. New York NY: Scientific American Library; 1996.
28. Linnaeus C. Flora Lapponica. Amsterdam: Salomon Schouten; 1737.
29. Svanberg I. Etnobiologen Linné. In: Jacobsson R, editor. Så varför reser Linné? Perspektiv på Iter Lapponicum 1732. Umeå: Carlsson; 2005. p. 135–162.
30. Svanberg I. The use of lichens for dyeing candles: ethnobotanical documentation of a local Swedish practice? Svenska Landsmål och Svenskt Folkliv. 1998;324:133–139.
31. Svanberg I. Fabulat, plagiat och spökuppgifter: att använda berättande källor inom etnobiologi. In: Nycklar till kunskap: om människans bruk av naturen. Stockholm: Kungliga Skogs- och Lantbruksakademien; 2010. p. 123–136.
32. Svanberg I. Linneaner: Carl von Linnés lärjungar i Sverige. Stockholm: Wahlström & Widstrand; 2006.
33. Svanberg I. Källor till folklig kunskap. In: Tunón H, Pettersson B, Iwarsson M, editors. Människan och floran: etnobiologi i Sverige. Stockholm: Wahlström & Widstrand; 2005. p. 23–34. (vol 2).
34. Bergfors G. Från självhushållets dagar i ett ångermanländskt jordbrukarhem. Ångermanland. 1947;4:59–76.
35. Fridner G. Folkliga växtnamn i Västerbotten. Umeå: Språk- och folkminnesinstitutet; 1999.
36. Svanberg I. Folkbotaniska uppgifter från Borgsjö socken i Medelpad: Gullik Gulliksons anteckningar från 1955. Uppsala: Maral Production; 2004.
37. Modin E. Växtnamn samt folkliga bruk och föreställningar rörande växter i Härjedalen. In: Festskrift til H. F. Feilberg fra nordiske sprog- og folkemindeforskere på 80 års dagen den 6. August 1911. Kristiania: Bymaalslaget; 1911. p. 697–737.
38. Landsmålsarkivets frågelista. 19, Nödbröd ock nödfoder. Svenska Landsmål och Svenskt Folkliv. 1929;22:213–226.
39. Johansson L. Växtnamn m.m. Manuscript collection (cat. No. ULMA 19 857). Uppsala: Swedish Institute for Dialects and Folklore; 1949.
40. Bergvall F. Edseles folkliga flora: växtnamn (latin, dialekt, riksspråk jämte uppgifter om växternas användning. Manuscript collection (cat. No. ULMA 28 641). Uppsala: Swedish Institute for Dialects and Folklore; 1972.
41. Drake G. Linnés försök till inhemsk teodling. Svenska Linnésällskapets Årsskrift. 1927;10:68–83.
42. Räsänen L. Of all foods bread is the most noble: Carl von Linné (Carl Linneaus) on bread. Scand J Food Nutr. 2007;51:91–99.
43. Linnaeus C. Archiaterns och riddarens af Kongl. Nordstjerne Orden Carl Linnæi berättelse om the inhemska wäxter, som i brist af säd kunna anwändas til bröd- och matberedning. Stockholm: Kongl. Tryckeriet; 1757.
44. Svanberg I. Fattigmanskost. In: Broberg G, Lindell G, editors. Till livs med Linné: om mat hälsa och levnadskonst. Stockholm: Atlantis; 2007. p. 185–195.
45. Lundström E. Ätliga vilda växter: en dyrtidsbok. Stockholm: Norstedt; 1918.
46. Holmboe J. Gratis mat av vilda växter. Stockholm: Albert Bonnier; 1942.
47. Pettersson B. Nötter, bär och svamp. In: Pettersson B, Svanberg I, Tunón H, editors. Människan och naturen: etnobiologi i Sverige. Stockholm: Wahlström & Widstrand; 2001. p. 147–154. (vol 1).
48. Nordenstreng E. 72 sätt att tillaga de vitaminrika rönnbären jämte inledande kapitel om rönnbär i folktro och forskning. Stockholm: Albert Bonnier; 1934.
49. Kardell L. Utmarkens bär och svampar samt inmarkens nötter. In: Lindwall B, editor. Tjära, barkbröd och vildhonung. Stockholm: Nordiska museet; 1996. p. 138–155.
50. Ekvall U. Gotländska växtnamn: tradition och förnyelse i ett lokalt växtnamnskick. Stockholm: Almqvist & Wiksell International; 1990.
51. Turner NJ, Łuczaj ŁJ, Migliorini P, Pieroni A, Dreon AL, Sacchetti LE, et al. Edible and tended wild plants, traditional ecological knowledge and agroecology. Cr Rev Plant Sci. 2011;30(1-2):198–225.
52. Nelson MC. Bitter bread: the famine in Norrbotten 1867–1868. Uppsala: Acta Universitatis Upsaliensis; 1988. (Studia Historica Upsaliensia; vol 153).
53. Høeg OA. Har De spist blåveis- eller hvitveisblomster om våren? Blyttia. 1950;8:126–128.
54. Svanberg I. Field horsetail (*Equisetum arvense*) as a food plant. Fróðskaparrit: Annales Societatis Scientiarum Færoensis. 1997;45:45–55.
55. Edlund LE, Svanberg I. Rallarros, *Epilobium angustiforlium*. In: Tunón H, Pettersson B, Iwarsson M, editors. Människan och floran: etnobiologi i Sverige. Stockholm: Wahlström & Widstrand; 2005. p. 193–194. (vol 2).
56. Keyland N. Svensk allmogekost. Stockholm: Svenska Teknologförening; 1919. (vol 1).
57. Ståhlberg S, Svanberg I. Gathering food from rodent nests in Siberia. J Ethnobiol. 2010;30:184–202.
58. Hansson AM. Pre- and protohistoric bread in Sweden: a definition and a review. Civilisations. 2002;49:183–190.
59. Niklasson M. Bark som människoföda ur agrart och samiskt perspektiv. In: Lindwall B, editor. Tjära, barkbröd och vildhonung. Stockholm: Nordiska Museet; 1996. p. 107–125.
60. Zackrisson O, Östlund L, Korhonen O, Bergman I. The ancient use of *Pinus sylvestris* L. (Scots pine) inner bark by Sami people in northern Sweden, related to cultural and ecological factors. Veg Hist Archaeobot. 2000;9:99–109.
61. Vilkuna K. Das Kauharz, ein uraltes Genussmittel. In: Furumark A, Lagercrantz S, Nesheim A, Widegren G, editors. Lapponica: essays presented to Israel Ruong 26.5.1963. Uppsala: Uppsala University; 1964. p. 295–303. (Studia Ethnographica Upsaliensia; vol 21).
62. Ränk G. Vilken roll har tätörten spelat i mjölkhushållningen? Folktro kontra vetande. Folk-Liv. 1961;24–25:65–75.
63. Hofsten N. Pors och andra humleersättningar och ölkryddor i äldre tider. Uppsala: Lundequistska; 1960. (Acta Academiae Regiae Gustavi Adolphi; vol 36).
64. Berg G. Att tappa björklake. Gastronomisk kalender. 1968:35–57.
65. Nelson MC, Svanberg I. Coffee in Sweden: a question of morality, health, and economy. Food Foodways. 1993;5:239–254.
66. Bergfors G. Vad roade sig barnen med när farfars far var liten? Fjällsjö Krönika. 1966;17:33–54.
67. Scrimshaw NS. The phenomenon of famine. Annu Rev Nutr. 1987;7:1–22.
68. Holmboe J. Mura (*Potentilla anserina* L.) og dens bruk som matnyttnig plante i ældre og nyere tid. Svensk Botanisk Tidskrift. 1928;22:160–168.
69. Bringéus NA. The red gold of the forest. In: Lysaght P, editor. Food from nature, attitudes, strategies and culinary practices. Uppsala: Academiae Regiae Gustavi Adolphi; 2000. p. 225–239. (Acta Academiae Regiae Gustavi Adolphi; vol 71).
70. Rattfelt A, Lunøe Pihl S. Blåbär: minnen, mat och smak. Stockholm:

Balkong; 2008.
71. Kardell L. Forest berries and mushrooms: an endangered resource? Ambio. 1980;9:241–247.
72. Irestedt B, Persson H, Sjöberg G. Förgiftningstillbud bland barn är vanliga men oftast inte allvarliga. Läkartidningen. 2004;101:2256–2260.
73. Lindhagen A, Hörnsten L. Forest recreation in 1977 and 1997 in Sweden: changes in public preferences and behaviour. Forestry. 2000;73:143–153.
74. Genrup K. Skogslök, *Allium scorodoprasum*. In: Tunón H, Pettersson B, Iwarsson M, editors. Människan och floran: etnobiologi i Sverige. Stockholm: Wahlström & Widstrand; 2005. p. 271. (vol 2).
75. Svanberg I. Blåhallon, *Rubus caesius*. In: Tunón H, Pettersson B, Iwarsson M, editors. Människan och floran: etnobiologi i Sverige. Stockholm: Wahlström & Widstrand; 2005. p. 259. (vol 2).
76. Svanberg I. Knallgul saft på Fyris torg. Biodiverse. 1998;3(2):8–9.
77. Ingmansson I. Kan man äta sånt? Stockholm: Prisma; 1996.
78. Källman S. Vilda växter som mat och medicin. Västerås: Icaförlaget; 1997.
79. Łuczaj ŁJ. Changes in the utilization of wild green vegetables in Poland since the 19th century: a comparison of four ethnobotanical surveys. J Ethnopharmacol. 2010;128(2):395–404.
80. Country report: Sweden. Rome: FAO; 2005.

8

Wild plants used for food by Hungarian ethnic groups living in the Carpathian Basin

Andrea Dénes[1]*, Nóra Papp[2], Dániel Babai[3], Bálint Czúcz[4], Zsolt Molnár[4]

[1] Natural History Department, Janus Pannonius Museum, Box 158, 7601 Pécs, Hungary

[2] Department of Pharmacognosy, University of Pécs, Rókus 2, 7624 Pécs, Hungary

[3] Centre for Humanities, Hungarian Academy of Sciences, Országház 30, 1014 Budapest, Hungary

[4] Centre for Ecology, Hungarian Academy of Sciences, Alkotmány 2-4, 2163 Vácrátót, Hungary

Abstract

A list of plant species used for food in Hungary and among Hungarian ethnic groups of the Carpathian Basin during the 19th and 20th centuries was compiled from 71 ethnographic and ethnobotanical sources and a survey among contemporary Hungarian botanists. Species used as food, spice, beverage or occasional snacks were collected. Sources mention 236 plant species belonging to 68 families. Most wild fleshy fruits (mostly *Rosa*, *Rubus*, *Cornus*, *Ribes*, *Vaccinium* spp.), dry fruits and seeds (*Fagus*, *Quercus*, *Corylus*, *Castanea*, *Trapa* spp.), several green vegetables (e.g. *Rumex*, *Urtica*, *Humulus*, *Chenopodiaceae* spp., *Ranunculus ficaria*), bulbs and tubers (*Lathyrus tuberosus*, *Helianthus tuberosus*, *Chaerophyllum bulbosum*, *Allium* spp.) used for food in Europe, are also known to be consumed in Hungary. A characteristic feature of Hungarian plant use was the mass consumption of the underground parts of several marsh (e.g. *Typha*, *Phragmites*, *Sagittaria*, *Alisma*, *Butomus*, *Bolboschoenus* spp., as well as the endemic *Armoracia macrocarpa*) and steppe species (e.g. *Crambe tataria*, *Rumex pseudonatronatus*). Consuming wild food plants is still important among Hungarians living in Transylvania: even nowadays more than 40 species are gathered and used at some locations.

Keywords: ethnobiology, historical ethnobotany, wild green vegetables, wild edible plants, tree saps

Introduction

In the last two decades several comprehensive surveys and reviews were performed in many European countries on wild plant use. Wild plants formed important parts of our ancestors' diet in a multitude of ways (e.g. as bread ingredients, vegetables, fruits, spices, snacks or beverages); essential in severe times, supplementary otherwise [1–3]. Revitalization of traditional practices is timely for ecology, economy and nutrition biology. Nevertheless, even though Hungary is no poorer in traditions of wild plant use than other parts of Europe, no broad-scale review has been undertaken for Hungary in this respect, yet.

Information on the human consumption of wild plant species is widely scattered among the ethnographic and botanic literature, mostly in Hungarian language. Publications on gathering economy, traditional nutrition, shepherding, forest goods utilization, and on the hunting-fishing-gathering "pákász" lifestyle in the former wetlands of the floodplains of the large Hungarian rivers are known from the end of 18th century on (e.g. [4]), and they became frequent from the second half of the 20th century. Publications of traditional research on medicinal uses of plants [5–8] and ethnobotany [9–11] also mention wild food plants, but a review concentrating on wild food plants is still missing.

In this work we systematically review and compile information on the utilization of wild food plants in the Hungarian speaking regions of the Carpathian Basin from the Hungarian ethnographic and ethnobotanical publications. Our study aims at compiling a knowledge base on wild food plants consumed in Hungary and by the Hungarians living in other countries of the Carpathian basin, containing information on the species, as well as the modes of use. As the Carpathian Basin is dominated by alluvial floodplains located on the margin of the vast Eurasian steppe regions, Hungarian traditional plant use might also include previously undocumented usage of wetland and steppe plant species, which can potentially add some new aspects to the existing knowledge on traditional plant use in Europe.

* Corresponding author. Email: denes.andrea@jpm.hu

Material and methods

Flora, vegetation and history of the Carpathian Basin

The Carpathian Basin is the contiguous floodplain area of the Danube and Tisza Rivers encircled by the Carpathian

Mountains in Central Europe. Containing an extensive transitional zone between European deciduous forests and the Eurasian steppe biome the study area harbors a particularly diverse vegetation, which is also acknowledged by the European Union by classifying the majority of this region into a singular biogeographic zone, the Pannonian biogeographic zone.

The periphery of the Carpathian Basin including the Carpathians can be characterized mostly by and alpine and subalpine vegetation, coniferous forests, which turn into broadleaved deciduous forest at lower elevations. The central part of the basin is dominated by continental forest-steppes, although only remnants of salty and sand steppes had survived to date. Gallery forests and wetlands on the floodplains of the two large rivers, Danube and Tisza, and their tributaries played a determining role in vegetation development and also in people's lives till their regulations in the second part of the 19th century. Presently their former area, as well as the majority of former steppe vegetation, is dominated by agriculture [12]. The flora of Hungary consists of 2600 species including many steppe species, whereas the flora of Transylvania is also about 2600 with significant proportions of boreal and alpine species [13]. The Carpathian Basin altogether harbors 3360 species. In addition to the dominant Eurasian, continental and European species, southern, submediterranean and Balcanic elements amount to 20%.

As a consequence of Hungarian history, there are Hungarian ethnic groups living in all countries of the Carpathian Basin. Due to ecological, historical and economic reasons, ethnicities separated from the mother country often preserve their traditions better, even archaic ones, so researchers prefer to conduct ethnographic and ethnobotanic studies among Hungarians living in Romania, Slovakia, Ukraine, Serbia and Croatia.

Research history

Historical records on the use of wild plants in Hungary are known since the Medieval period (16th century). Herbal knowledge of wise women and monks, and works of former botanists as Clusius (1526–1609), Beythe (1532–1612), Kitaibel (1758–1817), Borbás (1844–1934), and Dégen (1866–1934) are usually reviewed by ethnographers studying the gathering economy of the Carpathian Basin. Detailed records on edible plants from the end of the 18th century, can be found in the manuscript diaries of Kitaibel, reviewed recently by Molnár [4]. Rapaics [14] had published an overview on the history of food plants, going back to pre-Medieval times.

A complete list of former publications can be found in the reviews of Gunda [15,16], a key person of ethnographical research of gathering economy in Hungary. Reviews of folk nutrition [17–19] generally include more or less detailed overviews of wild plants collected for food [9,17–19]. Plants consumed during famines were studied by Rapaics [14], Györffy [20], Gunda [15,16], and Molnár [21]. The published studies collectively cover almost the entire Carpathian Basin, the most thoroughly explored regions being Transylvania [sensu lato, including e.g. Gyimes (Ghimeş) and Máramaros (Maramureş)] and the southern foothills of the Northern Carpathians [Nógrád, Heves, Borsod, and Zemplén counties as well as the Gömör (Gemer, Slovakia) region]. Nevertheless, not all papers provide a comprehensive ethnobotanical "wild food plant" survey of the studied regions – there are several papers, where the focus is on other aspects of traditional living, with only sporadic mention of a few wild plants.

Methods

We compiled a list of plant species used for food based on 71 ethnographic and ethnobotanical publications in Hungarian language. Most of the studied papers describe traditional ecological knowledge of Hugarian speaking ethnic groups living in the Carpathian Basin, even though there might be exceptions, e.g. some papers included wild food plants used by other ethnic groups (e.g. Slovaks) living in Hungary, while other papers include traditional plant use of Hungarians living even beyond the Carpathians (e.g. Hungarian refugees from Bukovina, whose plant use included ancient elements from their old home country as well as newer ones from their new home, Hungary). The studied publications explore the following regions: Dunántúli dombvidékek (SW Hungary: Őrség, Somogy, Zala, Tolna, Baranya) [22–29]; Kisalföld (NW Hungary and SW Slovakia) [30,31]; Déli-Alföld (SW Hungary and N Croatia) [32–35]; Dunántúli középhegység (mountains in W Hungary) [36,37]; Északi középhegység (Carpathian foothills in N Hungary) [38–47]; Gömör (Gemer, S Slovakia) [48–52]; Kárpátalja (Zakarpatska Oblast, W Ukraine) [53,54]; the Great Hungarian Plain [55–62]; Vajdaság (Voivodina, N Serbia) [63]; Erdély (Transylvania, Romania) [64–75]; and Bukovina (N Romania – collected from refugees settled down in Hungary) [76,77] (Fig. 1). This literature review was complemented by a survey among 34 Hungarian botanist on the wild food plants which they had collected and consumed in their childhood [78].

Fig. 1 Map of the study area including the geographic names used in the text.

The oldest source is from the end of the 18th century [4]; the most recent ones are up-to-date ethnobotanical surveys collected in Ghimeş, Transylvania and Hortobágy, Hungary [11,74,75]. In addition to local traditional plant names, most publications also mention the official Hungarian and/or the scientific (genus or species) Latin names of the food plants, even though these identifications can be easily mistaken if the local traditional names are similar to the official Hungarian name of another taxon. Such misidentifications are unfortunately typical in a part of the ethnography literature, nevertheless a good botanical knowledge, the descriptions of the plants discussed (if supplied), and the comparison of the different sources can effectively help to correct misidentifications [79]. In our work we corrected all obvious misidentifications which could be easily corrected, and omitted all records which were clearly invalid, but no unambiguous correction was available. Nevertheless, most of the species are mentioned in several publications, which reduces the uncertainties, particularly in the case of plant uses which were also documented by botanists.

In the reviewed papers we focused on plants which were collected from the wild and consumed as food or food ingredient (including spices, beverages, occasional snacks, etc.). This definition excludes medical plants (consumed only for their health impacts), and plants collected only for non-food use (e.g. dye, timber, etc.). On the other hand, we included gathering from spontaneous populations of escaped and naturalized cultivated plants and non-native invasives.

Results

We found altogether 235 plant species belonging to 67 families which were mentioned from the study area. We found 216 species in the literature survey, whereas contemporary Hungarian botanists (34 data providers) mentioned 91 taxa (the overlap between the two sets was 71 species). The species are listed in detail in Tab. 1 with their local names as mentioned in the sources, and their documented modes of usage. In order to provide an indicator for record uncertainty, species data published only by ethnographers are marked with "*" in Tab. 1.

We found that it is the species of the Rosaceae family (36 species) which are consumed most often; other frequently used families include Asteraceae, Lamiaceae, Liliaceae (sensu lato, before the APG division of the family), Fabaceae and Apiaceae. The list contains 36 trees, 27 shrubs, 4 dwarf shrubs and 169 herbaceous species. Green aboveground parts (leaves, young shoots, buds, and sometimes the whole plant) of 98 species were consumed (mainly Apiaceae, Lamiaceae, Liliaceae). Flowers of 39 species (mainly Asteraceae, Boraginaceae, and Lamiaceae) and fruits/seeds of 74 species (mainly Rosaceae, Grossulariaceae, and Ericaceae) were eaten. Underground parts – roots, rhizomes, tubers, bulbs – of 23 species (mainly Liliaceae, Apiaceae, Asteraceae, and Brassicaceae) were used. Saps of 8 species and dried saps (resins, gums) of 4 species were consumed or chewed. In several cases two or three parts of the same species were consumed, e.g. *Rosa gallica*: flowers, fruits, leaves; *Fagus sylvatica*: fruits, leaves and sap; *Sambucus nigra*: flower, fruit; *Taraxacum* spp.: leaves, flowers; *Fragaria* spp.: fruits and leaves.

Green vegetables

Green parts – mainly young spring shoots or young leaves – of 51 species were prepared raw for salad, or cooked for use in soup or sauce. Species used in most regions include *Rumex acetosa*, *Urtica dioica*, *Humulus lupulus*, *Ranunculus ficaria*, and *Allium* spp. Larger leaves are often used in regional dishes as a wrapping for some meaty stuffing, e.g. *Armoracia rusticana*, *Tussilago farfara* and *Fallopia* spp. Young shoots of *Typha* spp. and *Phragmites australis* used to be consumed as a salad in the Sárköz region of Hungary, a tradition abandoned long time ago. Sixteen species including *Anthriscus cerefolium*, *Glechoma hederacea*, *Thymus* spp., and *Verbena officinalis* were used as spice, pickling or preservative. *Satureja alpina* as a spice is mentioned from the early 18th century. The green parts of several plants were used to prepare a refreshing tea or as syrup (e.g. a "pine honey" made from the buds of Pinaceae species); more interesting uses include making candy from *Melittis melissophyllum* [76], or *Fagus* leaves. Babies unable to suck were fed with pressed nettle (*Urtica*) sap (I. Németh personal communication). Some species (e.g. *Oxalis acetosella*, *Galium verum*) were used in the past as curdling agents. Herdsmen chewed *Plantago lanceolata* leaves in order to clean their teeth (I. Németh personal communication.). Eating Lemnaceae species in famine is mentioned only from the Drava floodplain.

Flowers

Even though sucking nectar and eating flowers or inflorescences is a widespread and delightful occupation of children, major food or beverage products are rarely made from flowers. As an exception, flowers of *Sambucus nigra* or *Robinia pseudacacia* are commonly used for making refreshing drinks or fried into pancakes they can serve as popular dishes. Unique is the use of inflorescences of *Carlina acaulis* as a green vegetable. Flowers of *Humulus lupulus* and *Robinia pseudacacia* were added to sourdough in many places. In Gömör, there used to be a tradition of making "flower wines" from *Robinia* or *Taraxacum* flowers ("pimpóbor") [50].

Fruits and seeds

Eating and processing wild fruits of many species is a widespread, still a living tradition in Hungary. *Fragaria*, *Sambucus*, *Rubus*, *Crataegus*, *Vaccinium*, *Ribes*, *Rosa* spp. and *Cornus mas* have been eaten fresh, baked into cakes, prepared as beverages, or dried for a later use for long time. With the advent of cheap sugar additional preservation techniques became available including syrup and jam production [17]. Some fruits (*Malus*, *Pyrus*, *Vaccinium vitis-idaea*, *Streptopus amplexifolius*) were collected unripe; they were ripened in the attic, in hay, or in the sun. Formerly cider and vinegar were fermented from the fruits of *Malus sylvestris*. A traditional fruit brandy called "pálinka" is traditionally distilled after fermentation from many fruits even today. To improve taste and color fruits and spices can also be added to "pálinka" after distillation ("ágyaspálinka"). Fruit cider and brandy production was particularly important in regions without extensive grape productions. In Gömör cider was fermented from almost all wild fruits. Rarely consumed fruits include *Streptopus amplexifolius*, *Viburnum* spp. and *Cornus sanguinea*, which are considered slightly toxic by some sources but were still documented as consumed by trustworthy publications. Several dry fruits (nuts and seeds) were also widely collected. In addition to the nuts being still economically significant (*Corylus avellana*, *Juglans regia*, and *Castanea sativa*), *Fagus sylvatica* seeds were also eaten raw or roasted. Ground *Fagus acorns*, as well as *Quercus acorns* after leaching, were also used as coffee substitutes and flour in famine times. In addition to acorns, famine flour ingredients also included dried and ground *Crataegus*, *Rosa* and *Trapa* fruits, *Glyceria* seeds and *Corylus* buds. According to Kitaibel's data from the end of the 18th century, oil was pressed from *Sisymbrium altissimum*, *Brassica nigra*, and *Fagus sylvatica* seeds. *Staphylea pinnata* nuts used to be a kids' snack; cooked unripe fruits of *Daphne mezereum* were used as black pepper substitute in the time of Kitaibel, at the beginning of the 1800s. *Carum carvi* used to be a widespread spice not only for dishes, but also for palinka and tea. Food and wine were coloured, e.g. with *Phytolacca americana*; while shiny seeds, e.g. *Vicia* spp. and *Lathyrus aphaca* were used for decorating cakes.

Underground parts

Underground parts of of several wild plants were among the most important staple foods during famines. Bulbs, tubers or rhizomes of *Alisma plantago-aquatica*, *Bolboschoenus*

Tab. 1 List of wild food plants used by Hungarians in the Carpathian Basin.

Scientific name	Local names mentioned	Parts used	Mode of use	Reference No.
Abies alba Mill.	fehérfenyő	VEG, FLO	brandy (palinka) was distilled from buds and young cones	[74]
Acer campestre L.	kokasfa, kokastorufa, juhar	FRU	immature fruits were sucked as a snack	[23,78]
Acer pseudoplatanus L.	jávor, jávorfa, hegyi juhar	SAP	beverage	[19]
Acer sp.	juhar, jávor, jávorfa	SAP	beverage	[9,52,69,72]
Acorus calamus L.*†	kálmos	SUB	spice for liqueur	[29]
Aesculus hippocastanum L.†	vadgesztenye	FRU	seeds as coffee substitute	[19,78]
Agrimonia eupatoria L.	párlófű, tüdőfű, bojtorján, bojtorván, bojtorvány, repcsik	VEG	tea	[19,43,67]
Alchemilla spp.	palástfű, harmatfű	VEG	raw as salad	[74]
Alisma plantago-aquatica L.	type of bengyele, bakacs	SUB	food in famine and for herdsmen	[16,21]
Allium atroviolaceum Boiss.*	mezei fokhagyma	VEG, SUB?	raw	[57]
Allium obliquum L.	turkesztáni hagyma	VEG, SUB?	raw	[65]
Allium oleraceum L.	érdes hagyma	VEG, SUB?	raw and as spice	[21]
Allium rotundum L.	n.d.	VEG, SUB?	eaten	[65]
Allium scorodoprasum L.	hagyma, vadhagyma, kígyóhagyma, vad fokhagyma	VEG	young leaves were eaten; also used as spice like garlic	[4,19,41,57,78]
Allium ursinum L.	vadfokhagyma, medvehagyma, medvefokhagyma, vad hagyma, medvesósdi, sorhajma, salama	VEG	salad, vegetable, spice (put into sausage)	[27,41,65,70, 73–75]
Allium victorialis L.	győzedelmes hagyma	VEG, SUB?	eaten	[4]
Allium vineale L.	n.d.	VEG, SUB	leaves and bulbs were eaten	[78]
Allium sp.	hagyma	FLO	small onions in inflorescence were children's snack	[78]
Alopecurus pratensis L.	gombos ecsetpázsit	VEG	stem was children's snack	[78]
Anchusa officinalis L.	n.d.	VEG	eaten as salad, scalded with vinegar	[65]
Anthriscus cerefolium (L.) Hoffm. subsp. *trichosperma* (Spr.) Arc. (*Chaerophyllum trichosporum*) & *A. sylvestris* (L.) Hoffm. (*Ch. sylvestre*)	zamatos turbolya, turbolya	VEG	green spice for soups and vegetables; rarely mentioned in folk literature, but in a cookbook from the end of the 19th century *A. cerefolium* was mentioned as soup and spice	[65,78,86]
Amaranthus spp.	paraj	VEG	whole plant was consumed	[78]
Arctium lappa L.	bojtorján, keserűtorzsa, burusztujlapi, keserűtorzsa lapi, büdös lapi, parti lapi, büdös levél, parti fű, lapu, bogáncs	SUB	peeled and eaten	[77,78]
Armoracia macrocarpa (W. et K.) Baumg.	torma (debreceni torma)	VEG, SUB	VEG: soup, vegetable; SUB: spice for pickles	[14,59,62]
Armoracia rusticana G. Gaertnn. B. Mey. et Schreb (*A. lapathifolia* Usteri)	torma, tormalapu	VEG, SUB	SUB: side dish for meat, VEG: leaves: soup, sauce; meaty stuffing filled into them	[9,23,36,42,52, 53,62,76]
Arrhenatherum elatius (L.) J. et C. Presl.	fű	VEG	stem was children's snack	[78]
Artemisia absinthium L.†	fehér üröm, fejér üröm	VEG	spice for brandy (palinka) and wine	[30,62,77]
Asperula odorata L.*	szagos müge	VEG	spice for liqueur	[30]
Atriplex patula L.	vad laboda, sós paréj, fodros paréj	VEG	soup	[77]
Atriplex tatarica L.	fehér laboda	VEG	raw: salad, cooked: soup, former herdsmen prisoners of war put it into hot soup in Siberia	[11,71]
Berberis vulgaris L.	fajisóska, nyúlsom, sóska, sóskaborbolya	VEG, FRU	VEG, FRU: snack; FRU: substitute for vinegar	[36,67,78]
Betula pendula Roth (*B. verrucosa* Roth)	nyír, nyírfa, májfa	SAP, VEG	SAP: fresh beverage; wine and vinegar were fermented from it; herdsmen of Transylvania and Zemplén used it for inoculation of milk. VEG: substitute for tea	SAP: [9,22–24, 29,30,38,41–44, 48,50,52,53, 56,58,63,65,66, 69,72,74]; VEG: [7]
Bolboschoenus maritimus (L.) Palla	csatak, zsiók, zsiku, zsióka	SUB	food in famine and for herdsmen	[14,16]

Tab. 1 *(continued)*

Scientific name	Local names mentioned	Parts used	Mode of use	Reference No.
Brassica nigra (L.) Koch (*Sinapis nigra* L.)	franciamustár	FRU	oil was pressed from it	[4]
Bromus sterilis L	vadzab	VEG	stem was children's snack	[78]
Bunias orientalis L.	borsoslenke, borsos lenkő, rákányéc, szümcső	VEG	soup and salad in spring	[9,74,75,78]
Butomus umbellatus L.	alacs, elecs, elecske	SUB	food in famine times and for herdsmen	[14,16,21]
Campanula persicifolia L.	kupa, tőcsérvirág, vadcsengő	FLO	snack for children	[9]
Capsella bursa-pastoris (L.) Medik.	pásztortáska	VEG, FLO	young flowering shoots were snacks for children	[74,78]
Cardamine amara L.	vízitorma	VEG	snack in early spring	[70]
Cardamine pratensis L.	n.d.	VEG	n.d.	[73]
Carlina acaulis L.	bábakalács, bábakonty, kontybába, kenyérvirág	FLO	inner part of inflorescence was a raw snack	[9,19,74,75]
Carpinus betulus L.	gyertyán	VEG, SAP	SAP: fresh beverage. VEG: leaves as a spring snack	SAP: [9,23,39]; VEG: [22]
Carex elata ALL. & *Carex* spp.	sás, limbus	VEG	leaves and stems were children's snack	[78]
Carum carvi L.	keménmag, kemény, kömény, kömén, kümén, köminy, kömin, kömind, köménd,	FRU	spice for rye bread, soups and roasted meat; brandy (palinka) with honey and cumin was a traditional drink in Transylvania	[7,9,19,36,37, 41,52,67,70,71, 74–78]
Castanea sativa Mill.	szelidgesztenye, geszkenye, geszkönye	FRU	roasted or cooked for sweets and cakes	[23,28–30,36, 37,78]
Celtis occidentalis L.	ostorfa, zsidómeggy, zsidócseresznye, madárbogyó, gelegenye	FRU	children's snack	[78]
Centaurium erythraea Rafn.	cintória, ezerfű	VEG	raw as an appetizer; spice for brandy (palinka); tea	[26,30,57,74]
Cerasus avium (L.) Mönch (*Prunus avium*)	cseresznye, vadcseresznye, vadcserösznye	VEG, FRU, SAPs	VEG: leaf was spice for pickles. FRU: eaten raw and dried; brandy (palinka) was distilled from it; SAPs: gum chewing	[19,22,23,26, 27,29,37,38,40, 47,52,57,68,70, 71,74,75,78]
Cerasus vulgaris Mill. subsp. *acida* (Dumort.) Dostal (*Prunus cerasus* L.)	vadmeggy	FRU	raw snack	[52,78]
Chaerophyllum bulbosum L.	baraboj, bubályka, bobályka, bubolyicska, buboicska, bóbiska, mogyorófű, turbolya, csemegebürök, trombujka, mogyorós baraboly, baraboi, fődibarabój, mihályka, mihálka, Mihályka monya	SUB	eaten raw by children, also by adults, like carrot	[9,14,42,53,58, 59,62,65,67, 70,75]
Chenopodium album L.	laboda, cigara	VEG	salads and vegetable	[19,65,78]
Cichorium intybus L.	katáng, vad cikória, katángkóró	SUB	grated and roasted as a coffee substitute; used even today in Transylvania	[14,74,78]
Cirsium arvense (L.) Scop.	tüvis, tövis	VEG	young spring shoots gathered for soup; peeled stems eaten raw	[9,78]
Cirsium canum (L.) All.	bojhos tövis, ökörlevél, ökörnyelv	VEG	young leaves cooked for soup	[9,73]
Cirsium oleraceum (L.) Scop.	káposztás acat	VEG	young leaves were eaten	[21]
Convolvulus arvensis L.	szulák, győtény	FRU, VEG	FRU: seeds were famine food; VEG: children's snack	[16,21,78]
Cornus mas L.	som, sum	FRU	sour jam eaten with meat; syrup; dried; brandy (palinka) distilled from it	[19,26,29,30, 36–42,46–52, 57,62,65,70, 72,78]
Cornus sanguinea L.†	somfa	FRU	jam, syrup and brandy was made from it (even though some consider it poisonous)	[19]
Corylus avellana L.	magyarófa, mogyorófa, mogyoró, fáin magyaró, fájimogyoró, fájimagyaró, monyaru, monyaró, magyaru	FRU, VEG	FRU: baked into cakes, snack; VEG: buds: ground for flour substitute in famine	[19,22,26,27,30, 36–38,42,49–53, 65,67,69,72, 75,78]
Crambe tataria Sebeók	tátorján	SUB	famine and herdsmen's food	[4,14,16,21,55]

Tab. 1 *(continued)*

Scientific name	Local names mentioned	Parts used	Mode of use	Reference No.
Crataegus monogyna Jacq., *Crataegus oxyacantha* L. and related taxa	Istengyümölcs, Istengyümölcsfa, Istengyümöcse, édeskés galagonya, galaginya, gelegenye, gyümölcsény, Jézus Krisztus tövise	FRU, VEG	fruits and sometimes leaves were raw snack; in few places cooked into jam; dried flour substitute in famine	*C. monogyna*: [14,19,22,25, 27,34,35,39,51, 59,62,77,78]; *C. oxyacantha*: [22,30,36–38]; C. sp: [29,40,46, 50,76]
Crataegus nigra W. et K.	savanykás galagonya	FRU	raw snack	[35]
Crocus banaticus L. Gay (*Crociris iridiflorus* Heuff.) & *Crocus balcanicus* Janka	sáfrány & balkáni sáfrány	SUB	spread by wandering herdsmen; bulbs can be eaten	[16]
Crocus variegatus Hoppe & Hornsch	vadsáfrány	FLO	spice; flowers were collected and sold	[16]
Dactylis glomerata L.	ebír	VEG	stem was children's snack	[78]
Daphne mezereum L.†	farkasboroszlán	FRU	unripe fruits were used as a pepper substitute after cooking	[4]
Daucus carota L. subsp. *carota*	murok	SUB	added to soups and vegetables	[74]
Echium vulgare L.	édesfű	FLO	flowers (perhaps fruits also) eaten by children	[36]
Elaeagnus angustifolia L	olajfa, olajbogyó	FRU	eaten by children	[60,78]
Elymus repens (L.) Gould.	tarackbúza	VEG, SUB	VEG: raw stem as a snack; SUB: famine food	[19,78]
Equisetum arvense L.	mezei zsurló	FRU	fertile young shoots were eaten by Gypsies	[87]
Eryngium campestre L.	csipke, bikacsöke, bikacsipke, macskatövis, széhajtitövis	VEG	young shoots cooked for soup; stems and fresh leaves were eaten fresh, as salad	[18,77]; as *Xanthium spinosum*: [59,62]
Fagus sylvatica L.	bükk, bikfa, májusfa	FRU, VEG, SAP	FRU: raw and roasted for snack; for confectionery as a walnut substitute; coffee substitute; from seeds oil was pressed; famine food: ground seeds mixed into flour (even in 1957); VEG: slightly acidic leaves eaten raw by children in the spring, sometimes with sugar; SAP: children tapped and drank	FRU: [19,22,23, 33,36–39,42,50, 53,57,65,68,69, 72,74,75,78]; VEG: [19,36]; SAP: [9]
Fallopia baldschuanica (Regel) Holub	sóskafa, tőtike	VEG	leaves as a vegetable with a meaty stuffing filled into them; leaves were used as a children's snack	[78]; B Czúcz pers. comm.
Filipendula vulgaris L.	koloncos legyezőfű, bányavirág	SUB	sap pressed from roots were eaten; famine food	[9,16,21]
Fragaria moschata Duch.	piroseper, berkeeper, eper	FRU	raw and conserved as jam	[75,78]
Fragaria sp.	szamóca, bakkeper, fődi eper, földi eper, szimóca, SK: jahoda	FRU	raw and conserved as jam	[22,25,26,29,39, 42,47,50,52,62, 68,76,78]
Fragaria vesca L.	fődi eper, erdei eper, eper, vereseper, bekeeper, szamóca, vad eper, piroseper, berekeper, madár-epörgye, lányeper, bagóeper, erdei szamóca,	FRU, VEG	raw on site; jam, syrup and put into brandy (palinka)	[19,27,37,40,49, 52,57,72,74,75, 77,78]
Fragaria viridis Duch.	szamóca, csattogó eper, tokoseper, csattogó-epörgye, lányeper, bagóeper, csattogó, vadeper, réti szamóca	FRU	raw, jam, fruit and liqueur	[27,36,37,40,45, 53,78]
Galeobdolon luteum Huds. (syn. *Lamium galeobdolon* (L.) Ehrend & Polatschek.)	árvacsalyán, árvacsanál, árvalánycsanál, árvacsihán, szelidcsanál, szopóka	FLO	flowers were used as a children's snack	[19]
Galium verum L.	tejoltó galaj	VEG	used for milk inoculation	[66]
Gentiana cruciata L. & *Gentiana lutea* L.	epeburján, epefű, horecska, kösörűgyökér,	SUB	put into brandy (palinka), appetizer	[19]
Glechoma hederacea L.	katonapetrezselyem, vadpetrezselyem	VEG	green spice: parsley substitute	[19,41]

Tab. 1 *(continued)*

Scientific name	Local names mentioned	Parts used	Mode of use	Reference No.
Gleditsia triacanthos L.	koronatüsökfa, kruskuli, glédicstüsök, vad szentjánoskenyér, koronaakác, kreditsia, gleditse, édeske, szejjánoskönyérfa, ledics, lepényfa, lackószar, zsidótakony, glédicstüsök	FRU	children eat the spongy part of the fruit in autumn, or suck its sap	[11,26,27,29,36, 38,59,62,63,78]
Glyceria maxima (Hartm.) Holm. & *Glyceria fluitans* (L.) R. Br.	harmatkása	FRU	flour substitute in famine	[16,20,28,60,62]
Glycyrrhiza echinata L. & Glycyrrhiza glabra L. & Glycyrrhiza sp.	édesgyökér, idesgyökér	SUB	sweetener; children's snack; pressed juice drunk	*G. echinata*: [11, 22,28,59,62,78]; *G. glabra*: [14, 16,28,57,59,78]; *G.* sp.: [5,60,62]
Helianthus tuberosus L.	csicsóka, tótrépa, taknyos pityóka, picsóka, cicoski, mikóka, árvapityóka, árvapijó, árpapityóka, csókapityóka, csókapicsóka, disznópityóka, édespityóka, fingópityóka, fingóspityóka, picsócsa, pityójka	SUB, FLO	stands escaped from cultivation were gathered; SUB: pickled for winter; children and adult eat it raw, roasted or cooked, FLO: nectar was sucked out of flowers	[9,43,46,50,52, 57,58,62,63,71, 74,77,78]; FLO: [52]
Heracleum sphondylium L.	bojtorján	VEG	sour soup and refreshing drink was made from the leaves	[16]
Hippophaë rhamnoides L. subsp. *carpatica* Rousi	homoktövis	FRU	whole fruit was conserved in honey in Transylvania; syrup, jam	[74]
Hordeum murinum L.	vadárpa, fű, ragcsos fű, kalász	VEG	stem, spikes were children's snack	[78]
Humulus lupulus L.	komló, vadkomló	VEG, FLO	FLO: put into bread and beer sourdough (widespread); VEG: shoots fried (mainly in Transylvania); soup and vegetable like French beans	[7,9,19,27,30,36, 38,57,59,67,71, 75,77,78,86]
Juglans regia L.	dió, vad dió, dijófa	FRU, SAP	FRU: into sweets, confectionery; green husks cooked with sugar to make syrup or tea; from unripe, soft fruits pickles, jam, sweets and brandy (palinka) was made. SAP: children tapped; fresh drink	FRU: seeds in many sources; soft fruit: [68]; green husk: [19]; SAP: [9]
Juniperus communis L.	borsika, borsukafenyő, borókafenyő, bucsfenyő, sillő, süllő, borosán, borosánfenyő	FRU	spice for brandy (palinka), sour cabbage, ham marinade and pepper substitute for meat dishes; used for meat smoking: it gives a pleasant flavour to meat	[19,22,27,36–38, 50,52,68,69, 74,75]; meat smoking: [27]
Koelrauteria paniculata Laxm.	n.d.	FRU	children's snack	[78]
Lamium album L. & *L. maculatum* (L.) L. & *L. purpureum* L.	árvacsalán, árvacsalyán, árvacsanál, árvalánycsanál, árvacsihán, szelidcsanál, szopóka	FLO	children's snack	[19,70,78]
Larix decidua Mill.	fenyő	SAPs	gum was chewed for cleaning teeth	[72,78]
Lathyrus aphaca L.†	csiriborsó, csicsiriborsó, feketeborsó, fényesborsó, finyesborsó, vadborsó	FRU	nice shiny seeds were used for decorating confectionery	[9]
Lathyrus tuberosus L.	borsó viloja, borsó viola, vadborsó, csunya, csuma, csunyavirág, julisztavirág, jurisztavirág, zsírgaz, földimogyoró, fődimagyaró, földimagyaru, földi zsír, kutyulló fije, borsój, borsó, borsóvirág, dobra, csicsiriborsó, csicseriborsó, barabój, jenyestye, unalomvirág, vadborsóvirág	SUB, FRU	SUB: children and adults ate it almost everywhere; at ploughing it was ploughed out or was dug out by pigs; roasted at open fire (delicious); it has disappeared when deep ploughing came into practice and because of herbicides; FRU: seeds were eaten less often	[9,11,14,22,26, 27,35,49,52,59, 60,62,65,67,74, 75,77,78]
Lemna sp.*	fulencse	VEG	eaten by poor people	[33]
Lepidium perfoliatum L. & *L. ruderale* L.	borsika, cigánypaprika	VEG	eaten by Gypsies instead of horseradish and hot peppers	[11]
Lilium martagon L.	n.d.	SUB	bulbs were put into wine to make vinegar; herdsmen›s children ate it dried or was ground into flour	[4,65]
Linaria vulgaris Mill.*	sarkantyúvirág	FLO	flowers were snack	[59]

Tab. 1 (continued)

Scientific name	Local names mentioned	Parts used	Mode of use	Reference No.
Lolium perenne L.	fű	VEG	stem was children's snack	[78]
Lotus corniculatus and/or *L. glaber*	mogyoró, földimogyoró	SUB	roots (nodules) were children's snack	[78]
Lycium barbarum L.†	licium, kutyafa	FRU, FLO	eaten by children, but in some places forbidden because it was known as poisonous; FLO: pistil from the flowers were children's snack	[19,62,78]
Malus sylvestris (L.) Mill.	vadalma, vadóma, vadóka, madárbogyó, SK: plánka	FRU	generally everywhere vinegar was made from it; it was post-ripened, rarely eaten raw; it was dried; compote, wine, brandy (palinka) and "cibere" soup was made of it; as a spice it was put into sour cabbage	[19,22,23,26,27, 29,36,38–40, 42, 43,47–53,57–59, 62,63,67–69, 72,74,76,78]
Malva neglecta Wallr. & *Malva sylvestris* L. & *Malva* sp.	kalácska virág, papsajt, papsajtmályva, papsajtja, taknyozófű, kerekmályva, papkalács, papkenyérke, vadmályva, papsajtlevél	FRU	immature fruit was and is eaten raw by children	[9,11,14,21,22, 30,42,50,51,59, 60,62,70,75, 77,78]
Matricaria recutita L. (*M. chamomilla* L.)	libavirág, szíkfűvirág, szíkfű, kamilla	FLO	refreshing tea and medicinal infusion; flowers were children's snacks	[29,43,57,78]
Medicago sativa L.	lucwrna, lucerna	VEG	soup was made from the young shoots	[77]
Melissa officinalis L.	citromfű, citromszagú méhfű	VEG	modes of use not documented; today used as a tea and a spice, the plant occurs spontaneously	[9]
Melittis melissophyllum L.*	mecsekháti tea, mecseki-téja,	VEG	"most popular herb tea"; gathered basketful for sale; it was made into candies	[27,29]
Mentha sp., *M. aquatica* L., *M. arvensis* L., *M. spicata* cv. *crispa*, *M. longifolia* (L.) Nath., *M. pulegium* L.	balzsamka, vad fodormenta, fodorminta, menta, vízimenta, szagos menta	VEG	spice for brandy (palinka); tea	[7,11,19,43,67, 70,74,78]
Morus sp., *M. alba* L. (& *M. nigra* L.? = *M. alba* f. *nigra*)	epörgyefa, epörgye, szederfa, faszeder, eperfa, fekete eper, fehér eper, eperfa, eper	FRU	eaten raw by travellers; brandy (palinka), wine, jam, syrup made of it, without sugar	[19,26,27,29,30, 63,78]
Muscari botryoides (L.) Mill.	Szent-györgy virág	VEG, FLO	whole plant was a children's snack	[78]
Onopordum acanthium L.	bogáncs	VEG	leaves were a children's snack	[78]
Origanum vulgare L.	ezerjófű, lebetka, lebetkevirág, szúrfű, szurokfű	VEG, FLO	tea; appetizer	[19,73,74]
Oxalis acetosella L.	medvesósdi, madársósdi, medvesóska, erdei sósdi, bikksóska, kereksósnya, nyúlsaláta, nyúlsóska, nyúlsósnya, nyúlsózsnya	VEG, FLO	pressed juice was used like vinegar; chewed; raw leaves were children's snack; milk rennet	[9,16,19,38,70, 74,75,78]
Oxalis corniculata L. (*O. europaea* Jord.)	galambsóska, sárga madársóska, kakukksóska, madársóska	VEG	raw snack	[19]
Padus avium Mill.	zelnica, szelence, vadszelence	FRU	eaten by children; put into brandy (palinka)	[22,67,74]
Papaver rhoeas L.	pipacs	FLO	petals were a children's snack	[78]
Parthenocissus tricuspidata (S. et Z.) Planch.*†	vadszőllő	VEG, FRU	fruits and thin shoots were put into pickled cucumbers (even though fruits are considered toxic)	[27]
Pastinaca sativa L.	peszternák, vad pasztinák	SUB	added to soups, children's snack	[74,78]
Phragmites australis (Cav.) Steud.	nád	SUB, VEG	SUB: cooked rhizomes are herdsmen's and famine food; VEG: salad made from fresh shoots, inner part of the shoots was spring-summer snack	[11,14,23,28,78]
Phytolacca americana L.†	alkörmös	FRU	jam; food colouring	[19]
Picea abies (L.) Karsten	lucsika, vörösfenyő, veresfenyő, szëmërke szëmërcefenyő, havasifenyő, csetenyefa, lukszfenyő, parasztfenyő	VEG, FLO, SAPs	VEG, FLO: jam, syrup and preserve made from buds and juvenile cones; SAPs: gums were teeth cleaner, chewing gum substitute	[19,67,70,74, 75,78]
Pimpinella anisum L.	ánizs	FRU	spice for brandy (palinka) and confectionery	[77]
Pinus cembra L.	cirbolya	VEG, FLO	use not documented; presumably like *P. sylvestris*. Seeds: children snack	[16,78]

Tab. 1 *(continued)*

Scientific name	Local names mentioned	Parts used	Mode of use	Reference No.
Pinus sylvestris L.	lucfenyő, lucsfenyő, csëmëtefa, csetnye, lúcfenyő, lukszfenyő, veresfenyő, vörösfenyő, répafenyő	VEG, FLO	syrup was made from young buds and cones	[9,19,50,67, 69,78]
Plantago lanceolata L.	úti lapi, keskenylevelű útilapi	VEG	preservative was cooked from it; snack	[74,78]
Plantago major L. & *Plantago media* L.	úrfilapi, útilapi, útifű	VEG	raw plant put into butter against anaemia	[67]
Poa angustifolia L. & *Poa pratensis* L. & *Poa* spp.	fű	VEG	stems were used as a children's snack (chew for the sweet sap)	[78]
Polygonum lapathifolium L. & *P. aviculare* L.	savanyú keserűfű, lapulevelű keserűfű	VEG	slightly sour leaves were used as snacks	*P. lapathifolium*: [9,22,57,59]; *P. aviculare*: [78]
Polypodium vulgare L.	kőméz, édesgyükér	SUB	sweetener fresh and dried; snack; lapped into a rag it was given to babies to chew at teething; "sweet water" was made by soaking ground roots in water	[38,42,49,50,52, 67,70,74,75,78]
Portulaca oleracea L.	porcsin	VEG, FRU	eaten; leaves were children's snack; seeds were famine food	[16,78]
Potentilla anserina L.	földi mogyoró, libapimpó	VEG	young leaves were eaten	[57,58,62]
Primula veris L.	kukukvirág	FLO	flowers were cooked into a sweet syrup	[75]
Primula vulgaris Huds.	zsibavirág	VEG	whole plant was consumed	[78]
Prunus sp. & *P. cerasifera* Ehrh. & *P. domestica* L. subsp. *insititia*	fosóka szilva, fosóka, kökényszilva, korkodus, márabora, ringló, vadszilva, macskaszemű szilva	FRU SAPs	brandy (palinka), preserves, syrup, jam was made from the fruit; also dried; SAPs: gum chewing	[50,68,77,78]
Prunus spinosa L. & *P. spinosa* L. subsp. *fruticans* (Weihe) R. et Cam.	kökin, kökényszilva, kükény, kükényszilva, kökörcsönszilva, kükürcsönszilva, kökén, kükén. kükénfa, kökönye, porumbar	FRU	eaten mainly raw "after frost-bitten"; wine, soaked "kökényvíz" (blackthorn water); brandy (palinka) was made of it; also dried	[5,9,11,19,22,23, 25,29,30,36–40, 46–48,50–53,57, 59,60,62,63,65, 70,72,74,76–78]
Pulmonaria mollis Wulf & *Pulmonaria officinalis* L.	dungóvirág, dungófű, tüdőfű	FLO	children suck the nectar out of the flowers	[67,70]
Pyrus pyraster Burgsd. (*Pyrus achras* Gaertn.)	vadkörte, vackor, erdei vackor, vadvackor, malina, SK: cernice, dicka	FRU	harvested raw, ripened in the attic; stewed fruit, wine ("csügör"), vinegar, brandy (palinka) made of it; dried; soup was made from the dried fruit ("cibereleves")	[11,19,22,23, 25–27,29,36, 38–40,42,43, 46–53,57–59, 62,63,65,68,69, 72,74,76–78]
Quercus cerris L.	cserelfa, cserfa, cser	VEG, SAP, FRU	VEG: leaves were put into barrelled cucumber as a spice; SAP: herdsmen tapped it from spring to late autumn, "brown beer" for them; FRU: coffee substitute after leaching and roasting, famine food	SAP: [24,27,28, 38,39,44]; VEG: [19,77]; FRU: [16]
Quercus pubescens Willd.	magyal	FRU, VEG	FRU: herdsmen roasted the acorns on ember; VEG: leaves were put into barrelled cucumber as a spice	[14,16,77]
Quercus robur L. & *Q. rubra* L. & *Q. petraea* (Matt.) Liebl. & *Quercus* sp.	cserefa, tölgy, tölgyfa, töfa, tőfa, töljfa csepefa	VEG, FRU	VEG: leaves were put into barrelled cucumber as a spice; bark: famine flour substitute FRU: coffee substitute, children's snack and famine flour substitute: boiled acorns lost their bitter taste, after boiling it was dried, ground and mixed into flour	VEG: [77]; FRU: [19,23,27,33,35, 37,48,77,78]; Bark: [14]
Ranunculus ficaria L.	györgysaláta, papsaláta, szentgyörgybúza, madársaláta, galambsaláta, vadsaláta, búzasaláta, bükki saláta, erdejisaláta, harangversengő, kereksajáta, kakuksaláta, kukuksaláta, kukuksajáta, mezejisaláta, nyúlsaláta, nyúlsajáta, salátavirág, szaronkőtt saláta, vadsajáta, pipirisaláta	VEG	it was consumed mainly raw, with a sauce as a spring salad; roasted in speck lard with garlic; cooked for soup and vegetable (tradition survived in Transylvania); in Hungary it re-appears in markets, gathered by Gypsies	[9,27,31,35,58, 62,65,70,71,74, 77,78]
Ribes alpinum L.	leánykafüge, vad ribizli	FRU	eaten raw, wine was made from it	[70,75]

Tab. 1 *(continued)*

Scientific name	Local names mentioned	Parts used	Mode of use	Reference No.
Ribes aureum Pursch	aranyribizli	FRU	snack	[78]
Ribes nigrum L.	fekete ribizli, fekete ribizli, fekete szöllő, fekete vërësszőlő	FRU	eaten raw (also in gardens)	[9,50,74]
Ribes petraeum Wulf.	borfüge	FRU	eaten raw	[75]
Ribes rubrum L.	piros ribizli, vad ribizli	FRU	eaten raw (also in gardens)	[38,74]
Ribes uva-crispa L.	egres, egris, füge, szőrös füge, vad egres, agris, piszke, büszke	FRU	eaten raw; gathered for preserves and soup	[9,38,70,74, 75,78]
Robinia pseudacacia L.	akác, fehér akác, agacsi, ágác, mézvirág, kukucka, agáca	FLO, VEG	FLO: whole flower or just the nectar was children's snack; in many places it was fried in pancake dough; tea; less often wine was made of it; sometimes added to bread sourdough. VEG: young leaves were eaten by children	FLO: [19,23,27, 29,31,36,38,50, 59,60,70,74,77, 78]; to bread: [17]; VEG: [59]
Rosa canina L. agg. & *Rosa corymbifera* Borkh. (*Rosa dumetorum* Thuill.) & *Rosa* sp.	csipkerózsa, hecseli, hecsedli, hecserli, rózsabogyó, pecs, bucske, bücske, bütyke, seggvakaró, seggvakarcs, istengyümőcs, vadrózsa, csipkefa, szaragógya, szagrógya, bicskerózsa, csitkenye, csitke, csipka, hecse, hecse-pecse, hecsebokor, tüvisfa hecsempecs, csicskenye, csipkebokor	FRU, FLO	FRU: syrup, jam, preserves, stewed fruit, soaked drink; with yeast it was made into wine; "cibere" soup, enriched with bread; dried: tea; ground: flour substitute in famine; FLO: refreshing drink, vinegar and preserves were made from petals	*R. canina*: [9,11, 19,22,25,26,29, 30,36–40,42,43, 45–50,52,53,57, 60,65,68,71, 74–78]; *R. corymbifera*: [19, 22]; *Rosa* sp.: [22,59,62,74]
Rosa gallica L.	selyemrózsa, fátyolrózsa, rózsa, csicskenye nagylevelű csipkerózsa	FRU, FLO, VEG	FLO: syrup and jam were made from the petals with sugar and citric acid. VEG: tea from leaves. FRU: tea, jam, syrup.	[19,22,23,74,76]
Rubus caesius L. & *Rubus* sp. (also as *Rubus tomentosus* Borkh)	szeder, seder, fődi szeder, erdei szeder, fődi szödörnye, szödör, csuszkor, promber, futó szëdër, szëdër, földi szeder, vad szeder, fekete szeder, szederincs, szederéncs, szedernye, szeder, vadszeder, tüskeszeder szödörnye, kódis szeder, víziszeder, SK: carnica	FRU, VEG	FRU: eaten raw in the past and also today; jam, preserves, wine, brandy (palinka), syrup, thick jam ("dulcsesz") was made of it. VEG: tea from leaves	*R. caesius*: [9,11, 19,22,26,27,30, 35,36,45,47,49, 52,72,74,76–78]; *Rubus* sp.: [22,25,29,30,36, 38,42,43,48,52, 57–60,62,68]
Rubus fruticosus agg.	szeder, fekete szeder, fás szödörnye, tüskeszeder, lószamóca	FRU	FRU: eaten raw in the past and also today; wine, brandy (palinka) was made of it; sold in markets	[19,22,27,38,42, 65,75,77,78]
Rubus idaeus L.	málna, mána, malina, mánafa, málnafa	FRU, VEG	FRU: eaten raw; jam, syrup, wine, brandy (palinka) was made of it; also put into brandy (palinka); fruits were conserved in syrup. VEG: tea was made from leaves.	[19,23,38,40,42, 43,45,47–52,65, 67,70,72, 74–76,78]
Rumex acetosa L. & *Rumex* sp.	sóska, réti sóska, vadsóska, sóslorjum, sóslórum, vadsóska, papsaláta, sósdi, tavaszi sóska, sóski, sósnya, vadsósnya, lósóska	VEG	mentioned by almost all sources. It was an important spring vegetable everywhere; field snack, eaten raw mainly by children; cooked for soup and sauce; also used for inoculation of milk	*R. acetosa*: [9,19, 22,23,27,35,36, 38,39,42,43,46, 49–53,57–59,62, 65–67,70,77,78]; *Rumex* sp.: [33, 47,60,71,78]
Rumex acetosella L.	madársóska	VEG	mainly snack; sometimes eaten after being boiled in salty water; also used for inoculation of milk	[43,49,50,52,65, 66,78]
Rumex patientia L.	lósóska, lósósdi	VEG	cooked for sauce and soup	[5,31,43,76,78]
Rumex pseudonatronatus Borb.*	lósóska	VEG	stuffed with meat and rice like cabbage	[59]
Sagittaria sagittifolia L.	alacs, elecs, elecske	SUB	famine food	[14,16]
Salvia pratensis L. & *S. nemorosa* L.	bárányláb & vadorgona	FLO	flowers and nectar were snacks	[75,78]

Tab. 1 *(continued)*

Scientific name	Local names mentioned	Parts used	Mode of use	Reference No.
Sambucus ebulus L.†	fődi bodza, földi bodza, gyalogbodza, gyalog bédza, fődi-boza, borzang, fekete borzag, csete	FRU FLO	FRU: jam and syrup was cooked, brandy (palinka) was distilled from it; jam was made mixed with pumpkins. It was not eaten raw, and was used less often than *S. nigra*. FLO: tea.	FRU: [19,26,27, 30,37–39,45,69, 77]; FLO: [77] CZB
Sambucus nigra L. & *Sambucus* sp.	fabodza, borza, bodzavirág, gyalogbodza, bodza, fehér bozza, bozda, bozzafa, bojzafa, fekete bojza, bodzafa, bozdafa, borzag, borza, borzafa	FLO, FRU	FRU: jam was cooked; brandy (palinka) was distilled from it; used for colouring wine. FLO: tea, refreshing drink, syrup, "elder champagne"; fried in pancake dough: "elder doughnut"	[5,9,19,27,30,37, 42,43,49,50,52, 57,59,60,62,67, 69,74,75,78]
Sambucus racemosa L.*	n.d.	FRU	red fruits were eaten	[37,69]
Satureja alpina (L.) Scheele (*Thymus alpinus* L.)	n.d.	VEG	mountain herdsmen used it as a spice	[4]
Scorzonera hispanica L.	fekete gyökér	SUB	eaten	[36,37]
Scorzonera purpurea L. subsp. *rosea* (*Scorzonera rosea* W. et K.)	bakceka	FLO	children's snack	[74]
Setaria spp.	muhar	VEG	stem was used as a children's snack (chewed for the sweet sap)	[78]
Sinapis arvensis L.	repce, rabcsont, rebcsont	VEG, FRU	FRU: oil pressed; VEG: soup was made from young spring shoots	[4,74 76,77]
Sisymbrium altissimum L. (as *Sisymbrium pannonicum* Jacq.)	magas zsombor	FRU	oil pressed	[4]
Smyrnium perfoliatum L.	őzsaláta	VEG	salad was made from raw leaves	[19]
Solanum nigrum L.†	fekete szöllő, fekete vad szöllő, káposztaszöllő	FRU	sweet fruits were eaten by children (it is considered poisonous, but the source notes: "1–2 pieces do not make any harm")	[77]
Solidago gigantea Ait.	vadkender	VEG	children's snack	[78]
Sonchus arvensis L.	n.d.	VEG	n.d.	[65]
Sorbus aucuparia L.	belekenyér, istenkenyere	FRU	fruits eaten	[19,69]
Sorbus domestica L. & *Sorbus* sp.	berkenye, berkenyi	FRU	dried	[22,28,36,37]
Sorbus torminalis (L.) Cr.	barkóca, vadbarkóca	FRU	sweet, pulpy fruits were eaten by children, herdsmen	[37–40]
Staphylea pinnata L.*	klokocs,	FRU	eaten by children	[38]
Stellaria media (L.) Vill.	galambbegy, tyúkhúr, korpafű	VEG	eaten as salad and vegetable	[11,78]
Streptopus amplexifolius (L.) DC	nyúleper	FRU	fruits ripen in hay by winter, eaten raw	[75]
Symphytum officinale L.†	feketenadály, fekete nadálytő, dongóvirág	VEG, FLO	VEG: young leaves were eaten fried in harsh times; FLO: children's snack	FLO: [9]; VEG: [14], SZGY
Taraxacum officinale Weber & *T. laevigatum* (Willd.) DC.	öregapám pogácsája, láncvirág, lánclapi, cikória, tyúkvirág, pitypang, pipevirág, marcivirág, láncfű, kákics, kutyavirág, pimpó, békavirág, bíkavirág, cikornya, csikāra, hóttok virága, kotlóvirág, láncoslapu, nyúlsaláta, pipefű, pipevirág, tejesbúrjány, tejesbúrján, tejesfű, tejefű, fűgörhe, tejesgaz, tejes vadsaláta, tyúksegge, csorbóka, libuskavirág	VEG, FLO	VEG: young leaves were eaten as salad in spring; FLO: syrup was cooked from soaked flowers; or "cikoria honey" was made from flowers with thick sugar syrup; in Gömör "pimpó wine" was fermented from the flowers; in some places it was regarded as poisonous	[9,19,22,28,50, 59,63,65,70,74, 75,78]; *T. laevigatum*: [59,62]
Taxus baccata L.†	tiszafa	FRU	arils were eaten as children's snack	[78]
Thymus serpyllum L. & *Thymus* sp.	csombor, kakukkbora, kakukkfüj, vadborsfüj, vadcsombor	VEG	spice	[19,30,60,62,78]
Tilia platyphyllos Scop. & *Tilia cordata* Mill. & *Tilia* sp.	hársfa	FLO	tea; condiment for brandy (palinka) and wine	[19,27,29,38,40, 43,61,78]

Tab. 1 *(continued)*

Scientific name	Local names mentioned	Parts used	Mode of use	Reference No.
Tragopogon orientalis L. (as *Tragopogon pratensis* L.)	bakceka, bakszaka, baszkata, csuka, édesfű, idesfű, tefesfű, kukukté, szasza, szekeboka, tejesbúrján, tejesbúrjány	VEG	in the past young sweet stems were eaten peeled; in autumn curly leaves were chewed; in some places it was cooked in whey	[9,50,65,67,75]
Trapa natans L.	sulyom, suly, sójom	FRU	cooked and roasted; flour substitute in famine; till the 1930s it was sold in large quantities	[5,11,14,16,30, 33–35,58–60,62]
Trifolium pannonicum JACQ	nagy fehér vad here	FLO	syrup was cooked from the flowers; children's snack	[74]
Trifolium pratense L.	vörös lóhere, lucerna, istenke cipókája	FLO	children's snack	[16,35,78]
Trifolium repens L.	lóhere	VEG	leaves and stem were used as a children's snack	[78]
Tripleurospermum perforatum (Mérat) M. Lainz	kamilla	FLO	flowers were children's snacks	[78]
Tussilago farfara L.†	fodbájlapi, podbánlapi, podbállapi, martilapi, martilapu, martivirág, partilapu, pipevirág, ciberelapu, fehérhátu lapu, tejfölös lapu, pitypang, gyermekláncfű	VEG, FLO	VEG: stuffed with meat or mush; soup, vegetable, salad, in spring salad soup with bones (first cooking water was poured off, because it was bitter); FLO: flowers cooked into sugar syrup	[9,19,65,67, 71,74]
Typha latifolia L. & *Typha angustifolia* L.	csella, elecske, bengyele, pintér gyékény, nádi botikó	SUB, VEG	SUB: rhizomes were eaten raw or cooked by herdsmen; famine food; VEG: salad from young leaves	[4,14,28,30,34, 59,61,62,78]
Ulmus glabra Huds. ? = ? *Ulmus minor* Mill. & *Ulmus* sp. = *U. laevis* L.*	szilfa	SAP	fresh drink	[27,33]
Urtica sp., *Urtica dioica* L. & *Urtica urens* L.	csijány, csián, csiján, csalán, csihán, csallán, csollán, csohán, csonár, csajánt, csojánt, csoján, csojány, csohány, csípős csajánt, csípős csojánt, csípős csojány, csípős csollán, széleslevelü csoján, csípős csohán	VEG	soup, vegetable and salad was made from fresh shoots; weak children were fed with nettle; eggs with nettle: eastern dish; tea	[9,19,23,28,33, 36,48,50,52,53, 63,65,71,74–78], NI; *U. urens*: [52,53,65,77]
Vaccinium gaultherioides Bigelow	takonkokojza	FRU	eaten raw	[75]
Vaccinium myrtillus L.	feketekukujza, feketekokojza, kakóca, kakojza, fekete áfonya, fekete kukojsza, áfinya, áfinyála, háfonya, háfinya, feketemeggy, boronyica, brusnyica	FRU, VEG	FRU: raw; as jam, liqueur, syrup, preserves; preserved in rum; put into brandy (palinka); VEG: tea from leaves	[9,16,19,30, 48–52,70,72, 74–76,78]
Vaccinium vitis-idaea L.	piroskokojza, piroskukujza, fojminc, piros kukojsza, piros kakojza, havasi meggy, vörös áfonya, piros áfonya, ménisora, fásmeggy, botonyica, brusnyica	FRU	eaten raw, fresh and as pickles; preservation: pickled; sweet preserves; put into brandy (palinka) and wine; cooked into wine: glee; preserved in alcohol; dried; harvested semi-ripe and ripened in the sun	[19,48–52,70,74, 75,78]
Valerianella locusta (L.) Latterade	vadsaláta, nyúlsaláta, galambbegy saláta, madárka saláta	VEG	eaten in spring as salad, poured with sauce	[19,27,38,53,78]
Valerianella olitoria (L) Poll. or *V. dentata* (L.) Poll.	madársaláta, papsaláta, vadsaláta	VEG	eaten as salad with oil and vinegar	[22]
Verbena officinalis L.	vasfű, vasfűj, vasfi, szaporagaz	VEG	put into pickled cucumbers; spice and preservative	[9,38,67]
Viburnum lantana L.† & *Viburnum* sp.	nyomittó, nyomtató, ostorménfa, barátsza, barátszar, farkascseresnye, gusfa, bangita, korbácsnyélfa, ostórminya, ostornyélfa, szentfa, Szent-Ilona szőlő, szereputyka	FRU	fruits and pressed juice; snack for children	[9,19,26,67,74]
Viburnum opulus L.†	kányafa, kánya, veres kánya, gána, kálenka, kalina, kalinafa, kalinka	FRU	raw when frost-bitten but bitter: jam and syrup was made of it	[19,74]
Vicia spp.	csicseriborsó, vadborsó	FRU	cakes were decorated with its shiny seeds	[9]
Vitis sylvestris C. C. Gmel. (*V. vinifera* L. subsp. *sylvestris*)	vadszőlő	FRU	children's snack; vinegar and brandy (palinka) was fermented from it	[22,27,68,78]

Tab. 1 *(continued)*

Further poisonous species, tasted and snacked by some Hungarian botanists in their childhood include *Chelidonium majus* L, VEG *Conium maculatum* L. (VEG: piece of internode), *Hedera helix* L. (FRU), *Ligustrum vulgare* L. (FRU), *Parthenocissus quinquefolia* (L.) Planch (FRU), *Salix alba* L. (VEG: leaves), *Syringa vulgaris* L. (VEG: branch).

FLO – flowers, inflorescences, sometimes only petals or nectar; FRU – fruits or seeds; n.d. – no data about local names in the sources; SAP – liquid sap of trees; SAPs – dried forms of saps (resin or gums); SUB – subterranean parts (rhizomes, roots, bulbs, tubers); VEG – leaves, shoots, buds, sometimes the whole plant. * Uncertain identification (species inclusion in the list is based only on ethnographic sources). † Plant (the consumed part) is mentioned to be poisonous by some sources.

maritimus, Butomus umbellatus, Crambe tataria, Filipendula vulgaris, Phragmites australis, Sagittaria sagittifolia and *Typha latifolia* were known as important famine foods. *Crambe tataria* and *Typha latifolia* were reported to be preferred to bread as a staple food by some herdsmen even in normal times [60]. Collecting and eating *Lathyrus tuberosus* tubers used to be a widespread practice after ploughing. *Polypodium vulgare* and *Glycyrrhiza* rhizomes were generally consumed as kids' snacks and used as a sweetener. Tubers of *Chaerophyllum bulbosum* and *Helianthus tuberosus* were important wild vegetables eaten raw and cooked. *Acorus calamus* and *Gentiana roots* were used as pálinka spices; whereas *Cichorium intybus* roots can serve as a coffee substitute, still in use Transylvania [74].

Tree saps

Tapping trees for sap used to be a common practice in the forested regions of Eurasia. Saps from *Betula pendula, Quercus cerris, Carpinus betulus, Ulmus* spp. and *Acer pseudoplatanus* trees were generally drunk raw. *Carpinus betulus, Fagus sylvatica* and *Juglans regia* trees were also tapped in Transylvania. *Betula pendula* sap used to be an important commercial item sold in larger quantities at markets as a refreshing beverage, a medicine or a curdling agent. *Quercus cerris* sap was an important water source for herdsmen where spring water was scarce. Oaks giving the best sap were tapped permanently with built-in elder tubes; some trees were visited for drinking for 20 years. The dried resin of *Picea abies, Larix decidua, Prunus cerasifera* and *Cerasus avium* was used for chewing. This was not only useful for cleaning teeth, but also for stimulating the production of saliva, which was needed for spinning.

Social aspects of the gathering activities

Wild food plants, especially fruits, were collected both for own use or for sale. Unwritten laws regulated gathering. Signing an area rich in blueberries or strawberries, or a wild fruit tree, or arriving first in the gathering day, anyone could reserve the fruits for himself [16]. According to a documented tradition [48], families without horses were allowed to do gathering in closer to the village than families possessing horses. Horses knew their job, they followed their owners with baskets on their backs. Gathering traditions went from generation to generation in communities, sometimes in families. Jam making from *Sambucus ebulus* was a community activity among German-speaking people till the 1930s, like jam-making from *Rosa canina* in some villages; this latter tradition exist even today (e.g. in the village Szarvaskő), but more as a tourist attraction.

It was also noted that some people were ashamed that they need to eat green leaves, and were unwilling to speak about it, since gathering ("nettle-eating") was often regarded as a sign of poverty in some communities [9]. Teachers coming from an urban environment also influenced children to give up snacking on wild plants: in a documented case children have to write down 500 times "grass is eaten only by ruminant beasts" as a punishment [50].

Discussion

The diverse natural vegetation of the Carpathian Basin was a rich source of wild food plants. The consumption of wild plants was of different magnitude and economic significance in different periods and different regions, depending both on the natural characteristics of the landscape and the socio-economic background and tradition-preserving ability of the families or communities. The importance of gathering activities changed in parallel with the areal loss of natural vegetation and with socio-economic changes. The greatest changes occurred in the lowland floodplains, where traditional gathering, fishing and hunting activities provided livelihood for many people. After the extensive drainage of the floodplains in the late 19th and early 20th centuries, the utilization of wild food plants dropped dramatically. The two main reasons behind this transition were the reduced availability of the most important wetland plants and the new economic opportunities in the transformed landscape dominated by arable fields, which rendered gathering unnecessary, and left no time for such activities [16,32,35]. The tradition of consuming wild food plants could survive for a longer time in forested and mountain regions unsuitable for agricultural production, where the proportion of natural vegetation was larger. In such regions there were several documented cases of people actively consuming 20–30 species even in the middle of the 20th century (e.g. [27,36,50,74,75]). Snacking of wild fruits, flowers, and raw wild vegetables survived longest among herdsmen and children.

Today, the gathering and consumption of wild plants is increasingly becoming popular and fashionable activity again. Factors behind this process include health-conscious nutrition as well as the worsening economic situation experienced by many. Some species e.g. *Allium ursinum, A. scorodoprasum, Sambucus nigra, Ranunculus ficaria, Rosa canina, Prunus spinosa, Rubus, Crataegus,* and *Urtica* spp., are appearing in markets as sources for fashionable "nature products". Gathering wild plants for sale is generally a seasonal subsistence activity of the poorest people – e.g. gypsies or elderly people. Under harsh economic conditions the consumption of wild goods becomes a matter of survival. There always have been (and there are still) places, where indigent families made use of everything "presented by the Creator in his endless good will since the beginning of times" [63].

More traditional forms of gathering activities can also be a living tradition today. Among Hungarians living in Transylvania the traditional use of wild plants have continued till today in many places. There are settlements, where more than 30–40 wild species are used in everyday life [74,75], and wild fruits (e.g. *Fragaria*, *Vaccinium* spp.) are commonly sold in markets.

Comparing Hungary to Mediterranean (e.g. Spain and Italy [1,2]) and to the countries north of Hungary (Poland [3,80,81], Slovakia [82], Estonia [83]) there are some striking similarities and differences. Use of most wild fruits and seeds, as well as the little importance of wild greens is similar to that of Poland, Slovakia and Estonia. Similar to several Mediterranean countries, the consumption of the members of Liliaceae is high in Hungary.

A specific feature of the traditional plant use of the Hungarians is the mass consumption of the underground parts of some wetland species, particularly *Typha* spp. but also *Phragmites*, *Sagittaria*, *Alisma*, *Butomus*, and *Bolboschoenus*. Another Hungarian characteristic is the widespread consumption of some continental steppe species, reaching the Great Hungarian Plain from the east, most notably *Crambe tataria*, *Rumex pseudonatronatus* and the endemic *Armoracia macrocarpa*. The traditional knowledge of these plants might have arrived into the Carpathian Basin with the Hungarians coming from the Eurasian steppes in 896 AD. The Turkic origin of the common names of several species (e.g. "tátorján": *Crambe tatarica*, "bojtorján": *Arctium lappa*, "gyékény": *Typha latifolia*, "som": *Cornus mas*, "katáng": *Cichorium intybus*) suggests that these species might have been consumed before the Hungarians settled in the Carpathian basin. Interestingly, there are no records of Hungarians using some species (e.g. *Aegopodium podagraria*, *Alliaria petiolata*, *Sonchus oleraceus*, *Lactuca serriola*, *Stachys palustris*) common in the Carpathian Basin, which are important wild vegetables in other European countries, Poland, Germany, Spain or Italy [1–3,79–84].

Conclusions

We reviewed 71 papers and manuscripts summarizing information on the use of 236 species. Although we did our best to find all relevant papers, it is still necessary to search for further literature, and, particularly, to perform additional field data collection in order to record the traditions still surviving in some regions. One might ask: is it not too late? Does the knowledge on wild food plants and their traditional use still exist? As a few contemporary field studies testify, there is some hope left. There are places where this kind of traditional knowledge still exists, and even new species or new uses for old species can be documented. There are hints that this knowledge may exist also in other regions, and some communities consciously revitalize ancient traditions for economic or other reasons (e.g. the production of rosehip jam as a community activity in Szarvaskő village). But the erosion of traditional knowledge on wild food plants is very fast.

There are several regions where no research on this topic was ever pursued, (e.g. Hanság, Zselic, Mecsek, Balaton region); and the traditional wild plant use of non-Hungarian ethnic groups living in Hungary (e.g. Germans) would be also worth to explore. Exploring and documenting this form of traditional ecological knowledge is an important part of conserving cultural heritage. Lessons on sustainable interactions between nature and human communities can become an important source of information in an uncertain and energy scarce future [85]. In addition to serving as a basis for novel business opportunities, traditional forms of wild plant use can also improve the cohesion and resilience of local communities.

Acknowledgements

We are grateful to István Burján and Katalin Sárközi (Ethnography Department of Janus Pannonius Museum, Pécs) for their assistance in tracking down literature sources. We are also grateful to the 32 Hungarian botanists who responded to our query on their childhood snacks: Lajos Balogh, János Bölöni, Anikó Csecserits, Anna Mária Csergő, Ágnes Csomós, Áron József Deák, Tamás Exner, Sándor Farkas, Alexander Fehér, Gábor Fekete, Ferenc Gyulai, Eszter Illyés, Melinda Juhász, Árpád Kenéz, Géza Kósa, András Kun, Imre Majláth, Ákos Malatinszky, András Máté, József Nagy, Miklós Óvári, Róbert Pál, Tamás Pócs, Szilvia Rév, Dénes Saláta, Imelda Somodi, Klára Szabados, István Szabó, László Gyula Szabó, Attila Takács, Tamás Tóth, Gábor Turcsányi.

References

1. Tardío J, Pardo-De-Santayana M, Morales R. Ethnobotanical review of wild edible plants in Spain. Bot J Linn Soc. 2006;152(1):27–71.
2. Ghirardini M, Carli M, Del Vecchio N, Rovati A, Cova O, Valigi F, et al. The importance of a taste. A comparative study on wild food plant consumption in twenty-one local communities in Italy. J Ethnobiol Ethnomed. 2007;3(1):22.
3. Łuczaj Ł. Dziko rosnące rośliny jadalne użytkowane w Polsce od połowy XIX w. do czasów współczesnych. Etnobiologia Polska. 2011;1:57–125.
4. Molnár V. A. Kitaibel Pál élete és öröksége. Biatorbágy: Kitaibel Kiadó; 2001.
5. Oláh A. Zöld varázslók, virág-orvosok: népi gyógynövényismeret Békés megyében. Békéscsaba: Békés Megyei Tanács V. B. Tud.-Koordinációs Szakbizottság; 1987.
6. Kóczián G, Pintér I, Gál M, Szabó I, Szabó L. Etnobotanikai adatok Gyimesvölgyéből. Botanikai Közlemények. 1976;63(1):29–35.
7. Kóczián G, Szabó I, Szabó L. Etnobotanikai adatok Kalotaszegről. Botanikai Közlemények. 1977;64(1):23–29.
8. Papp N. Népi gyógynövény-ismereti kutatások a kolostori gyógyászatban és Erdélyben (2007–2010). Journal of History of Culture, Science and Medicine. 2011;2(2):76–88.
9. Péntek J, Szabó A. Ember és növényvilág: kalotaszeg növényzete és népi növényismerete. Bucharest: Kriterion; 1985.
10. Molnár Z, Babai D. Népi növényzetismeret Gyimesben I. Növénynevek, népi taxonómia, az egyéni és közösségi növényismeret. Botanikai Közlemények. 2009;96(1–2):117–143.
11. Molnár Z. A traditional ecological knowledge of herders on the flora and vegetation of the Hortobágy. Debrecen: Hortobágyi Természetvédelmi Közalapítvány; 2012.
12. Fekete G, Varga Z, editors. Magyarország tajainak növényzete és állatvilága. Budapest: MTA Társadalomkutató Központ; 2006.
13. Borhidi A. Magyarország növényföldrajzi képe. A magyar flóra származása és elemei. In: Fekete G, Varga Z, editors. Magyarország tajainak növényzete és állatvilága. Budapest: MTA Társadalomkutató Központ; 2006. p. 27–38.
14. Rapaics R. A kenyér és táplálékot szolgáltató növényeink története. Budapest: Királyi Magyar Természettudományi Társulat; 1934. (Népszerű Természettudományi Könyvtár; vol 16).

15. Gunda B. A magyar gyűjtögető és zsákmányoló gazdálkodás kutatása. Budapest: Néptudományi Intézet; 1948. (Magyar Népkutatás Kézikönyve; vol 2).
16. Gunda B. A vadnövények gyűjtése. In: Paládi-Kovács A, editor. Magyar néprajz II. Gazdálkodás. Budapest: Akadémiai Kiadó; 2001. p. 11–40.
17. Kisbán E. Táplálkozáskultúra. In: Paládi-Kovács A, editor. Magyar néprajz IV. Anyagi kultúra 3. Életmód. Budapest: Akadémiai Kiadó; 1997. p. 417–584.
18. Morvay J. Népi táplálkozás. Budapest: Magyar Nemzeti Múzeum; 1962. (Útmutató füzetek a néprajzi adatgyűjtéshez).
19. Kóczián G. A hagyományos parasztgazdálkodás termesztett, a gyűjtögető gazdálkodás vad növényfajainak etnobotanikai értékelése [PhD thesis]. Mosonmagyaróvár: ATE Mezőgazdasági Kar; 1985.
20. Györffy L. A nagy ínség 1863-ban. Túrkeve: Daróczy Antal Könyvnyomdája; 1931. (A Túrkevei Hírlap melléklete).
21. Molnár V. A. Növényi ínségeledelek. Természet Világa. 2010;141(11):514–516.
22. Kardos L. Az Őrség népi táplálkozása: tanulmányok az Őrség monográfiájához. Budapest: Államtudományi Intézet Táj- és Népkutató Osztálya; 1943.
23. Bödei J. Adatok Zalabaksa gyűjtögető gazdálkodásához. Néprajzi Értesítő. 1943;2:69–96.
24. Takáts G. Somogyi pásztorvilág. Kaposvár: Somogy Megyei Múzeumok Igazgatósága; 1986.
25. Füvesy A. Gyűjtögetés és erdőélés Mozsgó környékén. Tiscium. 1997;10:207–217.
26. Müller G. Gyűjtögető életmód szőkédi hagyományai. Budapest: Baranya Megyei Tanács V. B. Művelődésügyi Osztálya; 1973. (Baranyai Művelődés; vol 3).
27. Nagy R. Adatok a Baranyamegyei nagyváty növényekkel kapcsolatos szokásaihoz és néphagyományaihoz. Magyar Népnyelv. 1942;4:268–308.
28. Szigeti A. Népi konyha. Dunántúli tájak ételei. Ételkészítés alapanyagai tájanként. Budapest: Planétás Kiadó; 1999.
29. Táplálkozás adattári adatok Zengővárkony, Bonyhád, Szigetvár. Inv. No.: 7–75; 613–82; 255–80. Database of the Ethnography Department of the Janus Pannonius Museum, Pécs.
30. Pethő M. Győr-Moson-Sopron megye XX. századi népi táplálkozásának rövid jellemzése. Arrabona. 36(1–2):137–150.
31. Dantér I. Zsákmányoló gazdálkodás Farkasdon és Negyeden. In: Dantér I, editor. Hagyományos gazdálkodás a Kisalföld északi részén: Farkasd, Negyed. Galánta: Galántai Honismereti Múzeum; 2005. p. 51–73.
32. Zentai J. Egy letűnt életforma képviselője az Ormánságban. Janus Pannonius Múzeum Évkönyve. 1965;11:181–203.
33. Kiss Z. G. Zsákmányoló foglalkozások: Halászat gyűjtögetés. In: Kiss Z. G, editor. A régi Vajszló: 1244–1849. Pécs: Vajszló Nagyközség Önkormányzata; 1994. p. 174–179.
34. Andrásfalvy B. A Duna mente népének ártéri gazdálkodása : ártéri gazdálkodás Tolna és Baranya megyében az ármentesítési munkák befejezése elott. Budapest: Ekvilibrium; 2007.
35. Lábadi K. Kopács, a víz melletti falu. Budapest: Horvátországi Magyarok Demokratikus Közössége; 1994.
36. Vajkai A. A gyűjtögető gazdálkodás Cserszegtomajon. Néprajzi Értesítő. 1941;23:231–258.
37. Hegyi I. Gyűjtögető gazdálkodás az északkeleti Bakonyban. Ethnographia. 1970;2–4:442–452.
38. Fehér J. Adatok Bernecebaráti gyűjtögető és zsákmányoló gazdálkodásához. Néprajzi Értesítő. 1957;2(3–4):267–292.
39. Gunda B. A gyűjtögető gazdálkodás emlékei egy gerecsehegységi tót faluban. Ethnographia. 1938;49(1–2):213–214.
40. Barsi E. Sály : egy bükkalji falu a hagyományos gazdálkodás idején. Miskolc: Herman Ottó Múzeum; 1987. (Néprajzi Kiadványa; vol 17).
41. Kóczián G. Etnobotanikai vizsgálatok Répáshután. Répáshuta, egy szlovák falu a Bükkben. Miskolc: Herman Ottó Múzeum; 1984. (Néprajzi Kiadványa; vol 8).
42. Ujváry Z. A vadontermő növények szerepe a táplálkozásban az abaúj-zempléni hegyvidéken. Néprajzi Értesítő. 1957;39:231–243.
43. C. Schwalm E. Gyűjtögetés, mint nyersanyagkiegészítő. In: Bakó F, editor. Palócok III. Eger: Heves Megyei Múzeumok Igazgatósága; 1989. p. 417–423.
44. Paládi-Kovács A. A Barkóság és népe. Miskolc: Herman Ottó Múzeum; 2006.
45. Pálosné Nagy R. Az erdő hatása a gyöngyössólymosi emberek életére. Mátrai tanulmányok. 2000;
46. Sinkó R. Adatok Füzesgyarmat népi táplálkozáshoz. Békés Megyei Múzeumok Közleményei. 1996;16:267–288.
47. Petercsák T. Hegyköz. Miskolc: Herman Ottó Múzeum; 1978. (Borsodi Kismonográfiák; vol 6).
48. Márkus M. Gyűjtögetés a Csermosnya völgy. Néprajzi Értesítő. 1941;33(2):173–177.
49. Újváry Z. Vadon termő növények a táplálkozásban. In: Újváry Z, editor. Népi táplálkozás három Gömöri völgyben. Debrecen: Kossuth Lajos Tudományegyetem Néprajzi Tanszéke; 1991. p. 33–60. (Gömör néprajza; vol 29).
50. Zsupos Z. Dél-Gömör gyűjtögető gazdálkodása. Debrecen: Kossuth Lajos Tudományegyetem Néprajzi Tanszék; 1987. (Gömör Néprajza; vol 10).
51. Bodnár M. Adatok Tornaújfalu népi táplálkozásához. Miskolc: Herman Ottó Múzeum; 1988. (Gömör néprajza; vol 25–26).
52. Bődi E. A vadon flórája a gömöri magyarok táplálkozási kultúrájában. Tiscium. 1999;11:73–82.
53. Gunda B. Egy kárpátaljai magyar falu ethnobotanikája. Nyíregyházi Jósa András Múzeum Évkönyve. 1980;
54. Takács P, Udvardi I. Adalékok a Bereg, Ugocsa és Ung vármegyék lakóinak 18. századi erdőélési szokásaihoz. Hermann Ottó Múzeum Évkönyve. 1996;33–34:213–247.
55. Hanusz I. Fűben, fában. Budapest: Atheneum Kiadó; 1905.
56. Kiss L. A nyírvíz. A Földgömb. 1930;1:1–6.
57. Fazekas M. Karcag népi táplálkozása. Debrecen: KLTE; 1994. (Studia Folkloristica et Ethnographica).
58. Ecsedi I. A debreceni és tiszántúli magyar ember táplálkozása. Debrecen: A Debreceni Déri Múzeum Évkönyve; 1934.
59. Bencsik J. A gyűjtögető gazdálkodás emlékei a Tisza mentén. A Hajdúsági Múzeum évkönyve. 1973;1:111–126.
60. T. Bereczky I. Gyűjtögető élelemszerzés. In: Barna G, editor. Csépa. Tanulmányok egy alföldi palóc kirajzás népéletéből. Eger-Szolnok: Damjanich János Múzeum; 1982.
61. T. Bereczky I. Népi táplálkozás Szolnok megyében. Kossuth Lajos Tudományegyetem Néprajzi Tanszékenek; 1986. (Studia Folkloristica et Ethnographica; vol 19).
62. Varga G. A népi táplálkozás Hajdú-Bihar megyében a XX. század elso felében. Debrecen: Hajdú-Bihar Megyei Múzeumi Igazgatóság; 1993. (A Hajdú-Bihar Megyei Múzeumok Közleményei; vol 52).
63. Palatinus A. "Kész az étel, gyé kend má önni". Hagyományos táplálkozás a muzslyai házban. Népi táplálkozás a XX. században. Létünk. 2002;32(3–4):126–136.
64. Györffy I. Viricselés a Székelyföldön. Ethnographia. 1937;48(2):205–220.
65. Gunda B. A gyűjtögető életmód emlékei a Gyalui havasokban. Budapest: Tankönyvkiadó; 1960. (Muveltség és hagyomány).
66. Gunda B. Tejoltó növények a Kárpátokban. Ethnographia. 1967;78:161–175.
67. Gub J. Adatok a Nagy-Homoród és a Nagy-Küküllő közötti terület népi növényismeretéhez. Néprajzi Látóhatár. 1993;1–2:95–110.
68. Székely F. Népi gyűjtögetés Vadasdon. Kriza János Néprajzi Társaság Évkönyve. 1998;6:281–285.
69. Tarisznyás M. A gyűjtögető gazdálkodás hagyományai Gyergyóban. In: Népismereti dolgozatok. Bucharest: Kriterion; 1978. p. 25–33.
70. Rab J. Népi növényismeret a Gyergyói-medencében. Csíkszereda: Pallas-Akadémia Kiadó; 2001.

71. Vita Z. Tápláléknövények gyűjtése Nagyenyed környékén. In: Népismereti dolgozatok. Bucharest: Kriterion; 1994. p. 44–47.
72. Kruzslits I. Az erdő adta javak a kalotaszegi Kiskapus gyűjtögető gazdálkodásában. Tisicum. 2007;16:115–119.
73. Mihalescu L, Vosgan Z, Rosca OM, Danci O. The ornamental, aromatic and edible vegetal species from Lapus region (Transylvania). In: 19th EVS Workshop. Flora, vegetation, environment and landuse at large scale Congress. 29 April–2 May, 2010, Univeristy of Pecs, Hungary. Pécs: Univeristy of Pécs; 2010. p. 66.
74. Papp N. Etnobotanikai adatok Erdélyből 2007–2012 között. Úz-völgye (Csinód Egerszék), Gyimes, Erdővidék (Nagybacon, Kisbacon, Uzonka) Homoródkarácsonyfalva, Lövéte [Manuscript]. 2012.
75. Babai D. Vadon élő tápláléknövények Gyimesben – Etnobotanikai gyűjtések [Manuscript]. Budapest: MTA Research Centre for the Humanities; 2012.
76. F. Várózi Z. A bukovinai székelyek táplálkozása 1958–1961 közötti gyűjtés alapján. [Manuscript]. Inv. No.: I./ 736-80. Database of the Ethnography Department of the Janus Pannonius Museum, Pécs. 1961.
77. Grynaeus T, Szabó LG. A bukovinai hadikfalvi székelyek növényei. Kriza János Néprajzi Társaság Évkönyve. 2002;10:153–246.
78. Molnár Z, editor. 34 magyar botanikus gyerekkori vadnövény csemegéi [Manuscript]. Vácrátót: Centre for Ecology Hungarian Academy of Sciences; 2012.
79. Łuczaj ŁJ. Plant identification credibility in ethnobotany: a closer look at Polish ethnographic studies. J Ethnobiol Ethnomed. 2010;6(1):36.
80. Łuczaj Ł. Changes in the utilization of wild green vegetables in Poland since the 19th century: a comparison of four ethnobotanical surveys. J Ethnopharmacol. 2010;128(2):395–404.
81. Łuczaj ŁJ, Kujawska M. Botanists and their childhood memories: an underutilized expert source in ethnobotanical research. Bot J Linn Soc. 2012;168(3):334–343.
82. Łuczaj Ł. Ethnobotanical review of wild edible plants of Slovakia. Acta Soc Bot Pol. 2012;81(4):245–255.
83. Kalle R, Sõukand R. Historical-ethnobotanical review of wild edible plants of Estonia, 1770s–1960s. Acta Soc Bot Pol. 2012;81(4):271–281.
84. Łuczaj ŁJ, Svanberg I, Köhler P. Marsh woundwort, *Stachys palustris* L. (Lamiaceae): an overlooked food plant. Genet Resour Crop Evol. 2011;58(5):783–793.
85. Czúcz B, Gathman JP, McPherson GR. The impending peak and decline of petroleum production: an underestimated challenge for conservation of ecological integrity. Conserv Biol. 2010;24(4):948–956.
86. Zilahy Á. Valódi magyar szakácskönyv. Budapest: A Magyar nők Lapjának Kiadóhivatala; 1892.
87. Szabó György bükki füvesember weboldala [Internet]. 2012 [cited 2012 Aug 1].

9

Wild food plants and wild edible fungi of Heihe valley (Qinling Mountains, Shaanxi, central China): herbophilia and indifference to fruits and mushrooms

Yongxiang Kang[1], Łukasz Łuczaj[2*], Sebastian Ye[3], Shijiao Zhang[1], Jin Kang[1]

[1] College of Forestry, Northwest A&F University, Yangling 712100, China

[2] Department of Botany and Biotechnology of Economic Plants, University of Rzeszów, Werynia 502, 36-100 Kolbuszowa, Poland

[3] Skargi 11/70, 39-300 Mielec, Poland

Abstract

The aim of the study was to investigate knowledge and use of wild food plants and fungi in Han (i.e. Chinese) nationality villages in central China, including famine plants used in the respondents' childhood. A valley adjacent to the extremely species-rich temperate forest vegetation of the Taibai Nature Reserve was chosen. Eighty-two people from 5 villages took part in the study. Altogether, 159 wild food plant species and 13 fungi folk taxa were mentioned by informants. The mean number of freelisted wild foods was very high (24.8; median - 21.5). An average respondent listed many species of wild vegetables (mean - 17, median - 14.5), a few wild fruits (mean - 5.9 and median - 6) and very few fungi (mean - 1.9, median - 1), which they had eaten.

Over 50% of respondents mentioned gathering the young shoots or leaves of *Celastrus orbiculatus*, *Staphylea bumalda* and *S. holocapra*, *Caryopteris divaricata*, *Helwingia japonica*, *Pteridium aquilinum*, *Pimpinella* sp., *Amaranthus* spp., *Matteucia struthiopteris*, *Allium* spp., *Cardamine macrophylla* and *Chenopodium album*. Only one species of fruits (*Schisandra sphenanthera*) and none of the mushrooms were mentioned by over half of the respondents. Although very diverse, it can be noted that the use of wild vegetables has decreased compared to the second half of the 20th century, as informants listed several plants which they had stopped using (e.g. *Abelia engleriana*) due to the availability of cultivated vegetables and other foodstuffs. On the other hand, the collection of the most well-known wild vegetables is maintained by selling them to tourists visiting agritourist farms, and restaurants.

Keywords: ethnobotany, ethnomycology, wild edible plants, non-timber forest products, mycophobia, food security

Introduction

Wild food plant and fungi use in the world is very diverse. In many areas, mainly in urban centers or in the richest countries, few wild species are used, usually just a few wild fruits, mushrooms or green vegetables. However, in less developed countries wild food is still a significant component of nutrition. The use of wild greens and fungi shows a particularly interesting pattern as in some areas of the world these components of the diet are either avoided or little used (herbophobia and mycophobia, respectively), whereas in other areas they constitute an everyday part of nutrition [1]. For example in Amazonia and Eastern Europe wild greens are little used, in contrast to East Asia, India and parts of Africa [2]. Wild mushrooms are widely gathered in Italy, France, Catalonia, Slavic countries, Turkey, Mexico and parts of Africa, but traditionally little used in England [3].

The country the most renowned for its wide contemporary use of wild components in the human diet is China. This stems from a few factors [4,5]:

(*i*) China frequently suffered severe food crises up until the 1960s, so the use of any natural resources was important to the inhabitants,

(*ii*) wild plants and animals are believed to contain more "qi" ("life energy"), so they are regarded as more nutritious and healthy,

(*iii*) Chinese culinary art likes indulging in a large number of, often strange, ingredients.

Although the research on potentially edible wild plants has been well developed in China (e.g. [5–10]), studies focused on documenting local traditions of wild plant and fungi use using the methodology of modern ethnobotany are relatively few, and the publications in international papers solely concern ethnic minorities, i.e. Mongolians [11,12], Shaxi in Sichuan [13,14], Miao in Hunan [15] and a variety of ethnic groups in Yunnan [9,10,16–18]. Two papers (in Chinese) concerning wild food resources in the Qinling Mounatins were published, but from a different part of these mountains [19,20]. The relatively short list of species given in them [19,20] suggests that only the most

* Corresponding author. Email: lukasz.luczaj@interia.pl

commonly used species were included. The ethnomycology of edible mushrooms in China is also under-developed (in spite of the extensive literature on the economic use of mushrooms in China), and particularly lacking are studies simultaneously oriented towards wild vegetables, mushrooms and fruits.

Paradoxically, little is known about wild food plant use patterns in north-central, central and eastern China where the dominant Han (i.e. "Chinese") population lives. The aim of our study was to fill this gap and to record wild food species in one little-developed, mountainous, rural area in the province of Shaanxi. For our study we chose probably the best-preserved forest complex in central China – the vicinity of the Taibai Nature Reserve, where the local population has a particularly rich choice of wild food plants and mushrooms.

The working hypothesis was that in an area of high biodiversity, in a country famous for the use of many species of wild foods, the farmers use a large number of wild vegetables, fruits and mushrooms in their nutrition, and this knowledge is widespread in the community. An additional aim was to elicit answers on the differences between present day use and use during the last widespread famine (1958–1960).

Study area

The study was located in the Heihe National Forest Park, on the southern edge of the Taibai Nature Reserve, with the highest peak of northern China in the center of the reserve (Mt Taibai 3767 m a.s.l.). The nature reserve protects a highly diverse flora – from warm temperate (with subtropical elements) to alpine at the top. The National Forest Park (a less strict protection regime) is the southern extension of it, and mainly protects species-rich forests. The area is completely covered by ancient forest vegetation and rocky outcrops. The river Heihe valley belongs to the Houzhenzi administrative unit [town, zhen (镇)], with an area of 822 km^2. It is a very isolated place, which has vehicular access to the county town of Zhouzhi (where the post-office and schools are located) only via a 2.5 h drive through a winding precipitous gorge, often blocked for days by falling rocks. The whole valley is inhabited by 3500 people – ca. a thousand in the main settlement of Houzhenzi, and the rest in the hamlets scattered in the forest (Fig. 1).

Fig. 1 A typical landscape in Houzhenzi valley – small fields in a valley surrounded by dense primary forest vegetation. Photograph by Łukasz Łuczaj.

The studied villages lie between 1000 and 1400 a.s.l. At these altitudes the climate is humid temperate, with daily temperatures in summer oscillating around 20–30°C and winter temperatures around 10°C to −10°C. The mean annual temperature in Houzhenzi is 8.2°C, with high rainfall of nearly 1000 mm, out of which 44% is concentrated in the summer months [21]. The dominant vegetation is the species-rich *Quercus variabilis* and *Q. aliena* var. *acuteserrata* forest, with an admixture of *Pinus tabulaeformis*, and many deciduous tree species (e.g. *Acer* spp., *Tilia* spp.).

The majority of the local population are subsistence farmers who grow maize, potatoes, wheat and beans [21,22]. Sources of cash income are the orchards of zaopi (*Cornus officinalis*), walnuts (*Juglans regia*) and northern Sechuan pepper (*Zanthoxylum bungeanum*). Digging out medicinal roots and collecting medicinal herbs for wholesale buyers is also a very popular activity [21,22].

Methods

The field research was conducted in June and July 2011, as well as in August 2012, using transect walks and semi-structured interviews with key informants, individual and group freelisting interviews (36 freelists were created), and cross-checking of the gathered herbarium specimens with key informants. Altogether, 82 people from five settlements (Houzhenzi, Diaoyutai, Huaerping, Jiangjiaping, Sanhe) took part in the study. The mean age of participants was 50 (from 16 to 83). The data were supplemented by participant observations by one of the authors (S. Y.) who had frequently visited the area, in 2007–2010, during all seasons of the year. During freelisting we separately asked, which species of wild vegetables (including underground organs), wild fruits and wild mushrooms were used. Making the three separate freelists enabled the comparison of the use of these categories and helped elicit answers from the respondents [23,24]. Freelists were made orally and written down on the spot by our team, including the Chinese-script version of the plant/fungi names.

The nine restaurants selling wild vegetables in Houzhenzi were also visited and menus were photographed in order to record the taxa sold and their price.

The study started from a few informants found using the snowball technique, but most interviewees were found by systematic walks through the village, visiting houses and asking the inhabitants if they wanted to take part in the study. We aimed at interviewing only one person from each household, only occasionally were two people from the same house interviewed, if there were signs that their knowledge differed (e.g. one of the spouses comes from another village, etc.). We also displayed some of our collections (mushrooms and some wild vegetables) in the main street of Houzhenzi, to observe the reactions and comments of people to the edible taxa shown.

Voucher specimens were collected from the specimens gathered during transect walks or supplied by informants, and are stored in the Department of Forestry, Northwest A&F University in Yangling.

Results

General figures

Altogether, 159 plant species from 59 families (classified as 139 folk taxa) and 13 fungi taxa were mentioned by the informants as eaten at least once, but only 128 plant species and 12 fungi species were confirmed as eaten by more than one person (Tab. 1, Tab. 2, Tab. 3). The leaves and green parts of 104 species have been used, roots/rhizomes/tubers/bulbs of 18 species , flowers of 4 species and fruits of 36 species. Respondents mentioned wild vegetables most eagerly, and generally had problems listing wild fruits and fungi, stating that they are unimportant and are collected rarely. The gathering of young shoots of *Celastrus orbiculatus* and *Staphylea* spp. (mainly *Staphylea bumalda*, also *S. holocarpa*) was mentioned by nearly every respondent. Over 50% of respondents also mentioned gathering the young shoots or leaves of *Caryopteris divaricata*, *Helwingia japonica*, *Pteridium aquilinum*, *Pimpinella* sp., *Amaranthus* spp., *Matteucia struthiopteris*, *Allium* spp., *Cardamine macrophylla* and *Chenopodium album* (Fig. 2, Fig. 3). Six wild vegetables are served in most local restaurants (Tab. 3). Only one species of fruits (*Schisandra sphenanthera*) and none of the mushrooms were mentioned by over half of the respondents.

The mean number of freelisted wild foods was 24.8 (median – 21.5). An average respondent listed many species of wild vegetables (mean – 17, median – 14.5), a few wild fruits (mean – 5.9 and median – 6) and very few fungi (mean – 1.9, median – 1) as eaten.

The domination of wild vegetables in foraging activities is also confirmed by the fact that they are the only category of wild food stored for winter. Drying is a very common preserving technique (Fig. 4). Households dry between 1–5 species each year, usually a few kg of dry shoots and leaves, but some households who host tourists [so called "nong jia le" (农家乐)] can even dry a few dozen kg of dry "ye cai" (wild vegetables). Particularly large amounts of *Chenopodium album* are dried, as they are often used as winter fodder for pigs as well. Other commonly dried foods include *Staphylea* spp., *Helwingia japonica*, *Celastrus orbiculatus*, *Toona sinensis* and *Cardamine macrophylla*. Also, plants which are usually not treated as "ye cai", but as pig food (e.g. *Artemisia subdigitata*), are dried for the animals. In 2011 and 2012 dried *Staphylea* shoots were sold at the local food shop in Houzhenzi at 40 ¥/kg. Dried *Chenopodium* was, in 2011, sold to tourists a few km before entering the village, along with dried *Auricularia* and *Lentinula* mushrooms. Most families dry 1–5 species of wild vegetables outside on the concrete, on mats or inside the house on newspapers. Formerly wild vegetables were lacto-fermented, but now this is done very rarely.

Forest species, species of grasslands and forest edges as well as ruderal species are well represented in the list of collected taxa (however typical forest species dominate with 44% taxa, only 15% taxa are ruderal species, the rest are ubiquitous species or species of intermediate successional stages). The ruderal species are collected near homesteads. Their growth is often promoted. For instance when a farmer sprays their *Cornus officinalis* plantation with glyphosphate, they leave a clump of wild vegetables unsprayed. One of the most protected species is *Chenopodium album*, which is harvested and dried at the turn of May and June. Some forest species are harvested up to 5 km from the villages, up to the altitude of 1800 m a.s.l. At even higher altitudes, wild plants are only harvested while collecting medicinal herbs, which grow even higher.

All the older informants were asked about plants eaten during the severe food shortages that plagued China until the last case of famine in 1958–1960. The usual response was that they ate the same species but in larger quantities. They said that they were lucky having so many wild vegetables around, as some people from other, more populated areas had to take refuge in their mountains to avoid starvation. However, asking questions about the last time the particular species was used revealed that there is a group of taxa which could clearly be called famine plants – species, which were used in this area until the mid-20th century and are not used any more. These include: the leaves of *Abelia engleriana*, the rhizomes of *Pueraria lobata*,

Tab. 1 Most commonly freelisted species in the study.

Species	Category	*N* = 36
Celastrus orbiculatus	v	35
Staphylea spp.	v	33
Caryopteris divaricata	v	27
Helwingia japonica	v	27
Pteridium aquilinum	v	26
Pimpinella sp.	v	24
Amaranthus spp.	v	24
Matteucia struthiopteris	v	23
Cardamine macrophylla	v	22
Schisandra sphenanthera	f	22
Chenopodium album	v	22
Allium spp.	v	20
Toona sinensis	v	18
Akebia trifoliata	f	18
Rubus spp.	f	17
Prunus salicina	f	15
Chrysosplenium biondianum	v	15
Saussurea dolichopoda	v	14
Decaisnea fargesii	f	14
Adenophora spp.	v	13
Cantharellus cibarius	m	13
Allium paepalanthoides	v	12
Sedum amplibracteatum	v	11
Eleagnus umbellata	f	11
Tricyrtis macropoda	v	11

f – fruit; m – mushroom; *N* – number of freelists; v – green vegetable.

Tab. 2 Wild food plants sold in the restaurants in Houzhenzi.

Species	*N*
Staphylea spp. (mainly *S. bumalda*)	9
Toona sinensis	8
Chenopodium album	7
Matteucia struthiopteris	6
Pteridium aquilinum	6
Celastrus orbiculatus	6
Helwingia japonica	2
Medicago sativa	1
Cardamine macrophylla	1
Caryopteris divaricata	1
Pimpinella sp.	1
Lychnis senno	1

N – No. of restaurants selling wild food plants (total = 9).

Tab. 3 The list of wild edible plants and fungi confirmed by at least two respondents.

Scientific name	Family (according to APGIII [31])	Part used	Frequency	Local name in pinyin	Local name in Chinese
Vascular plants					
Abelia engleriana Rehder	Linneaceae (formerly in Caprifoliaceae)	aerial parts	***	shen xian dou fu	神仙豆腐
Achyranthes bidentata Blume	Amaranthaceae	aerial parts	*	niu xi	牛膝
Actinidia chinensis Planch.	Actinidiaceae	fruit	***	ye mi hou tao	野猕猴桃
Adenophora spp. (*A. capillaris* Hemsl., *A. polyantha* Nakai)	Campanulaceae	whole plant	***	nai jiang cai	奶浆菜
Akebia trifoliata (Thunb.) Koidz.	Lardizabalaceae	fruit	****	ba yue gua, ba yue zha	奶浆菜, 八月炸
Allium ovalifolium Hand.-Mazz., *A.* cf. *victorialis* L.	Amaryllidaceae (formerly in Liliaceae)	whole plant	*	ge jiu, ye jiu	茖韭, 野韭
Allium paepalanthoides Airy Shaw	Amaryllidaceae (formerly in Liliaceae)	whole plant	***	tian suan	天蒜
Allium spp. (*A.* cf. *senescens* L., *A. macrostemon* Bunge)	Amaryllidaceae (formerly in Liliaceae)	aerial parts	****	ai jiu cai, ai suan, ye suan, yong bao tou, luo er jiu, zong bao tou, ye jiu cai	崖韭菜, 崖蒜, 野蒜, 罗儿韭, 棕包头, 野韭菜
Amaranthus caudatus L.	Amaranthaceae	aerial parts	**	tian xi mi	甜菥蓂
Amaranthus retroflexus L., *A. paniculatus* L., *A. viridis* L.	Amaranthaceae	aerial parts	****	han cai, ren han cai	汉菜, 人汉菜,
Anaphalis aureopunctata Lingelsh et Borza	Asteraceae	aerial parts	*	shi qu cao	鼠曲草
Anaphalis margaritacea Benth. & Hook. f.	Asteraceae	aerial parts	*	qing ming cai	清明菜
Aralia chinensis L.	Araliaceae	tender buds	**	ci long pao	刺龙袍
Arctium lappa L.	Asteraceae	leaf stalks, underground parts	*	niu bang zi	牛蒡子
Artemisia subdigitata Mattf.	Asteraceae	aerial parts	*	ai	艾
Asarum himalaicum Hook. f. & Thomson ex Klotzsch	Aristolochiaceae	whole plant (as spice)	*	mao xi xin	毛细辛
Asarum sieboldii Miq.	Aristolochiaceae	whole plant (as spice)	*	xi xin	细辛
Begonia sinensis A. DC.	Begoniaceae	aerial parts	*	yi kou xie	一口血
Berchemia sinica Schneid.	Rhamnaceae	fruit	*	ya gu teng, gou er cha	亚古藤、勾儿茶
Boehmeria gracilis C. H. Wright	Urticaceae	aerial parts	*	hong he ma	红河麻
Boehmeria tricuspis Makino	Urticaceae	aerial parts	*	he ma	河麻
Broussonetia papyrifera (L.) Vent.	Moraceae	leaves	*	gou shu guo, gou ye	构树果, 构 叶
Cacalia roborowskii (Maxim.) Y. Ling	Asteraceae	aerial parts	*	xiong er duo	熊耳朵
Capsella bursa-pastoris Medik.	Brassicaceae	aerial parts	***	di di cai	地地菜
Cardamine macrophylla Willd.	Brassicaceae	aerial parts	***	shi jia cai	石夹菜
Cardamine spp. (other smaller species e.g. *C. flexuosa* With., *C. hirsuta* L.)	Brassicaceae	aerial parts	*	xiao shi jia cai	小石夹菜
Caryopteris divaricata Maxim.	Lamiaceae (formerly in Verbenaceae)	aerial parts	****	chou lao han, lao han xiang	臭老汉/老汉香
Castanea mollissima Blume	Fagaceae	fruit	**	ye mao li, ye ban li	野毛栗, 野板栗
Celastrus orbiculatus Thunb.	Celastraceae	aerial parts	****	bai wan ye	白蔓叶
Cephalotaxus sinensis (Rehder & E. H. Wilson) H. L. Li	Cephalotaxaceae	pseudo-fruit	*	bai gai guo, bi zi shu, shui bai, sun guo	白盖果、篦子树、水柏.松果
Chenopodium album L., also *C. giganteum* D. Don	Amaranthaceae (formerly Chenopodiaceae)	aerial parts	****	hui hui cai	灰灰菜
Chrysosplenium biondianum Engl.	Saxifragaceae	aerial parts	***	hong jin cai	红筋菜
Cirsium arvense var. *setosum* (Willd.) C. A. Mey	Asteraceae	aerial parts	**	hong hua miao, ci ji	红花苗, 刺蓟
Cirsium spp. eg *Cirsium botryoides* Petrak ex Hand.-Mzt.	Asteraceae	aerial parts	*	xiaoji	小蓟
Clerodendrum trichotomum Thunb.	Lamiaceae	aerial parts	*	chou mu dan, chou lao han	臭牡丹、臭老汉
Commelina benghalensis L.	Commelinaceae	aerial parts	*	dan zhu ye, zhu ye cao, mian da zi	淡竹叶、竹叶草、面达子
Cornus kousa Bürger ex Miq.	Cornaceae	fruit	***	shi zao	石枣

Tab. 3 *(continued)*

Scientific name	Family (according to APGIII [31])	Part used	Frequency	Local name in pinyin	Local name in Chinese
Corylus heterophylla Fisch. ex Besser	Corylaceae	fruit	*	zhen zi, mao li, mao li zi shu, xiao xiang zi shu	
Crataegus hupehensis Sarg.	Rosaceae	fruit	*	ye shan zha	野山楂
Cryptotaenia japonica Hassk.	Apiaceae	aerial parts	*	ya jiao ban	鸭脚板
Cynanchum giraldii Schltr.	Asclepiadiaceae	underground parts	*	ge shan xiao	隔山消
Decaisnea fargesii Franch.	Lardizabalaceae	fruit	***	mao shi gua, ye xiang jiao	猫屎瓜, 野香蕉
Descurainia sophia (L.) Webb ex Prantl (?)	Brassicaceae	aerial parts	**	yin chen, mi hao	因陈, 米蒿
Dioscorea batatas Decne.	Dioscoreaceae	underground parts	*	shan yao	山药
Elaeagnus umbellata Thunb.	Eleagnaceae	fruit	***	yang nai zi, niu nai zi	羊奶子, 牛奶子
Fragaria spp. (*F. corymbosa* Losinsk., *F. pentaphylla* Losinsk.)	Rosaceae	fruit	***	cao mei, di pao, di di pao xiang pao	草莓, 地泡, 地地泡, 香泡
Helwingia japonica (Thunb.) F. Dietr.	Helwingiaceae	leaves	****	ye shang hua	叶上花
Hemerocallis spp. (*H. dumortierii* C. Morren, *H. fulva* L.)	Xanthorrhoeaceae (formerly Liliaceae)	flowers	**	ye huang hua	野黄花
Hieracium sp.	Asteraceae	aerial parts	*?	ku mai cai	苦荬菜
Impatiens notolopha Maxim.	Balsaminaceae	aerial parts	*	dao laon nen	到老嫩
Ixeris sonchifolia Hance	Asteraceae	aerial parts	***	ku mai cai	苦荬菜
Juglans cathayensis Dode	Juglandaceae	fruit	**	ye he tao	野核桃
Kalimeris pinnatifida (Maxim.) Kitam.	Asteraceae	aerial parts	*	ma lan tou	马兰头
Kochia scoparia (L.) Schrad.	Amaranthaceae (formerly Chenopodiaceae)	aerial parts	*	tie sao ba	铁扫把
Lactuca serriola L.	Asteraceae	aerial parts	**	xiao bai jiang, xiao ku ma cai, ku ma cai	小苦荬菜, 苦荬菜
Leontopodium japonicum Miq.	Asteraceae	aerial parts	*	shi qu cao	石曲草
Ligusticum sinense Oliv., Chuanhsiung	Apiaceae	aerial parts	*	chuan xiong	川芎
Lilium giganteum Wall.	Liliaceae	underground parts	**	shui bai he	水百合
Lilium spp. (*L. leichtlinii* Hook. f., *L. tigrinum* Ker Gawl.)	Liliaceae	underground parts	**	ye bai he	野百合
Lonicera standishii Carr.	Caprifoliaceae	fruit	*	ku tang pao	苦糖泡
Lychnis senno Siebold & Zucc.	Caryophyllaceae	whole plant	*	huang hua cai	黄花菜
Lycopus lucidus Turcz. ex Benth.	Lamiaceae	aerial parts	*	ye bai cai, ze lan	野白菜、泽兰
Matteucia struthiopteris (L.) Tod.	Onocleaceae	tender shoots	****	ji tou cai	鸡头菜
Medicago sativa L.	Fabaceae	aerial parts	*	mu xi cai	苜蓿菜
Morus australis Poir.	Moraceae	fruit	*	sang pao, sang shu	桑泡、桑树
Oxalis spp. (*O. griffithii* Edgew. & Hook. f., *O. corniculata* L.)	Oxalidaceae	aerial parts	*	suan cao, suan cai, suan suan cao	酸草、酸菜、酸酸草
Picris hieracioides L.	Asteraceae	aerial parts	*?	ku mai cai	苦荬菜
Pilea mongolica Wedd.	Urticaceae	aerial parts	*	dao lao nen	到老嫩
Pimpinella sp. (probably a new taxon related to *Pimpinella arguta* Diels and *P. rhomboidea* Diels)	Apiaceae	aerial parts	****	shui qin cai, sha qin cai	水芹菜, 沙芹菜
Plantago asiatica L.	Plantaginaceae	aerial parts	*	kai men ye, che qian cao, che qian zi	开门叶、车前草、车前子
Polygonatum megaphyllum P. Y. Li and *Polygonatum odoratum* L.	Asparagaceae (formerly in Liliaceae)	underground parts	*	yu zhu, yu zhu shen	玉竹, 玉竹参
Polygonum ciliinerve (Nakai) Ohwi	Polygonaceae	underground parts	*	qiao mai tou	荞麦头
Potentilla arbuscula D. Don var. *veitchii* (E. H. Wilson) Liou	Rosaceae	aerial parts	*	guan yin cha	观音茶
Prunus armeniaca L.	Rosaceae	fruit	**	ye xing	野杏
Prunus canescens Bois, *P. pilosiuscula* Koehne	Rosaceae	fruit	**	ye ying tao	野樱桃
Prunus cf. *polytricha* Koehne	Rosaceae	fruit	*	chuan tao	川桃
Prunus persica (L.) Batsch	Rosaceae	fruit	**	ye tao zi	野桃子
Prunus salicina Lindl.	Rosaceae	fruit	***	ye li zi, ze maili, huo li zi, huo li, ye mai li	野李子, 火李子, 火李, 野麦李
Pteridium aquilinum (L.) Kuhn	Dennstaedtiaceae	tender shoots and underground parts	****	jue cai, jue gen, long zhua	蕨菜, 蕨根, 龙爪菜

Tab. 3 *(continued)*

Scientific name	Family (according to APGIII [31])	Part used	Frequency	Local name in pinyin	Local name in Chinese
Pueraria lobata (Willd.) Ohwi	Fabaceae	underground parts	**	ge gen	葛根
Pyrola decorata Andres	Ericaceae	aerial parts (spice and infusion)	*	hong ru, shou cha	红茹、寿茶
Pyrola rotundifolia L.	Ericaceae	aerial parts (spice and infusion)	*	bairu, shou cha	白茹、寿茶
Pyrus xerophila T. T. Yu	Rosaceae	fruit	***	ye li, ma li, shan li	野梨, 麻梨, 山梨
Rhus verniciflua Stokes	Anacardiaceae	aerial parts	*	qi shu	漆树
Robinia pseudoacacia L.	Fabaceae	flowers	*	huai hua	槐花
Rorippa montana Small	Brassicaceae	aerial parts	*	man jing cai, la la cai	蔓茎菜, 辣辣菜
Rosa sp.	Rosaceae	young shoots	*	ci mei hua	刺玫花
Rubus coreanus Miq.	Rosaceae	fruit	**	ci pao, di pao, fu pen zi	刺泡, 地泡、覆盆子
Rubus flosculosus Focke	Rosaceae	fruit	**	cai zi pao	菜子泡
Rubus pungens Cambess.	Rosaceae	fruit	**	huang ci pao	黄刺泡
Rubus spp.	Rosaceae	fruit	***	duan yang pao, xuan gou zi	端阳泡, 悬钩子
Rumex crispus L.	Polygonaceae	aerial parts	**	niu she tou, ye da huang	牛舌头
Sabia shensiensis H. Y. Chen	Sabiaceae	aerial parts	*	qing teng cai, teng er cai	青藤菜, 藤儿菜
Saussurea dolichopoda Diels	Asteraceae	aerial parts	***	kong tong cai, kong xin cai	空筒菜, 空心菜
Schisandra sphenanthera Rehder & E. H. Wilson	Schisandraceae	fruit	****	wu wei zi	五味子
Sedum aizoon L., *S. sarmentosum* Bunge, *S. pampaninii* Raym.-Hamet, S. lineare Thunb.	Crassulaceae	aerial parts	**	gou ya ban, gou za cai, machijie, da bu si, chui pen cao	狗牙瓣, 打不死
Sedum amplibracteatum K. T. Fu	Crassulaceae	aerial parts	***	hua qiao man, la zi miao, la jiao miao, ye la cai	花荞蔓, 野辣子苗苗, 辣椒苗, 叶辣菜
Silene conoidea L.	Caryophyllaceae	aerial parts	*	mai pian cai	麦片菜
Sinacalia tangutica (Maxim.) B. Nord.	Asteraceae	underground parts	*	shui luo bo	水萝卜
Smilacina japonica A. Gray, *Smilacina henryi* (Baker) Hara	Asparagaceae (formerly in Liliaceae)	aerial parts	*	pian tou cai	偏头菜
Stachys affinis Bunge	Lamiaceae	underground parts	*	di gu niu	地牯牛
Staphylea bumalda DC., *S. holocarpa* Hemsl.	Staphyleaceae	aerial parts, flowers	****	shu hua cai	树花菜
Stellaria media (L.) Vill.	Caryophyllaceae	aerial parts	*	e er chang	鹅儿肠
Taraxacum mongolicum Han.-Mzt	Asteraceae	aerial parts	**	pu gong ying, ku mai cai, da ku cai	蒲公英、苦荬菜、大苦菜
Thlaspi arvense L.	Brassicaceae	aerial parts	***	ji dan huang	鸡蛋黄
Toona sinensis (Juss.) M. Roem.	Meliaceae	tender leaf shoots	****	xiang chun	香椿
Tricyrtis macropoda Miq.	Liliaceae	aerial parts	***	huang gua cai	黄瓜菜
Ulmus bergmanniana C. K. Schneid., *U. propinqua* Koidz., *U. pumila* L.	Ulmaceae	leaves, bark, immature fruit	**	yu shu	榆树
Urtica fissa E. Pritz. ex Diels	Urticaceae	aerial parts	*	bai he ma	白河麻
Vicia cracca L.	Fabaceae	aerial parts	*	ye wan dou jian	野豌豆尖
Vitis ficifolia Bunge	Vitaceae	fruit	***	ye pu tao	野葡萄
Zanthoxylum bungeanum Maxim.	Rutaceae	fruits, aerial parts	*	ye hua jiao	野花椒
Fungi					
Boletus spp.	Boletaceae		**	niu gan jun, da jiao gu	牛肝菌, 大脚菇
Cantharellus cibarius Fr.	Cantharellaceae		***	huang si jun	牛肝菌
Grifola umbellata (Pers.) Pilát	Meripilaceae		***	zhu ling jun	猪苓菌
Hericium sp.	Hericiaceae		**	hou tou jun	猴头菌
Laetiporus sulphureus (Bull.) Murrill (??)	Polyporaceae		*	ji guan jun	鸡冠菌
Lentinula edodes (Berk.) Pegler (more often cultivated)	Marasmiaceae		*	ye xiang gu	野香菇
Morchella sp.	Morchellaceae		**	yang que jun	羊雀菌

Tab. 3 *(continued)*

Scientific name	Family (according to APGIII [31])	Part used	Frequency	Local name in pinyin	Local name in Chinese
Pleurotus sp.	Pleurotaceae		*	dong jun	冻菌
Ramaria spp.	Gomphaceae		**	shua ba jun	刷把菌
Unidentified terrestrial gilled mushroom	?		**	ban li jun	板栗菌
Unidentified mushroom	?		*	bao gu jun	包谷菌
Unidentified mushroom	?		*	qiao mian jun	荞面菌

Frequency: **** > 50% of respondents; *** > ¼ of respondents; ** > ⅛ of respondents; * ⅛ of respondents or less, but at least 2 respondents.

Pteridium aquilinum, *Polygonatum* spp., *Sinacalia tangutica*, the bulbs of *Lilium giganteum* and other *Lilium* species, and the bark of *Ulmus* spp. Nowadays the consumption of underground organs of plants has practically ceased and is restricted to the occasional use of *Lilium* spp. and *Stachys affinis*.

Wild vegetables are eaten in any of the daily three meals. The commonest preparation technique is boiling, then straining and sprinkling them with some oil in which Sechuan pepper, garlic, and sometimes ginger, was fried. Then they are served, warm or cold. This is a side dish, called "liang ban", accompanied by home-made wheat bread ("bing"), rice or other stir-fried foods. Sometimes wild vegetables are also put into broad, home-made noodles served in spicy and sour soup. They are also, rarely, lacto-fermented. Dried vegetables are first soaked in water for a few minutes or hours and then used like fresh vegetables. Some respondents say that in the case of *Staphylea* shoots, drying even improves their taste and digestibility. Wild vegetables are also sold in all the local restaurants (Tab. 2), and every agritourist farm has them on their menu.

Fruits are and have always been less appreciated than wild greens. They are sometimes collected for fun by children or grown-ups going to the forest to collect wild greens, medicinal herbs or wood. They have never been stored for winter and are not used in any dishes by anyone, apart from dried *Schisandra* and *Akebia* fruits, used medicinally. In spite of this, most people easily mentioned a few species of wild fruits they had eaten.

Few fungi species are used, as most of them are generally feared. We found only one inhabitant of the valley, who had a hobby of collecting wild edible mushrooms, which he developed after gathering *Boletus* species for sale a few years ago. Others never go to the forest with the purpose of collecting mushrooms, apart from going to collect cultivated *Auricularia* and *Lentinula edodes* grown on piles of logs located in the woods. The only mushrooms relatively more widely known and accepted are *Cantharellus cibarius*, an unidentified *Agaricales* (called "banlijun", i.e. "chestnut mushroom"), *Ramaria* spp. (and possibly *Clavaria* spp.) treated by locals as one folk taxon "shuabajun" (i.e. "brush mushroom") and *Grifola umbellata*, whose sclerotia are collected for medicinal purposes, although fruiting bodies are occasionally eaten as well. However, only a third of the respondents have ever eaten the most commonly listed mushroom, *Cantharellus cibarius*. More than half of the respondents had never collected wild fungi in the forest. A few people who eat *Boletus* spp. species started doing so after taking part in commercial mushroom collecting a few years ago.

Discussion

The large number of used wild greens in this study is one of the highest recorded on such a small scale in the history of ethnobotanical studies. The only comparable study, by Zou [15], recorded the use of 335 taxa of wild vegetables, belonging to 87

Fig. 2 *Staphylea bumalda* flowering shoots, boiled, strained and sprinkled with oil, Houzhenzi, 2011. Photograph by Łukasz Łuczaj.

Fig. 3 *Pteridium aquilinum* shoots, boiled, strained and sprinkled with oil, Houzhenzi, 2011. Photograph by Łukasz Łuczaj.

Fig. 4 Drying wild vegetables (*Chenopodium album* and *Staphylea bumalda*) on a farm in Diaoyutai in early June 2011. Photograph by Łukasz Łuczaj.

families and 119 genera in 10 villages of Hunan, however the latter study concerned a larger and more heterogenous area. Ghorbani [16] recorded the use of 173 wild food plants from 485 informants of four ethnic groups of Xishuanbanna valley, out of which only around a third were wild greens. However, his study concerned an area which was very heterogenous in terms of elevation, inhabitants and vegetation. The average number of wild food plants listed by one informant was only around 10 species, whereas in this study we documented a much higher rate of use per person (mean - over 24 species), with probably the highest average number of wild green vegetables listed per person (mean - 17 species) using freelisting techniques in any ethnobotanical study to date.

Knowledge of wild vegetables in China is additionally encoded in the language. Most wild vegetables have the word "cai", i.e. vegetable, so it is enough to know the name and to be able to recognize the plant to be able to presume its use, i.e. if it is a "cai" (vegetable) or "cao"/"yao" (medicinal plant).

The relatively high geographical diversity of the use of "ye cai" in China should be emphasized. For example none of the articles on wild vegetables of Qinling Mts mention the use of *Staphylea* spp. nor *Pimpinella* sp. - locally important vegetables [19,20].

The widespread phenomenon of drying wild vegetables is worth attention. This ancient preservation technique is nowadays rarely used for wild vegetables across the globe. Storing a particular food for winter may mean that this is a culturally significant item.

Several taxa are semi-domesticated, and undergo varying degrees of active protection. Some are increasingly brought from the forest to be planted in gardens (*Toona sinensis*, *Staphylea bumalda*, *Asarum sieboldii*). The two species of mushrooms widely cultivated using tree logs left in the garden or in the forest, *Auricularia* sp. and *Lentinula edodes*, were mentioned by some respondents as wild vegetables as well. This shows that, similarly to other geographic areas, the distinction between the wild and the cultivated is not usually sharp [2,25,26].

The local population utilizes a large proportion of the local edible flora. However, strong cultural biases can be seen even in such a herbophilous community. Some edible wild vegetables, e.g. *Galinsoga* sp., *Reynoutria japonica*, most *Polygonum* species and *Lamium barbatum* are not used, the common *Stellaria media*, is also eaten rarely and only by a few individuals.

Over half wild vegetables come from the forest. This is in contrast with some studies showing that human populations, even in wooded areas, tend to over-utilize the ruderal flora [27,28]. Here, at least two explanations are possible:

(*i*) the fields' area is relatively small, so in the past the volume of wild vegetables from ruderal sites may not have been sufficient,

(*ii*) many edible ruderal taxa are regarded here as pig food, and we observed some degree of separating wild vegetables for human consumption and pig food (the exception here is *Chenopodium*, eaten both by humans and pigs).

What is interesting is the large domination of wild greens over fruits and fungi. A typical answer of a respondent to the question about what wild veg they had eaten was: "ye cai hen duo" (there are many wild vegetables). On the other hand, the same question about fruits or fungi resulted in the opposite answer, e.g. "ye mogu/junzi hen shao" (there are very few wild fungi).

Wild vegetables are intentionally collected here, they are well known to everyone and dried for winter in most houses. Fruits are something unimportant, play objects, something one finds on a forest walk to collect herbs, something only eaten raw. They are never cooked or dried or added to any dishes. The only exceptions are the fruits of *Schisandra*, which are dried and sold as medicine. The lack of interest in mushrooms is puzzling, as China is usually regarded as a mycophilous part of the world [3,29,30].

Conclusions

The studied community displays one of the highest levels of herbophilia known in human cultures. On the other hand the community shows relative indifference to wild fruits and fungi, which are rarely collected, and only as an additional activity.

The results of this study show that further in-depth ethnobotanical research is needed to determine patterns in wild food plant and fungi use in different parts of China, as locally these patterns may be extremely variable.

Acknowledgments

Many thanks to the inhabitants of the studied villages for their generous help in sharing information on the use of the species. The program was financially supported by the Forestry Research Foundation for the Public Service Industry of China (2009,04004) and by the University of Rzeszów (Institute of Biotechnology and Basic Sciences, as well as a special grant from the rector of the University). We are also grateful to Prof. Khasbagen (Hohhot, China) for his help in literature search.

References

1. Łuczaj Ł. Changes in the utilization of wild green vegetables in Poland since the 19th century: a comparison of four ethnobotanical surveys. J Ethnopharmacol. 2010;128(2):395–404.

2. Turner NJ, Łuczaj ŁJ, Migliorini P, Pieroni A, Dreon AL, Sacchetti LE, et al. Edible and tended wild plants, traditional ecological knowledge and agroecology. Crit Rev Plant Sci. 2011;30(1–2):198–225.

3. Boa ER. Wild edible fungi: a global overview of their use and importance to people. Rome: FAO; 2004.
4. Anderson EN. The food of China. New Haven: Yale University Press; 1988.
5. Hu SY. Food plants of China. Hong Kong: Chinese University Press; 2005.
6. Long CL. Strategies for agrobiodiversity conservation and promotion: a case from Yunnan, China. Biodivers Conserv. 2003;12(6):1145–1156.
7. Wu X. "Turning waste into things of value": marketing fern, Kudzu, and Osmunda in Enshi Prefecture, China. J Dev Soc. 2003;19(4):433–457.

8. Wang X, Du X. Recent status of the development and strategies of exploitation of non wood forest products in China. Lin Ye Ke Xue Yan Jiu. 1997;10(2):199–205.
9. You-Kai X, Guo-Da T, Hong-Mao L, Kang-La Y, Xiang-Sheng D. Wild vegetable resources and market survey in Xishuangbanna, southwest China. Econ Bot. 2004;58(4):647–667.

10. Zhang XP, Wu JL, Li Y, Liu F, Wang JQ. Investigation on species resources and utilization of wild vegetable in nabanhe watershed nature reserve, Xishuangbanna. Journal of Southwest Forestry College. 2004;24:21–24.
11. Huai KHY, Pei SJ. Wild plants in the diet of Arhorchin Mongol herdsmen in inner Mongolia. Econ Bot. 2000;54(4):528–536.

12. Wujisguleng W, Khasbagen K. An integrated assessment of wild vegetable resources in Inner Mongolian Autonomous Region, China. J Ethnobiol Ethnomed. 2010;6(1):34.
13. Weckerle CS, Huber FK, Yongping Y, Weibang S. Plant knowledge of the Shuhi in the Hengduan mountains, Southwest china. Econ Bot. 2006;60(1):3–23.

14. Huber FK, Ineichen R, Yang Y, Weckerle CS. Livelihood and conservation aspects of non-wood forest product collection in the Shaxi Valley, southwest China. Econ Bot. 2010;64(3):189–204.

15. Zou X, Huang F, Hao L, Zhao J, Mao H, Zhang J, et al. The socio-economic importance of wild vegetable resources and their conservation: a case study from China. Kew Bull. 2011;65(4):577–582.

16. Ghorbani A, Langenberger G, Sauerborn J. A comparison of the wild food plant use knowledge of ethnic minorities in Naban River Watershed National Nature Reserve, Yunnan, SW China. J Ethnobiol Ethnomed. 2012;8(1):17.
17. Ghorbani A, Langenberger G, Liu JX, Wehner S, Sauerborn J. Diversity of medicinal and food plants as non-timber forest products in Naban River Watershed National Nature Reserve (China): implications for livelihood improvement and biodiversity conservation. Econ Bot. 2012;66(2):178–191.
18. Jin C, Yin-Chun S, Gui-Qin C, Wen-Dun W. Ethnobotanical studies on wild edible fruits in southern Yunnan: folk names; nutritional value and uses. Econ Bot. 1999;53(1):2–14.
19. Ma X, Zhang J, Lu S, Cui Z, Zhao H, Zheng J. The survey of the distribtion of wild vegetables in the northern slope of the Qinling Mountains. Quaterly of Forest By-Product and Speciality in China. 2002;61(2):49–50.
20. Ma X, Hu P, Wu J, Zhang J, Cui Z. The habitat, development and utilization of wild vegetables in the northern slope of the Qinling Mountains. Special Economic Animal. 2002;2:37–38.
21. Economic survey of the Zhouzhi County Houzhenzi town, protected areas from 2007 [Internet]. 2007 [cited 2012 Sep 20].

22. Kang Y, Łuczaj ŁJ, Ye S. The highly toxic *Aconitum carmichaelii* Debeaux as a root vegetable in the Qinling Mountains (Shaanxi, China). Genet Resour Crop Evol. 2012;59(7):1569–1575.

23. Quinlan M. Considerations for collecting freelists in the field: examples from ethobotany. Field Methods. 2005;17(3):219–234.

24. Sutrop U. List task and a cognitive salience index. Field Methods. 2001;13(3):263–276.
25. McClatchey WC. Wild food plants of Remote Oceania. Acta Soc Bot Pol. 2012;81(4):371–380.
26. Łuczaj Ł, Pieroni A, Tardío J, Pardo-de-Santayana M, Sõukand R, Svanberg I, et al. Wild food plant use in 21st century Europe: the disappearance of old traditions and the search for new cuisines involving wild edibles. Acta Soc Bot Pol. 2012;81(4):359–370.
27. Voeks RA. Disturbance pharmacopoeias: medicine and myth from the humid tropics. Ann Assoc Am Geogr. 2004;94(4):868–888.

28. Sõukand R, Kalle R. Change in medical plant use in Estonian ethnomedicine: a historical comparison between 1888 and 1994. J Ethnopharmacol. 2011;135(2):251–260.
29. Yamin-Pasternak S. Ethnomycology: fungi and mushrooms in cultural entanglements. In: Anderson EN, Pearsall D, Hunn E, Turner N, editors. Ethnobiology. Hoboken NJ: Wiley and Sons; 2011. p. 213–230.

30. Arora D. The houses that matsutake built. Econ Bot. 2008;62(3):278–290.

31. The Angiosperm Phylogeny Group. An update of the Angiosperm Phylogeny Group classification for the orders and families of flowering plants: APG III. Bot J Linn Soc. 2009;161(2):105–121.

10

Ethnobotanical review of food uses of *Polygonatum* (Convallariaceae) in China

Wujisguleng Wujisguleng[1,2], Yujing Liu[1], Chunlin Long[1]*

[1] College of Life and Environmental Sciences, Minzu University of China, Beijing 100081, China

[2] Mongolian Medicine College, Inner Mongolia Medical University, Hohhot 010110, China

Abstract

According to our literature survey, fifteen species in the genus *Polygonatum* (Convallariaceae) have been used as food in China. Five main categories of use (drinks, snacks, vegetables, staple food and seasoning) were found. The plants from this genus are used mainly as vegetables. Rhizomes are the main parts used. The categories of use and diverse preparation methods have some similarities with those in other countries. Among all the recorded species, *Polygonatum sibiricum* and *P. odoratum* are the most commonly consumed species in China. Local people from different places have their own unique preparation methods and customs. However, the use of *Polygonatum* still remains at the level of family consumption, and the resources have not been fully used yet. We proposed ideas for further exploitation and conservation of edible *Polygonatum* species, and suggested further research, e.g. on the nutritional properties of all recorded edible species. It is believed that there is still plentiful traditional knowledge about food uses of *Polygonatum* to be documented in China.

Keywords: *Polygonatum*, food uses, ethnobotany, China

Introduction

China is the third largest country in the world. Its diverse climate and geography, varied terrain and range of soil types harbor an enormous diversity of plants, with 31142 species (of higher plants). At least 1156 species and 274 varieties and/or cultivars of food plants have been used by the inhabitants of China, who belong to 56 ethnic groups, including the dominant Han Chinese nationality [1]. Over thousands of years, they have accumulated very rich and precious traditional knowledge of using plants. That is why Prof. Richard I. Ford, a renowned ethnobotanist at University of Michigan at Ann Arbor, once said "the root of ethnobotany is in China". Since the 1980s, ethnobotanical studies have been carried out to document traditional knowledge of plant use in China, including wild food species [2–12]. It seems that using wild plants for food is more popular in southern China than in its northern part. Ethnic groups in some areas in southern China, like Yunnan Province, which harbors the richest diversity of plants in the country, consume more wild vegetables than cultivated ones. In some villages, which are far from the markets, wild vegetables are the local people's main sources of vegetable dishes [7].

* Corresponding author. Email: long@mail.kib.ac.cn

The plants in genus *Polygonatum* Mill. are perennial rhizomatous herbs belonging to the family Convallariaceae (also named Liliaceae). Up to June 2012, 63 species are accepted by the "World checklist of selected plant families" in the genus *Polygoantum* [13]. Plants in the genus are widely distributed in temperate regions of the northern hemisphere, mainly from the Himalayas to Japan, and especially in China and Japan, where approximately 40 different species of *Polygonatum* have been reported. They are also found in India, Korea, Afghanistan, Bhutan, Nepal, Pakistan and Russia. *Polygonatum* also grows in the temperate climate zones of northern and eastern America and Europe [14].

China is one of the richest countries in *Polygonatum* species, with 39 described species (61.9% of the whole genus), and some studies suggest that China might be the distribution and differentiation center of the genus *Polygonatum* [15,16]. Among them, 20 species (31.7%) are endemic. They are distributed throughout the country, and they can be found mainly in the northeast, Yunnan, Guizhou, Anhui, Guangxi, Hebei, Inner Mongolia, Henan, Shandong, Shanxi, Zhejiang, and Jiangxi provinces of China. Yunnan and Guizhou are the richest in *Polygunatum* species. The typical natural habitat of these plants is wet slope below 2000 m above sea level. Most of them grow in wet places and forests or bushes with thick and fertile soil [17].

Polygonatum species have a long history of traditional use, both as medicine and food, in China and other countries, like the northern and eastern United States, south-central Canada, India and Korea [18–20]. There are numerous ancient books describing medicinal and food uses of *Polygonatum* in China. It was recorded for the first time in the Mingyi Bielu (a pharmaceutical work, written at the end of Han Dynasty, AD 220), and ranked to the top grade of herbal medicine. Thereafter, all Chinese medicine scholars studied and documented it. "Got the essence of soil, be the first in tonics", a description of *Polygonatum*, was recorded in the famous Chinese medicinal book "Bencao gangmu" (i.e. "Compendium of materia medica", written by the famous medical expert Li Shizhen in Ming Dynasty, in 1590).

Polygonatum was recorded as a food item in the ancient medical works Shiliao Bencao and Jiuhuang Bencao as early as in the Tang Dynasty (AD 618–907). However, folk usage of *Polygonatum* in China should be much earlier than these ancient published works. Also in recent years, some ethnobotanical studies have documented the food uses of *Polygonatum* among the rural population (including various ethnic groups) [21–35].

This paper aims to review the traditional knowledge of *Polygonatum* used as food in China. The objective is to gather and document information that will lead to a better understanding of the knowledge of traditional uses of *Polygonatum* in China.

Results

Fifteen species of *Polygonatum* are traditionally used as food in China through literature studies (Tab. 1). Five main categories of use (drink, snack, vegetable dish, staple food and seasoning) were found.

Vegetables dishes ('cai')

The largest category (14 species) is vegetable dishes ("cai"). In China "cai" is usually a plate of mainly one kind of vegetable eaten as a side-dish accompanying a meat dish and rice. Young leaves, stems and rhizomes are the main parts used, and are consumed raw or after frying, decocting, stewing, roasting or steaming. Different places have their own unique mode of preparation and consumptions.

One example is Simao, a tropical county in Yunnan, southwest China, with many minorities, e.g. Yi, Dai, and Lahu. The Dragon-boat festival is one of the important traditional holidays in China. In Simao, it is considered that this festival is the turning point of the wet and dry seasons in a year. Local herbal doctors believed that this time is when people are most vulnerable to disease, so people need to take some tonics to strength themselves and enhance immunity. They collected the rhizomes of *P. cirrhifolium* and *P. kingianum*, and braised them with chicken or pig's feet for use as tonics by people of all ages. The local people also considered that these dishes would strengthen the body, tranquilize the mind and increase appetite [6].

In Changbai Mountains, northeast China, because of the long and cold winter, the local people got used to eating steamed or roasted hot food. They usually use steamed or roasted rhizomes of *P. sibiricum*, *P. odoratum*, *P. filipes*, *P. acuminatifolium*, *P. stenophyllum*, *P. humile*, *P. involucratum* and *P. inflatum* as vegetables [23]. Also in northeast China, people like to eat some plants as salad, usually in the form of vegetables dressed with salty soybean paste. In northern China, fresh cultivated vegetables are not abundant in winter. During the summer or autumn, when plenty of green vegetables are harvested, the local people make some pickled vegetables (including wild vegetables), which can be stored for a long time. The sprouts of *Polygonatum odoratum* are salted for consumption in winter [31]. In some other places, the young leaves, stems, sprouts or rhizomes of *P. odoratum* are usually collected for use in fried dishes or salad, or for soup, after blanching [25].

Drinks

This group consists of three species used to prepare drinks, including wine or liquor and a substitute for tea. The rhizomes of *P. sibiricum* and *P. nakaianum* are soaked in rice wine for a month to make into herbal wine, which will make the drinkers' hair darker and brighter [35]. The rhizomes of *P. odoratum* are usually soaked with liquor and sugar for six months. Before use, the drink is strained with gauze. Drinking it frequently is supposed to eliminate fatigue, moisten skin, and make one beautiful. The rhizomes are also used as a substitute for tea. Some previous studies recorded that rhizomes of *P. sibiricum* were boiled with water for use as tea. It is believed that daily use of this tea will strengthen the body and is especially good for recovering from weakness [15].

Snacks, staple foods, and seasonings

Ten species are consumed as snacks. In the Inner Mongolian Autonomous Region, China, the rhizomes of *P. sibiricum*, *P. odoratum*, *P. macropodium*, were consumed as fresh "fruits" (snacks) by herdsmen. While the livestock were grazing on grassland, herdsmen could collect and eat edible plants any time they felt hungry or thirsty. Grassland areas have always been short on fruits. Because of their good taste, herdsmen usually took the rhizomes of *Polygonatum* as fresh "fruits" (snacks). They believed that people who eat the rhizomes of *Polygonatum* regularly will gain strength and live a long life [11]. In addition, the local Mongolians named *P. sibiricum* "imaan orhodai" and named *P. odoratum* "ebsen orhodai". Here "orhodai" means ginseng (*Panax ginseng*). It implied that the Mongolians have highly appreciated the values of these plants [34].

According to the nutritional study reported [35], the *Polygonatum* rhizomes are rich in starch (25.6–68.46% of dry mass). The starch from *P. sibiricum*, *P. odoratum*, *P. filipes*, *P. acuminatifolium*, *P. stencphyllum*, *P. humile*, *P. involucratum* and *P. inflalum* has traditionally been used in China for a long time.

The local people in northeast of China steam or roast the *Polygonatum* rhizomes and eat them as a snack. The rhizomes are also fried with sugar and honey to make a kind of "candy" [22].

When used as staple food, the local people usually sun-dry the rhizomes of *P. sibiricum*, *P. odoratum*, and *P. nakaianum*, grind them into powder with a stone roller, then sift them to mix into flour to make cakes.

The local people also prepare the rhizomes of *Polygonatum* as seasonings to make some sweet dishes. The pulped rhizomes of *P. sibiricum* are boiled for 4 hours, then filtered, and 2% barley flour is added, to obtain a sugar solution. The sugar solution will be placed into a pot and boiled for 4 hours over a gentle heat. When the sugar solution is boiled into filamentous bright yellow droplets, the *Polygonatum* seasoning named sugar diluents ("tangxi") was processed [36].

Polygonatum uses in other countries

In some other countries, some species in *Polygonatum* are believed to have medicinal properties [18], and have been used as a source of food too. For example, in India, *Polygonatum cirrifolium* and *P. verticillatum* are considered to be wild tasty leafy vegetables [19]. *Polygonatum biflorum* is distributed in the eastern United States and south-central Canada. Its starchy, edible rhizomes were formerly used by early Americans as a potato-like food or the starch was extracted and used to make bread, or mixed in with other foods such as soups. Young shoots can be added to salads, consumed raw or can be boiled

Tab. 1 Species of *Polygonatum* traditionally used as food in China.

Latin name	Used parts	Categories of use	Preparation methods	References
P. acuminatifolium Kom.	Rhizome	Snack	Cooked (steamed or roasted) as snack	[23]
	Sprout	Vegetable	Raw (salad, pickled, dressed with sauce) or cooked (fried or added to soups after blanching)	[31]
P. cirrhifolium (Wall.) Royle	Young stem	Vegetable	Cooked	[7]
	Tuber, rhizome	Vegetable	Raw (salad) or cooked as vegetable	[6,24,29,32]
P. cyrtonema Hua	Rhizome, sprout	Vegetable	Cooked	[25,26,32]
P. filipes Merr.	Rhizome	Vegetable, snack	Cooked as vegetable Cooked (steamed or roasted) as snack	[25]
	Sprout	Vegetable	Cooked as vegetable	[25]
P. humile Fisch. ex Maxim.	Rhizome	Vegetable, snack	Raw (salad, pickled, dressed with sauce) or cooked (fried, steamed, roasted or added to soups) as vegetable Cooked (steamed or roasted) as snack	[23,28]
	Sprout	Vegetable	Raw (salad, pickled, dressed with sauce) or cooked (fried or added to soups after blanching)	[28,31]
P. inflatum Kom.	Rhizome	Snack	Cooked (steamed or roasted) as snack	[23]
	Sprout	Vegetable	Raw (salad, pickled, dressed with sauce) or cooked (fried or added to soups after blanching)	[31]
P. involucratum (Franch. et Sav.) Maxim.	Rhizome	Vegetable, snack	Cooked as vegetable Cooked (steamed or roasted) as snack	[23,30]
	Sprout	Vegetable	Raw (salad, pickled, dressed with sauce) or cooked (fried or added to soups after blanching)	[28,31]
P. kingianum Coll. et Hemsl.	Tuber, rhizome	Vegetable	Raw or cooked (fried or stewed)	[6,7]
P. macropodium Turcz.	Rhizome	Vegetable, snack	Cooked as vegetable Raw as snack (fruit)	[12,23,28,34]
	Sprout	Vegetable	Raw (salad, pickled, dressed with sauce) or cooked (fried or added to soups after blanching)	[28,31,32]
Polygonatum megaphyllum P. Y. Li	Rhizome	Snack	Roasted on embers as famine food	[57]
P. nakaianum Ishidoga	Rhizome	Drink, staple food	Raw as drink (wine) Cooked as staple food (make into powder and mixed with flour)	[35]
P. odoratum (Mill.) Druce	Young stem and leave	Vegetable	Cooked (fried dish or added to soups after blanching)	[21,27–29]
	sprout	Vegetable	Raw (salad, pickled, dressed with sauce) or cooked (fried or added to soups after blanching)	[28,31]
	Rhizome	Vegetable, drink, snack, stale food	Raw or cooked (steamed or roasted) as snack Raw (wine) or cooked (boiled) as drink (substitute for tea) Cooked (steamed) as vegetable Make into powder and mixed with flour as staple food Roasted on embers as famine food in some areas	[11,23,27,28, 31,33–35,57]
P. sibiricum Delar. ex Red.	Sprout	Vegetable	Cooked (fried dish or added to soups after blanching)	[27,28,30]
	Rhizome	Vegetable, snack, drink, staple food, seasoning	Raw (fruit) or cooked (steamed or roasted) as snack Raw (pickled) or cooked (steamed, roasted, fried) as vegetable Raw as drink (wine/liquor) Boiled as substitute for tea Make into powder and mixed with flour as staple food. Cooked to make seasonings	[11,21–23, 27,28,31–35]
P. stenophyllum Maxim.	Rhizome	Snack	Cooked (steamed or roasted) as snack	[23]
P. verticillatum (L.) All.	Rhizome, sprout	Vegetable	Cooked	[24,28,29,32]

and served like asparagus [20]. Some species, in particular *P. sibiricum*, are used as a herbal tea called dungulle in Korea [13].

Medicinal effect

According to "China Pharmacopoeia" (2010), Rhizoma polygonati, prepared from the dried rhizomes of *Polygonatum sibiricum*, *P. kingianum* and *P. cyrtonema*, and processed *P. odoratum* products are the most commonly used medicines in Traditional Chinese Medicine (TCM). TCM considered that the former contains the properties of reinforcing "qi", nourishing "yin" and moistening the lungs, strengthening kidney and spleen, while the processed *P. odoratum* products have the properties of moistening heart and lungs, nourishing the stomach and enhancing bone marrow [37].

Different pharmacologically active groups of compounds have been isolated phytochemically, such as polysaccharides, saponins, phytohormones, glycosides, flavonoids and alkaloids. Pharmacological research has demonstrated that *Polygonatum* species produce effects, showing antihyperglycemic potential, anticancer, analgesic, antipyretic diuretic, antimalarial, antioxidant, antimicrobial, phytotoxic etc. [14]. Polysaccharides are the major bioactive compounds in these plants [38–41]. Phytochemical and pharmacological research on several species including *P. odoratum*, *P. verticillatum*, *P. cyrtonema*, *P. kingianum* have been reported. The research results have shown biological activities and possessed potentially medical applications [42–50]. In fact, based on our literature studies, we see that all the species recorded as used for food, also have medicinal effects. The local people in China believe that to take *Polygonatum* food enhances immunity and longevity, which are in full accord with the results of modern scientific research.

Discussion and conclusions

China is the richest country in *Polygonatum* resources, with 39 species (61.9%). The local inhabitants of China have developed a rich traditional knowledge about the food use of this genus. Based on our literature survey, we have found 15 species used for food, out of which *P. macropodium*, *P. cytonema*, *P. filipes* are endemic to China. There are 5 main categories of use in China, including wild vegetables, drinks, snacks, staple food and seasonings. All the species in the genus have been used for vegetable dishes, 10 species for snacks, 3 species for making drinks, 3 species as staple food and 1 species for seasonings. Some species can be classified into more than one category. For example, the rhizomes of *P. sibiricum* have traditionally been used as elements of vegetable dishes, drinks, snacks, and as a staple food.

Preparation methods are also diverse. Some of the species are eaten as salad, directly or after simple processing. Some of them can only be eaten after being well-cooked or processed. The cooking methods include frying, decocting, stewing, roasting and steaming.

There are some interrelations between the used parts and categories of use. Young leaves, stems and rhizomes are mainly used in vegetable dishes. Rhizomes are used to make drinks, snacks and seasonings. Therefore the rhizomes are the most commonly used parts. Some ancient publications recorded that the roots, leaves, flowers and fruits of *P. sibiricum* were edible. Nowadays, however, there are no records about the use of flowers or fruits of this species as food [51]. Meanwhile, some literature sources report that the fruits of *Polygonatum biflorum* (not distributed in China), contain anthraquinone, which is poisonous, but with a low toxicity, mostly causing vomiting and diarrhea [52]. Thus the fruits of this species are not edible. Based on this information we should be aware of the toxicity of *Polygonatum* fruits. It is possible that the toxicity of a *Polygoantum* species varies depending on the place where it grows or which species, subspecies or variety is taken into account. Therefore further research on the chemistry of *Polygonatum* fruits is necessary to determine whether the fruits of Chinese *Polygonatum* species are edible or poisonous. Some nutritional studies on *P. sibiricum* seeds have been done recently, and showed that the contents of methionine, lysine and isoleucine are higher than those of wheat or corn, and suggested that the seeds of *P. sibiricum* could be eaten [53].

Local Chinese people from all over the country have used *Polygonatum* as food sources. The different places have their own distinctive features. They believe that it would enhance immunity and longevity to take *Polygonatum* as food, which is in full accord with the results of modern scientific research. However, the usages of *Polygonatum* are still at the family level of consumption. Although China is rich in wild *Polygonatum* resources, they have not been fully used and exploited yet. To explore and develop new resources, the traditional preparation methods of *Polygonatum* can provide information. Considering that the plants from this genus have both nutritional and medicinal value, they can be classified into the health food category and are worth developing into more widely used functional foods by combining modern food processing technology and traditional preparation methods. The following points should be taken into account:

(*i*) The wild species of *Polygonatum* are usually pollution free, so they are attractive to food processing industries.

(*ii*) As the plants have medicinal functions, the rhizomes of *Polygonatum* can be developed into functional foods such as "*Polygonatum* paste", "*Polygonatum* porridge", "*Polygonatum* powder" [54]. These products have already appeared on the market, but are not very commonly used.

(*iii*) The rhizomes are the main parts used for the development of food products. The rhizomes are soft and contain starch, other carbohydrates, carotene, vitamins and low coarse cellulose. They are suitable for making various drinks, such as wine, tea substitutes or different flavored drinks. They can be processed into snacks, such as dried, canned, or puffed foods. The fruits of *P. sibiricun* can be exploited as a vitamin B1 source [15]. In addition, other parts, such as seeds, are worth using, but further investigation into the nutritional properties of all useful parts is necessary.

(*iv*) The most popular consumed species of *Polygonatum* in China were *P. sibiricum* and *P. odoratum* (Tab. 1). However, there are still some other species, which are little known and processed. For example, *P. cyrtonema*, which is rarely used for food by local people, has high nutritional quality, unique medicinal functions and plentiful wild resources [55]. P. filipes, which has a wide distribution in 8 provinces in China, i.e. Anhui, Fujian, Guangdong, Guangxi, Hunan, Jiangsu, Jiangxi and Zhejiang, was only used by the local people in Qiandaohu, Zhejiang Province. The *Polygonatum* plants in Anhui Province present another example. Anhui Province, located in eastern China, is one of the major production areas of *Polygonatum* in China, with 11 species of *Polygonatum* and the production of 20–100 tons every year [54]. However, the local people paid more attention to the medicinal value than the edible value so that a few species were collected and used for food. Further study and exploitation could be carried out regarding these species for product development.

(*v*) Most of the *Polygonatum* species are not protected. The local people mainly consume wild *Polygonatum* species for food, and the rhizomes are the main parts used. In addition, most of the *Polygonatum* plants used as medicines sold on the markets are also from wild specimens. All the above-mentioned factors have greatly damaged the local plant resources. There are some species, such as *P. macropodium*, *P. kingianum*, *P. verticillatum*, *P. cirrhifolium*, which have been frequently used both for food and medicine in some places. However, because of the increasing demand, their resources have shrunk. Given this situation, it is necessary to select the species which have the largest use potential for cultivation, or to conserve resources in situ to reduce threats to the rare species and their habitats.

In the past, the Chinese used wild plants for food in famine times when main crops and cultivated vegetables were not available. Nowadays, however, with the intensive development of agricultural and marketing techniques, it is easy to find various cultivated fruits and vegetables and other food in markets throughout the year. Therefore, the younger generation has little awareness of using wild food plants, and the traditional knowledge of using plants for food survives only in the memory of the elders. As a result, it is eroding due to changing social values and a lack of participation of the younger generation in collection and processing of wild plants [56]. However, we believe that there is still plentiful traditional knowledge of food usage of *Polygonatum* to be documented in China, which can provide important scientific clues to developing new food resources.

Acknowledgements

This study was supported by the Ministry of Education of China through its 111 and 985 projects (B08044, MUC 98503-001006 & MUC 98502-006004), the National Natural Science Foundation of China (31070288 & 31161140345), and the Ministry of Science and Technology of China (2012FY110300).

References

1. Hu SY. Food plants of China. Hong Kong: Chinese University Press; 2005.
2. Pei SJ. Preliminary ethnobotanical study of Xishuangbanna. In: Proceedings of tropical plants researches. Kunming: Yunnan People's Press; 1982. p. 16–32.
3. Dao ZL, Long CL, Liu YT. On traditional uses of plants by the Nu people community of the Gaoligong Mountains, Yunnan Province. Sheng Wu Duo Yang Xing. 2003;11(3):231–239.
4. Li QJ, Liu HM, Xu YK, Chen J, Xu ZF. Changes in species number and causes that used as wild vegetableby Dai people in Xishuangbanna, China. Acta Bot Yunnan. 2007;29(4):467–478.
5. Khasbagan, Soyolt, Man L, Enhebayar, Gerelt, Hu W. Traditional usage of wild plants for food by the Ejina Mongolians and its exploitation and ethnoecological significance. Journal of Inner Mongolia Normal University (Natural Science Edition). 2005;34(4):471–488.
6. Liu YT, Long CL, Dao ZL. Ethnobotanical survey on medicinal roots eaten by the local people in Simao, Yunnan Province, during the Dragon-boat Festival. Journal of Plant Resources and Environment. 2003;12(2):33–38.
7. Wang JR, Long CL. Ethnobotanical study of traditional edible plants of Jinuo nationality. Acta Bot Yunnan. 1995;17(2):161–168.
8. Long CL, Li H, Zhou YL, Dao ZL, Abe T. Ethnobotanical studies in Gaoligong Mountains: II. the Dulong ethnic group. Acta Bot Yunnan. 1999;11 suppl:137–144.
9. Long CL, Li H, Dao ZL, Zhou YL. Ethnobotanical studies in Gaoligong Mountains: I. the Dulong ethnic group. Acta Bot Yunnan. 1999;11 suppl:131–136.
10. Jin S, Tie J, Liu RX. Ethnobotanical study on folk wild edible plants in southeast region of Shanxi province. Journal of Jindongnan Normal College. 2004;21(2):13–15.
11. Khasbagen, Huai HY, Pei SJ. Wild plants in the diet of arhorchin mongol herdsmen in inner mongolia. Econ Bot. 2000;54(4):528–536.
12. Wujisguleng W, Khasbagen K. An integrated assessment of wild vegetable resources in Inner Mongolian Autonomous Region, China. J Ethnobiol Ethnomed. 2010;6(1):34.
13. World checklist of selected plant families [Internet]. Kew Royal Botanic Gardens. 2012 [cited 2012 Jul 7].
14. Haroon K, Muhammad S, Naveed M. Pharmacological and phytochemical updates of genus *Polygonatum*. Phytopharmacology. 2012;3(2):286–308.
15. Li JH, Zhou SB. Utilization of *Polygonatum* resources and its assessment of economic value. Anhui Agricultural Science Bulletin. 2006;12(5):64–65.
16. Yang CR, Zhang Y, Wang D, Zhang YJ. Molecular evolution of steroidal saponins in the genus *Polygonatum* (Convalleriaceae) and their chemotaxonomical significance. Acta Bot Yunnan. 2007;29(5):591–600.
17. *Polygonatum* [Internet]. eFloras.org. 2012 [cited 2012 Aug 25].
18. Youn YN, Lim E, Lee N, Kim YS, Koo MS, Choi SY. Screening of Korean medicinal plants for possible osteoclastogenesis effects in vitro. Genes Nutr. 2007;2(4):375–380.
19. Misra S, Maikhuri RK, Kala CP, Rao KS, Saxena KG. Wild leafy vegetables: a study of their subsistence dietetic support to the inhabitants of Nanda Devi Biosphere Reserve, India. J Ethnobiol Ethnomed. 2008;4(1):15.
20. *Polygonatum biflorum* [Internet]. Missouri Botanical Garden. 2012 [cited 2012 Aug 25].
21. Huang JS, Peng YL, Li M, Wang Y. Exploitation and utilization of wild vegetables in the north of Hunan. Chinese Wild Plants Resources. 2002;21(6):39–40.
22. Su SL, Ma B, Huang K. Research on the ethnobotany of *Panax notoginseng*. Anhui Agricultural Science Bulletin. 2012;18(1):56–58.
23. Zhou Y. The *Polygonatum* plant resources in Changbai Mountains and their exploitation. Chinese Wild Plants Resources. 2002;21(2):34–35.
24. Wang ZK, Wang ZH, Xing Z, Li RQ, Liu H, Deqingcuomu. Preliminary investigation and studies on wild vegetable resources in Nyingtri, Tibet. China Vegetables. 2010;2:75–78.
25. Feng BC, Yu MS, Xu GF, Yan SF, Chen AM, Liu WH, et al. Exploiting and utilizing edible wild plants in Thousand-Isle Lake. Central South Forest Inventory and Planning. 2010;29(1):41–44.
26. Xu LZ. Exploitation and utilization of the wild vegetable resource in northeastern Guangdong province. Anhui Agricultural Science Bulletin. 2006;3(12):2690–2694.
27. Zhu Q, Guo S, Wang J, Tian Y. Investigation on the wild edible plant resources in Ningxia. Journal of Ningxia University (Natural Science Edition). 2004;28(4):367–371.
28. Khasbagen, Soyolt, editors. Wild vegetable resources in Inner Mongolia and their ethnobotanical study. Beijing: Science Press; 2008.
29. Han YX, Zhao CH, Jing ZP, Li GF, Gong FL, Chen SY, et al. Hill wild vegetables investigation in Huzhubeishan forest region. Qinghai Prataculture. 2001;10(2):46–49.
30. Fan SJ, Zhao ZT. Wild vegetable blog in Shandong. Jinan: Shandong Science and Technology Press; 1996.
31. Dong R, Li GC, Fan SL. Wild vegetables in Changbai Mountains. Beijing: Science Press; 1997.
32. Zhao JG, Wei XB, Guo WC. Chinese potherb. Changchun: Jilin Science and Technology Press; 2004.
33. Dong SY, Wei CR, Yang CJ. Recipes of Chinese wild vegetables. Beijing: China Tourism Press; 1993.
34. Khasbagan, Pei SJ. Wild plants used for the folk dietotherapy in Arhorchin Mongolians. Zhong Yao Cai. 2001;24(2):83–85.
35. Bureau of Production & Consumption of Chinese Ministry of Commerce, Institute of Botany of Chinese Academy of Sciences. Flora of Chinese economic plants. Beijing: Science Press; 2012. (Flora of Chinese economic plants; vol 1).

36. Wang ZX. China resources plants utilization manual. Beijing: Chinese Science and Technology Press; 1989.
37. China Pharmacopoeia Committee. China Pharmacopoeia. Beijing: Chinese Medicine and Technology Press; 2010. (China Pharmacopoeia; vol 1).
38. Liu L. Purification, structure identification, biological activity and structure-activity relationship research of polygona polysaccharides. Shanghai: Shanghai Institute of Materia Medica; 2005.
39. Zhang TT, Xia XK, Chen CP, Wu AP, He M, Nie LW. Biological activities of polysaccharides from *Polygonatum sibiricum* Redoute. Chinese Journal of Experimental Traditional Medical Formulae. 2006;12(7):42–45.
40. Hu CY, Xu DP, Wu YM, Ou SY. Triterpenoid saponins from the rhizome of *Polygonatum sibiricum*. J Asian Nat Prod Res. 2010;12(9):801–808.
41. Deng Y, He K, Ye X, Chen X, Huang J, Li X, et al. Saponin rich fractions from *Polygonatum odoratum* (Mill.) Druce with more potential hypoglycemic effects. J Ethnopharmacol. 2012;141(1):228–233.
42. Janeczko Z, Jansson P, Sendra J. A new stereoidal saponin from *Polygonatum officinale*. Planta Med. 2007;53(1):52–54.
43. Yu HS, Ma BP, Kang LP, Zhang T, Jiang FJ, Zhang J, et al. Saponins from the processed rhizomes of *Polygonatum kingianum*. Chem Pharm Bull. 2009;57(9):1011–1014.
44. Hirai N, Miura T, Moriyasu M, Ichimaru M, Nishiyama Y, Ogura K, et al. Cardiotonic activity of the rhizome of *Polygonatum sibiricum* in rats. Biol Pharm Bull. 1997;20(12):1271–1273.
45. Lin HW, Han GY, Liao SX. Studies on the active constituents of the Chinese traditional medicine *Polygonatum odoratum* (Mill.) Druce. Yao Xue Xue Bao. 1994;29(3):215–222.
46. Khan H, Saeed M, Gilani AU, Khan MA, Dar A, Khan I. The antinociceptive activity of *Polygonatum verticillatum* rhizomes in pain models. J Ethnopharmacol. 2010;127(2):521–527.
47. Li XC, Yang CR, Makoto I, Hiromichi M, Ryoji K, Kazuo Y. Steroid saponins from *Polygonatum kingianum*. Phytochemistry. 1992;31(10):3559–3563.
48. Liu B, Cheng Y, Bian HJ, Bao JK. Molecular mechanisms of *Polygonatum cyrtonema* lectin-induced apoptosis and autophagy in cancer cells. Autophagy. 2009;5(2):253–255.
49. Liu F, Liu Y, Meng Y, Yang M, He K. Structure of polysaccharide from *Polygonatum cyrtonema* Hua and the antiherpetic activity of its hydrolyzed fragments. Antiviral Res. 2004;63(3):183–189.
50. Ye HC, Zhang XP, Gao GB, Jiang JH. Chemical constituents and biological activity of volatile oil from the tubers of *Polygonatum* filipes. Guihaia. 2009;29(3):417–419.
51. Materia medica for dietotherapy [Internet]. 2012 [cited 2012 Aug 25].
52. *Polygonatum* spp. [Internet]. Poisonous plants. 2012 [cited 2012 Aug 25].
53. Li YG, Ji JY, Zhang YJ. Measurement and analysis of the nutrients of *Polygonatum sibiricum* seeds. Xibei Zhiwu Xuebao. 2009;29(8):1692–1696.
54. Li JH, Zhou SB. Advances in *Polygonatum* from Anhui Province. Chinese Wild Plant Resources. 2005;24(5):17–19.
55. Sun Z. An investigation of three species of *Polygonatum* and quality assessment of *Polygonatum cirrhifolium* (Wall.) Royle. Beijing: Beijing University of Chinese Medicine; 2009.
56. Balemie K, Kebebew F. Ethnobotanical study of wild edible plants in Derashe and Kucha Districts, South Ethiopia. J Ethnobiol Ethnomed. 2006;2:53.
57. Kang Y, Łuczaj Ł, Zhang S, Kang J. Wild food plants and wild edible fungi of Heihe valley (Qinling Mountains, Shaanxi, central China): herbophilia and indifference to fruits and mushrooms. Acta Soc Bot Pol. 2012;81(4):405–413.

11

Wild food plants of Remote Oceania

Will C. McClatchey*

Botanical Research Institute of Texas, 1700 University Drive, Fort Worth, Texas 76107, USA

Department of Botany, University of Hawai'i at Manoa, Honolulu, Hawai'i 96822, USA

Environmental Studies Program, Texas Christian University, Fort Worth, Texas 76107, USA

Abstract

Agricultural societies partly depend upon wild foods. Relationships between an agricultural society and its wild foods can be explored by examining how the society responds through colonization of new lands that have not been previously inhabited. The oldest clear example of this phenomenon took place about 5000 years ago in the tropical Western Pacific at the "boundary" interface between Near and Remote Oceania. An inventory of wild and domesticated food plants used by people living along "the remote side of" that interface has been prepared from the literature. This was then assessed for the roles of plants at the time of original colonization of Remote Oceania. The majority of species are wild foods, and most of these are used as leafy vegetables and fruits. The wild food plants mostly serve as supplements to domesticated species, although there are a few that can be used as substitutes for traditional staples.

Keywords: Oceania, wild food plants, colonization, agriculture

Introduction

Human-plant co-evolutionary relationships have been documented for processes of wild plant food domestication into socially critical crops [1,2]. However, major "food" plants consumed by humans were first derived from plants that have co-evolved with other animals. The plants make use of the animal's food habits for pollination, seed dispersal, and other services. Other animals include grazers (such as cattle), browsers (such as deer), frugivores (such as monkeys and birds), and omnivores (such as pigs and rats).

Wild food plants contrast to those that were domesticated in that populations of domesticated plants have gone through a selection process, by humans, that has resulted in a change in their genetic structure that benefits humans [3]. Conversely, wild plants persist without need of any selection or tending by humans and have genetics that reflect their original co-evolutionary relationships. In many parts of the world, wild relatives of domesticated plants exist side by side, or within short distances of their domesticated kin. Speculation about human development of agriculture from wild plants has been based on observations that agriculture developed from seeds left in dump heaps [4] and/or wet river camp sites [5]. The implication is that humans brought wild plant materials to their homes or processing sites, much as squirrels or fruit bats might do, and then consumed less than all of the material collected, discarding material (seeds or other living parts) that could grow into new plants.

The process of transition from wild to domesticated remains vague. Some plants exist as persistent hangers-on in the shadow of humanity, while others are periodically brought near to the home to survive as captives. For the purposes of this paper, wild and domesticated are determined not by genetic changes in plants, but based on how humans interact with the plants. Those that are actively tended will be called domesticated, while those that are left to grow on their own, being ignored until needed for food, will be called wild. Plants that are generally cultivated, but sometimes survive on their own will be called feral when being ignored by humans.

This paper sets out to provide: (*i*) a brief inventory of wild foods used at the interface of geographic regions called Near and Remote Oceania and then; (*ii*) an assessment of the roles of these plants at the time of original colonization of Remote Oceania. Several background issues are presented first that set the context for understanding the environment, people, and plants. The colonization of Remote Oceania and the role of wild food plants is important because this is the most ancient known example of initial colonization of a part of the earth by an agricultural society.

Wild foods

Wild foods constitute the diet of traditional hunter-gathers although the details of how our ancestors lived and made decisions are controversial [6]. Human cultures that developed agricultural practices began to emphasize some plants (and animals) by investing more time in them. These plants became

* Email: wmcclatchey@brit.org

those that are cultivated. Uncultivated plants were still used for many purposes, including food. One of the uses of uncultivated plants has undoubtedly been for periods of insufficient production of cultivated plants. All cultures periodically experience some sort of food shortages [7] that require consumption of scarcity foods [8]. In the literature, these are often called "famine foods" even though starvation may not actually be associated very often with their consumption. These foods are typically produced from wild or feral plants. Probably it is better to sort foods into categories of preferred and secondary or alternative foods since this more accurately reflects what is seen in the world. In actual famine, starving people will eat almost anything. This paper is not about the foods eaten by starving people, but rather about the alternative foods, particularly those that are not cultivated and therefore may be called wild plant foods [9].

Cox [9] argues that alternative foods are consumed in patterns that mirror diets of normal times, serving as analogs of preferred foods. He uses two examples from Samoa (in Remote Oceania) to illustrate this. *Cyrtosperma* is a wild source of starchy rhizomes. It is an analog of cultivated *Colocasia* that also produces starchy rhizomes. Conversely, feral *Metroxylon* is a starch source (extracted from the stems) that is not eaten as an alternative food to *Colocasia* because it is dissimilar to the traditional diet pattern. He furthermore states that famine or alternative foods are likely to be introduced species and/or previous cultivars that were important in the past or in other prior locations.

Wild foods may be alternatives because they are less productive or require more difficult preparation. Preparation may involve longer cooking (higher fuel/time costs), removal of protective toxins (washing, cooking, peeling or abrading of toxic layers), or lengthy processing of more, smaller plant parts (such as seeds) to generate similar volumes of food. For example, aroids, such as *Alocasia* contain proteolytic enzymes that can cause severe irritation and tissue damage. These enzymes are denatured with heat so that cooking with steam is a common strategy for rendering the rhizomes and leaves edible [10] (it is common to find reports in the literature indicating that these must be cooked to destroy oxylate crystals. However, this is not supported by research [10]). Most varieties of *Colocasia* are easier to process than varieties of *Alocasia* rendering *Alocasia* as an alternative to *Colocasia.* A second example is processing of seeds from nuts. *Terminalia catalpa* L., *Canarium indicum* L., and *Barringtonia edulis* Seem., each have larger seeds that accumulate faster than their wild relatives of the same genera. Finally, among communities that do extract starch from plants such as *Amorphophallus, Metroxylon,* or *Tacca, Cycas* species are alternatives because they require multiple additional processing steps to remove toxins that are not present in the preferred species. Abbott [11] reports that Hawaiians pound, salt, or otherwise process algae to make it edible or palatable. Similar processes are required for some other wild species [8] that make them less desirable.

Oceania (Near and Remote)

The focus of this paper is on the wild foods that were traditionally eaten within the geographic area of Remote Oceania prior to colonization contact from Europeans. Oceania encompasses the islands of the tropical Pacific Ocean through New Guinea, Australia and New Zealand. The terrestrial biological core is a holdout for marsupials with only relatively recent invasions by mammals, including humans. Plant diversity is mostly a mixture of a series of invasions from the Gondwana Australian continent and SE Asian Borneo centers of diversity, with smaller contributions from mainland Asia. These converge within New Guinea and the Philippines forming rich floras. Subsequently, these floras have distributed eastward, island hopping into the Pacific basin with a resulting dramatic attenuation of species richness and diversity from west to east [12].

Biogeographically, Oceania has been divided into two functional areas, Near and Remote Oceania [13] (Fig. 1). Near Oceania was settled around 50000 years ago by pre-agricultural, hunter-gatherer people [14] with certain domesticates such as dogs [15] not arriving until 5000 years ago. Subsequently, they participated in domestication of many wild plant species within the region [16]. Around 5000 years ago, Austronesian language speakers invaded Near Oceania from areas of eastern Asia [17–19]. They introduced new agricultural practices, maritime technologies and perspectives, and alternative ways of perceiving the environment. Their relatively rapid expansion then led into Remote Oceania where they become the first settlers [20] expanding in pulses until reaching all of Eastern remote Oceania by 1000 years ago [21]. Remote Oceania includes islands to the north and south of "the geographic" Solomon Islands (Vanuatu, New Caledonia, Palau, Federated States of Micronesia), and the Pacific islands to the east beginning with the Santa Cruz Islands, almost until the Americas (all of Micronesia and Polynesia; Fig. 1).

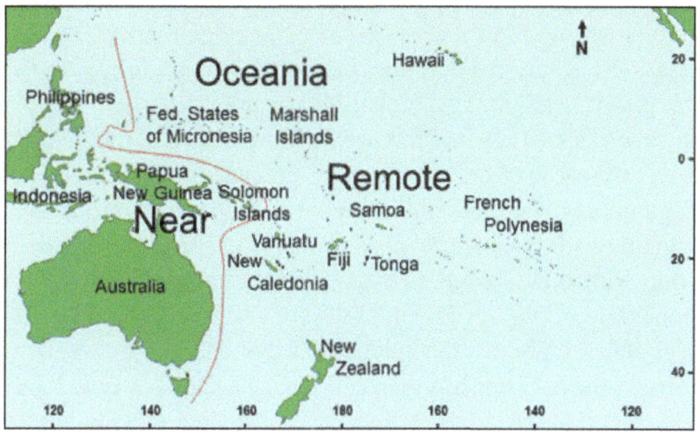

Fig. 1 Near and Remote Oceania separated by red line.

Large birds [22] and fruit bats [23] are important parts of the pre-human diversity in Oceania, especially Remote Oceania. These animals selected for plants with larger fruits and nuts, and were also the animals responsible for distribution of some species across and between islands. These plants would then have been good targets for possible human foods. For the most part, in Remote Oceania there are no terrestrial mammals and very few reptiles. Insectivorous and fruit bats are common in some but not all islands and atolls. Sea-birds and flightless birds, including some that were quite large, were once common. However, species diversity is generally low, based in part upon on strand organisms distributed by the oceans and common to the tropical Indian and Pacific Ocean regions. Other organisms are decedents of rare vicarious events, with some consisting of closely related adaptive radiations. In short, plant and animal diversity is lower than in most other locations where humans have settled. This results in fewer opportunities for wild foods, and because of limited animal life, many plants are not adapted for larger animal dispersal with edible fruits.

Colonization of Remote Oceania by Austronesians

About 5000 years ago, Near Oceania was invaded by Austronesian (a language family) speaking people from East Asia [17,24]. The Austronesian colonization of Near Oceania resembles the invasion of North America by Europeans. Austronesian settlers dominated interactions with the earlier inhabitants, introducing different technologies, including plant traditions, and adopting Oceanic cultivated and wild plants. The indigenous population and their languages were displaced or incorporated (we don't know how brutal or kind this was). Within a relatively short period of time, the colonists had reached the eastern edges of Near Oceania [25]. They did not stop. Leaving Near Oceania, they had to cross large ocean gaps to the next closest islands.

When they reached the islands of Remote Oceania, they found lands uninhabited by humans. Not only were there no people already in the islands but there were also not pre-established crops that could be taken over. There were, however, some of the same (or similar) wild plant resources that could serve as temporary sustenance until introduced crops could become established. In global history, this is probably one of only two examples of agriculturalists settling lands that had not been previously inhabited by hunter-gatherers [26] or other agriculturalists (the other example is the Viking settlement of Iceland [27]).

It is not entirely clear what the life ways were like of the first people to explore Remote Oceania [28]. However, it is almost certain that they practiced some form of agriculture, supplementing their diets with hunting and gathering of wild foods [29]. Kirch [30] describes first landfalls in parts of Remote Oceania as "Arriving in double-hulled canoes propelled by sails of woven mats, they had followed the flightways of the golden plover in search of unknown lands. The canoes held a precious cargo necessary to reestablish life on new landfalls: bundles of living taro shoots, breadfruit seedlings, and slips of sugarcane, along with other plants to establish new gardens. A floating menagerie of pigs, dogs, and chickens wallowed about in the bilge water. Little rats the Polynesians called kiore, the size of a house mouse, either hid themselves in the cargo or may have been taken along on purpose as a food source. A Polynesian Noah's Ark, bound for the unknown". Their diet possibly focused on *Colocasia* [31], or a combination of *Dioscorea*/*Musa* sp./*Amorphophallus paeoniifolius* (Dennst.) Nicolson/*Tacca leontopetaloides* (L.) Kuntze/*Pueraria montana* (Lour.) Merr. var. *lobata* (Willd.) Maesen & S. M. Almeida ex Sanjappa & Predeep [32]. Yen [33] has argued that the indigenous subsistence systems in the Western Pacific (interface region of Near/Remote Oceania) consisted of annual plant cultivation (with emphasis on root crops); arboriculture (trees and perennials); animal husbandry (pigs, chickens, dogs); and hunting and gathering (fish, birds, medicinals, raw materials). Agriculturalists would have been limited to wild foods during the initial stages of colonization, probably implying a recurring need to explore wild food resources (mostly marine) until agricultural crop cycles became established.

The combination of: (***i***) starch produced through swamp-based vegeculture of root crops; (***ii***) lipid/protein produced through arboriculture of large seeds; (***iii***) wild-collected vegetables, typifies the experience that colonists of Remote Oceania would have experienced in Near Oceania. Animal protein was likely limited with domesticated sources being pigs, chickens, dogs, and rats. Wild animal protein came primarily from fish and secondarily from birds, turtles, and bats (as well as native rats and possums in Near but not Remote Oceania).

Classification of food

The ways that people think about food, plan for food production, and classify plants and foods shape the results of their efforts. In order to develop a picture of how decisions were made, we must understand the emic classification system of Austronesian languages. Austronesian languages in general apply classifications recognizing plants, some with specific terms [34], and others with implied terms. A Proto Oceanic reconstructed language includes terms for many cultivated plants and some wild plants [35].

Two sets of Proto Oceanic terms are important for understanding "wild" plant foods in Oceania. The specific terms vary and are less important than the overall pattern [33]. First, there is a set of cognitive terms that are used to spatially and functionally arrange the world of plants. Natural or old primary vegetation is opposed to gardening areas. Swamps, lagoons, and riparian zones are important subsets of natural vegetation, while gardens and other locations where plants are tended may be divided into several categories following natural vegetation succession cycles that result from conversion of primary vegetation into gardens [36]. The products of these different vegetation categories may be termed as wild or not, with adjectives that often refer to source areas (e.g., from forests, swamps).

Second, meals of Oceanic peoples include two primary categories of ingredients: starches (mostly roots and rhizomes) and other (vegetables, meats) [37]. Food is often discussed as being an interchangeable term with starches [38]. The author has experienced many traditional meals in Oceania consisting entirely of one or more starch sources, confirming this perspective. Therefore, if primary starches were unavailable, Oceanic people would first look for wild starch sources before looking to other sources of sustenance. Ross [39] has in fact pointed out that in many Oceanic languages the verb "to eat" literally or implicitly refers to eating starchy foods. A tertiary category of non-meal foods is widely used as well. This category is raw fruit and raw fish [37].

An additional concern about classifications is the synonymy between circumscriptions of plants that are labeled with scientific names and the circumscriptions for plants in local languages [40]. There are undoubtedly many species that are recognized by science that are grouped into a single taxon by local people. Sometimes this is due to lack of exposure to more than one species at the same time (resulting in lumping of similar species), while in other cases, the uses or other culturally important indicators do not support a distinction with different names (and therefore only one name is applied to several). A different situation can occur with highly used, often cultivated plants. In this case, distinct species level names are applied by local people, and scientists lump together diversity that is not recognized by science [41].

Growing and using plants

Classification systems for plants and environments are bound into relationships with understandings of how and where plants grow. These forms of knowledge are reinforced through positive feedback cycles as plants are observed and used. Complicating the focus of this paper is the nature of cultivated plants that were introduced by the early settlers into Remote Oceania. Root crops and some vegetables are grown using vegetative cuttings. Tree seed/fruit crops in Remote Oceania are grown mostly by seeds but sometimes from root suckers or vegetative cuttings. We do not know if the colonists

Tab. 1 Cultivated food plants that were introduced into Remote Oceania prior to European contact.

Scientific name	English vernacular	Major part consumed							
		Fl	**F**	**L**	**Rh**	**R**	**Se**	**S**	**Sa**
Wild in Near and Remote Oceania (65+)									
Auricularia cornea Ehrenb.	Ear fungus		F						
Barringtonia spp.	Cut nut						Se		
Blechnum spp.	-			L					
Bruguiera gymnorhiza (L.) Savigny	Orange mangrove		F						
Bruguiera spp.	-		F						
Burckella obovata (G. Forst.) Pierre	Red silkwood		F						
Canarium spp.	-						Se		
Caulerpa racemosa (Forsskål) J. Agardh	-			"L"					
Claoxylon spp.	-							S	
Corynocarpus spp.	-		F						
Cyathea spp.	-			L				S	
Cycas circinalis L.	Cycad						Se		
Cyclosorus spp.	-			L					
Cyrtosperma merkusii (Hassk.) Schott	Swamp taro				Rh				
Dennstaedtia samoensis (Brack.) Moore	-			L					
Diplazium spp.	-			L					
Dracontomelon spp.	Dragon plum		F						
Drymophloeus spp.	-							S	
Ficus spp.	Wild fig			L				S	
Finschia chloroxantha Diels	-						Se		
Geniostoma rupestre J. R. Forst. & G. Forst.	-			L				S	
Gnetum gnemon L.	-			L			Se		
Gracilaria spp.	-			"L"					
Halymenia spp.	-			"L"					
Haplolobus floribundus (K. Schum.) H. J. Lam	-		F				Se		
Heritiera littoralis Aiton	Lookingglass tree						Se		
Hornstedtia scottiana (F. Muell.) K. Schum.	-		F						
Horsfieldia spp.	-		F					S	
Inocarpus fagifer (Parkinson ex Zollinger) Fosberg	Tahitian chestnut						Se		
Laportea spp.	-			L				S	
Maesa spp.	-		F						
Mangifera spp.	Wild mango		F						
Marsdenia tenacissima (Roxb.) Moon	-			L					
Metroxylon warburgii Becc.	Sago							S	
Morinda citrifolia L.	Indian mulberry		F	L					
Musa spp.	Wild banana		F						
Musa troglodytarum L.	-		F						
Myristica spp.	Wild nutmeg							S	
Nypa fruticans Wurmb	Nipa			L					
Ochrosia elliptica Labill.	-							S	
Omphalea spp.	-							S	
Pandanus spp.	Screw pine		F						
Pandanus tectorius Parkinson	Screw pine		F						
Parartocarpus beccarianus Baill.	-		F						
Pisonia grandis R. Br.	-			L					
Planchonella obovoidea H. J. Lam	-		F						
Pleocnemia spp.	-			L					
Polyscias scutellaria (Burm. f.) Fosberg	Balfour's aralia			L					
Pometia pinnata J. R. Forst. & G. Forst.	Island lychee		F						
Pouteria spp.	Egg fruits		F						
Pritchardia spp.	-			L					
Pseuderanthemum spp.	-			L					
Rhizophora apiculata Blume	-		F						
Rubus moluccanus L.	Molucca raspberry		F						
Schleinitzia spp.	-							S	
Solanum repandum G. Forst.	Pacific tomato		F						
Stenochlaena spp.	-			L					

Tab. 1 *(continued)*

Scientific name	English vernacular	Major part consumed							
Sterculia spp.	-		F						
Syzygium clusiifolium (A. Gray) Müll. Stuttg.	-		F						
Syzygium spp.	-		F						
Terminalia catappa L.	Indian almond						Se		
Terminalia spp.	-						Se		
Thelypteris spp.	-				Rh			S	
Ulva spp.	Sea lettuce			"L"					
Wollastonia biflora (L.) DC.	-							S	
Wild in Remote and not Near Oceania (1?)									
Rhizophora mangle L.	Red mangrove		F						
Often encountered in cultivation in Remote Oceania (16)									
Abelmoschus manihot (L.) Medik.	Slippery cabbage			L					
Citrus grandis (L.) Osbeck	Pomelo		F						
Cocos nucifera L.	Coconut						Se		Sa
Colocasia esculenta (L.) Schott	Taro			L	Rh				
Cucumis melo L.	Musk melon								
Curcuma longa L.	Turmeric				Rh				
Dioscorea alata L.	Greater yam					R			
Dioscorea esculenta (Lour.) Burkill	Lesser yam					R			
Dioscorea nummularia Lam.	-					R			
Homalomena pendula (Blume) Bakh.f.	-	Fl							
Ipomoea aquatica Forssk.				L				S	
Ipomoea batatas (L.) Lam.	Sweet potato			L		R			
Lablab purpureus (L.) Sweet	Hyacinth bean	Fl	F	L					
Musa acuminata × *balbisiana* Colla	Banana		F						
Saccharum edule Hassk.	Pitpit	Fl							
Saccharum officinarum L.	Sugar cane							S	
Often encountered as feral in Remote Oceania (18+)									
Aleurites moluccana (L.) Willd.	Candlenut						Se		
Alocasia macrorrhizos (L.) G. Don	Giant taro				Rh				
Alpinia novae-hiberniae B. L. Burtt & R. M. Sm.	-							S	Sa
Amorphophallus paeoniifolius (Dennst.) Nicolson	Elephant foot yam				Rh				
Archidendron spp.	-		F						
Canarium indicum L.	Pili nut						Se		
Cordyline fruticosa (L.) A. Chev.	Ti					R			
Dioscorea bulbifera L.	Air potato					R			
Dioscorea pentaphylla L.	Five-leaf yam					R			
Flacourtia rukam Zoll. & Moritzi	Rukam		F						
Metroxylon spp.	Sago							S	
Pueraria montana (Lour.) Merr. var. *lobata* (Willd.) Maesen & S. M. Almeida ex Sanjappa & Predeep	Kudzu					R			
Solanum viride Spreng.	Green nightshade		F						
Spondias dulcis Parkinson	Hog plum		F						
Syzygium corynocarpum (A. Gray) Müll. Stuttg.	-		F						
Syzygium malaccense (L.) Merr. & L. M. Perry	Mountain apple		F						
Tacca leontopetaloides (L.) Kuntze	Pacific arrowroot					R			
Zingiber zerumbet (L.) Roscoe ex Sm.	Shampoo ginger			L	Rh				
Domesticated in Remote Oceania (2)									
Artocarpus altilis (Parkinson) Fosberg	Breadfruit		F						
Barringtonia edulis Seem.	Cut nut						Se		

Only English names are given since most have many different local vernacular names. F – fruit or fruit-like reproductive structures; Fl – flowers; L – leaves or leaf-like structures; R – roots; Rh – rhizomes; S – stems or stem tips; Sa – stem sap; Se – seeds. See [35,43,44].

observed or understood the reproduction of the wild plants that they used both in Near and Remote Oceania. However, it would be consistent with their cultivation activities, if they considered starch sources to be reproduced via vegetative cuttings of herbaceous/rhizomatous plants and tree seed/fruit crops from seeds. If this were the case, then we would expect them to seek out root crops and tree seeds from among the plants of Remote Oceania. A propagation-based wild food plant search image such as this would focus their exploration efforts. Plants that did not fit the desired pattern might not have been used, or propagated, and would be unlikely to become domesticated. However, if discovered anyway, their use would persist as wild resources rather than entering a path toward domestication.

Cox [9] categorized wild plants used in Oceania into three different origins: (***i***) those predating arrival of humans; (***ii***) those introduced deliberately or inadvertently during the colonization process; (***iii***) those introduced subsequent to human colonization as part of voyaging/trading. To a certain extent, it is very difficult to distinguish when most of the introduced plants arrived and therefore, with a few exceptions, it is hard to tell which plants are parts of category (***ii***) verses (***iii***). Tab. 1 has been sorted into somewhat similar categories with the following distinctions: (***i***) those predating arrival of humans that are widely distributed (in Near and Remote Oceania) as wild plants (and therefore accessible to people prior to entering Remote Oceania) – indigenous wild plants; (***ii***) those distributed in Remote Oceania but not in Near Oceania (implying that these would be new) – endemic wild plants; (***iii***) those distributed in Near but not Remote Oceania prior to human colonization of Remote Oceania (implying that these were deliberately or inadvertently introduced) – introduced wild plants, and three categories of cultivated plants that are important to distinguish within this process; (***iv***) those domesticated in Near Oceania (or elsewhere outside of Remote Oceania) that were introduced prior to European contact and are largely grown as non-wild, cultivated food plants; (***v***) those that have become feral and are often not actively cultivated; (***vi***) cultivated plants that were domesticated from wild plants within Remote Oceania. The last category is essential for understanding the ingenuitive nature of the colonizers of Remote Oceania, and is difficult to distinguish because of the fine line between cultivation and wild. There is surely a continuum of plants from those receiving no tending whatsoever, to those that have either some level of human selection pressure and cultivation, to those that are clearly cultivated.

Inventory of wild plants in Remote Oceania

Names and distributions of cultivated and wild food plants that are used on islands at the boundary between Near and Remote Oceania have been brought together using studies of food plants from the region. While published literature [8,35,42–44] serves as the basis for much of the following discussion, the author's own published and unpublished interviews with indigenous communities in a wide variety of Oceanic locations inform specific details that are not easily found in the literature.

Species in genera with overlapping common names (more than one species with the same common name) have been lumped and only listed under the scientific genus. When multiple common names were reported for several species but there was partial overlap, then these were also lumped under the generic scientific name.

Results

More than 100 taxa are conservatively estimated to have been involved in the successful colonization of Remote Oceania. The majority of these (more than 65%), are wild plants with broad distributions across Oceania spanning the boundary between Near and Remote Oceania. More than 34% of plants were cultivated at one time or another. Only two food plants appear to have been domesticated within Remote Oceania.

In addition to those listed in Tab. 1, many modern introduced cultigens may also appear to be semi-cultivated or feral and therefore used as "wild" resources today. These include: *Allium cepa* L., *Capsicum annuum* L., *Carica papaya* L., *Mangifera indica* L., *Manihot esculenta* Crantz, *Passiflora edulis* Sims, *Passiflora foetida* L., *Psidium cattleyanum* Sabine, *Psidium guajava* L., *Psophocarpus tetragonolobus* (L.) DC., *Solanum granuloso-leprosum* Dunal, *Solanum melongena* L., *Tamarindus indica* L., *Trichosanthes cucumerina* L., *Vigna unguiculata* (L.) Walp., *Xanthosoma* spp. While not part of this assessment, the selections made are often consistent with ancient food choices: trees with edible fruit and starchy root crops.

Discussion

Near versus Remote Oceanic wild plant use

Wild plants used in Near Oceania are dramatically more diverse [8,35] than those in Remote Oceania and even when present in Near and Remote, fewer are used in Remote. This phenomenon can easily be observed within Solomon Islands where part of the islands are Near and part Remote. As an example, *Stenochlaena laurifolia* Presl. is widely considered as a food in Near Oceania (Malaita, Makira, Guadalcanal, Papua New Guinea [45]) but is not eaten in Remote Oceania (Santa Cruz Islands) even though it grows there.

Some species are wild in western parts of Remote Oceania and introduced in the east. There is, in fact, an attenuation with more wild and cultivated species in the west than east. This implies that all were not moved together, or did not equally persist, and possibly that they were discovered or introduced at different times and places. Certainly there are few that show evidence of being moved from east to west, the notable exception being *Ipomoea batatas* (L.) Lam.

Two food plants that were domesticated in Remote Oceania are *Artocarpus altilis* (Parkinson) Fosberg and *Barringtonia edulis*. In both cases, there are similar, related species of trees that are used as wild foods in Near Oceania. The case of *A. altilis* is particularly interesting because across Remote Oceania it is grown in a mixed range of almost wild (feral) to completely cultivated (dependent on humans for vegetative propagation). Furthermore, it has become a critical food resource for several cultural groups, perhaps enabling long-term settlement of some marginal islands.

The dominant cultivated foods of Near Oceania are herbaceous, starchy root/rhizome crops [e.g., *Alocasia macrorrhizos* (L.) G. Don, *Colocasia esculenta* (L.) Schott, *Cyrtosperma merkusii* (Hassk.) Schott, *Dioscorea* spp.] augmented with trees producing lipid/protein-rich seeds [e.g., *Barringtonia* spp., *C. indicum* L., *Cocos nucifera* L., *Inocarpus fagifer* (Parkinson ex Zollinger) Fosberg, *Terminalia* spp.]. These plants all thrive in wet, often swampy environments. Classic agriculture based on herbaceous grass and legume seeds was absent. However grasses (*Saccharum officinarum* L., *Saccharum edule* Hassk.), and legumes [*I. fagifer*, *Lablab purpureus* (L.) Sweet, and *Pueraria montana* (Lour.) Merr. var. *lobata* (Willd.) Maesen & S. M. Almeida ex Sanjappa & Predeep] were all grown and used differently than grass/legume agriculture in Asia, Europe, Africa, or North America, encouraging a different perspective on what food plants should look like.

Near Oceanic food production also included *Cordyline fruticosa* (L.) A. Chev. (roots and leaves) and *Musa acuminata* × *balbisiana* Colla (fruit and leaves), that were probably used as much for food processing (leaves) as they were for food sources (respectively roots and fruit). *Metroxylon* spp. trees were also harvested in some locations for their stem starch, however, their leaves served as superior thatching material making them more valuable in most locations for that purpose.

Leaves from *Abelmoschus manihot* (L.) Medik. and *C. esculenta* were the only commonly cultivated leafy green vegetables consumed. Most other vegetables and fruit were collected from mainly wild sources.

Nuts and seeds

Fruits of several trees were grown as staple nut/seed foods in parts of Oceania, feral or wild food in other parts, and are absent in others. This pattern is generally from west to east crossing the Pacific. Examples of trees that were domesticated in Near Oceania and then dispersed across Remote Oceania include: *Aleurites moluccana* (L.) Willd., some *Barringtonia* spp., *Canarium indicum*, *C. nucifera*, *Finschia chloroxantha* Diels, *Haplolobus floribundus* (K. Schum.) H. J. Lam, *Heritiera littoralis* Aiton, *I. fagifer*, and some *Terminalia* spp. The seeds of *Gnetum gnemon* L. and related members of this genus are also eaten as nuts but are technically not "fruits" (*Gnetum* leaves are also eaten). For many of these plants, it is difficult to tell a difference between tended (domesticated?) and untended wild populations. It is likely that these have continued to interbreed diluting any human selection activities. It is also possible that other species of *Canarium*, and *Barringtonia* were possibly domesticated within Remote Oceania but in modern times have become feral with disuse. Surely early recognition of these as possible foods was based upon experience eating related members of the same genera in Near Oceania.

Other fruit

Fleshy fruited species were likely the most biologically diverse wild food resource available to settlers of Remote Oceania. *Citrus grandis* (L.) Osbeck has been argued as being an introduced species [31] however, linguistic data supports possible ancient distribution of it or another *Citrus* species [35]. *Burckella obovata* (G. Forst.) Pierre, *Flacourtia rukam* Zoll. & Moritzi, *Pometia pinnata* J. R. Forst. & G. Forst., *Syzygium clusiifolium* (A. Gray) Müll. Stuttg., *Syzygium malaccense* (L.) Merr. & L. M. Perry, and *Spondias dulcis* Parkinson are fruit trees that are widely grown and consumed regularly. Many *Pandanus* fruit are edible, having some amount of starch or sweet juice on the inner part of each compound fruit "key". Some *Pandanus* (and *Freycinetia*) spp. may also have more fleshy fruit that is entirely edible, but these are not often mentioned. Some edible fruit are either rarely consumed or only discussed as famine foods (e.g., *Corynocarpus* spp., *Parartocarpus beccarianus* Baill.).

Rhizophora mangle L. has a complex and controversial history. It appears to be the only wild food plant in Oceania that originated in the Americas. Steele [46] has argued that it was naturally distributed into Remote Oceania with possible subsequent movement by people toward the west. However, other species of *Rhizophora* are widely distributed across Oceania, but were only rarely reported as eaten.

While not "fruit", the reproductive structures of wild fungi were harvested and eaten from several kinds of trees that serve as indicators of edibility. Some trees (e.g., *Metroxylon*) are felled and damaged to promote production of edible fungi. A wide variety of mushrooms (Ascomycota and Basidiomycota) are eaten in Near Oceania (e.g. [8]), but few are consumed in Remote Oceania except in New Zealand [47]. The only widely eaten fungus is *Auricularia cornea* Ehrenb. Ethnomycological studies in the Western Pacific region are limited and there is very little information identifying the taxa that are being consumed.

A large number of the wild food plants are fruits and leafy green vegetables. It would be very interesting to examine the nutritional contributions that these make, as well as the roles that they might play in childhood development and disease prevention (e.g., [48]).

Vegetables

Ross [39] has provided an excellent overview of green vegetables consumed in Oceania. The most commonly consumed are young leaves of *Abelmoschus manihot* and *Colocasia esculenta*. The latter includes wild and cultivated populations in Near Oceania, and cultivated and feral populations in Remote Oceania. *Ipomoea aquatica* Forssk. was likely domesticated in Near Oceania, but it is difficult to determine if it was introduced into Remote Oceania in ancient times. It is an important leafy green vegetable in some communities and would appear to be an excellent candidate for distribution. It may have been displaced more recently by the introduction of *Ipomoea batatas* from the Americas which has been eaten as much for its leaves (much like *I. aquatica*) as for its starchy roots.

Ferns were an important secondary category of edible leaves. Fern species are frequently endemic to particular islands or archipelagos, but genera are widespread and probably served as functional ethnotaxa (e.g., *Blechnum*, *Cyathea*, *Cyclosorus*, *Dennstaedtia*, *Diplazium*, *Thelypteris*). All of these leaves are generally consumed as young, often unfurled leaves that are steamed or boiled. "Salads" do not appear to have been part of traditional practices. All of the ferns are wild harvested.

Marine (as terrestrial) species richness drops off from Near through Remote Oceania [49]. However, there appears to be greater use of algae as a food resource within Remote than Near Oceania [50]. A few algae were widely consumed as foods or condiments [e.g., *Caulerpa racemosa* (Forsskål) J. Agardh, *Halymenia* spp., *Gracilaria* spp., *Ulva* spp.; e.g., [51]].

Leaves of many plants are incidentally edible at some growth stage, probably being eaten primarily when more desirable foods were in short supply. *Ficus* spp., *Gnetum* spp., *Morinda citrifolia* L., *Polyscias* spp., and *Pseuderanthemum* spp. are commonly reported as eaten. In each case, it is usually younger leaves that are preferred, sometimes still in bud. Of these, *Gnetum* is most common in little disturbed forests, while edible *Ficus* and *M. citrifolia* are common in coastal strand forests and adjacent open areas. *Polyscias* and *Pseuderanthemum* are common garden and village plants that may represent old settlements when found outside of these areas. *Laportea* and *Pisonia* leaves are commonly reported as food for pigs, and less often for humans. Although *M. citrifolia* is widespread in Near and Remote Oceania, reports of eating leaves are spotty, with some people even denying that they are edible (the same is incidentally true for the horrid smelling *M. citrifolia* fruit).

Most palms have an edible "palm heart" consisting of the growing apical meristem and adjacent young leaf material. This is true of all species native to Remote Oceania. However, only a few are regularly eaten, such as *Metroxylon*, *Nypa* and *Pritchardia* because the trees are cut down and used for other purposes (e.g., leaf thatch). *Cocos nucifera* hearts may also be eaten but because of the value of the tree for other purposes, they were not commonly consumed.

Alpinia novae-hiberniae B. L. Burtt & R. M. Sm., *Cyathea* spp., *Metroxylon* spp., *Wollastonia biflora* (L.) DC. each have partially edible stems. Traditional consumption levels of these are higher in New Guinea (Near Oceania) and taper off to the east and north with a lack of recognition that they are edible in the many locations where they grow. *Saccharum officinarum* is a stem crop domesticated in New Guinea that maintains its use across its distribution into Remote Oceania. However, its leaves became important as thatch in parts of Remote Oceania while this use is not mentioned in Near Oceania where there were many superior thatch sources.

Insects collected from plants

Several plants served as the food source for edible insect larvae. Many palms including *Metroxylon* spp., *Pritchardia pacifica* Seem. & H. Wendl., and *Veitchia* spp. are damaged so that insects will lay their eggs in them. Timing is important if the lipid-rich larvae are to be collected as food before they mature. *Caryota rumphiana* Mart. Stems host insect larvae that are eaten. It is unclear when this palm may have been introduced from Near to Remote Oceania. Other plants are either grown, encouraged to persist, or periodically checked for the presence of edible insect larvae.

Wild food plants when cultivation fails

The colonists of Remote Oceania generally encountered environments that were both supportive of their cultivated plant tool kit and contained familiar wild plants. However, some environments were climatically or ecological restricted (atolls – e.g., [52,53]) and others climatically marginal (Hawaiian islands – e.g., [54,55]). One major settlement was the most extreme that has persisted. This was the colonization of the massive uninhabited area of Aotearoa (New Zealand). It exceeds in area all of the rest of Remote Oceania combined, but it is so far to the south that much of the plant and animal life is different from that previously encountered by Austronesian settlers. Crowe [56] attempted to identify all possible wild plant foods in Aotearoa. Crowe conducted a field experiment, attempting to feed himself for an extended period only with native plants. His report is not encouraging, noting that on some days he was unable to find any plants that were edible. Even after including about 160 plants, it is painfully obvious that survival on wild plants in New Zealand would be difficult, especially for people primarily interested in starchy roots and protein-rich seeds/nuts. Fruit, mushrooms and ferns appear to be common, but do not seem to be ready substitutes for traditional dietary preferences. In the end, cultural expectations may be what drives the selection of foods and the sharing of what is learned with others. The original New Zealand colonists were highly resourceful, but it must have been a very difficult process to survive on such unusual fare.

Problems with this assessment

Because it is difficult to determine if and when a plant was introduced, it is possible to identify plants as being important ancient wild food plants, that in fact were not present in the past. Two methods have been used to reduce the number of false leads: linguistic analysis of plant names (e.g., [35]), and analysis of plant relationships with wild relatives [57]. Both of these have problems. For example, it is possible to produce linguistic reconstructions for plants that have well known recent introductions such as *Psidium guajava* or *Carica papaya*. In each case, this is probably just an indication that the plants were introduced from related sources and arrived with names in common. Plants in Tab. 1 that could suffer from this problem include *Lablab purpureus*, *Ipomoea aquatica*, and *Citrus grandis*. In each case, there is reason to believe, based upon relationships with other species of plants that these may be more recent introductions into Remote Oceania. Likewise, *Terminalia catappa* and *Bruguiera gymnorhiza* (L.) Savigny may be examples of plants that have actually been introduced by humans over much or part of Remote Oceania even though they are related to other plants within the region and have mechanisms for naturally dispersing through the region.

Scientists have "learned" biases about the history of plants. Those working in the Pacific "know" certain biogeographic facts that are based on scientific tradition rather than experimental or other positive evidence. For example, several widely distributed strand species [*Aleurites moluccana*, *Cocos nucifera*, *Inocarpus fagifer*, *Morinda citrifolia*, *Pandanus tectorius* Parkinson, *P. pinnata*, *Polyscias scutellaria* (Burm. f.) Fosberg] are thought to be distributed in eastern and northern parts of Remote Oceania via humans, despite each having natural dispersal capabilities via marine currents or fruit bats. The simple fact is that for most of these, we do not know how or when they were dispersed from island to island and assumptions are made, usually on a combination of local comments, and early botanical collection information that is surely incomplete.

Conclusions

Wild food plants in Remote Oceania have a complex history. The presence or absence of taxa as well as selection of previously existing taxa, is driven by cultural factors including expectations about what food should be and aesthetics (taste, texture, smell) rather than potential nutrition.

As an unusual region, originally colonized by agriculturalists, it is likely that there were many untapped wild food plants that would have been discovered by hunter-gathers if colonization had proceeded differently.

The majority of species used as foods in the initial colonization of Remote Oceania were wild foods, and most of these were used as leafy vegetables and fruits. The wild food plants mostly served as supplements to domesticated species, although there were a few that can be used as substitutes for traditional staples. The implication of this situation is that although some plants are managed as reserves or substitutes for lean times, the majority of wild foods serve other than primary caloric nutritional needs.

Based on this brief assessment, it should be expected (hypothesized) that within a particular culture, the selection of plants for domestication from among those discovered will strongly reflect the characteristics of the earliest founding domesticated foods. Plants that are drafted into roles as "famine foods" or wild supplemental foods, will be of two types: those that provide similar food products but have some deficiency (such as taking longer to process), and those that do not provide similar food products, but have supplementary benefits that are realized even when there is no stress on the food supply.

Acknowledgements

Thanks to Piet Lincoln and Kawika Winter for many discussions of the colonization of Remote Oceania. Thanks to Valerie McClatchey for commenting on drafts of this document.

References

1. Jackson F. The coevolutionary relationship of humans and domesticated plants. Am J Phys Anth. 1996;101(23 suppl):161–176.
2. Winter K, McClatchey W. The quantum co-evolution unit: an example of 'awa (kava-piper methysticum G. Foster) in Hawaiian culture. Econ Bot. 2009;63(4):353–362.
3. Fehr WR. Principles of cultivar development. Theory and technique. Ames IA: Macmillan; 1991. (vol 1).
4. Anderson E. Plants, man and life. Boston MA: Little, Brown; 1952.
5. Sauer CO. Agricultural origins and dispersals. New York NY: American Geographical Society; 1952.
6. Bettinger RL. Hunter-gatherers: archaeological and evolutionary theory. New York NY: Plenum Press; 1991.
7. Huss-Ashmore R, Johnston SL. Wild plants as cultural adaptations to food stress. In: Etkin N, editor. Eating on the wild side: the pharmacologic, ecologic and social implications of using noncultigens. Tucson AZ: University of Arizona Press; 1994. p. 62–82.
8. Henderson CP, R HI. A guide to the useful plants of Solomon Islands. Honiara: Ministry of Agriculture and Lands; 1988.
9. Cox PA. Wild plants as food and medicine in Polynesia. In: Etkin N, editor. Eating on the wild side: the pharmacologic, ecologic and social implications of using noncultigens. Tucson AZ: University of Arizona Press; 1994. p. 102–113.
10. Paull RE, Tang CS, Gross K, Uruu G. The nature of the taro acridity factor. Postharvest Biol Tec. 1999;16(1):71–78.
11. Abbott IA. La'au Hawai'i: traditional Hawaiian uses of plants. Honolulu: Bishop Museum Press; 1992.
12. MacArthur RH, Wilson EO. The theory of island biogeography. Princeton: Princeton University Press; 2001. (Monographs in Population Biology; vol 1).
13. Green RC. Near and remote Oceania: disestablishing "Melanesia" in culture history. In: Bulmer R, Pawley A, editors. Man and a half: essays in Pacific anthropology and ethnobiology in honour of Ralph Bulmer. Auckland: Polynesian Society; 1991. p. 491–502. (Memoirs of the Polynesian Society).
14. Kayser M. The human genetic history of Oceania: near and remote views of dispersal. Curr Biol. 2010;20(4):R194–R201.
15. Oskarsson MCR, Klutsch CFC, Boonyaprakob U, Wilton A, Tanabe Y, Savolainen P. Mitochondrial DNA data indicate an introduction through Mainland Southeast Asia for Australian dingoes and Polynesian domestic dogs. Proc R Soc B. 2011;279(1730):967–974.
16. Harris DR, Bellwood P. The origins and spread of agriculture in the Indo-Pacific region: gradualism and diffusion or revolution and colonization. In: The origins and spread of agriculture and pastoralism in Eurasia. London: University College London Press; 1996. p. 465–498.
17. Blust RA. The Austronesian languages. Canberra: Australian National University Press; 2009.
18. Goff J, McFadgen BG, Chague-Goff C, Nichol SL. Palaeotsunamis and their influence on Polynesian settlement. Holocene. 2012;22(9):1067–1069.
19. Wilmshurst JM, Hunt TL, Lipo CP, Anderson AJ. High-precision radiocarbon dating shows recent and rapid initial human colonization of East Polynesia. Proc Natl Acad Sci USA. 2010;108(5):1815–1820.
20. Gray RD, Drummond AJ, Greenhill SJ. Language phylogenies reveal expansion pulses and pauses in pacific settlement. Science. 2009;323(5913):479–483.
21. Matisoo-Smith E. Origins and dispersals of Pacific peoples: evidence from mtDNA phylogenies of the Pacific rat. Proc Natl Acad Sci USA. 2004;101(24):9167–9172.
22. McConkey KR, Meehan HJ, Drake DR. Seed dispersal by Pacific pigeons (*Ducula pacifica*) in Tonga, Western Polynesia. Emu. 2004;104(4):369–376.
23. Cox PA, Elmqvist T, Pierson ED, Rainey WE. Flying foxes as strong interactors in South Pacific island ecosystems: a conservation hypothesis. Conserv Biol. 1991;5(4):448–454.
24. Bellwood P. A hypothesis for Austronesian origins. Asian Persp. 1988;26(1):107–117.
25. Oppenheimer S. The "express train from Taiwan to Polynesia": on the congruence of proxy lines of evidence. World Archaeol. 2004;36(4):591–600.
26. Jennings JD. The prehistory of Polynesia. Canberra: Australian National University Press; 1979.
27. Vésteinsson O, McGovern TH, Keller C. Enduring impacts: social and environmental aspects of Viking age settlement in Iceland and Greenland. Arch Island. 2002;2:98–136.
28. Burley D, Weisler MI, Zhao JX. High precision U/Th dating of first polynesian settlement. PLoS One. 2012;7(11):e48769.
29. Yen DE. Polynesian cultigens and cultivars: the question of origin. In: Cox PA, Banack SA, editors. Plants, islands, and Polynesians: an introduction to Polynesian ethnobotany. Portland OR: Dioscorides Press; 1991. p. 67–96.
30. Kirch PV. A shark going inland is my chief: the island civilization of ancient Hawai'i. Berkeley: University of California Press; 2012.
31. Barrau J. Les plantes alimentaires de l'Océanie, origines, distribution et usages. An Musée Col Marseille. 1962;7:3–9.
32. Guppy HB. Observations of a naturalist in the Pacific between 1896 and 1899: plant-dispersal. London: Macmillan; 1906. (vol 2).
33. Yen DE. Subsistence to commerce in Pacific agriculture: some four thousand years of plant exchange. In: Predergast HDV, Etkin NL, Harris DR, Houghton PJ, editors. Plants for food and medicine. Kew: Royal Botanic Gardens; 1998. p. 161–183.
34. Evans B. Ethnobotanical classification. In: Ross M, Pawley A, Osmond M, editors. The lexicon of Proto Oceanic: the culture and environment of ancestral Oceanic society. Plants. Canberra: Australian National University Press; 2010. p. 53–84. (vol 3).
35. Ross M, Pawley A, Osmond M, editors. The lexicon of Proto Oceanic: the culture and environment of ancestral Oceanic society. Plants. Canberra: Australian National University Press; 2010. (vol 3).
36. Jansen T, Sirikolo MQ. Petanigaki ta Siniqa ni Lauru. Honiara: Kastom Garden Association; 2010.
37. Pollock NJ. These roots remain: food habits in islands of the Central and Eastern Pacific since western contact. Laie: University of Hawaii Press; 1992.

38. Barrau J. Witnesses of the past: notes on some food plants of Oceania. Ethnology. 1965;4(3):282.
39. Ross M. Introducing Proto Oceanic plant names. In: Ross M, Pawley A, Osmond M, editors. The lexicon of Proto Oceanic: the culture and environment of ancestral Oceanic society. Plants. Canberra: Australian National University Press; 2010. p. 25–52. (vol 3).
40. McClatchey W. Ethnobiology - basic methods used to document dynamic relationships between peoples, biota and environments, and ways in which this knowledge is represented in languages. In: Thieberger N, editor. The Oxford handbook of linguistic fieldwork. Oxford: Oxford University Press; 2011. p. 281–297.
41. Winter K, McClatchey W. Quantifying evolution of cultural interactions with plants: implications for managing diversity for resilience in social-ecological systems. Func Ecosyst Commun. 2008;2(1 special issue):1–10.
42. Mcclatchey W, Thaman R, Vodonaivalu S. A preliminary checklist of the flora of Rotuma with Rotuman names. Pac Sci. 2000;54:345–363.
43. Verheij EWM, Coronel RE. Edible fruits and nuts. In: Faridah Hanum I, van der Maesen LJG, editors. PROSEA: plant resources of South-East Asia. Wageningen: Pudoc; 1991. (vol 2).
44. Whistler WA. Polynesian plant introductions. In: Cox PA, Banack SA, editors. Plants, islands, and Polynesians: an introduction to Polynesian ethnobotany. Portland OR: Dioscorides Press; 1991. p. 41–66.
45. Powell JM. Ethnobotany. In: Paijmans K, editor. New Guinea vegetation. Amsterdam: Elsevier; 1976. p. 106–183.
46. Steele OC. Natural and anthropogenic biogeography of mangroves in the Southwest Pacific [PhD thesis]. Honolulu: University of Hawaii; 2006.
47. Fuller R. Fungi and Polynesia: New Zealand and Cook Island Maori ethnomycology [PhD thesis]. Honolulu: University of Hawaii; 2009.
48. Etkin N. Pharmacologic implications of "wild" plants in Hausa diet. In: Etkin N, editor. Eating on the wild side: the pharmacologic, ecologic and social implications of using noncultigens. Tucson AZ: University of Arizona Press; 1994. p. 85–101.
49. N'yeurt A, South G. Biodiversity and biogeography of benthic marine algae in the southwest Pacific, with specific reference to Rotuma and Fiji. Pac Sci. 1997;51(1):18–28.
50. Abbott IA. Polynesian uses of seaweed. In: Cox PA, Banack SA, editors. Plants, islands, and Polynesians: an introduction to Polynesian ethnobotany. Portland OR: Dioscorides Press; 1991. p. 135–146.
51. Conte E, Payri C. Present day consumption of edible algae in French Polynesia: a study of the survival of pre-European practices. J Polyn Soc. 2006;115(1):77.
52. Merlin M, Capelle A, Keene T, Juvik J, Maragos J. Keinikkam Im Melan Aelon Kein. Plants and environments of the Marshall Islands. Honolulu: East-West Center; 1994.
53. National Biodiversity Team of the Republic of the Marshall Islands. The Marshall Islands: living atolls amidst the living sea. Santa Clara: St. Hildegard Publisher; 2000.
54. Cuddihy LW, Stone CP. Alteration of native Hawaiian vegetation: effects of humans, their activities and introductions. Honolulu: University of Hawaii; 1990.
55. Handy ESC, Handy EG, Pukui MK. Native planters in old Hawaii: their life, lore, and environment. Honolulu: Bishop Museum Press; 1972.
56. Crowe A. Native edible plants of New Zealand. Auckland: Hodder & Stoughton; 1990.
57. Vavilov NI. Origin and geography of cultivated plants. Cambridge: Cambridge University Press; 1994.

12

Niche conservatism of *Eulophia alta*, a trans-Atlantic orchid species

Marta Kolanowska*, Dariusz L. Szlachetko

Department of Plant Taxonomy and Nature Conservation, University of Gdańsk, Wita Stwosza 59, 80-308 Gdańsk-Oliwa, Poland

Abstract

The genus *Eulophia* embraces over 230 species distributed through the tropical and subtropical Africa, Asia, Australia and the Americas. In Neotropics it is represented by a sole species – *E. alta*. The aim of the presented study was to evaluate the difference between ecological niches occupied by American and African populations of this species based on the ecological niche modeling. The similarity between the glacial and present niches occupied by *E. alta* was calculated and the factors limiting the species occurrence were identified. Areas of seasonal tropical forest, tropical savanna and woodland served as refugia for the studied species during last glacial maximum and they were more widespread in Neotropics than in Africa. No significant niche shift after last glacial maximum was observed. The distribution of *E. alta* in its whole range is restricted mainly by temperature seasonality. The differences in the niches occupied by African and Neotropical populations of *E. alta* suggest preglacial disjunction of the species range and independent adaptation of both groups. Despite the significant range disjunction of *E. alta* the species is characterized by relatively high degree of niche conservatism.

Keywords: ecological niche modeling; last glacial maximum; Orchidaceae; trans-Atlantic disjunction

Introduction

The trans-Atlantic range disjunction is observed within about 110 angiosperm genera [1]. Despite over 40 years of biogeographical studies [2–4] the understanding of timing, direction and pattern of long-distance dispersal across the Atlantic is still very limited. Little is known also about the niche variation of the taxa characterized by remarkable distribution gap.

The niche conservatism of numerous organisms was recently intensively studied using ecological niche modeling (ENM) methods [5–7], but so far this analysis was not apply to evaluate niche variation of species exhibiting trans-Atlantic disjunction. Based on the incoming research results it is difficult to formulate any explicit principle about the tendency of such species to retain characteristics of their fundamental niche over time. The degree of niche conservatism varies among groups of species (small-ranged and specialist [6]) and it is related to the evolutionary history of each taxon [8]. While some authors [9] considered evolutionary and biogeographic patterns of species diversity as completely separated processes, the concept of niche conservatism offers a bridge between them [10–13]. Studies on ecological niche variation are particularly important to recognize fundamental distribution factors and evolutionary models of species characterized by the disjunctive range. While in some taxa the geographical discontinuity resulted in adaptive [14,15], other adjusted evolutionary stable strategy [16].

The object of our study was *Eulophia alta* (L.) Fawc. & Rendle, a sole representative of the genus in Neotropics where its range extends from USA (Florida) south to Argentina. Except Americas *E. alta* occurs in the Africa, from Senegal to Zimbabwe and it is assumed that it originates in the Old World where high specific diversity of the genus is observed. However, no clear evidence of this hypothesis was presented so far. *Eulophia alta* is terrestrial, large plant producing a subglobose, subterranean corm and loosely many-flowered, racemose inflorescence. The species is found in open areas, along roadsides and on cleared land reverting to bush lower montane forest, usually in heavy, moisture-retentive soils. Populations of *E. alta* were found in lowland and premontane areas, at the altitudes up to 1500 m. The main flower visitors and most effective pollinators of this species are anthophorid bees. Six Hymenopteran and two Lepidopteran families were observed visiting the flowers [17].

The aim of this study was to evaluate the similarity of the niches occupied by African and Neotropical populations of *E. alta* using ENM tools. Because adaptation to local climatic conditions is a significant force driving morphological evolution and speciation, the availability of the suitable niches of the studied species during last glacial maximum (LGM; 26 500–19 000 years ago) was evaluated to estimate the possible postglacial niche shift and geographical range changes of *E. alta*. In the studies on Orchidaceae the ENM was applied so far mainly in the research on invasive species [18,19] and

* Corresponding author. Email: martakolanowska@wp.pl

Handling Editor: Aleksandra Samecka-Cymerman

a sole analysis [20] concerned the potential distribution of orchid during last glacial maximum.

Material and methods

Localities and georeferencing

The database of *E. alta* localities was prepared based on the examination of herbarium specimens deposited in herbaria AMES, BM, BR, COL, HUA, K, MO, NY, P, UGDA, VALLE, W, and WAG. Herbaria acronyms are cited according to "Index Herbariorum" [21]. To enlarge the dataset, the information obtained from the electronic database of the Missouri Botanical Garden (available at http://www.tropicos.org) was included in the analysis.

The georeferencing process followed Hijmans et al. [22] and only the localities, which could be precisely placed on the map were used in the study. The information about the latitude and longitude provided on the herbarium sheet labels were verified. If no geographic coordinates were indicated by the plant collector, they were assigned based on the data about the collection place. The Google Earth (v. 6.1.0.5001, Google Inc.) application was used to validate all gathered information.

In total 54 localities were included in the database (Fig. 1, Tab. 1), 8 African and 46 Neotropical, which is more than the minimum number of localities (>5) required by Maxent to obtain reliable predictions [23].

Fig. 1 Locations of *E. alta* used in the ecological niche modeling.

Maximum entropy analysis

The maximum entropy method implemented in Maxent version 3.3.2 [24–26] was used to create models of the distribution of the ecological niches of *E. alta*. Because Maxent is relatively robust against collinear variables [27,28], all available climatic factors (Tab. 2) in 2.5 arc-minutes developed by Hijmans et al. [29] as well as the altitudinal data were used as an input data. This was also justified by the insufficient data on habitat requirements of the studied species that could lead to premature exclusion of the correlated variables. The analogical bioclimatic data for the last glacial maximum period was mapped by Paleoclimate Modelling Intercomparison Project Phase II [30].

To assess high specificity of the modeling, the maximum iterations was set to 10 000 and convergence threshold to 0.00001. For each run 20% of the data were used to be set aside as test points [31]. The "random seed" option which provided random test partition and background subset for each run was applied. The run was performed as a bootstrap with 100 replicates, and the output was set to logistic. All operations on GIS data were carried out on ArcGis 9.3 (ESRI).

Niche similarity

The geographical overlap of the niches calculated based on two different datasets (African and Neotropical locations only) as well as the similarity of niches occupied by the populations from the New World and Africa were defined based on the test implemented in ENMTools application. The Schoener's *D* [32] and *I* [33] statistics were calculated. In Schoener's *D* statistic the local species density measures are compared with each other. "I" statistic is based on Hellinger distance and measures the ability of the model to estimate the true suitability of the habitat. Both metrics range from 0 (no similarity) to 1 (overlapping).

Results

Models evaluation

All repeated ecological niche models for the present time received high area under the curve (AUC) scores of 0.972–0.984 (Tab. 3) that indicate very high reliability of the analysis.

Bioclimatic limiting factors

The results of the niche modeling indicated three main bioclimatic variables limiting the distribution of suitable habitats for *E. alta*. The crucial factor is the temperature seasonality (bio4), which influenced significantly models of both African and Neotropical populations of the studies species. While the range of latter geographical group is also related with the amount of precipitation in the coldest quarter (bio19), the African plants are more dependent on the mean temperature of the coldest quarter (bio11). The estimates of relative contributions of the environmental variables to the Maxent model for both present time and LGM are presented in Tab. 4.

Potential glacial refugia

The model of habitats available in Neotropics during the LGM indicate few possible glacial refugia of *E. alta* (Fig. 2). In Mesoamerica the suitable niches were located in the Gulf of Honduras, Mosquitia lowlands, Mosquito Coast, Cordillera de Talamanca, the Darién Gap as well as in Jamaica. In South America the possible glacial refugia were located in the northern Andes, eastern pre-Andean highland and partially Amazon basin, Guiana Highlands as well as western Brazilian coast. The African refugia were much more limited and there were distributed in the northern Grain Coast and around the Gulf of Guinea. Less suitable habitats were located in the Zanzibar Archipelago and northern Madagascar. The areas determined for both, African and Neotropical populations,

Tab. 1 Localities of *Eulophia alta* used in ENM analysis.

Country	Latitude	Longitude	Collector and number	Institution
Angola	−9.553333333	16.3475	Mechow 344	W-R
Angola	9.132777778	14.77194444	Welwitsch 664	BM, K, W-R
Belize	16.99666667	−88.40694444	Percy H. Gentle 9262	MO
Belize	17.09944444	−88.32388889	William A. Schipp S-171	AMES, MO
Belize	−17.74166667	−63.2	Michael H. Nee 48842	NY
Bolivia	−14.75055556	−61.14361111	A. Carrión, M. Castro & V. Ayala 503	MO
Bolivia	−17.65	−63.16666667	Nur Ritter 2782	MO
Bolivia	−14.75055556	−61.14361111	A.M. Carrión, M. Castro & V. Ayala 503	MO, USZ
Cameroon	3.8	10.11666667	Sanford 5177	K
Cameroon	4.966666667	8.85	Thomas 2316	K, P, MO
Cameroon	18.83333333	−69.56666667	Thomas A. Zanoni & Milcíades M. Mejía 16316	MO
Caribbean	18.15	−77.23333333	Alwyn H. Gentry & Valerie Kapos 28306	MO
Caribbean	6.666666667	−74.86666667	Ricardo Callejas, Julio C. Betancur B. & Omar D. Escobar 9020	HUA
Central African Republic	6.394444444	21.59888889	Le Testu 4779	BM, K
Colombia	3.840277778	−76.89777778	Kolanowska 233	UGDA
Colombia	3.881111111	−76.33888889	Paz 1476	VALLE
Colombia	26.01386	−72.76	D. Cárdenas & R. López 6632	COL
Colombia	8.73	−83.4	Álvaro Fernández 401	CR
Costa Rica	8.74	−83.56	Gerardo Herrera Ch. 4558	CR
Costa Rica	−2.883333333	−78.35	B. Løjtnant & Ulf Molau 14510	AAU
Ecuador	−0.466666667	−76.91666667	L. Broder, Holm-Nielsen, Jaramillo J. & Coello F. 19607	MO
Ecuador	−0.466666667	−76.91666667	L.B. Holm-Nielsen & et al. 19607	AAU, MO
Ecuador	0.033333333	−77.38333333	Benjamin Øllgaard 99651	AAU
Ecuador	15.50833333	−87.45	John M. MacDougal, Paul R. House & Ramón Zúñiga 3276	MO
Gabon	0.408055556	9.4475	Le Testu 8669	BM
Guinea	10.62722222	−9.713611111	Adam 12539	K
Guyana	3.35	−59.56666667	Jansen-Jacobs, ter Welle, Gorts-van Rijn & Ek 670	K
Honduras	19.13944444	−89.32944444	Esteban M. Martínez S., Demetrio Alvarez M. & Santiago Ramírez A. 28639	MO
Ivory Coast	4.95	−6.066666667	de Wilde 358	K, WAG
Mexico	17.35	−100.4	Otto Nagel 2045	MO
Mexico	12.01666667	−85.15	Alfonso H. Heller s.n.	MO database
Nicaragua	9.168611111	−79.85194444	Thomas B. Croat 7788	MO
Panama	9.168611111	−79.85194444	Thomas B. Croat 12809	MO
Panama	9.169444444	−79.85416667	Thomas B. Croat 4391	MO
Panama	8.6	−80.13333333	Paul H. Allen 1988	MO
Panama	7.75	−77.66666667	Gordon McPherson 15038	MO
Panama	7.73	−80.87361111	Barry E. Hammel 4243	MO
Panama	9.100555556	−79.27777778	James A. Duke 5930	MO
Panama	9.266666667	−78.95	H.W. Churchill 3814	MO
Panama	9.266666667	−78.95	H.W. Churchill 3815	MO
Panama	8.875277778	−79.78777778	Paul H. Allen 2080	MO
Panama	9.365555556	−78.95	Gordon McPherson 11863	MO
Panama	9.3	−78.975	Greg C. de Nevers & Heraclio Herrera 4353	MO
Panama	−26.07611111	−56.85	Elsa M. Zardini 12678	MO, PY
Paraguay	−5.118888889	−78.32083333	Philip J. Barbour 4356	MO
Peru	−10.75	−74.38333333	D.N. Smith 6875	MO
Peru	−4.568055556	−78.19805556	Rodolfo Vásquez & et al. 24292	MO
Peru	−3.8	−73.41666667	Rodolfo Vásquez & Nestor Jaramillo 4953	MO
Peru	−3.746944444	−73.40861111	Rodolfo Vásquez & Rocío Rojas 22731	MO
Peru	−5.083333333	−73.83333333	Rodolfo Vásquez & Nestor Jaramillo 4804	MO
Peru	−11.91666667	−77.3	Percy Núñez V., J. Terborgh & et al. 14246	MO
Peru	−10.75	−74.38333333	David N. Smith 6875	MO
Peru	−5.866666667	−77.21666667	David N. Smith 5963	MO
Peru	−8.75	−75.08333333	David N. Smith & et al. 1187	MO

Tab. 2 Variables used in the modeling.

Code	Variable
bio1	Annual mean temperature
bio2	Mean diurnal range = mean of monthly (max temp – min temp)
bio3	Isothermality (bio2/bio7) × 100
bio4	Temperature seasonality (standard deviation × 100)
bio5	Max temperature of warmest month
bio6	Min temperature of coldest month
bio7	Temperature annual range (bio5 – bio6)
bio8	Mean temperature of wettest quarter
bio9	Mean temperature of driest quarter
bio10	Mean temperature of warmest quarter
bio11	Mean temperature of coldest quarter
bio12	Annual precipitation
bio13	Precipitation of wettest month
bio14	Precipitation of driest month
bio15	Precipitation seasonality (coefficient of variation)
bio16	Precipitation of wettest quarter
bio17	Precipitation of driest quarter
bio18	Precipitation of warmest quarter
bio19	Precipitation of coldest quarter
Alt	Altitude

Tab. 3 The AUC scores for each model created during the study with information on standard deviation values (*SD*).

	All locations	African locations only	Neotropical locations only
Present time	0.974 (*SD* = 0.005)	0.972 (*SD* = 0.012)	0.984 (*SD* = 0.004)
LGM	0.979 (*SD* = 0.005)	0.973 (*SD* = 0.012)	0.982 (*SD* = 0.005)

were covered during LGM with seasonal tropical forest as well as with tropical savanna and woodland [34].

Current potential range

The combined model (95% confidence level summary grids, Fig. 3a–c) of suitable habitats created using all available location data shows the general decreasing of the available niches in Neotropics and increasing of the African ones in comparison with the analogical models for LGM. In South America the highest concentration of proper niches is observed near the estuaries of Orinoco and Amazon rivers as well as in Cachimbo and dos Gradaús mountains. Less suitable areas are located in the lowland between Andes and Amazon basin. The African niches seem extended within the tropical Africa to east, but their general distribution did not change.

Niches overlap and identity

The calculated statistics confirm the geographical differences in the distribution of the suitable niches for African and Neotropical populations of *E. alta* ($D = 0.319$; $I = 0.587$), however the niche identity test indicate moderate similarity between the preferred habitats [$D = 0.657$ ($SD = 0.0609$); $I = 0.883$ ($SD = 0.036$)]. The overlap of models created for LGM and present time created based all known locations of *E. alta* gave results of $D = 0.800$ and $I = 0.947$.

Discussion

Ecological niches distribution vs. current range

The known geographical range of *E. alta* corresponds to the distribution of its suitable niches estimated in ENM analysis. The only areas where there are no proper niches for the studied species, according to the ecological model, and where the specimens of the studied orchid were found, are southern Florida and northern Cuba. The first region is significantly affected by non-native plants, which are naturalized from the horticultures, and it can be hypothesized that the ornamental usage of *E. alta* is the source of its North American populations [35]. The most probable reason of the occurrence of this orchid in Cuba is the migration of *E. alta* from other Caribbean islands. As it was shown in the previous studies the invasive populations may shift their niches to invade new areas [18,36]. As the Neotropical populations of *E. alta* often grow in disturbed, open areas, such as roadsides, their ecological amplitude is rather wide and it was confirmed in the ENM analysis. Hereby the chances of the North American populations to survive in habitats

Tab. 4 Relative contributions of the environmental variables to the Maxent models.

	African locations only		Neotropical locations only	
Present time	bio11 (46.5%)	bio4 (32.6%)	bio4 (35.5%)	bio19 (14%)
LGM	bio4 (36.9%)	bio11 (36.1%)	bio4 (35.5%)	bio19 (14.7%)

Fig. 2 The Maxent projections of suitable habitats of *E. alta* onto climatic conditions of LGM derived from Paleoclimate Modelling Intercomparison Project Phase II.

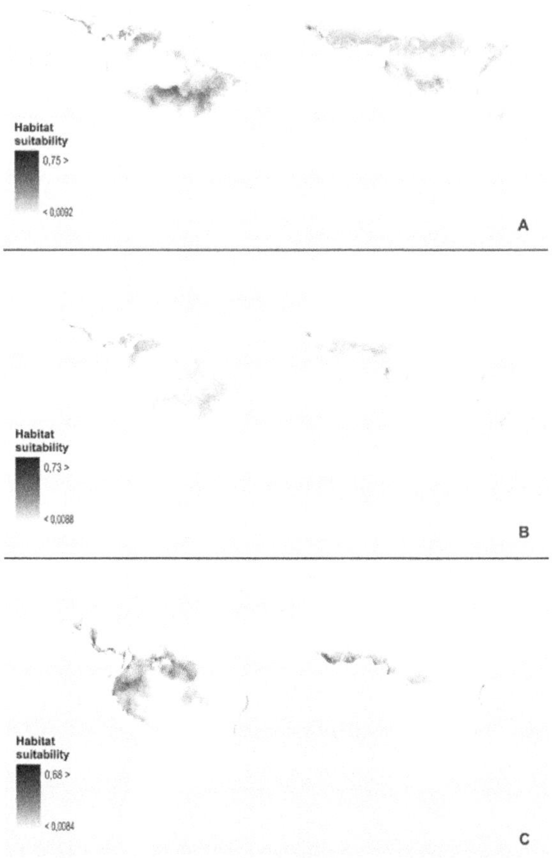

Fig. 3 The Maxent projections of suitable habitats of *E. alta* based on all localities (**a**), African localities only (**b**), Neotropical localities only (**c**).

less suitable for the studied orchid are high. Unfortunately, the number of the known locations of *E. alta* in this region is not sufficient to conduct reliable analysis of their ecology and to verify the possible niche shift of those populations.

African vs. Neotropical niches

The differences between niches occupied by African and Neotropical populations of *E. alta* suggest preglacial disjunction of its range and gradual, independent adaptation of both groups to slightly different climatic conditions. This timing is indicated by the consistency of the glacial refugia of this orchid in Africa and Neotropics with its current distribution that confirm the postglacial migration of this orchids on different continents from the refugia defined in the ENM analysis. Nevertheless, the occurrence of adaptation processes requires confirmation in the genetic studies. Apparently the climate changes during LGM were more harmful for African populations and while *E. alta* is now common in Neotropics, its African potential range declined.

Niche conservatism

While numerous species shift their niches in response to warming climate after the last glacial maximum [37–39], the habitats suitable for *E. alta* seem to be rather stable as indicated by the niche overlap test for LGM and present time. Considering also the relative similarity of the suitable niches and limiting factors for *E. alta* populations from different continents the phylogenetic niche conservatism theory seems to be applicable to the evolution history or this species. Despite the significant geographical disjunction of *E. alta* a high degree of its fundamental niche conservatism was found. This is also implied by the morphological uniformity of species representatives that suggest lack of substantial, abrupt adaptation to distinct habitat conditions.

Acknowledgments

The curators and staff of the cited herbaria are thanked for their kind hospitality and assistance during visits and for making specimens available on loan. Research was supported financially by the Ministry of Science and Higher Education of Poland (grant No. 8124/B/PO1/2011/40).

Authors' contributions

The following declarations about authors' contributions to the research have been made: research designing, conducting experiments, writing the manuscript (except "Introduction" section), appendices and figures preparation: MK; location database preparation, herbarium specimens examination, taxonomic information on the study object, writing the manuscript ("Introduction" section): DLS.

References

1. Renner S. Plant dispersal across the tropical Atlantic by wind and sea currents. Int J Plant Sci. 2004;165(S4):S23–S33.
2. Thorne RF. Floristic relationships between tropical Africa and tropical America. In: Meggers BJ, Ayensu ES, Duckworth WD, editors. Tropical forest ecosystems in Africa and South America: a comparative review. Washington, WA: Smithsonian Institution; 1973. p. 27–47.

3. Manchester SR. Biogeographical relationships of North American Tertiary floras. Ann Mo Bot Gard. 1999;86(2):472–522.

4. Thulin M, Thiede J, Liede-Schumann S. Phylogeny and taxonomy of *Tribulocarpus* (Aizoaceae): a paraphyletic species and an adaptive shift from zoochorous trample burrs to anemochorous nuts. Taxon. 2012;61(1):55–66.

5. Rödder D, Lötters S. Niche shift or niche conservatism? Climatic properties of the native and invasive range of the Mediterranean House-gecko *Hemidactylus turcicus*. Glob Ecol Biogeogr. 2009;18(6):674–687.

6. Cooper N, Freckleton RP, Jetz W. Phylogenetic conservatism of environmental niches in mammals. Proc R Soc B. 2011;278(1716):2384–2391.

7. Ryan Shipley J, Contina A, Batbayar N, Bridge ES, Peterson AT, Kelly JF. Niche conservatism and disjunct populations: a case study with Painted Buntings (*Passerina ciris*). Auk. 2013;130(3):476–486.

8. Crisp MD, Cook LG. Phylogenetic niche conservatism: what are the underlying evolutionary and ecological causes? New Phytol. 2012;196(3):681–694.

9. Algar AC, Kerr JT, Currie DJ. Evolutionary constraints on regional faunas: whom, but not how many. Ecol Lett. 2009;12(1):57–65.

10. Wiens JJ, Donoghue MJ. Historical biogeography, ecology and species richness. Trends Ecol Evol. 2004;19(12):639–644.

11. Wiens JJ, Graham CH. Niche conservatism: integrating evolution, ecology, and conservation biology. Annu Rev Ecol Evol Syst. 2005;36(1):519–539.

12. Wiens JJ, Ackerly DD, Allen AP, Anacker BL, Buckley LB, Cornell HV, et al. Niche conservatism as an emerging principle in ecology and conservation biology: niche conservatism, ecology, and conservation. Ecol Lett. 2010;13(10):1310–1324.

13. Losos JB. Phylogenetic niche conservatism, phylogenetic signal and the relationship between phylogenetic relatedness and ecological similarity among species. Ecol Lett. 2008;11(10):995–1003.

14. Lindqvist C, Albert VA. Origin of the Hawaiian endemic mints within North American *Stachys* (Lamiaceae). Am J Bot. 2002;89(10):1709–1724.

15. Spalik K, Downie SR. Intercontinental disjunctions in *Cryptotaenia* (Apiaceae, Oenantheae): an appraisal using molecular data. J Biogeogr. 2007;34(12):2039–2054.

16. Duchen P, Renner SS. The evolution of *Cayaponia* (Cucurbitaceae): repeated shifts from bat to bee pollination and long-distance dispersal to Africa 2–5 million years ago. Am J Bot. 2010;97(7):1129–1141.

17. Jurgens A, Bosch SR, Webber AC, Witt T, Frame D, Gottsberger G. Pollination biology of *Eulophia alta* (Orchidaceae) in Amazonia: effects of pollinator composition on reproductive success in different populations. Ann Bot. 2009;104(5):897–912.

18. Kolanowska M. Niche conservatism and the future potential range of *Epipactis helleborine* (Orchidaceae). PLoS ONE. 2013;8(10):e77352.

19. Kolanowska M, Konowalik K. Niche conservatism and future changes in the potential area coverage of *Arundina graminifolia*, an invasive orchid species from Southeast Asia. Biotropica. 2014;46(2):157–165.

20. Kolanowska M. Glacial refugia and migration routes of the Neotropical genus *Trizeuxis* (Orchidaceae). Acta Soc Bot Pol. 2013;82(3):225–230.

21. Thiers B. Index Herbariorum: a global directory of public herbaria and associated staff. New York Botanical Garden's Virtual Herbarium [Internet]. 2013 [cited 2013 Jan 10].

22. Hijmans RJ, Schreuder M, de la Cruz J, Guarino L. Using GIS to check co-ordinates of genebank accessions. Genet Resour Crop Evol. 1999;46(3):291–296.

23. Pearson RG, Raxworthy CJ, Nakamura M, Townsend Peterson A. Predicting species distributions from small numbers of occurrence records: a test case using cryptic geckos in Madagascar: predicting species distributions with low sample sizes. J Biogeogr. 2006;34(1):102–117.

24. Elith J, Phillips SJ, Hastie T, Dudík M, Chee YE, Yates CJ. A statistical explanation of MaxEnt for ecologists: statistical explanation of MaxEnt. Divers Distrib. 2011;17(1):43–57.

25. Phillips SJ, Dudík M, Schapire RE. A maximum entropy approach to species distribution modeling. In: ICML '04: proceedings of the twenty-first international conference on machine learning. New York, NY: ACM; 2004. p. 655–662.

26. Phillips SJ, Anderson RP, Schapire RE. Maximum entropy modeling of species geographic distributions. Ecol Modell. 2006;190(3–4):231–259.

27. Rodríguez-Robles JA, Jezkova T, Leal M. Climatic stability and genetic divergence in the tropical insular lizard Anolis krugi, the Puerto Rican 'Lagartijo Jardinero de la Montaña. Mol Ecol. 2010;19(9):1860–1876.

28. Kuemmerle T, Hickler T, Olofsson J, Schurgers G, Radeloff VC. Reconstructing range dynamics and range fragmentation of European bison for the last 8000 years: European bison range dynamics. Divers Distrib. 2012;18(1):47–59.

29. Hijmans RJ, Cameron SE, Parra JL, Jones PG, Jarvis A. Very high resolution interpolated climate surfaces for global land areas. Int J Clim. 2005;25(15):1965–1978.

30. Braconnot P, Otto-Bliesner B, Harrison S, Joussaume S, Peterchmitt JY, Abe-Ouchi A, et al. Results of PMIP2 coupled simulations of the mid-holocene and last glacial maximum – part 1: experiments and large-scale features. Clim Past. 2007;3(2):261–277.

31. Urbina-Cardona JN, Loyola RD. Applying niche-based models to predict endangered-hylid potential distributions: are neotropical protected areas effective enough. Trop Conserv Sci. 2008;1(4):417–445.

32. Schoener TW. The anolis lizards of bimini: resource partitioning in a complex fauna. Ecology. 1968;49(4):704–726.

33. Warren DL, Glor RE, Turelli M. Environmental niche equivalency versus conservatism: quantitative approaches to niche evolution. Evolution. 2008;62(11):2868–2883.

34. Olson JS, Watts JA, Allison LJ. Carbon in live vegetation of major world ecosystems. Report DOE/NBB-0037 for US Dept. of Energy. Washington, WA: Carbon Dioxide Research Division; 1983.

35. Stewart SL, Johnson TR, Dutra D, Kane M, Richardson L. Ecology, propagation and conservation of *Eulophia alta* L., a threatened Florida orchid. Gainesville, FL: Florida Cooperative Fish and Wildlife Research Unit; 2007.

36. Angetter LS, Lötters S, Rödder D. Climate niche shift in invasive species: the case of the brown anole. Biol J Linn Soc Lond. 2011;104(4):943–954.

37. Thuiller W. Patterns and uncertainties of species' range shifts under climate change. Glob Chang Biol. 2004;10(12):2020–2027.

38. Jezkova T, Olah-Hemmings V, Riddle BR. Niche shifting in response to warming climate after the last glacial maximum: inference from genetic data and niche assessments in the chisel-toothed kangaroo rat (*Dipodomys microps*). Glob Chang Biol. 2011;17(11):3486–3502.

39. Veloz SD, Williams JW, Blois JL, He F, Otto-Bliesner B, Liu Z. No-analog climates and shifting realized niches during the late quaternary: implications for 21st-century predictions by species distribution models. Glob Chang Biol. 2012;18(5):1698–1713.

13

The lichen family Parmeliaceae in Poland. III. *Parmelia serrana*, new to Poland

Emilia Ossowska[1], Rafał Szymczyk[2], Adam Bohdan[3], Martin Kukwa[1]*

[1] Department of Plant Taxonomy and Nature Conservation, University of Gdańsk, Wita Stwosza 59, 80-308 Gdańsk, Poland

[2] EKOPROJEKT Environmental Survey Laboratory, Nowica 24, 14-405 Wilczęta, Poland

[3] Society Workshop for All Beings, Świętojańska 22/1, 15-082 Białystok, Poland

Abstract

Parmelia serrana A. Crespo, M.C. Molina & D. Hawksw. is reported here for the first time from Poland. The species has been recorded from more than 20 localities and exclusively on the bark of trees. Its general distribution, habitat requirements, morphology, secondary chemistry are provided and the differences between this species and similar taxa, especially *P. saxatilis* and *P. ernstiae*, are discussed.

Keywords: parmelioid lichens; chemotaxonomy; *Parmelia serrana*; distribution

Introduction

The genus *Parmelia* Ach. occurs in the boreal-temperate Northern Hemisphere [1,2]. It is characterized by foliose thalli, which are loosely to closely adnate to the substrate. Upper surface is smooth to foveolate, grey to brown-grey or grey-green, occasionally pruinose, with protoplectenchymatous and non-pored epicortex and always with usually elongated pseudocyphellae. The lower surface is black and uniformly rhizinate up to the margins, with simple, branched or squarrose rhizines. Some species develop soredia, pustules or isidia as the vegetative propagules. Apothecia are rare, especially in the species, which reproduce by soredia or isidia, with brown or rarely blackish discs; ascospores are colorless, ellipsoidal to oval in shape [1–5]. All *Parmelia* species contain atranorin and chloroatranorin in the cortex, but a variety of secondary metabolites are present in the medulla, including lobaric, salazinic and protocetraric acids [3,5].

Although *Parmelia* species are easily observed in the field and occur in a variety of habitats, up to the end of the last decade its taxonomy was based mainly on morphological characters and rarely on chemical characters for species identification. However, as it was proved recently, this approach masked the presence of species representing distinct phylogenetic clades with similar or identical morphology and chemistry (so-called cryptic or semi-cryptic species) [6–12]. The taxonomy of the genus *Parmelia* appears to be still far from resolved as material from only a few regions has been investigated and the discovery of additional undescribed taxa is expected.

Previously in Poland, only six taxa have been reported: *Parmelia discordans* Nyl., *P. ernstiae* Feurer & A.Thell, *P. omphalodes* (L.) Ach., *P. saxatilis* (L.) Ach., *P. submontana* Nádv. ex Hale and *P. sulcata* Taylor [13–18]. Two of these, the isidiate *P. saxatilis* and the sorediate *P. sulcata*, commonly reported from Poland, were recently shown to represent a group of cryptic or semi-cryptic species [8–11,19]. Therefore, their previously known distribution and frequency in Poland should be treated rather as potential and are in need of re-evaluation.

During the revision of the material previously referred to *P. saxatilis*, *P. serrana*, a recently recognized species previously not reported from Poland, was discovered. The aim of this article is to present the first records of this species in the country. The morphology, chemistry and general distribution of the species and discussion on allied taxa are provided. This paper is the third in the series dealing with lichens of the Parmeliaceae in Poland [20,21], and one of numerous designed to aid recognition of the country's lichen biota (e.g. [22–29]).

Material and methods

Herbarium material loaned from the following Polish herbaria: GPN, KRAM, KRAP, OLS, WA, WRSL, UGDA and herbarium of the Society Workshop for All Beings from Białystok (in the locality section as herb. SWAB) has been revised.

* Corresponding author. Email: dokmak@ug.edu.pl

Handling Editor: Krzysztof Spalik

Morphology was studied with a stereomicroscope and following characters were examined: color and structure of the upper surface (pruinose, not pruinose or dull), shape and development of lobes (elongated, broad, short, oval, imbricate or not), type of rhizines and their shape (simple, squarrose or branched), the abundance, shape (narrow or oval) and distribution (marginal, laminal) of pseudocyphellae, the abundance, shape and distribution on thallus of isidia. The secondary lichen compounds were studied with thin layer chromatography (TLC) in solvent A and C, according to Orange et al. [30]. The localities in Poland were mapped according to the ATPOL grid square system [31].

Results and discussion

Parmelia serrana A. Crespo, M.C. Molina & D. Hawksw. in Molina et al., Lichenologist 36(1): 48 (2004).

ICONOGRAPHY. Molina et al. [9], Thell et al. [5].

MORPHOLOGICAL CHARACTERS. Thalli adnate to loosely adnate, 3–7 cm in diameter; lobes sometimes overlapping, rounded to sublinear, up to 2–4 mm wide; upper surface pale greenish-grey to whitish-grey, shiny mainly towards the margins, sometimes dull and reticulately cracked, epruinose (sometimes, few lobe tips have faint pruina); pseudocyphellae linear or irregular in shape; isidia simple, cylindrical or branched, up to 0.2–0.8 mm tall and 0.05–0.1 mm wide, laminal and marginal, in well developed specimens densely covering central parts of the thalli; lower surface black, with long, simple or branched (but not squarrose) rhizines, up to 0.2–1 mm long; apothecia rare in Polish material; disc brown and concave; ascospores broadly ellipsoidal, 16–18 × 11.5–13 μm.

SECONDARY CHEMISTRY. The species produces atranorin (moderate amounts), salazinic acid (major), consalazinic acid (minor), often protocetraric acid (trace amounts only) and fatty acids: lichesterinic and protolichesterinic acids. The chemistry agrees with that reported by Molina et al. [9] and Thell et al. [5,11].

NOTES. *Parmelia serrana* is morphologically and chemically very similar to *P. saxatilis* and apparently for many years considered as a chemical variant of the latter species [1]. However, recent molecular data supported the recognition of *P. serrana* as a distinct taxon [9]. Both taxa are very similar, but they can be easily separated by their chemistry since *P. serrana* produces fatty acids instead of lobaric acid, which is present in *P. saxatilis*. There are also some morphological differences [5,9,11], which are summarized in Tab. 1.

Parmelia serrana can also be confused with *P. ernstiae*, but the thallus of the latter is strongly pruinose and, in addition to all substances produced by *P. serrana*, *P. ernstiae* contains lobaric acid [5,9,11,32]. All differences are presented in Tab. 1.

Recently, another species similar to *P. serrana*, *P. mayi* Divakar, A. Crespo, M.C. Molina, has been described from North America. The differences between both species refer mainly to their chemistry. *Parmelia mayi* produces almost the same set of secondary metabolites as *P. serrana*, but protocetraric acid is absent while alectorialic acid is present instead. Until now *P. mayi* is known only from high montane areas of northern Appalachian Mountains, North America [8].

Parmelia serrana can be confused with *P. submontana*, which produces granular to nearly isidioid soredia. However, the thallus lobes of *P. submontana* are elongated and little branched with down-rolled margins and the species lacks fatty acids (only atranorin and salazinic acid are present) [3,5,15].

Other *Parmelia* species reported from Poland cannot be mistaken for *P. serrana* because *P. discordans* and *P. omphalodes* lack vegetative diaspores, while *P. sulcata* propagates through soredia [3,5].

HABITAT REQUIREMENTS. *Parmelia serrana* grows on coniferous and deciduous trees and rarely on mossy rocks [3,5]. The Polish specimens are mostly epiphytic and have been collected predominantly from bark of deciduous trees (*Betula* spp., *Alnus* spp., *Carpinus betulus*, *Fagus sylvatica*, *Fraxinus excelsior*, *Populus* spp., *Quercus* spp., *Salix* spp., *Tilia cordata*) in forests or roadside alleys and only rarely (one specimen) on bark of *Pinus sylvestris*. One sample was also found on sand dunes.

DISTRIBUTION IN POLAND. The currently known distribution of *Parmelia serrana* in Poland is presented in Fig. 1. Thr species is distributed in both the northern and southern parts of the country, and is probably much more common, since *P. saxatilis* s.l., under which name the species was previously recorded, was not always documented with herbarium vouchers and therefore some published records cannot be verified.

GENERAL DISTRIBUTION. *Parmelia serrana* appears to be rather widely distributed. So far, it has been found in Europe (Austria, Germany, Finland, Russia, Spain, Sweden and Ukraine) [5,8,9,11,16,33] and Africa (Canary Islands). It probably occurs in North America, since Hale [1] reported specimens of *P. saxatilis* s.l. without lobaric acid; however, this hypothesis needs to be confirmed and specimens studied by Hale must be re-examined.

NUMBER OF SPECIMENS EXAMINED – 28.

SELECTED SPECIMENS EXAMINED. Ac-37 – Mierzeja Helska Sandbar, 1 km from Władysławowo town, on *Pinus sylvestris*, 1986.07.10, B. Wojtyła (UGDA L-2899). Ac-43 –Wybrzeże Słowińskie Coast, Słowiński National Park, Mierzeja Łebska Split, forest section No. 8, on sand, 1996.09.08, W. Fałtynowicz (UGDA L-5061). Ac-56 – Pradolina Redy i Łeby Glacial Valley, Świetlino village, forest section No. 31, on bark of tree, 1972.07.27, T. Sulma (UGDA L-5851). Ac-76 – Pojezierze Kaszubskie Lakeland, Szczelina Lechicka Nature Reserve, on *Alnus glutinosa*, 2012.05.05, A. Kowalewska (UGDA). Ac-95 – Pojezierze Kaszubskie Lakeland, 1 km E of Klukowa Huta village, on *Tilia cordata*, 1994.06.16, W. Fałtynowicz (UGDA L-4375). Bc-08 – Pojezierze Kaszubskie Lakeland, Przywidz village, by Przywidz Lake, on *Fagus sylvatica*, 1985.10.25, W. Fałtynowicz (UGDA L-2581). Bc-25 – Pojezierze Kaszubskie Lakeland, by Sasnowicz Lake near Dziemiany village, on *Betula pendula*, 1986.07.25, J. Miądlikowska (UGDA L-3543). Bd-79 – Garb Lubawski, Wzgórza Dylewskie Landscape Park, near Wygoda village, roadside trees, on *Betula pendula*, 2003.08.23, R. Szymczyk (OLS L-339). Bf-03 – Pojezierze Ełckie Lakeland, Borki Nature Reserve, forest section No. 22, on *Carpinus betulus*, 2012.09.19, R. Szymczyk (OLS L-1284); ibid., forest section No. 15, on bark of tree, 2012.08.22, R. Szymczyk (OLS

Tab. 1 Diagnostic characters of isidiate *Parmelia* species occurring in Poland.

Characters	*P. serrana*	*P. ernstiae*	*P. saxatilis*
type of thallus	epruinose, rarely faint pruina at lobe tips	strongly pruinose	epruinose or partially pruinose
shape of lobes	broad and round, often overlapping	broad and round, not overlapping	sublinear, not overlapping
type of isidia	epruinose	pruinose	epruinose
atranorin	present	present	present
salazinic acid	present	present	present
lobaric acid	absent	present	present
lichesterinic and protolichesterinic acids	present	present	absent

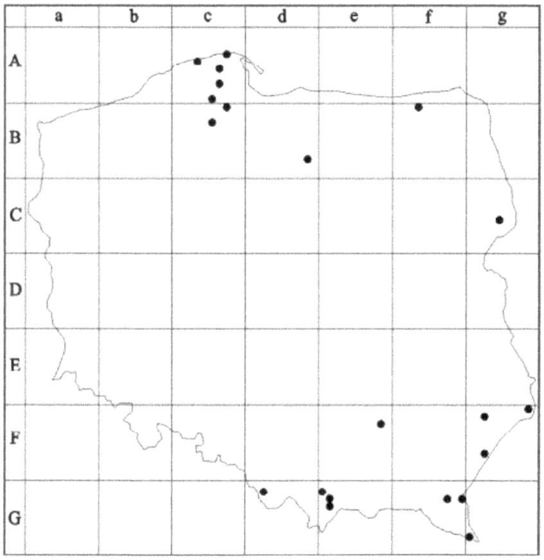

Fig. 1 Distribution of *Parmelia serrana* in Poland.

L-1243); Borki Nature Reserve, on *Quercus* sp., 1987.08.27, J. Zielińska (WA 28982). Cg-54 – Równina Bielska Plain, Białowieża Forest, Białowieża forest division, Zwierzyniec forest district, Natural Woodlands of Białowieża Forest Reserve, forest section No. 392A, wet alder forest, on *Fraxinus excelsior*, 2013.07.12, A. Bohdan (herb. SWAB); ibid., on branch of *Carpinus betulus*, 2013.07.12, A. Bohdan (herb. SWAB). Fe-28 – Niecka Połaniecka Basin, Przeczów Stara Wieś village, on *Betula* sp., 1990.09.04, J. Kiszka (KRAM L-60083). Fg-08 – Pogórze Przemyskie Foothills, Cisowa, forest section No. 30b, roadside, on *Quercus* sp., 1997.08.24, J. Kiszka (KRAM L-60081); road between Sanok to Przemyśl towns, on *Salix* sp., 1984.07.23, J. Kiszka & J. Piórecki (KRAM L-60091). Fg-12 – Roztocze Środkowe, Hamernia, on old *Quercus*, 1986.06.08, J. Kiszka & J. Piórecki (KRAM L-60092). Fg-62 – Płaskowyż Tarnogrodzki Plateau, S of Lubaczów town, Wielkie Oczy village, on *Betula* sp., 1984.07.24, J. Kiszka & J. Piórecki (KRAM L-60094). Gd-12 – Beskid Śląski Mts, Dolina Białej Wisełki valley, Fojtula village, on *Alnus* sp., T. Sulma (UGDA L-5330). Ge-10 – Gorce Mts, Poręba Wielka village, alt. 550 m, on bark of tree, 1993.11.23, P. Czarnota (GPN/1/94). Ge-21 – Gorce Mts, Gorce National Park, Turbacz Nature Reserve, below Czoło glade in Olszowy Potok valley, alt. 960 m, on *Abies alba*, 1994.06.23, P. Czarnota (GPN/653/94). Ge-31 – Pieniny Mts, Kramnica village, on *Salix* sp., 1998.08.15, J. Kiszka & J. Piórecki (KRAM L-64841). Gf-27 – Góry Słonne Mts, Rudenka village, on *Salix* sp., 1988.07.08, J. Kiszka & J. Piórecki (KRAM L-60085). Gf-29 – Pogórze Przemyskie Foothills, Liskowate, on *Populus tremula*, 1987.09.30, J. Kiszka & J. Piórecki (KRAM L-60080); ibid., on *Salix* sp., 1981.08.31, J. Kiszka & J. Piórecki (KRAM L-60077). Gg-70 – Bieszczady Zachodnie Mts, Połonina Bukowska, near the road to Bukowska Przełęcz pass, on *Alnus* sp., 1998.08.05, J. Kiszka & J. Piórecki (KRAM L-60097).

Acknowledgments

We are grateful to the Curators of herbaria for the loan of specimens and the reviewers for the constructive remarks. We thank also Professor Mark R.D. Seaward (Bradford) for checking the English. The research was supported by the Ministry of Science and Higher Education, project No. 2012/07/N/NZ8/00061 granted to EO.

Authors' contributions

The following declarations about authors' contributions to the research have been made: field and herbarium research: EO, RS, AB, MK; determination of specimens: EO, RS, MK, chromatographic analyses: EO, RS; writing of the manuscript: EO, MK, RS, AB; preparation of the distribution map: EO, MK.

References

1. Hale ME. A monograph of the lichen genus *Parmelia* Acharius sensu stricto (Ascomycotina: Parmeliaceae). Smithson Contrib Bot. 1987;(66):1–55.
2. Ferencova Z, Cubas P, Divakar PK, Molina MC, Crespo A. *Notoparmelia*, a new genus of Parmeliaceae (Ascomycota) based on overlooked reproductive anatomical features, phylogeny and distribution pattern. Lichenologist. 2014;46(1):51–67.
3. Louwhoff SHJJ, Purvis OW, James PW. *Parmelia* Ach. (1803). In: Smith CW, Aptroot A, Cooppis BJ, Fletcher A, Gilbert OL, James

PW, et al., editors. The lichens of Great Britain and Ireland. London: British Lichen Society; 2009. p. 651–654.

4. Crespo A, Kauff F, Divakar PK, del Prado R, Pérez-Ortega S, de Paz GA, et al. Phylogenetic generic classification of parmelioid lichens (Parmeliaceae, Ascomycota) based on molecular, morphological and chemical evidence. Taxon. 2010;59(6):1735–1753.
5. Thell A, Thor G, Ahti T. *Parmelia* Ach. In: Thell A, Moberg R, editors. Nordic lichen flora. Göteborg: Nordic Lichen Society; 2011. p. 83–90. (vol 4).
6. Crespo A, Pérez-Ortega S. Cryptic species and species pairs in lichens: a discussion on the relationship between molecular phylogenies and morphological characters. Jard Bot Madr. 2010;66(S1):71–81.
7. VondráK J, Říha P, Arup U, SøChting U. The taxonomy of the *Caloplaca citrina* group (Teloschistaceae) in the Black Sea region; with contributions to the cryptic species concept in lichenology. Lichenologist. 2009;41(6):571.
8. Molina MC, Del-Prado R, Divakar PK, Sánchez-Mata D, Crespo A. Another example of cryptic diversity in lichen-forming fungi: the new species *Parmelia mayi* (Ascomycota: Parmeliaceae). Org Divers Evol. 2011;11(5):331–342.
9. Molina MC, Crespo A, Blanco O, Lumbsch HT, Hawksworth DL. Phylogenetic relationships and species concepts in *Parmelia* s.str. (Parmeliaceae) inferred from nuclear ITS rDNA and tubulin sequences. Lichenologist. 2004;36(1):37–54.
10. Molina MC, Divakar PK, Millanes AM, SáNchez E, Del-Prado R, Hawksworth DL, et al. *Parmelia sulcata* (Ascomycota: Parmeliaceae), a sympatric monophyletic species complex. Lichenologist. 2011;43(6):585–601.
11. Thell A, Elix JA, Feuerer T, Hansen ES. Notes on the systematics, chemistry and distribution of European *Parmelia* and *Punctelia* species (lichenized ascomycetes). Sauteria. 2008;15:545–559.
12. Crespo A, Divakar PK, Hawksworth DL. Generic concepts in parmelioid lichens, and the phylogenetic value of characters used in their circumscription. Lichenologist. 2011;43(6):511–535.
13. Nowak J, Tobolewski Z. Porosty Polskie. Warsaw: Polish Scientific Publishers PWN; 1975.
14. Fałtynowicz W. The lichens, lichenicolous and allied fungi of Poland – an annotated checklist. Cracow: W. Szafer Institute of Botany, Polish Academy of Sciences; 2003.
15. Motiejūnaitė J, Kukwa M. *Parmelia submontana* in Lithuania and northern Poland. Graph Scr. 2003;14:62–64.
16. Hawksworth DL, Blanco O, Divakar PK, Ahti T, Crespo A. A first checklist of parmelioid and similar lichens in Europe and some adjacent territories, adopting revised generic circumscriptions and with indications of species distributions. Lichenologist. 2008;40(1):1–21.
17. Kukwa M, Łubek A, Szymczyk R, Zalewska A. Seven lichen species new to Poland. Mycotaxon. 2012;120(1):105–118.
18. Ossowska E, Kukwa M. Nowe stanowiska porostu *Parmelia ernstiae* Feuerer & A. Thell w Polsce. Acta Bot Cassub. 2012;11:195–199.
19. Crespo A, Molina MC, Blanco O, Schroeter B, Sancho LG, Hawksworth DL. rDNA ITS and β-tubulin gene sequence analyses reveal two monophyletic groups within the cosmopolitan lichen *Parmelia saxatilis*. Mycol Res. 2002;106(7):788–795.
20. Jabłońska A, Oset M, Kukwa M. The lichen family Parmeliaceae in Poland I. The genus *Parmotrema*. Acta Mycol. 2009;44:211–222.
21. Kukwa M, Pietnoczko M, Czyżewska K. The lichen family Parmeliaceae in Poland. II. The genus *Cetrelia*. Acta Soc Bot Pol. 2012;81(1):43–52.
22. Kowalewska A, Kukwa M. *Cladonia metacorallifera* (lichenized Ascomycota, Cladoniaceae) new to Poland and additional record from Slovakia. Biol Bratisl. 2004;59:433–434.
23. Flakus A. Three species of lichenized Ascomycota new to Poland. Biol Bratisl. 2006;61(1):15–17.
24. Kukwa M, Kubiak D. Six sorediate crustose lichens new to Poland. Mycotaxon. 2007;102:155–164.
25. Syrek M, Kukwa M. Taxonomy of the lichen *Cladonia rei* and its status in Poland. Biol Bratisl. 2008;63(4):493–497.
26. Krzewicka B. The "*Verrucaria fuscella* group" in Poland with some nomenclatorial remarks. Acta Soc Bot Pol. 2009;78(3):229–234.
27. Kukwa M, Jabłońska A. New records of two crustose sorediate lichens from central Europe. Mycotaxon. 2009;107(1):375–381.
28. Kubiak D. Distribution and ecology of the lichen *Fellhanera gyrophorica* in the Pojezierze Olsztyńskie Lakeland and its status in Poland. Acta Soc Bot Pol. 2011;80(4):293–300.
29. Śliwa L, Flakus A. *Lecanora microloba*, a new saxicolous species from Poland. Lichenologist. 2010;43(1):1–6.
30. Orange A, James PW, White FJ. Microchemical methods for the identification of lichens. London: British Lichen Society; 2001.
31. Cieśliński S, Fałtynowicz W. Note from editors. In: Cieśliński S, Fałtynowicz W, editors. Atlas of the geographical distribution of lichens in Poland. Cracow: W. Szafer Institute of Botany, Polish Academy of Sciences; 1993. p. 7–8. (vol 1).
32. Feuerer T, Thell A. *Parmelia ernstiae* – a new macrolichen from Germany. Mitt Inst Allg Bot Hamb. 2002;30-32:49–60.
33. Hawksworth DL, Divakar PK, Crespo A, Ahti T. The checklist of parmelioid and similar lichens in Europe and some adjacent territories: additions and corrections. Lichenologist. 2011;43(6):639–645.

14

No greens in the forest? Note on the limited consumption of greens in the Amazon

Esther Katz[1]*, Claudia Leonor López[2], Marie Fleury[3], Robert P. Miller[4], Valeria Payê[5], Terezinha Dias[6], Franklin Silva[7], Zelandes Oliveira[7], Elaine Moreira[8]

[1] Institut de Recherche pour le Développement, UMR 208 IRD/MNHN, Cuvier 57, 75005 Paris, France

[2] Museu Paraense Emílio Goeldi, Belém, Avenida Perimetral 1901, Box 399, 66077-530, Belém, Pará, Brazil

[3] Muséum National d'Histoire Naturelle, UMR 208 IRD/MNHN, Cayenne, French Guiana

[4] United Nations Development Program, Project GATI, Brasilia, DF, Brazil

[5] Institute of Social Sciences, University of Brasilia, Campus Darcy Ribeiro, Asa Norte, 70910-900 Brasilia, DF, Brazil

[6] Empresa Brasileira de Pesquisa Agropecuária, Recursos genéticos e biotecnologia, Parque Estação Biológica – W 5 Norte Final Asa Norte, Box 02372, 70700-900 Brasília, DF, Brazil

[7] Center of Sustainable Development, University of Brasilia, Campus Darcy Ribeiro, Asa Norte, 70904-970 Brasilia, DF, Brazil

[8] Department of Anthropology, Federal University of Roraima, Avenida Capitao Ene Garcez, 64000-000 Boa Vista, Roraima, Brazil

Abstract

The consumption of greens is reported as being very minor among Amazonian Indians. The authors of this article present a new review of this subject, based on fieldwork with Amerindians and other populations in different parts of the Brazilian Amazon and French Guiana. Written sources on Brazilian, Peruvian, Columbian and Venezuelan Amazon were also reviewed. The consumption of cultivated, semi-cultivated and wild species of greens was taken into account here, as the data specific to wild greens is very scarce. It is confirmed that greens are not commonly eaten among native Amazonians and that some ethnic groups do not consume them at all. The consumed species are usually young shoots of weeds or cassava leaves. Common in the Belém region are some specific aromatic plants, which have been diffused to other parts of the Amazon, together with introduced plants such as kale and coriander. Migrants from Northeastern Brazil settled in the Amazon consume some cultivated greens, especially aromatic plants. Maroons are the ones who use more greens in their diet. Native Amazonian people, who supplement agriculture with game and fish, follow a hunter-gatherer pattern, preferring wild fruit and tubers to greens.

Keywords: Amazon, greens, herbs, wild foods, weeds, ethnobotany

Introduction

Canadian Indians of the West Coast consume wild greens in the springtime [1], Mexican Indians in the rainy season [2–4], but what happens with the inhabitants of the Amazon, who benefit of a warm climate all year long, of rain for several months and are surrounded by forests? In an overview of the consumption of wild plants by Amazonian Indians, Dufour and Wilson [5] mentioned that very few greens were consumed. Taking this idea as a starting point, we present here data collected in the field in different parts of the Amazon as well as a review of the available literature. We confirm that consumption of greens is a rarity in the Amazon, but note that it does happen in a few places, in specific contexts or populations. As the wild greens used are very few, we will deal as well with ruderal species, weeds and cultivated greens and try to understand why this consumption represents such a minor component of diets.

* Corresponding author. Email: esther.katz@ird.fr

Material and methods

The authors are either social or natural scientists who have focused on ethnobotany at least in part of their research. They have conducted fieldwork in different parts of the Brazilian Amazon (and one author in French Guiana), working on use of the environment, agriculture, agroforestry and food with several Amerindian ethnic groups. They have also visited other parts of the Amazon where they gathered empirical observations. Three authors are Amerindian students at the University of Brasilia, in Social Sciences (Payê) and Sustainable Development (Silva, Oliveira). The last two students carry out research in their communities. The Amerindian peoples surveyed (see Tab. 1) are: Ethnic groups of the Carib linguistic family from the Northern Amazon: Wayana and Kali'ña in French Guiana (Fleury), Kaxuyana and Tiryo in the Brazilian State of Amapa,

Tab. 1 Quoted ethnic and socio-cultural groups.

Country and region	Localization	Ethnic group	Linguistic family	Reference*
Amerindian ethnic groups				
French Guiana (France)				
NW	lower Maroni, coast	Ka'liña	Carib	Fleury
SW	upper Maroni	Wayana	Carib	Fleury
Venezuela				
State of Amazonas	Upper Rio Negro	Baniwa	Arawak	[6]
Columbia				
Dpt of Vaupés		Tatuyo	Tukanoan (Eastern Tukanoan)	[5]
Dpt of Caqueta		Ingano (Quichua)	Quechuan	[8]
Peru				
Dpt of Amazonas	Cenepa river	Awajún	Jivaroan	[12,13]
Dpt of Amazonas/Loreto		Wampis	Jivaroan	K. Świerk (pers. com.)
Brazil				
State of Amazonas	Rio Negro	Baniwa	Arawak	Silva, Katz
	(NW Amazon)	Baré	Arawak, speakers of Nheengatu (Tupi)	Katz
		Tukano	Tukanoan (Eastern Tukanoan)	Katz
		Desana	Tukanoan (Eastern Tukanoan)	Katz
		Piratapuia	Tukanoan (Eastern Tukanoan)	Katz
	Solimões (upper Amazon river)	Ticuna	Ticuna	López
	middle Amazon river	Sateré-Mawé	Tupi	[7]
State of Roraima	NE (Guyana border)	Patamona	Carib	Oliveira
	N, NE	Makushi	Carib	Miller
	N, NE	Taurepang	Carib	Miller
	N, NE	Wapishana	Arawak	Miller
	S	Waimiri-Atroari	Carib	Miller
	NW (Venezuela border)	Ye'kuana	Carib	Moreira
State of Amapa	Surinam border	Tiryo	Carib	Payê
	Surinam border	Kaxuiana	Carib	Payê
State of Pará		Parakanã	Tupi-Guarani	Miller
	Middle Xingu river	Kayapó	Jê	López
State of Pará/Maranhão		Ka'apor	Tupi-Guarani	López
State of Tocantins		Krahô	Jê	Dias
State of Mato Grosso		Xavante	Jê	S. Silva, S. Juruna (pers. com.)
	National Park of Xingu	Kuikuro	Carib	M. Smith (pers. com.)
State of Acre	Peru border	Kaxinawa	Pano	[10,11]
Other ethnic/cultural groups				
French Guiana (France)				
SW	upper Maroni	Aluku (Maroons)		Fleury
Brazil				
State of Pará	Belém and surroundings	mestizos (caboclos)		López
	Tapajos River	mestizos (of Nordestino descent)		[20]
State of Acre	upper Juruá (Cruzeiro do Sul)	mestizos (of Nordestino descent)		Katz
	upper Juruá (Extractive Reserve)	mestizos (of Nordestino descent)		[19]
State of Amazonas	Manaus	Amerindians, mestizos, Nordestinos,…		[21]

* Names indicate personal observations of the authors of the article.

Tab. 2 Greens consumed in the Amazon.

Family	Species	Status	Preparation form	Local name	Ethnic group using it
Leaves (eaten cooked or as salad)					
Amaranthaceae	*Amaranthus spinosus* L.	cultivated	cooked		Aluku
Amaranthaceae	*Amaranthus dubius* Mart.	cultivated	cooked		Aluku
Apocynaceae	*Matelea rivularis*	?	cooked		Awajún, Wampis
Araceae	*Xanthosoma sagittaefolium* Schott.	cultivated	cooked	taioba	Wayana, Aluku, Patamona?
Araceae	*Xanthosoma* spp.	cultivated	cooked		Awajún
Araceae	*Caladium bicolor* (Aiton) Vent.	cultivated	cooked		Awajún
Araceae	*Philodendron* sp.	cultivated/spontaneous	cooked		Awajún
Araceae	*Anthurium* sp.	?	cooked		Awajún, Wampis
Asteraceae	*Acmella oleracea* (L.) R. K Jansen (syn. *Spilanthes oleracea*)	cultivated	cooked	jambú	Belém caboclos
Asteraceae	*Lactuca sativa* L.	cultivated	salad	salada	Acre Nordestinos
Asteraceae	*Spilanthes acmella* Murray	cultivated	cooked, raw	jambú	Belém caboclos, Acre Nordestinos
Basellaceae	*Basella alba* L.	cultivated	cooked	caruru	Rio Negro ethnic groups
Basellaceae	?	cultivated	cooked	caruru	Sateré-Mawé
Brassicaceae	*Brassica oleracea* L.	cultivated	raw, cooked	couve, repolho	Acre Nordestinos
Brassicaceae	*Eruca sativa* Mill.	cultivated	salad	rúcula	Acre Nordestinos
Euphorbiaceae	*Manihot esculenta* Crantz	cultivated	cooked	maniçoba	Belém caboclos, Rio Negro, Sateré-Mawé, Ingano, Kaxinawa, Aluku
Malvaceae	*Abelmoschus esculentus* (L.) Moench	cultivated	cooked	okra leaves	Aluku
Oxalidaceae	*Oxalis barrelieri* L.	spontaneous	salad	suwa wiwii (Aluku)	Aluku
Palmae	*Astrocaryum chambira* Burret	cultivated/spontaneous	cooked		Awajún
Phytolaccaceae	*Phytolacca rivinoides* Kunth & Bouché	spontaneous, in swidden fields	cooked	caruru	Rio Negro ethnic groups, Patamona, Makushi, Ka'apor, Aluku, Acre Nordestinos
Piperaceae	*Piper* sp.	cultivated/spontaneous	cooked, raw		Awajún
Piperaceae	*Peperomia pellucida* (L.) H. B. K.	spontaneous	salad	"soldier salad"	Aluku
Portulacaceae	*Talinum paniculatum* (Jacq.) Gaertn.	spontaneous, in swidden fields	cooked	caruru	Belém caboclos, Rio Negro ethnic groups
Portulacaceae	*Talinum triangulare* (Jacq.) Willd	spontaneous, in swidden fields	cooked	caruru	Rio Negro ethnic groups
Portulacaceae	*Portulaca oleracea* L.	cultivated	salad	pourpier (French)	Aluku
Herbs (aromatic plants)					
Apiaceae	*Eryngium foetidum* L.	cultivated		chicória	Belém caboclos, Acre Nordestinos, Kaxuiana/Tiryo (semi-wild), Rio Negro (cultivated, recent), Ka'apor, Kayapó, Sateré-Mawé
Apiaceae	*Coriandrum sativum* L.	cultivated		cheiro verde	Acre Nordestinos, Kaxuiana/Tiriyo (cultivated, recent), Rio Negro (cultivated, recent)
Apiaceae	*Petrosellinum sativum* (Mill.) Fuss	cultivated		salsa	Acre Nordestinos
Lamiaceae	*Ocimum* cf. *micranthum* Willd.	cultivated		alfavaca	Sateré-Mawé
Liliaceae	*Allium fistulosum* L.	cultivated		cebolinha	Acre Nordestinos, Rio Negro, Sateré-Mawé
Malvaceae	*Hibiscus sabdariffa* L.	cultivated		vinagreira	Belém caboclos, Acre Nordestinos
Solanaceae	*Capsicum* spp.	cultivated		folhas de pimenta	Makushi, Wapishana
?	?	spontaneous		nawanti (Kaxinawa)	Kaxinawa
?	?	spontaneous		xiwan (Kaxinawa)	Kaxinawa
?	?	spontaneous		sawiti (Sateré)	Sateré-mawé
Stems					
Cucurbitaceae	*Cucurbita* sp.	cultivated			Tiryo, Kaxuiana
Araceae	*Xanthosoma* sp.?	cultivated		taioba	Tiryo, Kaxuiana

close to Surinam (Payê), Waimiri-Atroari, Makushi, Taurepang (Miller), Ye'kuana (Moreira) and Patamona (Oliveira) in the Brazilian State of Roraima, as well as Wapishana (of the Arawak linguistic family) of the same region (Miller); Baniwa (of the Arawak linguistic family), Baré (people of the Arawak linguistic family who lost their language in favor of Nheengatu or Lingua geral, a Tupi language) and Tukano, Desana, Piratapuia (of the Tukanoan linguistic family, Eastern Tukanoan branch) from the middle Rio Negro (Katz); Baniwa of the upper Rio Negro (Silva); Parakanã of Pará (Miller) and Ka'apor of Pará/Maranhão (of the Tupi-Guarani linguistic family) and Ticuna (Ticuna linguistic family) of the Solimões (upper Amazon; López); Kayapo of Pará (López), and Krahô of Tocantins (Dias; of the Jê linguistic family). Fleury has also done fieldwork with the Aluku Maroons of Western French Guiana. Katz studied the market of Cruzeiro do Sul, in Acre (South-Western Amazon) and did fieldwork with farmers of the same region, with ethnic origins in the Northeast of Brazil (Nordestinos). López led a specific enquiry in and around Belém (Pará), in July 2012, on the management, use and consumption of local cultivated and semi-wild greens by the local mestizo population. Most authors have been revising literature on management and use of resources and food plants in the Amazon. Except for the recent enquiry led by López, the data presented have been collected over several years through observation and interviews that were not specifically oriented on greens but on food, agriculture and/or management of resources.

Results

Greens and the characteristics of the environment

First, what is called the "Amazon" is not a homogeneous forest area. Several types of environments must be distinguished. We will not enter here in ecological details, but roughly there are dense and more open forests, flooded and non-flooded forests ("terra firme"), and a wide range of vegetation formations and transition zones. Moreover, the South and Southeastern part of the legally defined "Amazon" region is actually covered by tree savanna ("cerrado"), also divided in several types of vegetation: "closed" and "open" "cerrado", gallery forests, and patches of various types of dry forests. In the northern portion of Amazonia, there is a corridor of savanna, called "lavrado" in the Brazilian regional terminology, contiguous with savannas of Guyana and Venezuela. In the "cerrado", rainy and dry seasons are strongly contrasted, with plant growth very limited in the dry season. As Dias observes, most leaves are very leathery, therefore not palatable, even if they are edible. In the dense forest, rainy and dry seasons are not so contrasted, with the forest maintaining a humid microclimate in its interior. As Fleury notes, the tree canopy is generally high, and the understory may be sparse. Tender leaves are not so easily available. Only a portion of tropical forest trees change their leaves on a seasonal basis, and therefore do not produce many young shoots, which are usually the consumed parts. Additionally, as highlighted by Miller, many plants contain secondary toxic compounds that protect leaves against potential predators. Available edible greens are more likely to grow as weeds in swidden fields, in fallow lands or along the paths.

Greens and Amerindian food habits: tortoise food?

As Miller observed among the Waimiri Atroari of Roraima and Amazonas states, leaves such as lettuce are jokingly considered "tortoise food" ("comida de jabuti"). Other researchers report that leaves are considered all over the Amazon as "animal food" ("comida de bicho"; Donald Sawyer, CDS-UnB, Brasilia, personal communication). According to Payê, when Carib Kaxuyana and Tiryo people of Amapa have wild animals as pets, they give them specific foods: monkeys are fed with fruit, agouti with nuts, birds with sweet potato and cassava bread, and tortoises ("jabuti" and "tartaruga") with leaves. Kaxuyana people who go to the city for the first time and are offered lettuce, perceive it as an "animal food" and consider it tasteless. In their community, they eat no leaves except for an aromatic herb, "chicória" (*Eryngium foetidum* L. - Apiaceae), used to flavour fish stew. They do not really cultivate "chicória", but rather favour it by throwing its seeds in the areas where it grows easily, close to the river. They also consume the stems of some plants. When a meal is prepared with leftovers of a fish stew, they add squash (*Cucurbita* sp. - Cucurbitaceae) and cocoyam or "taioba" (*Xanthosoma* sp.?, Araceae) stems, as well as sweet potato tubers [*Ipomea batatas* (L.) Lam. var *batatas* - Convolvulaceae]. Recently they have been adding vegetables introduced by Franciscan missionaries, such as "maxixe" (*Cucumis anguria* - Cucurbitaceae), now well appreciated, and coriander leaves ("cheiro verde", *Coriandrum sativum* - Apiaceae).

When greens are eaten in the Northern Amazon (Tab. 2), they are often cooked in fish stew, "to flavour it" ("temperar

Fig. 1 Wayana Indian cutting off the leafy parts of *Xanthosoma sagittaefolium* (M. Fleury; identification: M. Fleury).

Fig. 2 "Caruru" (*Phytolacca rivinoides*) in a swidden in Santa Isabel do Rio Negro (E. Katz; identification: L. Emperaire).

o peixe") or give it consistency (especially when there is not much fish), as noted by Silva; yams (*Dioscorea* spp. - Dioscoreaceae), peach palm flowers (*Bactris gasipaes* Kunth - Arecaceae) and other ingredients may also "flavour the fish". Wayana Indians of French Guiana use leaves of cultivated cocoyam in their fish stew, as observed by Fleury, picking the leafy part off from the veins (Fig. 1). In Santa Isabel, in the middle Rio Negro, as observed by Katz, Eastern Tukanoan and Arawak people use the leaves of weeds called "caruru" and of cultivated cassava (*Manihot esculenta* Crantz - Euphorbiaceae). Cassava leaves are called "maniçoba", in "Lingua geral" and local Portuguese. The botanist Laure Emperaire (personal communication) identified several species under the name of caruru in that region: *Phytolacca rivinoides* Kunth & Bouché (Phytolaccaceae), which usually grows in recently burnt swidden fields (Fig. 2), *Talinum paniculatum* (Jacq.) Gaertn. and *Talinum triangulare* (Jacq.) Willd (Portulacaceae), as well as an introduced species, *Basella alba* L. (Basellaceae), cultivated in the garden of the Salesian mission. As observed by Silva, in the upper Rio Negro, *Phytolacca* is also consumed in Baniwa communities of the Içana river, and *Talinum* is sold on the market in the town of São Gabriel da Cachoeira. Hoffmann [6] also mentions the consumption of caruru (*Phytolacca* cf. *rivinoides*) among the Baniwa of the upper Rio Negro in Venezuela, and Dufour and Wilson [5] among the Tatuyo, Eastern Tukanoan people of Columbia. "Caruru" and manioc leaves are usually boiled with chilli pepper, sometimes boiled manioc juice ("tucupi") also flavours the broth, then fish is added at the end. Manioc leaves may also be ground in a mortar before or after being cooked.

In the urban context of Santa Isabel do Rio Negro, not everybody eats these greens. Katz was shown by two older ladies how they are prepared, but most people referred to them as something that they used to eat more in the past, or when they lived in the upper Rio Negro (a more traditional area). Younger generations are not attracted by this type of food. Nevertheless, when demonstrated by a lady in a workshop on local food heritage (organised by the local Amerindian association, ACIMRN, and E. Katz), it raised curiosity and recalled the memories of some participants. In the Rio Negro towns, vegetables and aromatic plants have also been introduced in the last decades by missionaries, traders and public institutions: "chicória", which is not native to the region, coriander leaves, salad onions ("cebolhinha" or "cebola de palha"; *Allium fistulosum* - Liliaceae), "maxixe", cultivated locally by some people in the yard or home garden. Presently, people also consume onion ("cebola", *Allium cepa*), carrot [*Daucus carota* L. subsp. *sativus* (Hoffm.) Arcang. - Apiaceae] and tomato (*Lycopersicon esculentum* Mill. - Solanaceae), brought by boat from Manaus (Fig. 3).

The Sateré-Mawé of the middle Amazon also eat manioc leaves, and cultivate near the house "caruru" (identified as a Basellaceae species) which they use as a condiment, as well as other aromatic herbs used as condiment (salad onions, "chicórea") or medicine, such as basil - "alfavaca" - (*Ocimum* cf. *micranthum* Willd., Lamiaceae). They use a non-identified leaf from the abandoned swiddens, "sawiti", to flavour cassava fermented drinks [7]. More to the West, the consumption of cassava leaves is also mentioned for the Ingano, a Quichua-speaking ethnic group of the upper Columbian Amazon, in the Caqueta Department [8].

The Makushi and Wapishana who inhabit the savanna regions of Roraima (the "lavrado") commonly prepare a very spicy fish or meat stew called "damorida", in which main condiments are chili peppers (*Capsicum chinense* Jacq. and *Capsicum* spp. - Solanaceae) and the leaves of the chili pepper plant. Miller also observed a spontaneous plant in a swidden field (most likely *Phytolacca*) being collected for use in "damorida". According to Oliveira, the Patamona, who live in the highlands savannas above the Makushi and Wapishana communities, at the border of Guyana, eat leaves of at least three species of Araceae (including *Xanthosoma* sp.) generally called "taioba" ("João Gomes" in the local Portuguese, "er'oiá" in Patamona), and of a weed found in cassava swidden fields, after the burning, called in Patamona "aur'otsa" (very probably *Phytolacca rivinoides*).

As observed by Moreira, Carib Ye'kuana people of Northern Roraima use several types of leaves as medicines and for rituals, and wrap in large leaves (banana leaves, leaves of local wild Musaceae, etc.) food they want to cook or smoke-dry - a common practice all over the Amazon. While the Ye'kuana do not eat greens, they sprinkle some dry powdered leaves on cassava bread to speed growth of mould, in order to prepare fermented drinks, a practice also employed by other Carib people, the Tiryo, Kaxuyana (mentioned by Payê) and Wayana (Lucia van Velthem, personal communication). The Makushi reported to Miller the use of manioc leaves for this purpose, or leaves of the pioneer tree *Trema micrantha* (L.) Blume (Celtidaceae).

The Kuikuro, Carib ethnic group of the upper Xingu, do not consume any greens, according to the botanist Maira Smith (CDS-UnB, Brasilia, personal communication). As observed by López, the Ticuna of Amazonas also do not eat greens, the Ka'apor of Maranhão and the Kayapó of Pará only cultivate *Eryngium foetidum* to flavour their dishes, mainly beans in the case of the K'aapor. The consumption of *Phytolacca rivinoides* was also mentioned among the Ka'apor [5]. Among the Jê people of the "cerrado", Krahô children of Tocantins eat some flowers, but neither the children nor the adults eat greens, as observed by Dias, and the Xavante of Mato Grosso do not eat greens either (Sayonara Silva, Samantha Juruna, CDS-UnB Brasilia, personal communications).

In reviewing the food/agriculture/economic activities sections of the "Encyclopedia of the Indigenous People of Brazil" [9], as well as other sources on Amerindian food (which are actually not very numerous), Katz only found one reference on use of greens in the Amazon: according to Lagrou [10], the Kashinawa, a Pano people of the Southwestern Amazon, commonly use sweet manioc leaves as a condiment, as well as a herb that looks like coriander leaves. They make a gruel out of

Fig. 3 "Chicória", onion and salad onion on a kitchen table, Santa Isabel do Rio Negro (E. Katz).

cooked sweet cassava, flavoured with ground toasted peanuts and ground cassava leaves or "nawanti" or "xiwan", non-identified forest leaves with an acid taste similar to *Eryngium foetidum*. Pantoja et al. [11] also mention the consumption of "nawanti" and cassava leaves by the Kashinawa, as cooked together with manioc tubers. They note that women not only collect "nawanti" in the forest, but also cultivate it in "canteiros" (large pots) near the house.

Would the consumption of greens be more significant in the Western Amazon? Among all the cases we reviewed, the Awajún (or Aguaruna), a Jivaroan people from the Peruvian Amazon, are the Amerindians who consume the widest range of greens [12,13]. They cultivate several greens species: cassava leaves, *Caladium bicolor* (Aiton) Vent. (Araceae), *Xanthosoma* spp. (Araceae); other species are both cultivated and found in the wild: *Piper* sp. (Piperaceae), *Philodendron* sp. (Araceae), *Astrocaryum chambira* (Palmae), the only reference we found of the consumption of palm leaves. *Piper* sp. may be eaten cooked or raw, the other species are used cooked, in soups or in "patarashca" with fish (grilled food wrapped in leaves). According to the anthropologist Kacper Świerk (personal communication), the Awajún and Wampis (Huambisa/Peruvian Shuar, Jivaroan) also eat leaves of *Anthurium* sp. (Araceae; Wampis: "eep", regional Spanish: "repollo de monte", "wild cabbage") added to "patarashca" with fish, meat or with edible mushrooms called "esempu"; leaves of the small plant *Matelea rivularis* Woodson (Apocynaceae; Wampis: "tsamantsma", identified by the botanists David Neill and Camilo Kajekai) are also added to "patarashca".

Outside of the indigenous populations mentioned above, other inhabitants of the Amazon are mestizos of indigenous descent (called in Brazil "caboclos"), maroons and people who migrated from the Northeast of Brazil. We do not have data on other populations (of European origin or who migrated from other regions), except roughly on the city of Manaus.

Greens consumed by the mestizo population of the State of Pará

López conducted a short survey in the city of Belém (Pará; where she lives) and on the neighbouring Atlantic coast. Most inhabitants of the rural areas in Pará, and many of the city dwellers are "caboclos". Two greens enter in the composition of typical dishes of Belém region: ground manioc leaves and "jambú" [*Acmella oleracea* (L.) R. K. Jansen (syn. *Spilanthes oleracea* L.); *Spilanthes acmella* Murray - Asteraceae]; "Chicória" (*Eryngium foetidum*) and "vinagreira" (*Hibiscus sabdariffa* L. - Malvaceae) flavour several dishes. "Cariru" (or "caruru"; *Talinum esculentum*) is also consumed. Kale ("couve", *Brassica oleracea* L. - Brassicaceae), coriander leaves (*Coriandrum sativum*) and salad onion (*Allium fistulosum* L.), common in most regions of Brazil are also sold in Belém market (Fig. 4).

Ground manioc leaves are a culinary specialty of the Belém región (Fig. 5). In the State of Pará, manioc leaves are known under the name of "maniva", which refers in other regions to the manioc plant. When ground, they are called "maniçoba", name given to the leaves in other regions of the Amazon, whether fresh, cooked, entire or ground. "Maniçoba" is also the name of a dish prepared with the ground leaves, first mentioned in the literature in the 18th century by an Italian astronomer, Giovanni Agnelo Brunelli, who participated in the Comisión Demarcadora de Limites, organized to define the boarders between Spain and Portugal according to what had been stipulated in the 1750 Madrid Treaty [14]. In 1767

Fig. 4 Cultivated aromatic leaves sold on Belém market (kale, "chicória", coriander and salad onion; E. Katz).

he published a text called "About cassava" in the "Journal of the Sciences Academy of Bologna Institute", where he describes the dish:

"Mas também as folhas da planta da 'Maniva' costumam ser preparadas muitas vezes pelos índios na alimentação, como legumes. Amassam as folhas, seguramente no pilão, e cozinham ao fogo em água, com pedaços de carne ou de peixe, com gordura também adicionada ao caldo de mandioca, pimenta e tempero. Essa comida que chamam de 'maniçoba', embora pareça ser de bárbaros, eu não pude desprezar" [But the leaves of the 'Maniva' plant are very often prepared by the Indians in their food, as greens. They grind the leaves, probably in a mortar, and cook them in water, with pieces of meat or fish, with grease added to the cassava broth, chili pepper and condiments. They call this dish maniçoba. Although it looked like barbarian food, I could not refuse it] (Brunelli, apud. [14]).

At that time, there were still Indians in the regíón. Today many people probably descend from them but do not identify themselves as such. Presently "maniçoba" is offered in many food stands of regional dishes all over the city and is consumed in different parts of the State of Pará. Along with "pato no tucupi" (duck in "tucupi", the cooked juice extracted from ground manioc roots), it enters in the composition of the traditional "Lunch of the Círio de Nazaré" (the Candle of Nazareth), the main religious festival of the city, celebrated in September or October all over Pará, and which, in Belém, attracts hundreds of thousands of people and was declared National Cultural Heritage.

The leaves of "jambú" (*Acmella oleracea, Spilanthes acmella*) are one of the main greens consumed by the rural and urban population of Pará and Eastern Amazon. "Jambú" is either favoured or cultivated in the homegardens of the rural and periurban areas of Pará. It often grows spontaneously in the wet parts of the garden, close to where clothes are washed (Fig. 6). It is cultivated more intensively, albeit on a small scale, by entrepreneurs who sell it in local markets and supermarkets in Belém to supply urban dwellers. It is an important ingredient in regional dishes such as "pato no tucupi" and "tacacá". "Tacacá" is a broth made of "tucupi" flavoured with "chicória", dry shrimp and "jambú", and thickened by "goma" - manioc starch. It is sold by small food stands in the streets of Belém and other town of the States of Pará and Amazônas [15]. "Jambú" is also consumed in chicken dishes (Fig. 7). Recently it has been explored as a gourmet ingredient in the best restaurants of regional paraense cuisine, and a pizza of "jambú" and shrimp, inspired by "tacacá", has been created. The late chef Paulo Martins invented a dish called "rondelle de jambú" in his highly renowned restaurant. In the Eastern Amazon, "jambú" is definitely the most consumed green, with the most diverse culinary uses.

At least one type of "caruru" [*Talinum paniculatum* (Jacq.) Gaertn., erroneously identified in the quoted article as *Talinum esculentum*] is used in Pará regional food [16]. It is cultivated, together with "jambú", in humid parts of rural house yards (Fig. 8) and is also sold in urban markets (Fig. 9) and supermarkets. In Pará, "caruru" is one of the main ingredients

Fig. 5 Grinding manioc leaves on Belém market (E. Katz).

of dish called by the same name, prepared with dry shrimp, okra [*Abelmoschus esculentus* (L.) Moench – Malvaceae], and cassava "flour", and sold in the small food stands in Belém. It is also commonly served in the religious celebrations of the month of June ("festas juninas"), all over the State. Occasionally it is cooked with beans.

Roselle (*Hibiscus sabdariffa* L.; Fig. 10), a cultivated plant originally from Africa [17], is consumed in the Northeast of Brazil. It is very popular in the State of Maranhão (Eastern Amazon), where it enters in the composition of different regional dishes, the most famous being "arroz com cuchá" ("rice with roselle"). It is not as common in Pará, but it flavours a few dishes of grilled fish, especially on the coast.

Maroons and greens

The Aluku Maroons, who live close to the Wayana Amerindians, on the upper Maroni River, in French Guiana, were studied by Fleury. They practice slash-and-burn agriculture and cultivate dry land rice, which is the base of their diet. They usually eat the rice with oleaginous or mucilaginous sauces, made respectively of peanut (*Arachis hypogaea* L. – Fabaceae), sesame seeds (*Sesamum orientale* L. – Pedialaceae) or okra (*Abelmoschus esculentus*). They also cook the leaves of cultivated plants such as cassava, okra, cocoyam (*Xanthosoma sagittaefolium* Schott) and amaranth (*Amaranthus spinosus* L., *Amaranthus dubius* Mart. – Amaranthaceae), and pick the leaves of makoko (*Phytolacca rivinoides*; the "caruru" of Brazil) on the borders of swidden fields or in fallows. They also pick "pourpier" ("poseen": *Portulaca oleracea* L.), eaten raw or cooked, "four-leaves clover" ("suwa wiwii": *Oxalis barrelieri* L. – Oxalidaceae), with acidy leaves, that are consumed raw to give appetite, and "soldier salad" ("salade soldat", "konsaka wiwii", *Peperomia pellucida* (L.) H. B. K. – Piperaceae) eaten as a salad. They definitely eat much more greens than the Wayana Amerindians, with whom they share the same environment.

Fig. 6 "Jambú" (*Acmella oleacera*) growing in the yard of a house on the Atlantic coast of Pará (C. López; identification: L. Emperaire).

Fig. 7 Chicken with "jambú" (C. López).

Peoples from the Northeast in the Amazon: cultivated herbs and greens

Since the end of the 19th century, periodic droughts in the semiarid Northeast of Brazil have led people to migrate to other regions. Many such migrants went to the Amazon and were engaged in rubber tapping until the end of the 2nd World War. In the State of Acre, descendants from these migrants presently form the majority of the non-indigenous population. In the market of Cruzeiro do Sul, one of the main towns of Acre, Katz [18] observed the following greens being sold on the market: herbs: coriander leaves, "chicória", parsley [*Petroselinum crispum* (Mill.) Fuss] and salad onion (*Allium fistulosum*); plants for salads: lettuce (*Lactuca sativa* L.), "agrião" (*Spilanthes acmella*, Asteraceae), garden rocket (*Eruca sativa* Mill.), and kale ("couve", *Brassica oleracea*), eaten raw or in soups; cabbage ("repolho", *Brassica oleracea*), cooked. In the Extractive Reserve of the Upper Juruá (Acre), Emperaire [19] observed the consumption of green onions, "chicória", coriander leaves and kale, cultivated in "canteiros" (raised beds), *Hibiscus sabdariffa* ("vinagreira") cultivated in homegardens and *Phytolacca rivinoides* ("couve-manteiga" – "butter kale") picked in swidden fields. Along the Tapajos River in Pará, where part of the population is of Northeastern origin, Passos et al. [20] found the consumption of cultivated greens, mainly kale and coriander leaves.

Urban population in Manaus

In a nutritional study carried out in 1973–1974, Shrimpton and Giuliano [21] found Manaus to be the Brazilian city where fewest greens were consumed. We must mention that, at the level of the country as a whole, greens are already much less consumed than in Europe, for instance, and a more reduced range of vegetable species is available. Not surprisingly, Manaus, in the middle of the Amazon, had a lower consumption of greens than cities on or near the coast, such as São Paulo, Rio de Janeiro, Salvador, and Recife. Shrimpton and Giuliano also noted that fruit and vegetables were more expensive in Manaus. Coriander leaves, garlic and onion were not bought frequently by people with lower incomes, and only people with higher incomes were buying kale and lettuce. However, besides income, the question of social origin and food habits must also be considered. The consumption of fish, for instance, was higher in lower classes than in upper classes. The authors noted that the consumption of fruit was higher in towns and villages outside of the city, such as Codajás on the Solimões river, where another nutritional study was conducted in 1958. There, green onions and coriander leaves were not even mentioned. Possibly, nowadays, under national influence, inhabitants of Manaus may eat a little more of salads than they used to in the seventies. Comparatively, cultivated *Eryngium foetidum* was the only green found on the markets of the Columbian Amazon, in the cities of Leticia and Florencia, studied by Pulido and Cavelier [22].

Discussion

Despite a green environment, we find that actually there are not so many wild greens available in the Amazon (see Tab. 2). The few greens which are consumed are weeds, semi-cultivated or cultivated plants. The few spontaneous plants found are two types of leafy greens usually called "caruru" or "cariru": *Talinum paniculatum*, *T. triangulare* (Portulacaceae) and *Phytolacca rivinoides* (Phytolaccaceae), which seem to be widely distributed at least in the Northern Amazon. As observed by Emperaire (personal communication), *Phytolacca rivinoides* can actually grow up to the size of a shrub. The name "caruru" applies also to introduced *Basella alba*. According to Carney & Rosomoff [17], it may also refer in some regions to *Amaranthus* species. Among semi-wild herbs, *Eryngium foetidum*, *Acmella oleacera* and *Spilanthes acmella* seem to be native of the Northeastern Amazon. *Eryngium foetidum* (called "false coriander" in French Guiana) has been widely diffused as a cultivated plant in the rest of the Amazon, but not much in other regions of Brazil. *Spilanthes acmella* has been diffused only in some parts of the Amazon, but it was taken to Asia and Madagascar, where it is known as "brède mafane" [15]. Among cultivated plants, the leaves of manioc (*Manihot esculenta*) and cocoyam (*Xanthosoma*) are the most commonly consumed. Maroons who eat more greens than the Indians, actually do not use a much wider range of species. Leaves of *Xanthosoma* sp. are also consumed by some Mexican Indians, who also carefully separate out the veins [4].

Amerindian diet in the Amazon is based on starch and animal protein, more commonly bitter or sweet cassava, sometimes corn, and fish or game. All over the Amazon, many species of wild and cultivated tubers (yams, sweet potato, etc.) are used. These basic diets are complemented with fruit. The area is very rich in wild and cultivated fruit species, in particular palm fruits, oleaginous or starchy. The diet of indigenous peoples, richer in animal protein than many traditional diets in the world, and supplemented with forest fruits, is probably satisfactory without the need of greens.

What makes people choose to eat greens rather than other types of resources? When Katz asked inhabitants of the Rio Negro if they ate palm hearts (eaten in other parts of the Amazon), they answered: "why would we eat something like that? There are so many other tasty foods available, cassava, juice of palm fruit, sweet fruit from the forest or the gardens, fish, and so on". The same answer probably applies to greens. Young people in particular mentioned they perceived greens as bitter food. Many Rio Negro inhabitants do not appreciate the bitter taste at all, and consider bitter foods as not edible. It is the taste of remedies, and only the shamans look for bitter foods in specific diets. The Kaxuyana and Tiryo perceived lettuce as tasteless and, as we saw above, consider greens as "tortoise food". Supposedly the Touareg in the Sahara, who are probably far from having so many plants and animals available, do not eat greens either, because they perceive them as "camels' food" (Ph. Lefèvre-Witier, personal communication, 1984). This consumption is definitely cultural.

According to Turner et al. [23], hunters-gatherers collect mainly wild tubers and fruit; wild greens, or rather weeds, are more collected by cultivators. Most Amazonian Indians practice agriculture, but hunting, fishing and gathering are still important activities. We see that their gathering activities focus more on fruit and tubers, especially among "cerrado" people. Their attitude towards greens in the food could be described, using Łuczaj's term [24], as "herbophobia". The few greens they gather are more commonly weeds.

Amazonians with cultural origins in Northeastern Brazil (Nordestinos) eat a few cultivated greens as herbs or in soups. Maroons are the ones who consume them most. As Carney and Rosomoff [17] noted, greens were part of the traditional diets of African people, especially West Africans. As Fleury notes, the Aluku Maroon eat rice, which is a dry dish, and need to complete it with an oleaginous or mucilaginous sauce. Although living in the same environment as the Wayana, they definitely eat greens much more frequently, as mentioned above.

Carney & Rosomoff [17] found greens or dishes with greens eaten in different part of the Americas called "callalou", "calelu" and in Brazil "caruru", which refer to *Amaranthus* spp. (for instance in Jamaica), a genus of plants found in both continents, sometimes *Hibiscus sabdariffa* ("caruru azedo" – "sour caruru"

Fig. 8 "Caruru" (*Talinum paniculatum*) cultivated in a home garden (C. López; identification: L. Emperaire).

Fig. 9 "Caruru" (*Talinum paniculatum*) sold on Belém market (E. Katz; identification: L. Emperaire).

Fig. 10 Cultivated roselle (*Hibiscus sabdariffa*) on the coast of Pará (C. López).

- in some parts of Brazil) or to dishes with greens or with okra, like the "caruru" of Belém and of Bahia, in Brazil. The authors believe this name came from Africa with the slaves who applied it to species similar to the ones they knew in Africa. According to the data we collected in the Amazon, it also refers to at least three other weedy species. This name may have been taken from the coasts to the Amazon during the colonization, but we have no information on this process.

The same authors also consider the consumption of manioc leaves, common in Africa, boiled or ground, as an African innovation. Given the wide spatial range where it is consumed in the Amazon by Amerindian people, from the Northeastern Amazon to the Western Amazon, we are not totally convinced of an African origin for this culinary practice.

Presently, the cultivation of greens (lettuce and aromatic greens, such as kale and coriander) is being promoted in different development projects applied in the Amazon. However, little attention has been given to the fact that greens, salads, herbs are not part of the local people's traditional diet!

Conclusion

We confirm that greens are an extremely minor item in the diet of the inhabitants of the Amazon, with consumption lowest among Amerindians, with a few exceptions. Greens are very seldom mentioned in the bibliography as researchers usually focus on the main elements of the diet. Some of the authors here who have performed participant observation confirm that some Amazonian people do not eat any greens at all. In the other cases, we see that is not a very usual consumption, with few species being consumed. People with cultural origins in Northeastern Brazil consume more cultivated greens and Maroons are the ones who consume more species, either wild or cultivated, and in a more frequent pattern. Few species are wild or semi-wild, others are cultivated or semi-cultivated. We took the last ones into account here to understand the position of the different greens in this wild-cultivated gradient and in the diet. Greens are not totally absent from Amazonian diets, but they are not valued and preferably given as food for the tortoise.

Acknowledgements

Thanks to Laure Emperaire (IRD - UMR 208) for botanical identifications of the Rio Negro and Belém region and Kacper Świerk (University of Szczecin, Poland) for his comments on the article.

In the Rio Negro, Esther Katz led her research within two programs of bilateral cooperation CNPq/IRD No. 492693/2004-8 and 490826/ 2008-3 "Local populations, agrobiodiversity and associated traditional knowledge" (PACTA I and II), in partnership between the Institut de Recherche pour le Développement [IRD, UMR 208 PALOC ("Local heritage"), IRD/Muséum National d'Histoire Naturelle) and the Universidade Estadual de Campinas (Unicamp). The research was supported financially by CNPq, IRD, BRG (Bureau des Ressources Génétiques) and ANR (Agence Nationale de Recherche)/IFB - (Institut Français de la Biodiversité). The workshop on traditional food heritage was supported financially by the French Ministry of Foreign Affairs, through the French Embassy in Brasilia. The research of Esther Katz, Claudia López, Franklin Silva, and Zelandes Oliveira has partly been supported by the European project ENGOV ("Environment and governance in Latin America" (2011–2015); FP7-SSH-2010-3/SSH.2010.4.1-2).

References

1. Turner N, Harvey T, Burgess S, Kuhnlein H. "The Nuxalk food and nutrition" program, coastal British Columbia, Canada: 1981–2006. In: Kuhnlein H, Erasmus B, Spigelski D, editors. Indigenous people's food systems: the many dimensions of culture, diversity and environment for nutrition and health. Rome: CINE/FAO; 2009. p. 23–44.
2. Messer E. Patterns of "wild" plant consumption in Oaxaca, Mexico. Ecol Food Nutr. 1972;1(4):325–332.
3. Bye R. Quelites. Ethnoecology of edible greens. Past, present and future. J Ethnobiol. 1981;1(1):109–123.
4. Katz E. La cueillette des adventices comestibles au Mexique. Ecol Hum. 1992;10(1):25–41.
5. Dufour DL, Wilson WM. Characteristics of "wild" food plants used by indigenous populations in Amazonia. In: Etkin N, editor. Eating on the wild side: the pharmacologic, ecologic and social implications of using noncultigens. Tucson AZ: University of Arizona Press; 1994. p. 114–142.
6. Hoffmann SA. Subsistence in transition: indigenous agriculture in Amazonas, Venezuela [PhD thesis]. Berkeley CA: University of California; 1993.
7. Figueroa ALG. Guerriers de l'écriture et commerçants du monde enchanté : histoire, identité et traitement du mal chez les Sateré-Mawé (Amazonie central, Brésil) [PhD thesis]. Paris: EHESS; 1997.
8. Correal C, Zuluaga G, Madrigal L, Caicedo S, Plotkin M. Ingano traditional food and health, Phase 1. 2004–2005. In: Kuhnlein H, Erasmus B, Spigelski D, editors. Indigenous people's food systems: the many dimensions of culture, diversity and environment for nutrition and health. Rome: CINE/FAO; 2009. p. 83–108.
9. Ricardo FP. Povos Indígenas no Brasil [Internet]. 1997 [cited 2012 Aug 30].
10. Ricardo FP. Kaxinawa [Internet]. Povos Indígenas no Brasil. 2004 [cited 2012 Aug 30].
11. Pantoja M, Almeida MB, Conceição MG, Coffaci de Lima E, Aquino TV, Iglesias MP, et al. Botar roçados. In: Cunha MC, Almeida MB, editors. Enciclopédia da floresta. O Alto Juruá: práticas e conhecimentos das populações. São Paulo: Companhia das Letras; 2002. p. 249–283.
12. Carrasco M, Roche M, Tuesta I, Abad M, Ganoza L, Creed-Kanashiro H.

Sistema tradicional de alimentación de los Awajún del rio Cenepa. Lima: IIN/CINE/ODECOFROC; 2009.

13. Creed-Kanashiro H, Roche M, Tuesta I, Kuhnlein H. Traditional food system of an Awajún community in Peru. In: Kuhnlein H, Erasmus B, Spigelski D, editors. Indigenous people's food systems: the many dimensions of culture, diversity and environment for nutrition and health. Rome: CINE/FAO; 2009. p. 59–81.
14. Papavero N, Sanjad N, Chiquieri A, Overal WL, Mugnai R. Os escritos de Giovanni Angelo Brunelli, astrônomo da Comissão Demarcadora de Limites portuguesa (1753–1761), sobre a Amazônia brasileira. Boletim do Museu Paraense Emílio Goeldi. Ciências Humanas. 2010;5(2):433–533.
15. de Robert P, van Velthem L. L'heure du tacacá. Anthropology of Food [Internet]. 2008 [cited 2012 Aug 30].
16. Manhães LRT, Marques MM, Sabaa-Srur AUO. Composição química e do conteúdo de energia do cariru (*Talinum esculentum*, Jacq.). Acta Amazonica. 2008;38(2):307–310.
17. Carney J, Rosomoff RN. In the shadow of slavery. Africa's botanical legacy in the Atlantic world. Berkeley CA: University of California Press; 2009.
18. Katz E. Cruzeiro do Sul market (Acre, Brazilian Amazon): reflection of the regional culture and agricultural diversity. In: Pochettino ML, Ladio AH, Arenas PM, editors. Tradiciones y transformaciones en etnobotánica/Traditions and transformations in ethnobotany, Vth International Congress of Ethnobotany (ICEB 2009). San Salvador de Jujuy: CYTED – Programa Iberoamericano Ciencia y Tecnología para el Desarrollo; 2010. p. 525–533.
19. Emperaire L. Entre paus, palheiras e cipós. In: Cunha MC, Almeida MB, editors. Enciclopédia da floresta. O Alto Juruá: práticas e conhecimentos das populações. São Paulo: Companhia das Letras; 2002. p. 389–417.
20. Passos C, Mergler D, Gaspar E, Morais S, Lucotte M, Larribe F, et al. Caracterização do consumo alimentar de uma população ribeirinha na Amazônia Brasileira. Saúde e Ambiente. 2001;4(1–2):72–84.
21. Shrimpton R, Giugliano R. Consumo de alimentos e alguns nutrientes em Manaus, Amazonas 1973–4. Acta Amazonica. 1979;9(1):117–141.
22. Pulido MT, Cavelier J. Comercialización de productos vegetales no maderables en los mercados de Leticia y Florencia, Amazonia colombiana. In: Duivenvoorden JF, Balslev H, Cavelier J, Grandez C, Tuomisto H, Valencia R, editors. Evaluación de recursos vegetales no maderables en la Amazonía noroccidental. Amsterdam: IBED; 2001. p. 265–310.
23. Turner NJ, Łuczaj ŁJ, Migliorini P, Pieroni A, Dreon AL, Sacchetti LE, et al. Edible and tended wild plants, traditional ecological knowledge and agroecology. Cr Rev Plant Sci. 2011;30(1–2):198–225.
24. Łuczaj Ł. Changes in the utilization of wild green vegetables in Poland since the 19th century: a comparison of four ethnobotanical surveys. J Ethnopharmacol. 2010;128(2):395–404.

15

The lichen-forming fungi of the *Xanthoparmelia pulla* group (Parmeliaceae, Ascomycota) in Poland

Katarzyna Szczepańska[1]*, Maria Kossowska[2]

[1] Department of Botany and Plant Ecology, Wrocław University of Environmental and Life Sciences, pl. Grunwaldzki 24a, 50-363 Wrocław, Poland

[2] Institute of Environmental Biology, University of Wrocław, Kanonia 6/8, 50-328 Wrocław, Poland

Abstract

The paper presents the results of studies of *Xanthoparmelia pulla* group in Poland. The morphological and chemical analysis of herbarium materials confirmed the presence of four species of this group reported from Poland before. The study however, revealed considerable changes it the partiuclular species distribution. *X. verruculifera*, so far considered the most endangered species in the country, turned out to be the most frequent taxon. *X. loxodes*, regarded as common, has much less known localities than previously thought. This species was usually confused with *X. verruculifera*. *Xanthoparmelia pulla* is the rarest species and should be considered critically endangered in Poland. Most specimens of *X. pulla* collected in the country belong to *X. delisei*, which so far had only two historical records in Poland. All these taxa are characterized in terms of morphology, the content of secondary metabolites, ecology and distribution.

Keywords: parmelioid lichens; *Xanthoparmelia delisei*; *X. loxodes*; *X. pulla*; *X. verruculifera*; chemotaxonomy; ecology; distribution

Introduction

The genus *Xanthoparmelia* (Vain.) Hale, comprising ca. 800 species of worldwide distribution belongs to the largest family of lichenized fungi – Parmeliaceae [1,2]. The distinguishing features of the genus are: hyphal cell walls' polysaccharides with *Xanthoparmelia*-type lichenan [3–5], small ascospores with an arachiform vacuolar body [6], lack of pseudocyphellae, presence of a pored epicortex, eperforate apothecia, the presence of bifusiform conidia and usually simple rhizines [5]. This genus is further characterized by a considerable variation in cortical chemistry of individual species. Most taxa of this genus occur on the siliceous rocks in dry and well-sunlit places, in periarid, arid, semiarid and Mediterranean climates, mainly in the southern hemisphere [5].

Originally the genus *Xanthoparmelia* included exclusively the species containing atranorin and usnic or isousnic acids in the upper cortex, causing the yellowish tinge of thallus [7,8]. Lichens currently included to *Xanthoparmelia pulla* group but not containing these substances, with brown thalli and upper cortex stained by HNO_3 to blue-green, were initially grouped in the *Neofusca* subgenus (within the *Parmelia* genus) [9], which was then set up as a separate genus *Neofuscelia* [10]. Molecular studies have shown, however, that the genus *Neofuscelia* was polyphyletic, with its clades scattered within *Xanthoparmelia*. Consequently, the species of the genus *Neofuscelia* have been included in the *Xanthoparmelia* [11].

Xanthoparmelia pulla group includes about 25 taxa dispersed throughout the world, seven of which occur in Europe [12]. Due to the frequent absence of apothecia and the lack of obvious differences in their structure, these taxa are traditionally distinguished on the basis of morphological features of the thallus, such as the color of lower cortex, shape of lobes and the presence or absence of vegetative propagules [12]. However, chemical characteristics play the most important role in the identification of taxa [9,13]. Within the species of the genus *Xanthoparmelia*, ca. 90 secondary metabolites were identified, mainly phenolic compounds such as depsides, depsidones, antraquinones and monocilic compounds, as well as aliphatic acids [8,14], 15 of which were present within *Xanthoparmelia pulla* group [13].

So far four species of the *Xanthoparmelia pulla* group were reported from Poland: *X. delisei*, *X. loxodes*, *X. pulla* and *X. verruculifera* [15]. One of them, *X. loxodes*, was considered to be common throughout the country, while the *X. pulla* and *X. verruculifera* were admitted as more or less rare and endangered [16]. *X. delisei* in general was not distinguished by Polish lichenologists and its only records from present Polish territory were of historical nature [17].

* Corresponding author. Email: katarzyna.szczepanska@up.wroc.pl

Handling Editor: Joanna Zalewska-Gałosz

Proper identification of these species is relatively difficult and requires application of chemotaxonomic methods, e.g. thin layer chromatography (TLC) in addition to standard microscopic methods. Since these have not been widely used in the study of this group of lichens in Poland, data on the distribution, ecology and a status of threat for individual taxa are incomplete and require rewording. Therefore, the authors undertook a detailed review of herbarium material from Poland in order to verify data on lichens in this group.

Material and methods

A total of 304 herbarium specimens, deposited in the following herbaria: BSG, GPN, KRA, KRAM-L, KRAP, KTC, LOD-L, OLTC-L, POS-L, TRN, UGDA, WA, WRSL and private collections of M. Dimos-Zych, P. Grochowski, W. Gruszka, M. Kossowska, L. Lipnicki, K. Pietrzykowska and K. Szczepańska were investigated. Each specimen was analyzed for the presence of secondary metabolites in the thallus using thin layer chromatography (in solvent A and C), in accordance with the methods described by Orange et al. [18]. When the substance was present in all specimens in relative high concentration, it is marked (+), when it was only trace constituent (t), if it was absent in some samples (+/−) is used. In addition, the morphological features, especially the shape and size of lobes and isidia were investigated, using a stereoscopic microscope. Brief characteristics of each species are based on personal observations.

The list of examined specimens is available in Appendix S1. The names of physico-geographical mesoregions are given according to Kondracki [19]. The distribution of all species in Poland is shown on maps based on the ATPOL grid square system [20], modified by Cieśliński and Fałtynowicz [21].

Results

Xanthoparmelia delisei (Duby) O. Blanco, A. Crespo, Elix, D. Hawksw. & Lumbsch

Taxon 53(4): 967. 2004.≡*Parmelia olivacea* var. *delisei* Duby, Bot. Gall. 2: 602. 1830.≡*Neofuscelia delisei* (Duby) Essl., Mycotaxon 7: 50. 1978.

DIAGNOSTIC CHARACTERS. Thallus foliose, forming rosettes, loosely appressed to the substrate. The upper surface light yellowish-brown or slightly darker, often distinctly maculate, smooth or transversely wrinkled towards the center. Lobes irregular, flat or slightly convex, with overlapping edges, slightly shining at tips. Isidia absent. Apothecia usually present and numerous. Secondary metabolites detected by TLC: glomelliferic (+), glomellic (+), perlatolic (+), stenosporic (t) and gyrophoric (+/−) acids.

HABITAT. In Poland the species was recorded on siliceous boulders and stones in treeless, open places exposed to direct sunlight, often in areas used for agricultural purposes and on the roadsides.

GENERAL DISTRIBUTION. The species is widespread almost worldwide. It occurs in Africa, Australia, Asia, South America and Europe [12]. In Europe it was reported from Belgium [22], Germany [23], Great Britain and Ireland [24], Greece [25], Italy [26], Montenegro [27], Netherlands [28], Spain and Portugal [29], Sweden, Norway and Finland [30].

DISTRIBUTION IN POLAND. *X. delisei* has been considered an extremely rare species on Polish territory. It was only reported in historical works from the present Polish north and north-east [17,31]. This analysis showed that the taxon is relatively common and widespread in the lowland parts of the country, with a distinct concentration of localities in the north-east Poland (Fig. 1). The species reaches the highest altitude in the Sudety Mountains (about 600 m above sea level). In the Carpathians it has not been found to date.

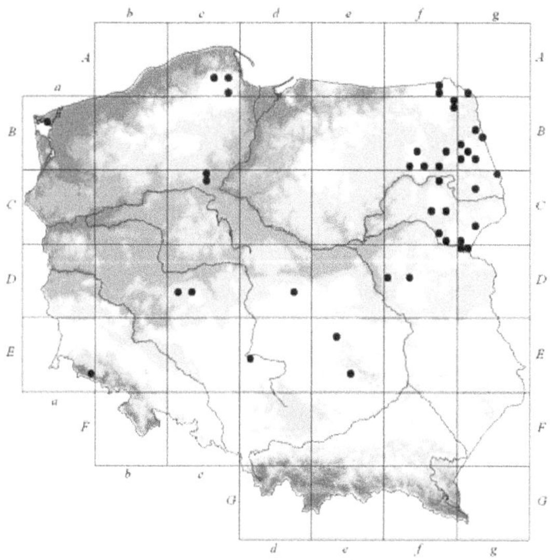

Fig. 1 Distribution of *Xanthoparmelia delisei* in Poland.

NUMBER OF SPECIMENS EXAMINED – 55.

COMMENTS. All examined specimens of *X. delisei* were originally identified as *X. pulla*. According to Esslinger [9] the taxa are morphologically almost identical, although in some cases they can be separated on the basis of morphological features, such as the color of thalli (paler and yellowish to gray-brown in *X. delisei*, darker brown in *X. pulla*) and the appearance and shape of the lobes (broader, thicker and more strongly maculate in *X. pulla*). In addition, these species are different in the reaction of the medulla with KC, which is red to orange in *X. delisei* and pink-red to red in *X. pulla* [9,23]. These features, however, may not be sufficient for proper determination of the species. Both species differ much in the content of secondary metabolites in thalli, and therefore should be identified based on TLC analysis. *X. delisei* contains glomelliferic, glomellic, perlatolic, loxodellic, stenosporic, anziaic acids and occasionally gyrophoric acid [9,13], while in *X. pulla* were found stenosporic, divaricatic, perlatolic, 4-O-demethylstenosporic and oxostenosporic acids, sometimes with atranorin and gyrophoric or lecanoric acids [32].

The chemistry of *X. delisei* is identical as in case of *X. loxodes*. However, mature and well-developed thalli of the latter species produce isidia, visible on the upper surface as cauliflower-like clusters. In the case of young and poorly developed specimens, where isidia may not be fully developed, the separation of these two species is not possible. Because of the identical chemistry of *X. loxodes* and *X. delisei*, some authors suggest that these two taxa may represent intergrading morphotypes of a single species [33]. In some sources *X. delisei* was treated as a chemotype of *X. pulla* and highlighted in a rank of variety [22,23,30].

Xanthoparmelia loxodes (Nyl.) O. Blanco, A. Crespo, Elix, D. Hawksw. & Lumbsch

Taxon 53(4): 968. 2004.≡*Parmelia loxodes* Nyl., Flora 55: 426. 1872.≡*Neofuscelia loxodes* (Nyl.) Essl. Mycotaxon 7: 51. 1978.

DIAGNOSTIC CHARACTERS. Thallus foliose, forming rosettes, loosely appressed to the substrate. Lobes rather flat, elongated and overlapping. The upper surface yellowish-brown to reddish-brown, smooth or wrinkled. Isidia present, coarse, more or less spherical and distinctly pustular. Apothecia rare. Secondary metabolites detected by TLC: glomelliferic (+), glomellic (+), stenosporic (+), perlatolic (t) and gyrophoric (+/−) acids.

HABITAT. *X. loxodes* is a species with a very wide range of habitat requirements. In the lowland part of Poland it was recorded on siliceous erratic boulders and stones, in well sunlit places on the edges of roads, fields and meadows. In mountain areas the species occurs on natural or artificially uncovered (e.g., abandoned quarries) rock outcrops of various chemistry and mineral composition (gneiss, sandstone, granite, basalt). Single records were also taken from anthropogenic calcareous substrates and from wood.

GENERAL DISTRIBUTION. The species is widespread, reported from North Africa, Asia, North America and Europe [12]. In Europe it occurs in Austria [34], Belgium [22], Czech Republic [35], Denmark [36], Germany [23], Great Britain and Ireland [24], Greece [25], Italy [26], Netherlands [28], Portugal [37], Spain [29], Sweden, Norway and Finland [30].

DISTRIBUTION IN POLAND. Until now *X. loxodes* was regarded as the most widespread and common member of the *X. pulla* group in Poland. In the checklist of Polish lichens [15] it was reported from numerous localities across the country. However, most of the 164 examined herbarium specimens of *X. loxodes* in fact represent another species, *X. verruculifera*.

The records of *X. loxodes* are concentrated in the northern part of the country (Pojezierze Kaszubskie and Pojezierze Mazurskie lakelands), within the limits of the last glaciation, which left a large number of erratic boulders (Fig. 2). Individual records are also from Wielkopolska district, Wyżyna Wieluńska Upland and the Góry Świętokrzyskie Mountains. In the southern part of the country it was found only in the lower parts of the mountains and the foothills of the Sudetes and the Przedgórze Sudeckie Foreland, up to about 800 m above sea level. It was not recorded in the Carpathians so far.

NUMBER OF SPECIMENS EXAMINED – 72.

COMMENTS. *X. loxodes* is usually confused with another species producing isidia, namely *X. verruculifera*. Both taxa can, however, be distinguished on the basis of morphological features. *X. loxodes* develops larger and paller thalli, thicker and wider lobes and larger and more strongly pustular isidia [9]. These species also slightly differ in the reaction of medulla with KC, which in *X. loxodes* is red to orange, and in *X. verruculifera* pink and disappearing [9]. Both species have different chemistry of thalli and in case of any doubts the specimens should be analyzed for the content of secondary metabolites. *X. loxodes* contains glomelliferic, glomellic, perlatolic acids and occasionally anziaic, stenosporic, loxodellic or gyrophoric acids [9,12,13] while in *X. verruculifera* divaricatic, nordivaricatic, oxostenosporic, stenosporic, subdivaricatic, perlatolic and 4-O-demethylstenosporic acids were found. Gyrophoric and lecanoric acids may also be present as accessory compounds [26].

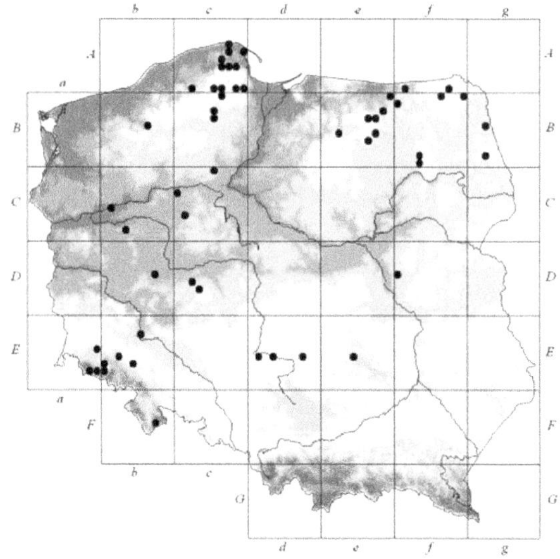

Fig. 2 Distribution of *Xanthoparmelia loxodes* in Poland.

The chemistry of *X. loxodes* is very similar to the *X. delisei*, therefore, when the thalli are young and poorly developed (without distinct isidia), proper identification of these taxa may be impossible.

In the studied herbarium materials *X. loxodes* was also confused with *Melanelixia fuliginosa* (Fr. ex Duby) O. Blanco, A. Crespo, Divacar, Essl., D. Hawksw. & Lumbsch, which, however, develops thalli with distinctly shiny surface, covered by fine, cylindrical isidia and the medulla C+ turning red. Another species confused with *X. loxodes* is *Melanelia disjuncta* (Erichsen) Essl. The last one may be distinguished by generally smaller thalli, upper cortex not reacting with HNO_3 and isidia transforming into convex, distinctly globose soralia. It is worth to note that the taxonomic affinity of *M. disjuncta* is uncertain and requires further studies (e.g. [2,38]).

Xanthoparmelia pulla (Ach.) O. Blanco, A. Crespo, Elix, D. Hawksw. & Lumbsch

Taxon 53(4): 970. 2004.≡*Parmelia pulla* Ach., Syn. Meth. Lich.: 206. 1814.≡*Neofuscelia pulla* (Ach.) Essl., Mycotaxon 7: 52. 1978.

DIAGNOSTIC CHARACTERS. Thallus foliose, forming rosettes or irregular, weakly appressed to the substrate. Lobes slightly convex, rounded, rather thick. Upper surface olive-brown, yellowish-brown or slightly darker, distinctly wrinkled, slightly shining at tips of the lobes, without isidia. Apothecia frequent. Secondary metabolites detected by TLC: stenosporic (+), divaricatic (+), perlatolic (t), oxostenosporic (t), 4-O-demethylstenosporic (t), and gyrophoric (+) acids.

HABITAT. *X. pulla* was confirmed in Poland only from two sites till now, hence it is difficult to define its habitat preferences. However, it seems that this taxon requires much more xeric habitats then the other species of the *X. pulla* group. In Poland it has been recorded in warm, sunny and dry places, preferably with a southwestern exposure. It occured on the acidic or neutral volcanic rocks, overgrown by xerothermic grasslands with *Jovibarba sobolifera, Sedum maximum, Thymus* sp. etc.

GENERAL DISTRIBUTION. The species occurs in Africa, Australia, New Zealand and Europe [12]. In Europe it was recorded in Austria [31], Belgium [21], Cyprus [39], Czech Republic [32], Denmark [33], Germany [22], Great Britain and Ireland [20], Greece [23], Italy [24], Netherlands [26], Portugal [24], Spain [27], Sweden, Norway and Finland [28].

DISTRIBUTION IN POLAND. Up to date, the species was thought to be relatively common, with many localities across the country [15]. As shown by the chemotaxonomic revision of the available herbarium material, most of the 86 specimens previously labeled as *X. pulla* in fact represent *X. delisei*, and only two of them belong to *X. pulla*. The known localities of the species in Poland are situated in the southwestern part of the country, within the Sudetes and their foreland (Fig. 3).

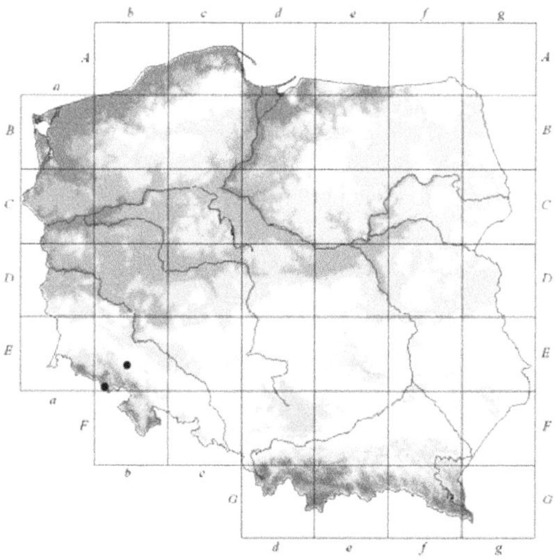

Fig. 3 Distribution of *Xanthoparmelia pulla* in Poland.

NUMBER OF SPECIMENS EXAMINED – 2.

COMMENTS. *X. pulla* is very similar to *X. delisei*. According to Esslinger [9], both species are slightly different in color, morphology of the lobes and reactions with KC (see comments on *X. delisei*). All morphological features, however, are so variable that they are not sufficient to separate these taxa properly. The identification of the species requires application of thin layer chromatography.

The chemistry of *X. pulla* is very similar to the chemistry of *X. verruculifera*. Therefore, morphological features of the thalli need careful examinations. In the case of young and poorly developed thalli without isidia or apothecia, the separation of these taxa can be problematic.

Another species very similar to *X. pulla* is *X. perrugata* (Nyl.) O. Blanco, A. Crespo, Elix, D. Hawksw. & Lumbsch, which has not been recorded in Poland so far. However, it may be possible to find this taxon in the country, because it was recorded in Central Europe [32]. Both species are slightly different in the chemistry of thalli. *X. perrugata* contains divaricatic acid as major and stenosporic acid as minor substances, in contrast to *X. pulla*, where the major substance is stenosporic acid [32]. In addition, *X. perrugata* has a markedly more rugose upper surface than the other members of *X. pulla* group [26].

***Xanthoparmelia verruculifera* (Nyl.) O. Blanco, A. Crespo, Elix, D. Hawksw. & Lumbsch**

Taxon 53(4): 972. 2004.≡*Parmelia verruculifera* Nyl., Flora 61: 247. 1878.≡*Neofuscelia verruculifera* (Nyl.) Essl., Mycotaxon 7: 53. 1978.

DIAGNOSTIC CHARACTERS. Thallus foliose, medium or loosely appressed to the substrate. Lobes flat, short and rounded to elongated, contiguous or imbricate, slightly shiny at tips. Upper surface olive-brown, reddish-brown to dark-brown, smooth or slightly rough. Apothecia rare. Isidia present, pustular, densely arranged, forming branched-coralloid structures. Secondary metabolites detected by TLC: divaricatic (+), stenosporic (+), perlatolic (t), oxostenosporic (t), 4-O-demethylstenosporic (t) and gyrophoric (+/–) acids.

HABITAT. This species occurs in both lowland and mountainous areas, usually in open and well-lit places. It was recorded mostly on siliceous boulders and rocks, on the roadsides, fields and meadows and on natural rock outcrops (andesite, gneiss, sandstone, quartzite, serpentinite, sporadically also calcareous rocks) as well as on anthropogenic substrates, e.g., on bricks, stone tombs and walls.

GENERAL DISTRIBUTION. *X. verruculifera* occurs in North Africa, North America and Europe [12]. In Europe it was recorded in Austria [34], Belgium [22], Czech Republic [35], Denmark [36], Great Britain and Ireland [24], Germany [23], Italy [26], Netherlands [28], Spain [29], Sweden, Norway and Finland [30].

DISTRIBUTION IN POLAND. The species was considered rare and endangered in Poland, having just a few locations in the country [15]. The analysis of the available herbarium materials has shown that this taxon was rarely distinguished and had often been referred to *X. loxodes*. Thus, *X. verruculifera* proved to be the most common species of the *X. pulla* group in Poland.

This species occurs in almost all territory of Poland, from the lowlands to the lower mountain areas (Fig. 4). As the only representative of the *X. pulla* group it occurs in the Carpathians and is particularly abundant in the Gorce

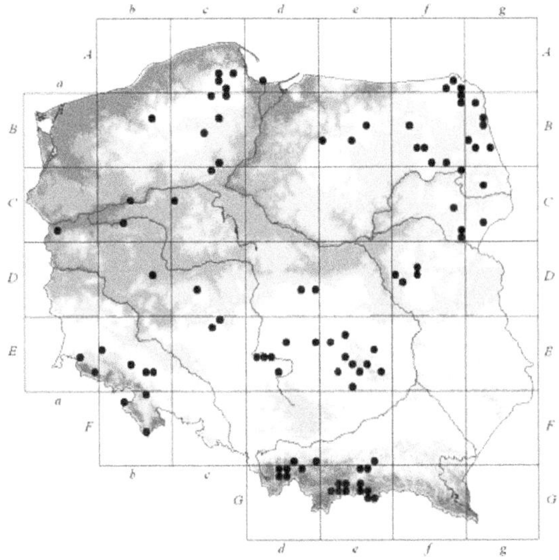

Fig. 4 Distribution of *Xanthoparmelia verruculifera* in Poland.

Mountains and the Beskid Sądecki Mountains. It reaches the highest altitude in the Pieniny Mountains at 760 m.

NUMBER OF SPECIMENS EXAMINED – 136.

Exiccates examined: Nowak, Lichens Poloniae Meridionalis Exsiccati 90 (LOD-L, sub *Parmelia isidiotyla*).

COMMENTS. *X. verruculifera* is most often confused with *X. loxodes*. Both species produce isidia on thallus surface, but their morphology is different. The isidia of *X. verruculifera* can be defined as higher and tinier. In addition, Esslinger [9] pointed at the other morphological features that allow to separate these species (see comments on *X. loxodes*). However, the most important feature is the chemical content of the thallus.

The chemistry of *X. verruculifera* is very similar to *X. pulla* and *X. perrugata*. Therefore the special attention should be paid to the presence or absence of isidia. In the case of young and poorly developed thalli separation of these species may not be possible.

X. verruculifera can be confused with species belonging to the genera *Melanelia* and *Melanelixia*, especially *Melanelia disjuncta* and *Melanelixia fuliginosa*. These species are, however, relatively easy to distinguish from the species of *X. pulla* group due to the lack of reaction of the upper cortex with HNO_3 and a generally different chemical content of the thalli.

It is worth to note that the old name of *X. verruculifera* – *Parmelia verruculifera* Nyl. has sometimes been erroneously used for an epiphytic species – *Melanelixia subargentifera* (Nyl.) O. Blanco, A. Crespo, Divacar, Essl., D. Hawksw. & Lumbsch. During this study, among the examined herbarium specimens, we identified also specimens of this species. The taxonomy and nomenclature of *P. verruculifera* were discussed in detail by Esslinger [40].

Discussion

The examined herbarium specimens confirmed the presence of all four species of lichens belonging to the *Xanthoparmelia pulla* group reported from Poland before, however their distribution and ecology seem to be other than previously thought. *X. verruculifera*, so far considered the most vulnerable and classified to the EN (endangered) category on the red list of Polish lichens [16], proved to be the commonest species of the group. This taxon had been often confused with *X. loxodes*, especially in case of young specimens with poorly developed isidia. In view of the obtained results it should be excluded from the Polish red list.

According to previous data [15], *X. delisei* was considered to be an extremely rare species, having just two historical localities in Poland. The revision has shown that the taxon is much more frequent in the whole country. In fact, the rarest of the studied species is *X. pulla*. For a total of 86 herbarium specimens labeled as *Xanthoparmelia* (*Neofuscelia*) pulla only two specimens were ascribed to this species. Other represent mainly *X. delisei*, but also *X. loxodes* and *X. verruculifera*. *X. pulla* is thus the only species actually endangered in Poland, so the change in the status of the species within the red list of lichens from the NT (near treatment) to CR (critically endangered) needs to be taken into account.

As mentioned above, the propper identification of the species of the *X. pulla* group, requires not only a morphological analysis, but also chemotaxonomic studies. The most important diagnostic features, which enable to identify all of the taxa occurring in Poland, are presented in a Tab. 1.

Tab. 1 Overview of the distinguishing features in the *Xantoparmelia pulla* group in Poland.

Species	Isidia	Apothecia	Chemistry
X. delisei	absent	common	glomelliferic, glomellic, perlatolic, stenosporic and occasionally gyrophoric acid
X. loxodes	present	rather rare	glomelliferic, glomellic, perlatolic, stenosporic and occasionally gyrophoric acid
X. pulla	absent	common	stenosporic, divaricatic, perlatolic and occasionally gyrophoric acid
X. verruculifera	present	rather rare	stenosporic, divaricatic, perlatolic and occasionally gyrophoric acid

Acknowledgments

We are deeply indebted to the all the Polish lichenologists and curators of Polish herbaria for the loan of the specimens for the study. We are also grateful to anonymous Reviewers for all valuable remarks and corrections. The study was funded by Department of Botany and Plant Ecology, Wrocław University of Environmental and Life Sciences, and Department of Biodiversity and Plant Cover Protection, University of Wrocław.

Authors' contributions

The following declarations about authors' contributions to the research have been made: determination of the specimens: KS, MK; taxonomic analyses: KS, MK; writing of the manuscript: KS, MK; preparation of the distribution maps: MK.

Supplementary material

The following supplementary material for this article is available online at http://pbsociety.org.pl/journals/index.php/asbp/rt/suppFiles/asbp.2014.004/0:

1. Appendix S1: the list of specimens examined.

References

1. Crespo A, Lumbsch HT, Mattsson JE, Blanco O, Divakar PK, Articus K, et al. Testing morphology-based hypotheses of phylogenetic relationships in Parmeliaceae (Ascomycota) using three ribosomal markers and the nuclear *RPB1* gene. Mol Phylogenet Evol. 2007;44(2):812–824.
2. Thell A, Crespo A, Divakar PK, Kärnefelt I, Leavitt SD, Lumbsch HT, et al. A review of the lichen family Parmeliaceae – history, phylogeny and current taxonomy. Nord J Bot. 2012;30(6):641–664.
3. Elix JA. Progress in the generic delimitation of *Parmelia* sensu lato lichens (Ascomycotina: Parmeliaceae) and a synoptic key to the Parmeliaceae. Bryologist. 1993;96(3):359.
4. Elix JA. Xanthoparmelia. Flora Aust. 1994;55:201–308.
5. Blanco O, Crespo A, Ree RH, Lumbsch HT. Major clades of parmelioid lichens (Parmeliaceae, Ascomycota) and the evolution of their morphological and chemical diversity. Mol Phylogenet Evol. 2006;39(1):52–69.
6. Del Prado R, Ferencová Z, Armas-Crespo V, Amo de Paz G, Cubas P, Crespo A. The arachiform vacuolar body: an overlooked shared character in the ascospores of a large monophyletic group within Parmeliaceae (*Xanthoparmelia* clade, Lecanorales). Mycol Res. 2007;111(6):685–692.
7. Hale ME. *Bulbothrix*, *Parmelina*, *Relicina* and *Xanthoparmelia*, four new genera in the Parmeliaceae. Phytologia. 1974;28:479–490.
8. Hale ME. A synopsis of the lichen genus *Xanthoparmelia* (Vainio) Hale (Ascomycotina, Parmeliaceae). Smithson Contrib Bot. 1990;(74):1–250.
9. Esslinger TL. A chemosystematic revision of the brown Parmeliae. J Hattori Bot Lab. 1977;42:1–211.
10. Esslinger TL. A new status for the brown Parmeliae. Mycotaxon. 1978;7:45–54.
11. Blanco O, Crespo A, Elix JA, Hawksworth DL, Lumbsch HT. A new classification of parmelioid lichens containing *Xanthoparmelia*-type lichenan (Ascomycota: Lecanorales) based on morphological and molecular evidence. Taxon. 2004;53(4):959.
12. de Paz GA, Cubas P, Crespo A, Elix JA, Lumbsch HT. Transoceanic dispersal and subsequent diversification on separate continents shaped diversity of the *Xanthoparmelia pulla* group (Ascomycota). PLoS ONE. 2012;7(6):e39683.
13. Culberson CF, Culberson WL, Esslinger TL. Chemosyndromic variation in the *Parmelia pulla* group. Bryologist. 1977;80(1):125.
14. Amo de Paz G, Raggio J, Gómez-Serranillos MP, Palomino OM, González-Burgos E, Carretero ME, et al. HPLC isolation of antioxidant constituents from *Xanthoparmelia* spp. J Pharm Biomed Anal. 2010;53(2):165–171.
15. Fałtynowicz W. The lichens, lichenicolous and allied fungi of Poland – an annotated checklist. Cracow: W. Szafer Institute of Botany, Polish Academy of Sciences; 2003. (vol 6).
16. Cieśliński S, Czyżewska K, Fabiszewski J. Red list of the lichens in Poland. In: Mirek Z, Zarzycki K, Wojewoda W, Szeląg Z, editors. Red list of plants and fungi in Poland. Cracow: W. Szafer Institute of Botany, Polish Academy of Sciences; 2006. p. 71–89.
17. Hillmann J. Zur Flechtenflora der Mark Brandenburg. V Verh Bot Ver Prov Brandenbg. 1936;76:6–21.
18. Orange A, James PW, White FJ. Microchemical methods for the identification of lichens. London: British Lichen Society; 2001.
19. Kondracki J. Regional geography of Poland. Warsaw: Polish Scientific Publishers PWN; 2002.
20. Zając A. Atlas of distribution of vascular plants in Poland (ATPOL). Taxon. 1978;27(5–6):481–484.
21. Cieśliński S, Fałtynowicz W. Atlas of the geographical distribution of lichens in Poland. Cracow: W. Szafer Institute of Botany, Polish Academy of Sciences; 1993. (vol 1).
22. Diederich P, Ertz D, Stapper N, Sérusiaux E, van den Broeck D, Ries C. The lichens and lichenicolous fungi of Belgium, Luxembourg and northern France [Internet]. 2013 [cited 2013 Sep 24].
23. Wirth V. Die Flechten Baden-Württembergs. Stuttgart: Ulmer; 1995. (vol 2).
24. Louwhoff SHJJ, James PW, Smith CW. *Xanthoparmelia* (Vain.) Hale (1974). In: Smith CW, Aptroot A, Coppins BJ, Fletcher A, Gilbert OL, editors. The lichens of Great Britain and Ireland. London: British Lichen Society; 2009. p. 963–967.
25. Grube M, Lindblom L, Mayrhofer H. Contributions to the lichen flora of Crete: a compilation of references and some new records. Stud Geobot. 2001;20:41–59.
26. Giordani P. The lichen genus *Neofuscelia* (Ascomycota, Parmeliaceae) in Italy. Lichenologist. 2003;35(5–6):377–385.
27. Knezevic B, Mayrhofer H. Catalogue of the lichenized and lichenicolous fungi of Montenegro. Phyton. 2009;48:283–328.
28. Aptroot A, van Herk K, Sparrius L, van den Boom P. Checklist van de Nederlandse lichenen en lichenicole fungi. Buxbaumiella. 1999;50:4–64.
29. Hafellner J. A new checklist of lichens and lichenicolous fungi of insular Laurimacaronesia including a lichenological bibliography for the area. Fritschiana. 1995;5:1–132.
30. Santesson R, Moberg R, Nordin A, Tønsberg T, Vitikainen O. Lichen-forming and lichenicolous fungi of Fennoscandia. Uppsala: Museum of Evolution, Uppsala University; 2004.
31. Lettau G. Flechten aus Mitteleuropa XII. Feddes Repert. 1957;59(3):192–257.
32. Elix JA. Chemical variations of the lichen *Neofuscelia pulla* (Ascomycotina: Parmeliaceae) sensu Esslinger. Australas Lichenol. 2002;51:7–9.
33. Coppins BJ, Seed L, Earland-Bennett PM. *Neofuscelia luteonotata*, new to the British Isles, and notes to the *N. pulla* group. Br Lichen Soc Bull. 2002;90:29–33.
34. Hafellner J, Türk R. Die lichenisierten Pilze Österreichs – eine Checkliste der bisher nachgewiesenen Arten mit Verbreitungsangaben. Stapfia. 2001;76:1–167.
35. Vězda A, Liška J. A catalogue of lichens of the Czech Republic. Průhonice: Institute of Botany, Academy of Sciences of the Czech Republic; 1999.

36. Søchting U, Alstrup V. Danish lichen checklist. Copenhagen: Botanical Institute, University of Copenhagen; 2002.

37. van den Boom PPG. Contribution to the flora of Portugal, lichens and lichenicolous fungi III. Nova Hedwig. 2003;76(1):157–171.

38. Crespo A, Kauff F, Divakar PK, del Prado R, Pérez-Ortega S, de Paz GA, et al. Phylogenetic generic classification of parmelioid lichens (Parmeliaceae, Ascomycota) based on molecular, morphological and chemical evidence. Taxon. 2010;59(6):1735–1753.

39. Litterski B, Mayrhofer H. Catalogue of lichenized and lichenicolous fungi of Cyprus. Stud Geobot. 1998;16:57–70.

40. Esslinger TL. Nomenclatural notes on some members of *Parmelia* section *Melanoparmelia*. Bryologist. 1973;76(2):306.

16

Ethnobotanical review of wild edible plants of Slovakia

Łukasz Łuczaj*

Department of Botany and Biotechnology of Economic Plants, University of Rzeszów, Werynia 502, 36-100 Kolbuszowa, Poland

Abstract

This paper is an ethnobotanical review of wild edible plants gathered for consumption from the 19th century to the present day, within the present borders of Slovakia. Twenty-four sources (mainly ethnographic) documenting the culinary use of wild plants were analysed. The use of 106 species (over 3% of the Slovak flora) has been recorded. Nowadays most of them are no longer used, or used rarely, apart from a few species of wild fruits. The most frequently used plants include the fruits of *Rubus idaeus*, *Fragaria* spp., *Rubus* subgenus *Rubus*, *Vaccinium myrtillus*, *V. vitis-idaea*, *Fagus sylvatica*, *Corylus avellana*, *Prunus spinosa*, *Pyrus* spp., *Malus* spp., *Crataegus* spp. and the leaves of *Urtica dioica*, *Rumex acetosa*, Chenopodiaceae species, *Cardamine amara*, *Glechoma* spp., *Taraxacum* spp. and *Oxalis acetosella*. The most commonly used wild food taxa are nearly identical to those used in Poland, and the same negative association of wild vegetables with famine exists in Slovakia, resulting in their near complete disappearance from the present-day diet.

Keywords: historical ethnobotany, ethnobiology, wild green vegetables, wild food plants, wild edible plants

Introduction

The growing interest in the use of wild food plant resources nowadays stems from efforts to find alternatives to the industrialization and globalization of agriculture and to provide food security in times of agricultural crisis. In the not so distant past many wild plants, instead of being eliminated from agricultural systems, constituted valuable supplementary sources of nutrition [1]. Within the last two decades, detailed ethnobotanical studies have been carried out in European countries to preserve the disappearing traditions of wild food plant use. Such studies were performed for instance on the Iberian Peninsula (e.g. [2–8]), in Italy [9–13], Greece [13,14], Turkey [15,16], Bosnia and Herzegovina [17], Albania [18] and Austria [19]. The phenomenon of foraging in Europe has been, however, studied from different perspectives for centuries. It was present in economic plant encyclopaedias [20–23] and later appeared as the subject of ethnographic studies. A separate branch of study concerns wild food plants as a means of alleviating food shortages during times of crop failures and wars [24,25]. Countries where ethnobotanical studies are most intensive now are usually places where little ethnobotanically oriented research has been done before. In contrast, there are a few European countries in which research on the ethnobotany of rural populations started at the end of the 19th century. Here we shall first of all mention two of them. One is Poland, where local ethnographic monographs, Józef Rostafiński's study of 1883, and the "Polish ethnographic atlas" all contributed to a large body of data concerning the use of wild food plants [26–29]. The other country is Estonia, where a similarly large number of ethnographic elaborations and queries is available- and it has been recently synthesized [30].

Łuczaj and Szymański [26] pointed out that the cross-cultural and geographical analysis of the patterns of plant use in Europe is hindered by the fact that most publications were written in national languages, mainly in small ethnographic journals and monographs. Thus English-language critical reviews in widely available journals can constitute "building blocks" for further international analyses. The studies from Poland show a gradual disappearance of traditions of wild food gathering, since the 19th century or even earlier [26–29]. A similar gradual decrease must have occurred in other European countries as well.

Slovakia has extensive published data on wild food plant use in the 19th and 20th century, but lacks a comprehensive review of them, apart from a short entry in a dictionary of folk culture [31] and three short essays with only a few literature references [32–34]. Hence a review of Slovak publications concerning this topic became the aim of the study.

* Email: lukasz.luczaj@interia.pl

Material and methods

Slovakia – its flora, geography and history

Slovakia covers an area of 49 thousand km^2. It has a population of 5.4 million inhabitants [35], and lies within the cold temperate climate zone, in the intermediate zone between the maritime and continental climate. The dominant natural vegetation is composed of deciduous and mixed forests, dominated, depending on the elevation, by *Quercus* spp., *Carpinus betulus, Fagus sylvatica, Abies alba* and *Picea abies* [35]. Slovak vascular flora is made up of ~3000 native and naturalized species of vascular plants [36,37].

Publications used in the review

The documentation of traditions of plant use in Slovakia was begun by a distinguished scholar, Jozef Ľudovit Holuby (1836–1923) in the 19th century. He was a very active researcher both in studying Slovak flora and documenting the ethnographic traditions of western Slovakia [38]. Most of his work concerns the area around the village of Zemianske Podhradie near Trenčin (W Slovakia), where he lived for more than half his life [39–44]. The depth of Holuby's ethnobotanical observations must be emphasized. He not only carefully studied the adult world but also devoted a separate article to children's ethnobotany, which is the first of its kind in the history of ethnobotany [41].

Another important individual in this area of study was Michal Markuš (1912–2004). He was the leading expert on the ethnography of traditional Slovak food, and devoted parts of some of his publications to foraging in central and eastern Slovakia [45–49] (Fig. 1).

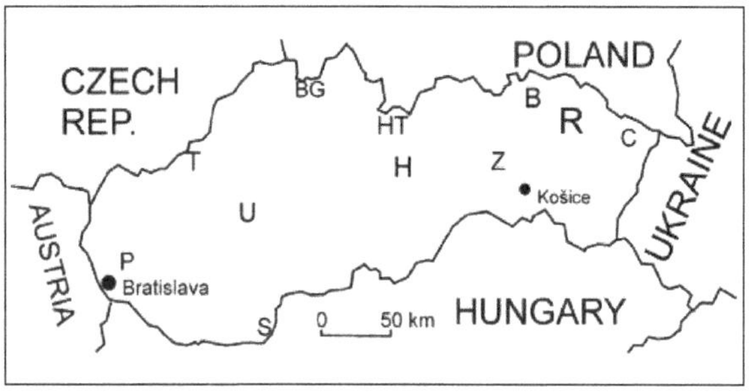

Fig. 1 Localities of the regional studies on Slovak wild food plants: B – Bardejov [47]; BG – south of Mount Babia Góra [52]; H – Horehronie region [46,48]; HC – Horna Cirocha [49,57]; HT – the High Tatra Mountains [54–56] (and K. Zlkovanová's letter); P – Pezinok [42]; R – Rusyn minority areas in general [51]; S – Z. Melečková's letter from Šturovo; T – Trenčin area [39–41,44]; U – Uhrovska Dolina [58]; Z – Žakarovce [45].

A few other ethnographers also devoted parts of their ethnographic publications to wild food provision, which leaves us with 24 publications altogether, usually containing both Latin and/or local plant names of wild edible plants used in Slovakia [31–34,39–58] (Fig. 1). From Stoličná's essays [32,33], only information not quoted from other sources analysed in the review was taken into account.

The review also includes two letters received in response to a query about wild food plants which I sent to 170 Slovak botanists:

(*i*) From Ms Zuzana Melečková, a PhD student from Bratislava, who listed 25 wild food snacks from her childhood in the vicinity of Šturovo (formerly Parkan, Párkány), and provided local Hungarian names (the area is predominantly Hungarian-speaking). Only 18 species, which were listed at least once by some other source than this review, were included.

(*ii*) From Ms Katarína Zlkovanová (Tatra National Park), who listed 14 wild food snacks from her childhood in the Tatra Mounatins. Only 9 species, which were listed at least once by some other source than this review, were included.

Methodology of plant identification

All the available publications containing data on wild food plant use were included. Most of them had Latin names provided along with local names. In a few cases data without Latin names of plants were also used for plants with widely known, unambiguous names. Latin names of plants were critically evaluated in order to avoid the introduction of "ghost data" sensu Svanberg [59,60]. Such ghost data is information about some use attributed to an incorrectly identified taxon, later repeated by other authors. Distribution of species was checked using standard Slovak floras. A list of taxa from all the publications was created, using a special code for the credibility of identification following Łuczaj [61] (Tab. 1). Plant names were given according to "Flora Europaea" [62].

Results

The use of at least 106 species of vascular plants as food, seasoning or beverages has been recorded in Slovakia since the 19th century (Tab. 1, Tab. 2). They belong to 79 genera from 36 plant families. The list includes 14 species of trees, 24 species of shrubs (including 2 species of dwarf shrubs), 54 species of perennials, 6 species of biennials and 8 species of annuals. The two largest categories are nearly of equal size: "ripe fruits" (both fleshy and dry fruit and seeds) – 41 species, and "green vegetables" – 39 species. Also the underground parts of 12 species, and flowers of 7 species were consumed. Various parts of 8 species were used as seasoning, and 4 taxa were used as bread ingredients. Eighteen species were collected for making beverages – 8 for liqueurs, 7 for herbal infusions used on an everyday basis, and three taxa were sources of tree sap.

Fruits and seeds

Out of 41 species whose use was recorded in the category of fruits and seeds (excluding species used only as seasoning), nearly half, i.e. 19 species, belonged to the Rosaceae family. The other best represented families are Caprifoliaceae and Grossulariaceae, with 4 species.

The fruits most commonly collected from the wild include *Rubus idaeus, Vaccinium vitis* idaea, *Rubus* subgenus *Rubus* spp. and *Fragaria* spp. (Tab. 2). Fruits are the most appreciated category of wild plants collected in Slovakia – they are both eaten raw and made into preserves (jams, juices, sauces). In Communist Slovakia large amounts of them were bought by special collection points, but even now they are sometimes collected for personal use. In the past rosehips, mainly from *Rosa* spp. were also widely used, mainly to make a type of jam.

In the past most fruits were eaten fresh or dried. In the latter half of the 20th century the preserving of soft fruits in the form of jams, wines and pasteurized compotes became popular (due to the decreasing prices of sugar). However over the last few years it has been in decline due to growing affluence.

Green vegetables

Green vegetables, including plants whose green parts such as leaves, stalks or unripe fruits are eaten raw or after special preparation (cooking, frying, etc.), and excluding plants used

Tab. 1 List of wild food plants used in Slovakia since the 19th century.

Family	Latin name	Cr.	Local names	Parts used	Mode of use	Reference No.
Aceraceae	*Acer campestre* L. (only this species is mentioned with a Latin name but probably *Acer pseudoplatanus* and *A. platanoides* were used as well)	A	klen	BEVsap	sap drunk fresh or frozen by children;	[40,41]
		L	javor	VEG	immature fruits sucked as a snack	[46]
	Acer sp.	O	javor	BEVsap	sap made into sugar in the early 19th century	[32,33]
Apiaceae	*Anthriscus cerefolium* (L.) Hoffm.	L	mačia madra	VEG SUB	whole plants as an ingredient of soup or other boiled dishes	[46,48]
	Angelica archangelica L.	N	(archangelika)**	VEGfam	famine food	[55]
	Carum carvi L.	L	rasca	VEG	aerial parts added to boiled dishes	[49]
				SEA	seeds as spice for bread, soups and vodka	[46], KZ
	Chaerophyllum bulbosum L.	A	repa, repka, krkoška	SUB	raw tubers eaten in spring "until the first thunder" mainly by children	[39,40,46]
	(?) *Daucus carota* L.	L?	poľna mrkva, knahenka	SUB	eaten in the fields after ploughing; this information may, at least partly, refer to Stachys palustris whose rhizomes were on the Polish side of the Carpathians (at the feet of the Tatras) gathered from fields after ploughing and called dzika marchew, i.e. wild carrot [26].	[46]
	(?) *Pastinaca sativa* L.	A?	paštrnák	SUB?	first year root, not clear if the reference reports actual use	[40], KZ?
	Pimpinella saxifraga L.	L	bedrenčok	BEVliq	roots formerly added to vodka	[46]
Asteraceae	*Arctium lappa* L.	L	lopuch, lopúch	VEGfam	boiled as famine food	[46,55]
				BEVliq	roots added to vodka	[46]
	Carlina acaulis L.	L	repka	VEG	inner part of inflorescences raw	[46,48]
	Cichorium intybus L.	A	čakanka, tuhovňik	VEG	young leaves raw	[40,46], KZ
	(?) probably *Cirsium oleraceum* (L.) Scop. or/and *C. rivulare* (Jacq.) All	N?	štrbak (as *Cichorium endivia*), bodiak	VEG	potherb	[46,55]
	Inula helenium L.	L	maškuľin, mačkuľin	BEVliq	spice for vodka	[46]
	Chamomilla recutita (L.) Rauschert	N	rum'janok	BEVher	infusion drunk until the mid-20th c.	[51]
	Petasites sp.	L	ďeviasel, ďiviasel	SUBfam	rhizomes as famine food	[46]
	Tanacetum sp. (probably *Tanacetum offcinale* L.)	L	boľačkovo žilja, boľačkovo ziľa, boľačkovoj ziľa	SEA	chopped as spice mainly for scrambled eggs	[49,57]
	Taraxacum sp. (as *T. officinale*)	A	pupava, popovňak, HU: pitypang	VEG	raw or fried – for a snack, salads and potherb; often the use was a new tradition introduced at the end of the 19th century by returning emigrants	[40,45,46], ZM, KZ
	Tragopogon orientalis L.	A	koria brada	VEG	stalks sucked by children	[40,41]
	Tussilago farfara L.	L	podbeľ, podmačka, podlieu, podbieľ, veľki podbeľ, žltie pierka	VEG	commonly made into soup until 1920s, VEG were chopped, spiced with flour, pig fat and vinegar, it was a popular food in Horehronie	[46,48]
				SUB	raw, eaten by children, supposedly tasted like cabbage head cores	[46]
Berberidaceae	*Berberis vulgaris* L.	A	nd, HU: sóskaborbolya	VEG	young shoots eaten raw in spring or added to sauces	[40]
				FRU	nd	[40], ZM
Betulaceae	*Betula pendula* Roth.	A	breza	BEVsap	sap drunk fresh or frozen by children	[40,41,45]
Brassicaceae	*Armoracia rusticana* P. Gaertn., B. Mey & Scherb.	L	chren	VEG	boiled, sometimes also after lactofermentation, mainly famine use	[46,55]
				SEA	roots as meat condiment, also added to vodka	[46], KZ

Tab. 1 *(continued)*

Family	Latin name	Cr.	Local names	Parts used	Mode of use	Reference No.
	Brassica nigra (L.) W. D. J. Koch	L	čierná kapusta, podlubki	VEG	soups and sauces	[46,58]
	Cardamine amara L.	A	žerucha, režucha, žerušnica, horká, šalvija, krasch, kreschbrunn, kressbrunnen	VEG	eaten mainly before St George's Day (24 Apr), raw with vinegar or oil, also added to soups and meat dishes	[39,40,43,44, 46,54,56], KZ
	(?) *Lepidium latifolium* L.	L?	žerucha	VEG	raw	[46]
	(?) *Nasturtium officinale* R. Br.	L?	žerucha	VEG	raw	[46,48]
Cannabaceae	*Cannabis sativa* L.	L	konope	FRU	seeds raw, crushed and sucked	[46,48]
Caprifoliaceae	*Sambucus ebulus* L.	A	podzemní bez; podzemný bez	FRU	fried into jam, used mainly as medicine against respiratory infections	[39,40]
	Sambucus nigra L.	A	bez, čierna baza, čierny bez, kozičky, chabs	FRU	as above	[39,40,45,49], KZ
	Viburnum lantana L.	A	nd	FRU	raw (by chidren)	[40]
	(?) *Viburnum opulus* L.	NO?	kalina	FRU	nd	[55]
Chenopodiaceae (recently moved to Amaranthaceae)	*Atriplex hortensis* L. and *Chenopodium bonus-henricus* L. mentioned but probably more species from these genera were used	L	loboda, ľebeda, natina, zeľina, gracka, mastná zelina	VEG	boiled and/or fried as potherb, soup or with potatoes and flatbread	[46–48,53, 55,58]
Cornaceae	*Cornus mas* L.	N	drienky	FRU	formerly mainly raw, in the 20th c. also made into vodka	[40] (mistakenly as *C. sanguinea* L.) [45,46]
Corylaceae	*Corylus avellana* L.	AO	lieska, fruits as oriešky	FRU	raw, or added to bread or cakes	[40,41,46,48, 49,58]
			catkins called brost, riasa	FLOfam	dried catkins ground into flour at the beginning of the 19th	[46,55,58]
Cupressaceae	*Juniperus communis* L.	AO	jadlovec, jalovec, borovka, borovňice	FRU	spice for vodka and sauerkraut	[40,45,46]
Equisetaceae	*Equisetum* sp. (probably *E. arvense* L.)	N	(praslička)**	VEGfam	famine food	[55]
Ericaceae	*Vaccinium myrtillus* L.	AO	čučoriedky, jafury, borovňice, borovki, HU: áfonya	FRU	raw, jams, wines, dried for infusions	[40,45,46,48, 49,55,56,58], ZM
	Vaccinium vitis-idaea L.	AO	brusnica, brušnica	FRU	preserves	[40,45,46,55, 56]
Fabaceae	*Lathyrus tuberosus* L.	A	repnica, repňica, orešíc, orešina, červené orešie, oresie	SUB	tubers raw or baked in ashes, eaten by shepherds and children	[39–41,46,48]
	Onobrychis viciifolia Scop. (as *Onobrychis*)	L	slaninka	FLO	eaten raw by children	[46]
	Robinia pseudoacacia L.	A	biely agát, HU: akác	VEG	fried in oil (coated in batter)	[39], ZM
				FLO	flowers as children snack	ZM
	Trifolium pratense L.	L	červeni chľebik	FLO	nectar from flowers sucked by children	[46]
	Trifolium sp.	L	lóhere	VEG	leaves as children snack	ZM
	Vicia lathyroides L.	L	ľadňik, ďivi hrach	FLO VEG	flowers and green pods raw (by children)	[46]
Fagaceae	*Fagus sylvatica* L.	AO	buk, fruit called bukvica	FRU	raw or dried, or pressed into oil	[39,40,49,58]
				OTHfam	wood shavings added to food during the 1847 famine	[52]
			buková voda	BEVsap	raw, also as medicine	[45]
	Quercus petraea Matt. (Liebl.) (as *Q. sessilis* Ehrh.) and probably *Q. robur* L. as well	L	žalude	FRUfam	famine food	[46,58]
	Quercus sp.	L	HU: tölgy	FRU	raw, children's snack	ZM

Tab. 1 *(continued)*

Family	Latin name	Cr.	Local names	Parts used	Mode of use	Reference No.
Gentianaceae	*Gentiana cruciata* L.	L	terlie, trlie, terleč	BEVher	roots gathered until recently as spice for tea	[57]
	Gentiana punctata L.	L	horec	BEVliq	highly prized addition to vodka	[46]
	Gentiana sp. (?; as *Gentianella*)	LD	goročka, gorička	BEVliq	"roots like fingers" used as an addition to vodka or as infusion	[46]
Grossulariaceae	*Ribes uva-crispa* L. (as *Ribes grossularia** and *Grossularia*)	L	egreš, gigadze, muchinki, muchiňak	FRU	raw, eagerly sought after by children	[40,46]
	Ribes nigrum L.	A	čierný rýbiz, smrdlenka	FRU	raw	[40]
	Ribes rubrum L. (maybe a related *R. spicatum* Robson)?	L	kvasňice	FRU	raw	[46]
	Ribes petraeum Wulfen	L	kvasňice	FRU	raw	[46]
Lamiaceae	*Glechoma hederacea* L. and *Glechoma hirsuta* Waldst. & Kit.	A	nádešník, nádešníček, oponka, kocurovo vajca/vajcia, kocurovo jajca, kocurovo jajička	VEG	leaves as spice for soups, sauces and potatoes; widely used until the 1940s	[39,40,43,44, 46,48,49,52, 57]
	Lamium album L.	L	hlucha pokriva	VEG	potherb	[45]
	(?) *Origanum vulgare* (as Melissa)	N?	lebitka	SEA	flowering shoots as a spice	[49]
	Mentha sp.	N	mjata kruta	SEA	spice	[49]
	Symphytum officinale L.	A	madunica, medunica	FLO	nectar from flowers sucked by children	[39,41]
	Thymus sp.	L	materina duška	SEA BEVher	flowering shoots	[49,51]
(?) Lamiaceae	unidentified (as *Nepeta* sp.)	?	madra	SEA	seasoning	[49]
Liliaceae (recently moved to Amaryllidaceae)	*Allium scorodoprasum* L.	A	divý česnek	SEA	used as a garlic surrogate	[40]
	Allium ursinum L.	A	remža	VEG	potherb	[42]
		L	HU: medvehagyma	FRU	fruits as children's snack	ZM
Lythraceae	*Trapa natans* L.	L	nd	FRU	eaten boiled in S Slovakia until the end of the 19th century	[32,33]
Malvaceae	*Malva pusilla* Sm. (as *M. rotundifolia* L. and *M. borealis* Wallr.)	A	pánbožkové koláčky, pánbozkové kolácky, pagačica, pánbíčkové koláčky, peniažková zelina	VEG	unripe (green) fruit eaten raw by children	[39–41,43]
	Malva sp.	L	sliz	VEG	added to cooked dishes	[49]
	Malva sylvestris L.	A	rindziki, sirjeňak, sirjenčak	VEG	soup	[46]
				VEG	unripe (green) fruit eaten raw by children	[46]
Orchidaceae	*Orchis morio* L.	L	žaludkovi koreň, vlaski orech	BEVliq SUB	tubers added to vodka and formerly used as food	[46]
Oxalidacaceae	*Oxalis acetosella* L.	A	zajačía ďetalinka, zajačková kapusta	VEG	mainly raw as children's snack, also added to sauces	[40,46], ZM, KZ
Pinaceae	*Abies alba* Mill.	L	jedľina	VEGfam	raw as famine snack	[46]
	Larix decidua Mill.	L	červeni smrek	OTH	resin chewed mainly by children	[46,48]
Plantaginaceae	*Plantago major* L.	A	babka, kolocier, kološija	VEG	mainly as potherb	[40,43,46,49]
	Plantago media L.		volovo ucho	VEG	potherb	[43]
Poaceae	*Elymus repens* (L.) Gould (as *Triticum repens* L.)	L		SUBfam	rhizomes, in 1847 (famine), dried, ground for bread	[39]
Polygonaceae	*Rumex acetosa* L.	A	štiav, kvasinka, šťava, štaveľ, šťovik, kozina, mokri križ	VEG	soups and sauces	[40,41,46,49, 55,58]
	Rumex sp. (some large-leaves species, given probably mistakenly as *R. hydrolapathum* Huds.)	LNR	konský šťav	VEGfam	boiled – only famine food	[46]

Tab. 1 *(continued)*

Family	Latin name	Cr.	Local names	Parts used	Mode of use	Reference No.
Polypodiaceae	*Polypodium vulgare* L.	A	sladič, slaďička, slaďica, soldečka, HU: páfrány	SUB	sucked by children, also used as sweetener	[40,46], ZM, Dr. N. Varchol – pers. comm.
Portulacaceae	*Portulaca oleracea* L. (as *P. sativa* L.)	A	nd	VEG	salad	[40]
Myrsinaceae (recently moved to Primulaceae)	*Cyclamen purpurascens* Mill.	ND	cyclaminy	SUBfam	tubers as famine food at the beginning of the 19th century	[45]
Ranunculaceae	*Ranunculus ficaria* L. (*Ficaria verna* Huds.)	A	blyskáč, blýštek, blyšček, liži kvet	VEG	salad	[39,40,42–44, 46]
				SUBfam	bulbils as famine food, especially in 1844	[42,43,46]
				BEVher SEA	formerly drunk as infusion, also added to dishes as spice	[46]
Rosaceae	*Aruncus dioicus* (Walter) Fernald	N	parilo	BEVher	infusion as beverage	[51]
	Crataegus spp. (*C. monogyna* Jacq., *C. laevigata* (Poir.) DC. and other species)	L	hloh, HU: galagonya	FRU	mainly as children's snack	[40] (as *C. oxyacantha* L. and *C. monogyna* Jacq.), [46] (as *C. oxyacantha**), ZM
	Fragaria spp. (mainly the commonest *Fragaria vesca* L.)	AO	lesné jahody, jahoda, vtáčenička, sunyca, HU: eper	FRU	raw, very commonly eaten	[40,46,49,51, 55,58], ZM
	Fragaria moschata Duchesne (as *Fragaria elatior* Ehrh.)	A	smokva	FRU	raw	[40]
	Fragaria viridis Duchesne (as *Fragaria collina* Ehrh.)	A	truskavec, druzgavec	FRU	raw	[40]
	Geum urbanum L.	L	prestriť, pristriť	BEVliq	until the 1950s used as spice for wine	[57]
	Malus sp. (partly as *M. silvestris**, *Pirus malus* L.)	AO	plánky, planki, HU: vadalma	FRU	raw or added to sauerkraut; also as famine food	[40,46,49,56, 58], ZM
	Potentilla erecta L.	L	červienkovi koreň	BEVliq	roots added to vodka	[46]
	Sanguisorba minor Scop. (as *Poterium sanguisorba* L.)	A	žabacia zelinka	VEG	soups	[39,40]
	Prunus avium L.	A	nd	FRU	raw	[40]
	(?) *Prunus fruticosa* Pall. (mistakenly as *P. chamaecerasus* Jcq.)	DR?	nd	FRU	raw, eaten by shepherds	[40]
	Prunus spinosa L.	A	tereň, tŕnky, tarňik, psí trn, trnka, HU: kökény	FRU	eaten raw after frosts (esp. by children) or used to make plum brandy	[39,40,45,46, 49,55,56], ZM
	Pulmonaria officinalis L.	L	čmeľik, pľúcna zelina, medunica, včelka, cicalka	FLO	nectar from flowers sucked by children	[46]
	Pyrus spp. (as *P. communis* L., *P. piraster**)	A	planki	FRU	added to sauerkraut, eaten raw	[40,45,46,58]
	Rosa spp. (mainly *R. canina* L., *R. rubiginosa* L.)	A	šverboritki, šverboritky, šip, šipová ruža, šipki, sviboritki, HU: csipkebogyó	FRU	mainly made into fried jam (lekvar), also made into wine (since 20th century) and extensively used as famine food (unsweetened jam)	[39] (as *R. canina* L.and *R. rubiginosa* L.), [40,45] (as *R. canina**), [46] (as *R. canina**), [47], ZM

Tab. 1 *(continued)*

Family	Latin name	Cr.	Local names	Parts used	Mode of use	Reference No.
	Rubus subgenus *Rubus*	AO	čornica, černica, čierna malina, ostružina, užyna, HU: rina (for *R. caesius*), szeder (for the other species from the subgenus)	FRU	mainly raw, also made into jam, or brandy and drunk as infusion	[39] (as *Rubus candicans* Whe, *R. corylifolius* Sm., *R. hybridus* Vill. *R. fruticosus* L.), [45,46] and ZM (both as *R. fruticosus** and *R. caesius**), [49,58,51]
	Rubus idaeus L.	AO	malina, malyna, červená malína, HU: málna	FRU	raw or jam	[40,49,51,55, 58] ZM
				VEG	infusion	[49,51]
	Sorbus aucuparia L.	A	skorucha, skoruša, karušniak	FRU	eaten mainly as famine food, dried for winter for food, also added to vodka	[40,46] (mistakenly as *S. domestica**), [55]
	Sorbus aria L.	A	skorucha, skoruša, karušniak	FRU BEVliq	dried for winter for food, also added to vodka	[40,46]
Scrophulariaceae (recently moved to Plantaginaceae)	*Veronica beccabunga* L.	A	bobovník	VEG	soup	[40]
Tiliaceae	*Tilia* sp. (*Tilia cordata* Mill. and *Tilia platyphyllos* Scop.)	O	lipa	BEVher	infusion from flowers, until the mid-20th century	[51]
Urticaceae	*Urtica dioica* L.	AO	žichľava, pokriva, kopriva, veliká žihlava, veľká žihľava, pŕhľava, pŕhlava	VEG	young shoots in soups, potherb, sauces, used mainly until World War I	[39,45,46,48, 49,50,51,55, 58], KZ
Valerianaceae	*Valeriana* sp. (probably *Valerianella* sp.?)	LD	čortove zilja	VEG	shoots added to cooked dishes	[49]
	Valerianella locusta (L.) Laterr. (as *Valeriana olitoria* Poll.)	A	nd	VEG	young plants in salads with vinegar	[40]

Cr. – credibility of identification: ? and (?) – identification uncertain; A – Latin names identified by a botanical expert; L – Latin name given by other researchers. Botanical name identified using: D – species description; N – local name; R – analysis of the range of related species. Local names: HU – a local Hungarian name; KZ – Zlkovanová's letter; nd – no data on local Slovakian names; ZM – Melečková's letter. Parts used: BEVher – herbal infusions; BEVliq – alcoholic beverages; BEVsap – tree sap; fam – used only in times of famine; FLO – flowers (their nectar eaten raw or flowers added in larger quantities to dishes); FRU – fruits (raw or in preserves); OTH – other; SEA – only small amounts added to dishes as seasoning; SUB – subterranean parts (rhizomes, roots, bulbs, tubers) – raw as a snack or added to boiled dishes; VEG – green vegetables: aerial parts (leaves and shoots), raw, boiled or fried.

in small quantities only as seasoning, constitute the second largest use category, with 39 species recorded. The most represented families are Asteraceae (7 species), Brassicaceae (5 species) and Apiaceae (3 species). Most of the recorded green vegetables are plants, which were eaten in times of scarcity, usually as mixed potherb or soup, often with an admixture of potatoes, cereals, butter, milk or cream. The only widely used green vegetable, which is still commonly used in nutrition is *Rumex acetosa* (leaves used to make soup). In some parts of Slovakia *Cardamine amara* leaves were also used relatively frequently, in a similar fashion to cultivated cress – raw or added to boiled dishes (Tab. 1). Up until the beginning of the 20th century the shoots of *Urtica dioica* and Chenopodiaceae species were often used with a variety of cooked and fried dishes, particularly in the years of bad harvest. As these plants were associated with poverty, their use is practically extinct nowadays (Tab. 1).

Underground parts

The use of underground parts of plants (roots, rhizomes, bulbs) was recorded only for 12 species. The sweet rhizomes of *Polypodium vulgare* were particularly widely used. Other species were used less frequently, as children's snacks (e.g. *Chaerophyllum bulbosum*, *Lathyrus tuberosus*) or famine food (*Elymus repens*, *Ranunculus ficaria*).

Seasoning and preservatives

Out of the ten species used as seasoning, only four have been used commonly: *Glechoma hederacea* and *G. hirsuta* (leaves of both as spice for soups and sauces), *Carum carvi* (seeds) and *Armoracia rusticana* (grated roots, as spice for meat). The use of *Glechoma* species gradually decreased in the 20th century and by the 1970s was only practiced by the older generation.

Beverages

A few species of mainly medicinal plants were drunk locally on an everyday basis (e.g. *Tilia* spp., *Chamomilla recutita*). In spring fresh tree sap was drunk, mainly by children (*Betula pendula* and *Acer* spp.). Making juices, wine and liqueurs out of wild fruits seems to be mainly a 20th century fashion, rarer in earlier times (due to the high price of sugar), which became an extremely widespread activity in the countryside in the second half of the 20th century in the Communist period, then diminished due to the increasing affluence of society. On the other hand adding wild plants to liqueurs as flavouring was commonly practiced both in the 19th and 20th century.

Bread ingredients

The reviewed article contains surprisingly few references to wild food bread ingredients, even in times of famine. Altogether only 4 such taxa are mentioned (*Fagus sylvatica*, *Elymus repens*, *Rosa* spp. and *Crataegus* spp.).

Children's snacks

A large proportion of the presented species list is made up by children's snacks. The use of some wild snacks, such as the unripe seeds of *Malva* spp. or *Polypodium vulgare* rhizomes may be a relic of use of food plants by adults, whereas using the nectar sucked from flowers is a part of pan-European children's ethnobotanical folklore [2,28].

Tab. 2 Wild food species reported by at least three authors.

Latin name	No. of authors
Urtica dioica L.	7
Rubus idaeus L.	6
Fragaria spp., *Malus* sp., *Rubus* subgenus *Rubus*, *Vaccinium myrtillus* L., *Rumex acetosa* L.	5
Atriplex and *Chenopodium* spp., *Cardamine amara* L., *Glechoma* spp., *Oxalis acetosella* L., *Polypodium vulgare* L., *Prunus spinosa* L., *Pyrus* sp., *Taraxacum* spp.	4
Armoracia rusticana P. Gaertn., B. Mey & Scherb., *Cichorium intybus* L., *Corylus avellana* L., *Crataegus* spp., *Fagus sylvatica* L., *Quercus* spp., *Rosa* spp., *Sambucus nigra* L., *Sorbus* spp., *Vaccinium vitis-idaea* L.	3

Discussion

Comparison with other countries

The pattern of species use in Slovakia is very similar to that Polish, Ukraine and Hungary [26–28,63,64]. In these areas, usually only a few species of fruits and wild vegetables were used within one village or small region, and a larger number was used only in times of famine. The use of fruits in both countries is nearly identical – nearly the same species (mainly various Rosaceae and *Vaccinium* spp.) have been collected and eaten raw or turned into jams, juices and alcoholic beverages. Both in Poland and Slovakia the larger scale processing and preserving of fruits with the addition of sugar is a relatively new phenomenon, which was originally restricted to the aristocracy, but throughout the 19th and 20th century it became a part of everyday cuisine [29]. This similarity to plant use in the Polish and Ukrainian parts of the Carpathians is caused by the linguistic and habitat similarities between these areas: all of them inhabited by people speaking related languages and dialects with very similar vegetation and rural economies.

Nowadays Slovaks use relatively few species of wild food plants, however in the olden times people knew many species of wild green vegetables, which could be used to supplement nutrition during spring food shortages. Markuš [48] wrote that up to 50 species of wild greens may have been gathered in times of poverty (without specifying all of them). However due to the negative reputation of wild greens they have not been preserved the in modern diet, apart from the use of sorrel (*Rumex acetosa*). Similarly to Poland, the most commonly used wild greens were *Rumex acetosa*, *Urtica dioica* and Chenopodiaceae species. There are actually more parallels with the use of plants in the Polish Carpathians: using the leaves of *Glechoma* spp. as soup seasoning and the use of the rhizomes of *Polypodium vulgare* as a children snacks and sweeteners [29].

A specific feature of some parts of Slovakia is the use of *Cardamine amara*, not reported from other surrounding countries. Another particular feature of Slovakia is the widespread use of medicinal herbs as seasoning for liqueurs. These species are relatively frequently mentioned in the literature on wild food plants, which suggests that they were culturally salient [40,46,57].

On the whole, Slovak culinary culture could positioned on the herbophobous side of the "herbophilia – herbophobia" spectrum. These terms, introduced by the author of this article in 2008, distinguish cultures where large amounts of wild greens are consumed directly as food or medicine (herbophilous cultures) from those which are "herbophobous", in which wild greens are used almost exclusively as famine food, and even in herbal medicine are not consumed directly but in the form of infusions, decoctions or liqueurs [26,27]. The lack of interest in eating wild greens in contrast to calorie-rich dairy and cereal products was summed up by Holuby in 1872 with a popular saying: "Nemcom zelina, Ma'arom slanina, a Slovákom kaša s mliekom" which can be translated as "greens for Germans, lard for Hungarians and kasha and milk for Slovaks" [39].

It is more difficult to compare the use of wild food plants in Slovakia, with that of other neighbouring countries (Czech Republic, Austria and Romania) as they lack detailed ethnobotanical reviews. However in all of them wild fruits seem to be more appreciated and used than wild green vegetables [19,65,66]. It is only in the Hungarian ethnographic literature

[64], e.g. in the works of Bella Gunda [67], about the eastern part of the Pannonian Plain (eastern Hungary), that we get a description of the use of water and aquatic plants in nutrition, e.g. *Glyceria* and *Typha* sp. This category of plants (apart from *Cardamine amara*) is hardly present among Slovak food plants, mainly due to the natural conditions of Slovakia, which is a country composed mainly of hills, mountains and, to a lesser extent, fertile flatlands taken up by intense agriculture, with little area covered by marshy habitats.

The proportion of families in the wild plants consumed in Slovakia is similar to that of Poland [28,29], as in both countries Rosaceae dominate among fruits, Lamiaceae in the category of seasoning and Asteraceae among wild vegetables. Also, similarly to Poland and in contrast to the Mediterranean countries [1], hardly any wild Liliaceae were used in nutrition.

Credibility of presented material

The reviewed material was critically evaluated in search of possible identification mistakes. As shown by Łuczaj [61], ruling out identification mistakes is not possible without voucher specimens. On the other hand the credibility of identification can be assessed indirectly, for example by the profession and expertise of the author of the publication, and by analysing the data from other publications, particularly those concerning local plant names and distribution maps of the analysed taxa. Expert botanists tend to report a much lower number of incorrect ethnobotanical taxa [61]. It seems that the publications of Holuby, the leading 19th century Slovak botanist, are highly credible (although even he confused the Latin names of the edible *Cornus mas* and inedible *Cornus sanguinea*). The works of Markuš, an ethnographer by profession, can however contain a few erroneous identifications (marked in Tab. 1). There is no problem with the identifications of other authors as they reported only common, widely known species.

Scope for future studies

Slovakia is a modern country which underwent intense industrialization processes in the Communist period. The loss of ethnobotanical knowledge and local tradition was enhanced by the collectivization and destruction of private farming, causing the abandonment of "old ways". However there is still much scope for local studies on traditional plant uses in some parts of the country, including the large ethnic minority groups of Hungarians, Roma and Carpatho-Rusyns. How much of the knowledge is still preserved can be seen from Varchol's study on the beliefs about plants among the Carpatho-Rusyns of the Prešov area [68]. A study of the use and traditional knowledge of wild food plants by Roma communities in Slovakia could be a particularly interesting issue, as only scraps of their ethnobotanical knowledge are recorded [69]. Return studies to the areas where Holuby and Markuš researched wild food plant use could also be of great interest, as well as a survey of wild food snacks among children parallel to the one carried out in Poland [70].

Conclusions

It must be stressed that the analysed literature documents an absolutely dramatic decrease in the use of wild plants as food. Most species formerly used as green vegetables are not used any more. Less of the traditional heritage has been lost in the case of fruits. The majority of the fruits reported in most publications are still gathered. Actually their use increased in the 20th century, due to the use of sugar as a preservative.

Slovak use of wild food plants is very similar to that of southern Poland and can be characterized by a high appreciation of wild fruits and low appreciation of wild green vegetables, which are regarded mainly as famine plants. There is still scope for local ethnobotanical studies in Slovakia, particularly as Slovakia is a country with well-preserved natural vegetation and a few large ethnic minorities.

Acknowledgments

Many thanks to Dr. Nadia Varchol (Museum of Rusyn-Ukrainian Culture in Svidnìk), Prof. Rastislava Stoličná (Slovak Academy of Sciences, Institute of Ethnology, Bratislava) and Paweł Kamiński (Vienna) for their help in the literature search, Dr. Ingvar Svanberg (Uppsala) for his helpful comments and to Ms Zuzana Melečková and Ms Katarína Zlkovanová for their data on wild food plant use in Slovakia.

References

1. Turner NJ, Łuczaj ŁJ, Migliorini P, Pieroni A, Dreon AL, Sacchetti LE, et al. Edible and tended wild plants, traditional ecological knowledge and agroecology. Cr Rev Plant Sci. 2011;30(1–2):198–225.
2. Tardio J, Pardo-De-Santayana M, Morales R. Ethnobotanical review of wild edible plants in Spain. Bot J Linn Soc. 2006;152(1):27–71.
3. Bonet MÀ, Vallès J. Use of non-crop food vascular plants in Montseny biosphere reserve (Catalonia, Iberian Peninsula). Int J Food Sci Nutr. 2002;53(3):225–248.
4. Tardío J, Pascual H, Morales R. Wild food plants traditionally used in the province of Madrid, central Spain. Econ Bot. 2005;59(2):122–136.
5. Rivera D, Obon C, Inocencio C, Heinrich M, Verde A, Fajardo J, et al. The ethnobotanical study of local Mediterranean food plants as medicinal resources in Southern Spain. J Physiol Pharmacol. 2005;56(1 suppl):97–114.
6. Parada M, Carrió E, Vallès J. Ethnobotany of food plants in the Alt Emporda region (Catalonia, Iberian Peninsula). J Appl Bot Food Qual. 2011;84(1):11–25.
7. Menendez-Baceta G, Aceituno-Mata L, Tardío J, Reyes-García V, Pardo-de-Santayana M. Wild edible plants traditionally gathered in Gorbeialdea (Biscay, Basque Country). Genet Resour Crop Evol. 2011;59(7):1329–1347.
8. Pardo-de-Santayana M, Tardío J, Blanco E, Carvalho A, Lastra J, San Miguel E, et al. Traditional knowledge of wild edible plants used in the northwest of the Iberian Peninsula (Spain and Portugal): a comparative study. J Ethnobiol Ethnomed. 2007;3(1):27.
9. Pieroni A, Nebel S, Quave C, Münz H, Heinrich M. Ethnopharmacology of liakra: traditional weedy vegetables of the Arbëreshë of the Vulture area in southern Italy. J Ethnopharmacol. 2002;81(2):165–185.
10. Guarrera PM. Food medicine and minor nourishment in the folk traditions of Central Italy (Marche, Abruzzo and Latium). Fitoterapia. 2003;74(6):515–544.
11. Pieroni A, Nebel S, Santoro RF, Heinrich M. Food for two seasons: culinary uses of non-cultivated local vegetables and mushrooms in a south Italian village. Int J Food Sci Nutr. 2005;56(4):245–272.

12. Guarrera PM, Salerno G, Caneva G. ood, flavouring and feed plant traditions in the Tyrrhenian sector of Basilicata, Italy. J Ethnobiol Ethnomed. 2006;2:37.
13. Leonti M, Nebel S, Rivera D, Heinrich M. Wild gathered food plants in the European Mediterranean: a comparative analysis. Econ Bot. 2006;60(2):130–142.
14. Della A, Paraskeva-Hadjichambi D, Hadjichambis AC. An ethnobotanical survey of wild edible plants of Paphos and Larnaca countryside of Cyprus. J Ethnobiol Ethnomed. 2006;2:34.
15. Dogan Y, Baslar S, Ay G, Mert HH. The use of wild edible plants in western and central Anatolia (Turkey). Econ Bot. 2004;58(4):684–690.
16. Ertuğ F. Wild edible plants of the Bodrum area (Mugla, Turkey). Turk J Bot. 2004;28(1–2):161–174.
17. Pieroni A. Local plant resources in the ethnobotany of Theth, a village in the Northern Albanian Alps. Genet Resour Crop Evol. 2008;55(8):1197–1214.
18. Redzic S. Wild edible plants and their traditional use in the human nutrition in Bosnia-Herzegovina. Ecol Food Nutr. 2006;45(3):189–232.
19. Schunko C, Vogl CR. Organic farmers use of wild food plants and fungi in a hilly area in Styria (Austria). J Ethnobiol Ethnomed. 2010;6(1):17.
20. Couplan F. Le régal végétal: plantes sauvages comestibles. Flers: Equilibres; 1989. (vol 1).
21. Hedrick UP, editor. Sturtevant's notes on edible plants. New York: Dover Publications; 1919.
22. Tanaka C. Tanaka's cyclopedia of edible plants of the world. Tokyo: Keigaku Publishing; 1976.
23. Kunkel G. Plants for human consumption: an annotated checklist of the edible phanerogams and ferns. Koenigstein: Koeltz Scientific Books; 1984.
24. Maurizio A. Pożywienie roślinne i rolnictwo w rozwoju dziejowym. Warsaw: Kasa Mianowskiego; 1926.
25. Redzić S. Use of wild and semi-wild edible plants in nutrition and survival of people in 1430 days of siege of Sarajevo during the war in Bosnia and Herzegovina (1992–1995). Coll Antropol. 2010;34(2):551–570.
26. Łuczaj Ł. Archival data on wild food plants used in Poland in 1948. J Ethnobiol Ethnomed. 2008;4(1):4.
27. Łuczaj Ł. Changes in the utilization of wild green vegetables in Poland since the 19th century: a comparison of four ethnobotanical surveys. J Ethnopharmacol. 2010;128(2):395–404.
28. Łuczaj Ł, Szymański WM. Wild vascular plants gathered for consumption in the Polish countryside: a review. J Ethnobiol Ethnomed. 2007;3(1):17.
29. Łuczaj Ł. Dziko rosnące rośliny jadalne użytkowane w Polsce od połowy XIX w. do czasów współczesnych. Etnobiologia Polska. 2011;1:57–125.
30. Kalle R, Sõukand R. Historical ethnobotanical review of wild edible plants of Estonia (1770s–1960s). Acta Soc Bot Pol. 2012;81(4):271–281.
31. Slavkovský P. Zberné a koristné hospodárstvo. In: Botík J, Slavkovský P, editors. Encyklopédia ľudovej kultúry Slovenska. Bratislava: Veda; 1995. p. 342–343.
32. Stoličná R. Alternatívne zdroje rastlinnej stravy v strednej Európe. Slovenský Národopis. 1997;45(3):285–294.
33. Stoličná R. Alternative sources of food in Central Europe. In: Lysaght P, editor. Food from nature. Uppsala: The Royal Gustavus Adolphus Academy for Swedish Folk Culture; 2000. p. 195–203. (Acta Academiae Regiae Gustavi Adolphi LXXI).
34. Dillnbergerová S. Domestic and commercial uses of the products of wild nature in Slovakia. In: Lysaght P, editor. Food from nature. Uppsala: The Royal Gustavus Adolphus Academy for Swedish Folk Culture; 2000. p. 204–208. (Acta Academiae Regiae Gustavi Adolphi LXXI).
35. Počet obyvateľov SR k 30. septembru 2011 [Internet]. 2012 [cited 2012 Nov 5];
36. Dostál J, editor. Flóra Slovenska. Bratislava: Slovenská Akadémia Vied; 1966.
37. Čeřovský J, Feráková V, Holub J, Maglocký Š, Prochazká F. Červena kniha ohrozených a vzácnych druhov rastlín a živočíchov SR a ČR. Vyššie rastliny. Bratislava: Príroda; 1999. (vol 5).
38. Mjartan J, editor. Jozef Ľudovít Holuby. Národopisné práce. Bratislava: Vydavateľstvo Slovenskej Akadémie Vied; 1958.
39. Holuby JL. Domáce lieky ľudu slovenského. Letopis Matice Slovenskej. 1872;9–10:41–55.
40. Holuby JL. Die gewöhnlichsten wildwachsenden Genusspflanzen des Trencsiner Comitates. Verhandlungen des Vereins für Natur-und Heilkunde zu Pressburg, Neue Folge. 1891;7:91–105.
41. Holuby JL. Aus des Botanik slowakischer Kinder des Trentschiner Komitates in Ungarn. Deutsche Botanische Monatsschrift. 1896;19(8–9):126–131.
42. Holuby JL. Z vojnových čias v Pezinku. Kalendár Slov týždenníka na rok 1918. Budapest: 1917.
43. Holuby JL. Rastlinné domáce lieky a poviery slovenské. Národopisný Všetnìk Československi. 1921;15:68–96.
44. Holuby JL. Lubosti a mrzutosti botanikove. Slovenské pohľady. 1923.
45. Markuš M. Ľudová strava. In: Mjartan J, editor. Banícka Dedina Žakaro. Bratislava: Vydavateľstvo Slovenskej Akadémie Ved; 1956. p. 283–324.
46. Markuš M. Zberné hospodárstvo na Horehroní. Slovenský Národopis. 1961;9(2):190–243.
47. Markuš M. Interetnické vzťahy v ľudovej strave na okolí Bardejova. Nové Obzory. 1973;15:393–439.
48. Markuš M. Stravá. In: Mjartan J, editor. Horenhronie. Bratislava: Veda; 1974. p. 119–158.
49. Markuš M. Ľudová strava. In: Podolák J, editor. Horna Cirochá: vlastivedná monografia zátopovej oblasti. Košice: Východoslovenské vydateľstvo; 1985. p. 351–374.
50. Babylon J. Prvá kuchárska kniha w Slovenskej reči. 1870.
51. Dillnberger S. Roslynni produkty v stravi shidnoslovac'kogo naselennâ. Annales Musei Culturae Ukrainiensis Svidnik. 1988;15:125–133.
52. Kurjaková E. Czym żywili się mieszkańcy wsi na południe od Babiej Góry. In: Kociołek J, editor. Kalendarz 2004 – z informacjami co się dawniej jadło i piło i z czego się pod Babią Górą strawę robiło oraz z przepisami kulinarnymi dawnymi i nowszymi. Zawoja: Stowarzyszenie Gmin Babiogórskich; 2004. p. 14–26.
53. Mišík S. Dačo z ludového herbára na Hnilci. Časopis Museálnej Slovenskiej Spoločnosti. 1905;8:90–92.
54. Olejník J. O niektorých liečivých rastlinách v oblasti Vysokých Tatier. Nové Obzory. 1968;10:365–385.
55. Olejník J. Ľud pod Tatrami. Martin: Vydateľstvo Osveta; 1978.
56. Olejník J. Strava lesných robotníkov v podmienkach Vysokých Tatier v minulosti a v podmienkach správy Tatranského Národného Parku v súčasnosti. Nové Obzory. 1991;32:149–161.
57. Šipka M. Liečivé rastliny Hornej Cirochy. Annales Musei Culturae Ukrainiensis Svidnik. 1985;12:457–488.
58. Urbancová V, Burlasová S, editors. Zo života a bojov ľudu Uhrovskej doliny. Bratislava: Slovenská Akadémia Vied; 1977.
59. Svanberg I. Fabulat, plagiat och spökuppgifter: att använda berättande källor inom etnobiologi. In: Tunón H, Dahlström A, editors. Nycklar till kunskap: om människans bruk av nature. Stockholm: Kungl. Skogs- och Lantbruksakademien; 2010. p. 121–136.
60. Svanberg I. The use of lichens for dyeing candles: ethnobotanical documentation of a local Swedish practice? Svenska Landsmål och Svenskt Folkliv. 1998;324:133–139.
61. Łuczaj ŁJ. Plant identification credibility in ethnobotany: a closer look at Polish ethnographic studies. J Ethnobiol Ethnomed. 2010;6(1):36.

62. Tutin TG, Heywood VH, Burges NA, Moore DM, Valentine DH, Walters SM, et al., editors. Flora Europaea. Cambridge: Cambridge University Press; 1964. (vol 1–5).
63. Łuczaj Ł. Dzikie rośliny jadalne używane w okresach niedoboru żywności we wschodniej części Karpat (powiaty Krosno, Sanok, Lesko, Nadwórna, Kosów i Kołomyja) według ankiety szkolnej z 1934 roku. In: Łuczaj Ł, editor. Dzikie rośliny jadalne - zapomniany potencjał przyrody. Bolestraszyce: Arboretum i Zakład Fizjografii w Bolestraszycach; 2008. p. 161–181.
64. Denes A, Papp N, Babai D, Czúcz B, Molnár Z. Wild plants used for food by Hungarian ethnic groups living in the Carpathian Basin. Acta Soc Bot Pol. 2012;81(4):381–396.
65. Butură V. Enciclopedie de etnobotanică românescă. Bukareszt: Editura Ştiinţifică şi Enciclopedică; 1979.
66. Péntek J, Szabó A. Ember és növényvilág. Kalotaszeg növényzete és népi növényismerete. Bukarest: Kriterion Könyvkiadó; 1985.
67. Gunda B. Plant gathering in the economic life of Eurasia. Southwestern Journal of Anthropology. 1949;5(4):369–378.
68. Varhol N. Roslini v narodnih povir'âh Rusinìv-Ukraïncìv Prâšìvščini. Prešov: Excol; 2002.
69. Čajánková E. Život a kultúra rožkovianskych Cigáňov. Slovenský Národopis. 1954;2(3–4):285–308.
70. Łuczaj Ł, Kujawska M. Botanists and their childhood memories: an underutilized expert source in ethnobotanical research. Bot J Linn Soc. 2012;168(3):334–343.

Thirteen years on the hunt for *Buxbaumia viridis* in the Czech Republic: still on the tip of the iceberg?

Eva Holá[1,2]*, Jan Vrba[2], Renata Linhartová[2], Eva Novozámská[3], Magda Zmrhalová[4], Vítězslav Plášek[5], Jan Kučera[1]

[1] Department of Botany, University of South Bohemia, Branišovská 31, 370 05 České Budějovice, Czech Republic

[2] Nature Conservation Agency of the Czech Republic, Kaplanova 1931/1, 148 00 Prague 11, Czech Republic

[3] Prague Botanic Garden, Nádvorní 134, 171 00 Prague 7, Czech Republic

[4] Regional Museum in Šumperk, Hlavní tř. 22, 787 31 Šumperk, Czech Republic

[5] Department of Biology and Ecology, University of Ostrava, Chittussiho 10, 710 00 Ostrava, Czech Republic

Abstract

Rare epixylic moss *Buxbaumia viridis*, which is one of the "Annex II" species of the European "Habitat directive", has been recorded at 124 localities in the Czech Republic in course of the last 13 years. Most of them were discovered in the last five years following a dedicated search at both historical sites of occurrence and new localities with putatively suitable habitat conditions. The recent and historical area of occupancy and extent of occurrence are not obviously different, although most of the recent localities are concentrated in the Western Carpathians and the Hrubý Jeseník Mts. Surprisingly, only 38% of the localities are located in natural forest habitats, of which herb-rich and acidophilous beech forests were among the most commonly inhabited ones. The rest of occurrences were recorded in non-natural forests (habitats strongly influenced or created by man), particularly the coniferous forest plantations. Sufficient amount of decaying wood of the advanced decay stages, as well as sufficient and constant humidity are crucial prerequisites for the occurrence of *B. viridis* in both types of forests. True epixylic moss *Herzogiella seligeri* and the liverwort *Chiloscyphus profundus* were recorded as the most common associated species, while the rest of commonly co-occurring species were facultatively epixylic, ground or ubiquitous bryophytes; other specialized and rare epixylic mosses or liverworts were only rarely recorded. We also designed a potential distribution model for *B. viridis* based on the distribution of habitats most commonly occupied by the studied moss. Despite the simplicity of the model, its close match with the recent distribution in the Western Carpathians, the Hrubý Jeseník Mts. and Šumava Mts. supports its relevance for the real distribution of *B. viridis*.

Keywords: *Buxbaumia viridis*; Czech Republic; epixylic bryophyte; the "Habitat directive"; threatened bryophyte

Introduction

Although the Central European landscape is extensively forested, most of these forests are strongly impacted by human activity. In the Czech Republic, the area of forest has been reported to account for about 35% but only a third of them (about 10% of the country) are considered to have near-natural composition [1]. Nature-like old-growth forests became extremely rare and heavily fragmented in whole Europe [2]. The degradation and fragmentation of natural forests poses a significant threat to biodiversity, as shown in several comparative studies [3,4]. One of the key factors affecting the biodiversity in managed production forests is the complete removal of wood during harvesting [5,6]. Decaying wood in different stages of decay is a substratum for many invertebrates, fungi, lichens and bryophytes. The threatened nature of epixylic bryophytes has been reflected in the red lists of most European countries [2,7].

A "flagship species" of rare and endangered epixylic mosses is *Buxbaumia viridis*, which received Europe-wide attention as one of the "Annex II" species of the "Habitat directive" [8]. It is a tiny and inconspicuous moss, which can be traced only in its sporophytic stage due to the extremely reduced gametophyte, and thus makes its recording hard even for an experienced bryologist. Its dioicy, resulting in low fertilization rate together with low establishment and poor competitive abilities of the minute gametophyte have been speculated to be among the main reasons for the rarity of *B. viridis* [9]. In addition, the species is sensitive to desiccation of its substratum and the insufficient air humidity or low precipitation during the summer months restrict spore production and germination [9–11].

* Corresponding author. Email: eva.neurazy@gmail.com

Handling Editor: Zygmunt Kącki

The occurrence of *B. viridis* is generally restricted to decaying wood, especially in advanced phases of decay [12]. Its detailed ecological requirements have however only been investigated in a few studies [9,10,13], focused mainly on substratum preference. The species is assumed to grow mainly in old-growth (primeval) forests [13].

We have summarized the information about the current occurrences of *B. viridis* in the Czech Republic, which has accumulated over the last thirteen years (2001–2013) and compared the recent and historical distribution pattern. The ecological information noted by the recorders, together with data from habitat mapping were used to develop a model predicting its distribution over the whole area of the country.

Material and methods

Buxbaumia viridis has been mapped by the Nature Conservation Agency of the Czech Republic since 2001. Positively mapped localities (i.e. localities with *B. viridis*) were visited at least once (see Appendix S1; number of visits is dependent on the availability of sites, the number of surveyor and on the budget for the respective year). The mapping was usually carried out during autumn season, because sporophytes of *B. viridis* were easily visible. Presence of dead wood, forest type, and forest age were considered as main features for selection of areas. In addition, the vast majority of historical localities [13–15] were included in the survey. In the case of the positive discovery at a locality, neighboring logs (typically within ca. 500 m) were particularly thoroughly searched for additional occurrences (microlocalities). At each microlocality, the number of capsules was recorded together with substratum data (the identity of decaying wood where detectable, decay degree) and co-occurring species. The decay degree was measured using a semi-quantitative scale modified from Söderström [16] (Tab. 1). Co-occurring species were defined as those occurring within 15 cm from the capsules of *B. viridis*; the nomenclature of bryophytes follows the current checklist [17]. Bryophytes were divided into four ecological groups: epiphytes (species commonly found on living trees), ground flora (most often recorded on the ground), true epixylic species (exclusively found on rotting wood) and facultative epixylic species (rotting wood is only sub-optimal substratum). The assignation of the species to these groups followed other studies [18,19] and own experience.

Tab. 1 Scale for identification of decay degree [16].

Decay degree	Description of decay degree
1	freshly fallen log, hard, whole, bark intact, branches present
2	log solid, wood hard, more than 50% of bark present
3	some patches of soft wood, more than 50% of bark missing
4	wood softening, outline retained, bark missing
5	log outline slightly deformed
6	pieces of soft wood, log outline still distinguishable
7	only log core remains
8	log outline indeterminable, humification 100%, no evidence of hard wood

Localities were defined as one or more logs of *B. viridis* within the radius of 500 m. Except for a few exceptions (Appendix S1), only one habitat type occurred at one locality. For each locality, mean altitude, and for each microlocality distance from the nearest watercourse were assessed from topographic maps. The distance from watercourse was classified into 4 categories: (***i***) up to 10 m; (***ii***) 11–100 m; (***iii***) 101–500 m; (***iv***) >500 m, i.e. no watercourse.

Another type of data assigned to localities was the habitat information. These data were extracted from the geographic information system (GIS) layer of the habitat mapping of the Czech Republic [20]. The habitats were classified according to habitat catalogue [21]. This classification utilizes primarily the formation–vegetation (physiognomic) approach and, secondarily, the floristic approach. The habitats of supposedly higher importance for nature conservation are termed natural habitats. In contrast, the non-natural habitats are those of presumably lower importance for biodiversity or nature conservation and are strongly influenced or created by humans [21]. In addition, syntopic combinations (mosaics) of natural and non-natural habitats can occur. If one of the habitats from the mosaic was non-natural, then the entire mosaic was considered as a non-natural habitat. In the areas of supposedly low frequency of natural habitats, only the natural habitats were recorded and mapped (so-called contextual mapping; [22]), whereas the non-natural habitats were not mapped and were merged in the resulting GIS layer in a single category, labelled "−1".

For the analysis of distribution, we defined the area of occupancy as the number of the occupied grid squares of the Central European floristic grid [23], with the squares 5 × 3 geographical minutes. Extent of occurrence is defined as the area contained within the shortest continuous imaginary boundary, which can be drawn to encompass all the known, inferred or projected sites of present occurrence of a species [24]. The distance of the farthest points was set to be 40 km [25]. The area of the occupied grid squares and extent of occurrence were calculated using ArcGis ver. 9 [26].

We combined the GIS layer of habitat mapping [20] and the Corine land cover layer [27] with our results of habitat association frequency and altitudinal distribution of *B. viridis* for the construction of the map of its potential distribution. The Corine land cover refers to the European programme establishing a computerized inventory on land cover of the 27 European Union member countries and other European countries, at an original scale of 1:100 000, using 44 classes with five main categories – artificial surfaces, agricultural areas, forest and other semi-natural areas, wetlands, and water bodies. Forests are subdivided into three categories – broad-leaved, coniferous, and mixed forests. Most frequently occupied habitats (i.e., habitats of interest, habitats in which the occurrence of *B. viridis* was more frequent than five times, see "Results") and their mosaics (only mosaics wherein the representation of habitats of interest was higher than 50%) within altitudes 300–1100 m a.s.l. were included in the model. The potential distribution map was constructed as the relative proportion of habitat area (only habitats of interest) per each square. For the non-natural habitats merged in the category "−1" (one of the habitats of interest) in the layer of the habitat mapping, the information from the Corine land cover layer was used to select coniferous and broad leaved-coniferous forests. Only these two types of "−1" habitats were used in the modeling of potential distribution of *B. viridis*. All other "−1" habitats were regarded as completely uninhabitable for *B. viridis*.

The general linear model, calculated by R version 2.12 [28], was used to test the influence of the altitude, habitat type (natural vs. non-natural), habitat unit, degree of decay, and distance from watercourse on the sporophyte production, measured as the mean number of capsules per each locality. The mean number of sporophytes per each locality was log-transformed prior to analysis to improve normality.

Results

Historical and recent distribution

Buxbaumia viridis has to date been recorded on 167 localities in the Czech Republic (Fig. 1, Tab. 2, Appendix S1). One hundred twenty four localities (and 192 microlocalities) were recorded in the recent 13 years (2001–2013), of which 89 localities were newly found occurrences. Of the 78 historical localities (recorded before 2001 [13–15]), 35 were checked with positive results, while at 21 localities *B. viridis* was not re-found and 22 historical records were not verified due to insufficient capacities, inaccurate localization etc. Since 2001, 242 sites were explored in total (i.e., both revision of historical localities and survey of newly selected ones), *B. viridis* was not found at 118 of them (Fig. 2).

Although the number of recent localities is higher than the number of historical localities, the area and extent of occurrence of historical localities reached higher values (Tab. 2). Altitudinal distribution of the localities spanned 250–1200 m, of which most were recorded at the altitudes between 600–900 m a.s.l. (Fig. 3). 42% of all microlocalities (N = 192) were situated within 10 m from a watercourse, 32% microlocalities within 11–100 m from a watercourse, 23% microlocalities within 101–500 m from a watercourse and only 4 microlocalities were farther than 500 m from watercourse (Fig. 4).

Habitat requirements and association, co-occurring species

The recorded substratum of *B. viridis* was primarily the decaying wood of conifers in late decay phases (degree 5, 6, 7 or 8; Fig. 5). Of the 148 cases (microlocalities), when the substratum could be determined, 88% were conifers (mainly *Picea abies* and *Abies alba*) and 12% were broad leaved trees, mainly beech (*Fagus sylvatica*; 8%), 1–2 microlocalities were

Tab. 2 The area of occupancy (AOO) and the extent of occurrence (EOO) for recent and historical localities were summarized.

Recent localitis	*N*	AOO	EOO (km²)
	124	74	4445
Historical localitis	*N*	AOO	EOO (km²)
	78	66	4359

The grid squares are defined by the project "Kartierung der Flora Mitteleuropas" (KFME), with the squares of 5′ longitude and 3′ latitude.

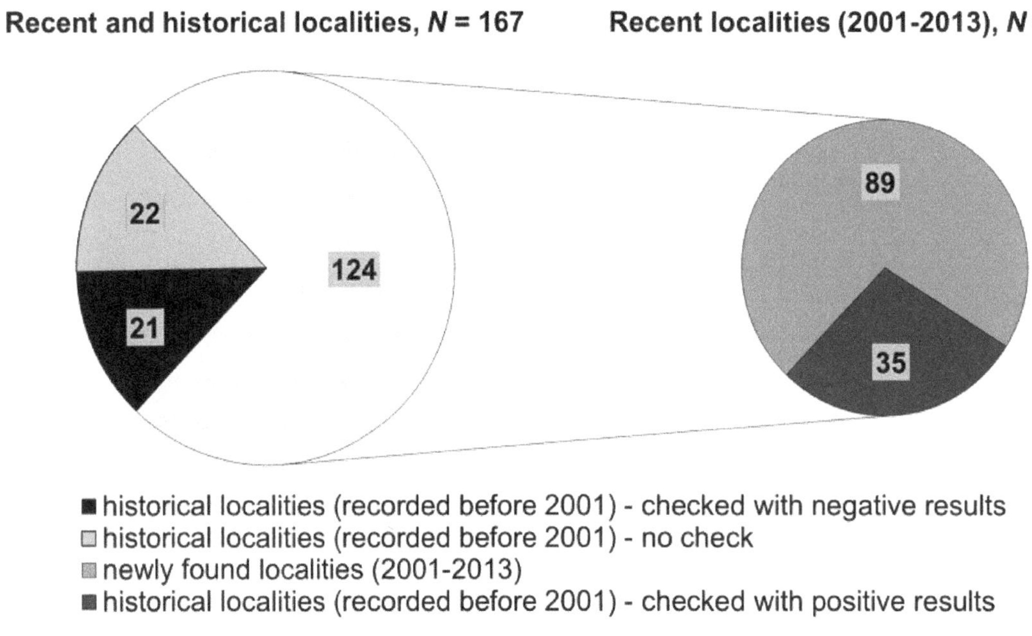

Fig. 1 The number of historical and recent localities.

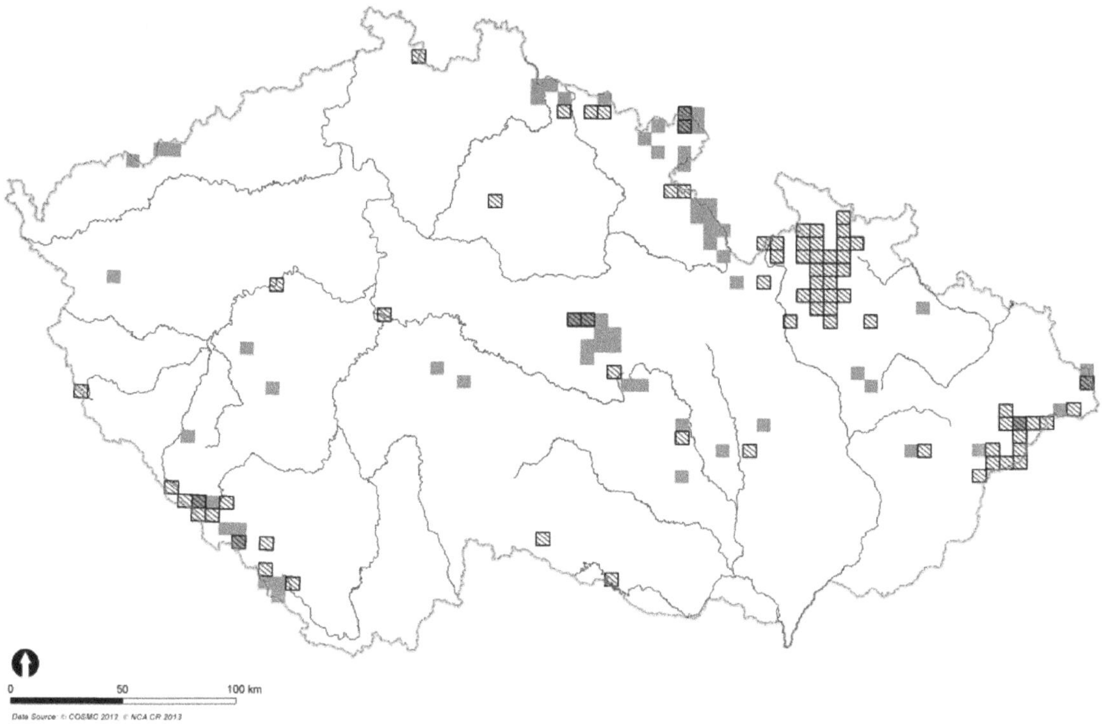

Fig. 2 Map of explored area. Hatched squares show positively explored area; grey squares are negatively explored area. The grid squares are defined by the project "Kartierung der Flora Mitteleuropas" (KFME), with the squares of 5′ longitude and 3′ latitude.

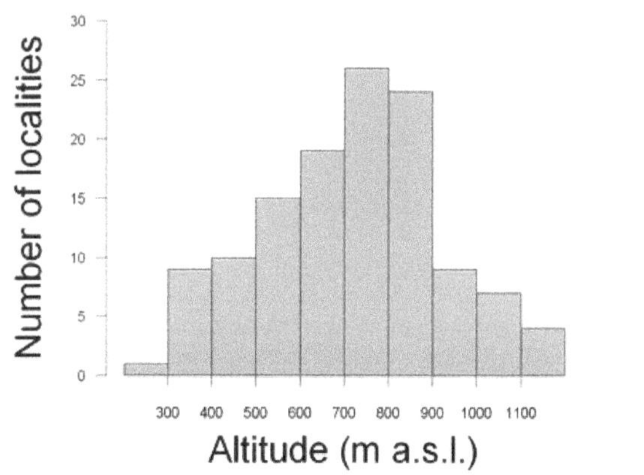

Fig. 3 Altitudinal distribution for recent localities of *B. viridis* from the Czech Republic; $N = 124$.

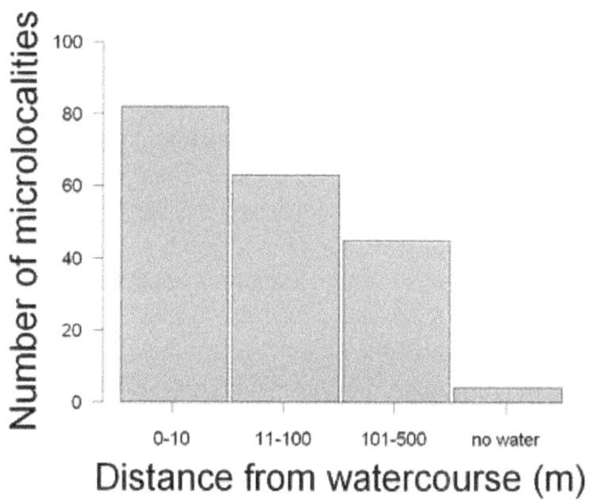

Fig. 4 Distance of *B. viridis* microlocalities from a watercourse; $N = 192$.

recorded for alder (*Alnus glutinosa*), ash (*Fraxinus excelsior*) and birch (*Betula pendula*).

The co-occurring species and their frequency at recent microlocalities are summarized in Tab. 3. Altogether, 68 species were recorded (24 liverworts and 44 mosses), of which the most common (growing at more than 40% of microlocalities) were liverwort *Chiloscyphus profundus* and mosses *Herzogiella seligeri*, *Rhizomnium punctatum*, *Dicranum scoparium* and *Tetraphis pellucida*. Forest ground species dominated among co-occurring species (40 species), whereas only 9 species were classified as true epixylic species and 17 as facultative epixylic species, 3 as the epiphytes. Nevertheless, the most common accompanying species (see above) were true or facultative epixylic species.

The mean number of sporophytes per one microlocality varied from 1 to 108 capsules (mean = 8, median = 3), most microlocalities (80%) and localities (70%) had only up to 10 sporophytes. However, number of sporophytes

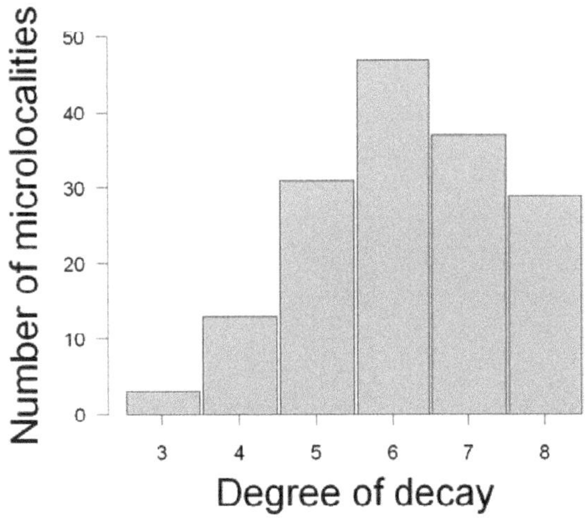

Fig. 5 Frequency of different decay stages of logs (microlocalities) with presence of *B. viridis*; *N* = 160 (modified from Söderström [16] – (3) some patches of soft wood, more than 50% of bark missing; (4) wood softening, outline retained, bark missing; (5) log outline slightly deformed; (6) pieces of soft wood, log outline still distinguishable; (7) only log core remains; (8) log outline indeterminable, humification 100%, no evidence of hard wood).

on microlocalities varied among years, e.g. at the locality Halenkov (195 sporophytes in 2001, 6 sporophytes in 2011), Senice – Val. Kyčera (7 sporophytes in 2002, 64 sporophytes in 2011), Vrchovištní pot. (94 sporophytes in 2001, 14 sporophytes in 2009). The numbers for all localities are listed in the Appendix S1. The average number of sporophytes per locality showed no relation to altitude (*F* = 2.3924, *P* = 0.1254), habitat type (*F* = 0.0432, *P* = 0.8357), habitat unit (*F* = 1.6006, *P* = 0.1463), decay degree (*F* = 3.4437, *P* = 0.0667), or watercourse (*F* = 2.2432, *P* = 0.1377). It seems that variation among years is an important predictor of persistence of *B. viridis* on locality. However, our data do not allow us this type of analysis because we have not accumulated sufficient amount of repeated observations from each locality.

Only 38% of recent localities were situated in 6 habitats units and 17 subunits, which are classified as natural according to habitat mapping GIS layer (Tab. 4). The remaining localities were recorded in 3 units and 8 subunits of habitats classified as non-natural. The most common natural habitats were beech forests, particularly herb-rich and acidophilus beech forests, the most common non-natural habitats were non-specified habitats of contextual mapping and habitats strongly influenced or created by man, particularly the forest plantations of coniferous trees.

Potential distribution map

The potential distribution model is shown in Fig. 6. The model and recent distribution show a close match mainly in the Western Carpathians and the Hrubý Jeseník Mts., where there are frequent records of *B. viridis* as well as the

Tab. 3 Co-occurring species at microlocalities of *B. viridis*.

Species	% of co-occurring species (*N* = 169)	Ecological category
Liverworts		
Chiloscyphus profundus	56	TX
Cephalozia bicuspidata	31	FX
Lepidozia reptans	27	TX
Blepharostoma trichophyllum	16	FX
Plagiochila porelloides	11	G
Scapania nemorea	8	FX
Mosses		
Herzogiella seligeri	70	TX
Rhizomnium punctatum	50	FX
Dicranum scoparium	45	FX
Tetraphis pellucida	40	TX
Dicranum montanum	28	E
Hypnum cupressiforme	21	FX
Sanionia uncinata	17	FX
Plagiothecium curvifolium	12	G
Dicranella heretomalla	12	G
Polytrichum formosum	12	G
Pohlia nutans	11	G
Atrichum undulatum	9	G
Dicranodontium denudatum	7	TX
Plagiomnium affine	6	G
Plagiothecium laetum	6	G
Sciuro-hypnum reflexum	6	G

Bryophytes with low frequency (less than 5%). Liverworts: *Bazzania trilobata* (ecological category: G), *Calypogeia azurea* (G), *C. integristipula* (G), *Crossocalyx hellerianus* (TX), *Chiloscyphus coadunatus* (G), *Ch. cuspidatus* (G), *Diplophyllum obtusifolium* (G), *Liochlaena lanceolata* (FX), *Lophozia ascendens* (TX), *Lophozia ventricosa* (FX), *Marchantia polymorpha* (G), *Nowellia curvifolia* (TX), *Orthocaulis attenuatus* (FX), *Ptilidium pulcherrimum* (E), *Riccardia latifrons* (TX), *Riccardia palmata* (TX), *Schistochilopsis incisa* (FX), *Tritomaria exsectiformis* (FX). Mosses: *Aulacomnium androgynum* (FX), *Brachythecium rivulare* (G), *B. rutabulum* (G), *B. salebrosum* (FX), *B. velutinum* (FX), *Bryum moravicum* (E), *Buxbaumia aphylla* (G), *Cirriphyllum piliferum* (G), *Dicranella subulata* (G), *Dichodontium pellucidum* (G), *Eurhynchium angustirete* (G), *Hylocomium splendens* (G), *Mnium hornum* (G), *M. spinosum* (G), *Plagiomnium cuspidatum* (FX), *P. undulatum* (G), *Plagiothecium curvifolium* (G), *P. nemorale* (G), *P. succulentum* (G), *Pleurozium schreberi* (G), *Pogonatum urnigerum* (G), *Polytrichum juniperinum* (G), *P. longisetum* (G), *Rhynchostegium riparioides* (G), *Rhytidiadelphus loreus* (G), *Sciuro-hypnum curtum* (G), *Sphagnum girgensohnii* (G), *Thuidium tamariscinum* (G). Ecological categories: E – epiphyte, FX – facultative epixylic species, G – ground species, TX – true epixylic species and.

high modeled probability of its occurrence (cf. Fig. 2 and Fig. 6). The high probability of occurrence according to the model is also reflected in at least several recent records of *B. viridis* in Moravian Karst, Giant Mts. and Šumava Mts. On the other hand, the relatively high modeled probability of occurrence contrasts with only two positive recent records

Tab. 4 Representation of habitats on recent localities of *Buxbaumia viridis* (*N* = 121).

Forest habitat	*N*	Habitat unit	*N*	Habitat subunit	*N*
Natural	46	Alluvial forests (L2)	2	Ash-alder alluvial forests (L2.2)	2
		Oak-hornbeam forests (L3)	2	Hercynian oak-hornbeam forests (L3.1)	2
		Ravine forests (L4)	3	Ravine forests (L4)	3
		Beech forests (L5)	24	Herb-rich beech forests (L5.1)	12
				Montane sycamore-beech forests (L5.2)	3
				Acidophilous beech forests (L5.4)	9
		Spruce forests (L9)	4	Montane *Calamagrostis* spruce forests (L9.1)	2
				Bog and waterlogged spruce forests (L9.2)	1
				Montane *Athyrium* spruce forests (L9.3)	1
		Mosaic	11	Ash-alder alluvial forests (L2.2) + ravine forests (L4)	1
				Ravine forests (L4) + chasmophytic vegetation of siliceous cliffs and boulder screes (S1.2)	2
				Ravine forests (L4) + chasmophytic vegetation of calcareous cliffs and boulder screes (S1.1)	1
				Herb-rich beech forests (L5.1) + chasmophytic vegetation of siliceous cliffs and boulder screes (S1.2)	2
				Herb-rich beech forests (L5.1) + ravine forests (L4)	2
				Herb-rich beech forests (L5.1) + chasmophytic vegetation of calcareous cliffs and boulder screes (S1.1)	1
				Acidophilous beech forests (L5.4) + herb-rich beech forests (L5.1)	1
				Herb-rich beech forests (L5.1) + forest springs without tufa formation (R1.4)	1
Non-natural	75	Contextual mapping (−1)	41	Contextual mapping (−1)	41
		Habitats strongly influenced or created by man (X)	27	Anthropogenic areas with sparse vegetation (X6)	1
				Forest plantations of coniferous trees (X9A)	22
				Forest clearings (X10)	4
		Mosaic	7	Ash-alder alluvial forests (L2.2) + forest plantations of coniferous trees (X9A)	3
				Forest plantations of coniferous trees (X9A) + herb-rich beech forests (L5.1)	2
				Forest plantations of coniferous trees (X9A) + forest clearings (X10)	1
				Forest clearings (X10) + forest springs without tufa formation (R1.4)	1

Mainly microlocalities of one locality belonged to same habitat type. Only a few microlocalities of three localities differed in habitat type (for more details see Appendix S1), this localities were not included. Most frequented habitat subunits (habitats of interest) are highlighted.

in the Orlické hory Mts., although we paid an adequate effort to the recent focused search for the species (cf. Fig. 2 and Fig. 6). The military area Brdy remains among regions with the high-modeled probability of occurrence and inadequate recent exploration.

Discussion

Historical and recent distribution

The considerable amount of recent localities, especially in the Hrubý Jeseník in the last five years, cannot be interpreted in terms of the recent spread of the species. Rather, we could see the increasing intensity of the exploration, together perhaps with the increasingly better search strategy and knowledge of ecology, resulting in the more focused search of the surveyors. In addition, *Buxbaumia viridis* was re-found at nearly a half of the historical localities (44%), which convincingly shows that the species may persist at the localities for a long time, provided that sufficient amount of suitable deadwood is retained, as most of the historical localities date back to or precede the 1950s. Even though, *B. viridis* may be still under-recorded, as the rate of its potentially suitable habitats, visible only in sporophyte stage to the number of surveyors, is still far too high. On the other hand, the survey density and success rate in finding is probably highest in the Czech Republic in comparison with neighboring countries. About 95 recent localities have been found in significantly larger Germany [29] and 85 were reported from Austria [30], and only 15 localities were recorded in Slovakia [31]. In neighboring Poland, the current state of the population is less known. The species was reported from tens of localities over a wide geographic range, discovered mainly in the 19th century [32]. Most occurrences were reported from the Western Carpathians, as well as in the Sudety Mts. and the Sudety foothills [32]. Nevertheless, even in Poland, the trend of recent discoveries of new localities has been observed [33–36].

Fig. 6 Potential distribution map of *B. viridis*, the relative proportion of habitat area (only habitats of interest) per each square are given. The grid squares are defined by the project "Kartierung der Flora Mitteleuropas" (KFME), with the squares of 5′ longitude and 3′ latitude.

Most of the recent localities are situated in the Western Carpathians and the Hrubý Jeseník Mts., while the historical localities were distributed more evenly in the Czech Republic. However, the recent and historical area of occupancy and extent of occurrence are not obviously different. The highest density of recent occurrences in the mountains of the northeastern part of Czech Republic agrees well with our model of potential distribution (Fig. 6). The reasons for this pattern might include the highest proportion of forests [37], ideal span of altitude, and generally more humid conditions, resulting from high precipitation averages [38] and favorable geomorphology. At the same time, the Western Carpathians and the Hrubý Jeseník Mts. are the best-explored areas for the species in the recent period (Fig. 2), in contrast to the historical state. Other well-researched mountain regions of Czech Republic (Orlické hory Mts., Giant Mts.) demonstrably support a much smaller number of recent localities (Fig. 2), which implies that the Western Carpathians and the Hrubý Jeseník Mts. indeed represent the core area for the occurrence of *B. viridis* in the Czech Republic. In addition, many new records in the Šumava Mts. during the two last years of focused survey suggest that the real situation here might be closer to that of Western Carpathians and the Hrubý Jeseník Mts. than previously thought. However, additional survey in the Šumava Mts. is needed to confirm this trend.

Habitat requirements and co-occurring species

The most important abiotic factor, which affects the occurrence of *B. viridis* is probably the humidity. The insufficient moisture was speculated to hamper the sporophyte production by the inhibition of gametangia and fertilization, or death of the protonema [9]. *B. viridis* also prefers the wood in advanced phases of decay, with high and fairly constant water content. Indeed, the majority of sites were located within 100 m from a stream. The exact pattern of the yearly course of air humidity at the sites, which is probably crucial for the successful establishment of gametophytes and development of sporophytes, is nevertheless much more difficult to trace. Our data unfortunately do not allow any scholarly analysis in this way, particularly because the GIS layer used for the retrieval of watercourses might not have included smaller streams, which nevertheless provide the sufficient source of humidity for the development of *Buxbaumia sporophytes.*

Epixylic bryophytes do not seem to be restricted to specific wood species [3,9]. This was confirmed for *B. viridis*, although Spruce logs were the most frequently occupied substratum. This may be due to the better holding capacity of brown-rot fungi that grow in conifer wood [9,18].

B. viridis tends to occur with rather common moss species. It prefers mosses to liverworts, and only two of the most frequently recorded mosses (*Herzogiella seligeri*, *Tetraphis pellucida*) are rather specifically associated with dead wood (true epixylic mosses) in the Czech Republic. True epixylic liverworts are generally good indicators of the high air-humidity, being moisture-demanding or intolerable of prolonged drought periods. However, *B. viridis* rarely if ever grows associated with such liverworts (*Crossocalyx hellerianus*, *Lophozia ascendens*, *Nowellia curvifolia*), which seems to be contradictory to the general demand on air humidity discussed above. Protonematal growth of *B. viridis* is very slow (0.5–3 mm after 70 days in cultivation) and does not seem to expand rapidly to the empty space [10]. Therefore, *B. viridis* is probably not successful in competitive interaction with bryophytes during the juvenile stage of the life cycle. The pleurocarpous mosses, which are commonly associated with the sporophytes of *B. viridis* (*Herzogiella seligeri*, *Hypnum cupressiforme*, *Sanionia uncinata*) may provide shadier and damper shelter for germination and protonematal growth of *B. viridis* than tiny liverworts.

Habitat association and potential distribution

B. viridis is generally considered to be an indicator species of old-growth (primeval) forests (e.g. [13,39]). Nevertheless, only 38% of the Czech localities are situated in natural forest habitats, while the rest is situated in managed forests, most typically in spruce plantations. It seems that the crucial factor is not habitat association (type of forest) but probably suitable humidity, typically near a watercourse with a sufficient amount of decayed wood. These conditions seem to be crucial for the occurrence of *B. viridis*. However, it is exactly these factors, which are difficult to retrieve, analyze and include to the distribution model. The microclimatic humidity is influenced by topography and substratum characteristics of the logs. The amount of dead wood is not specified in any of the existing databases (cf. habitat mapping [20] and Czech national phytosociological database – ČNFD [40]). Although only the frequency of the most commonly occupied habitat (which is definitely also linked to certain extent to these parameters) was included to the model, its match with the recent distribution shows that the model may be used as a basis for focused search for the new localities of *B. viridis*.

Acknowledgments

We thank Jan Košnar (University of Hradec Králové), Veronika Oušková, Iva Hönigová (Nature Conservation Agency) for heplful discussion of the manuscript and to all bryologists mainly Štěpán Koval, who participated in the monitoring. We acknowledge the financial support of the Nature Conservation Agency of the Czech Republic, Grant Agency of the University of South Bohemia in České Budějovice 138/2010/P, the project Institute of Environmental Technologies, reg. No. CZ.1.05/2.1.00/03.0100, supported by the "Research and development for innovations" operational program and project of the "National feasibility programme" of the Czech Republic LO1208.

Authors' contributions

The following declarations about authors' contributions to the research have been made: responsible for coordination of monitoring *Buxbaumia viridis* (since 2008), field work, data and GIS analysis, paper writing and manuscript preparation: EH; GIS analysis: JV; field work, manuscript preparation: EN, MZ, VP; coordination of monitoring *Buxbaumia viridis* (2000–2007), field work: RL; field work, paper writing and manuscript preparation: JK.

Supplementary material

The following supplementary material for this article is available online at http://pbsociety.org.pl/journals/index.php/asbp/rt/suppFiles/asbp.2014.015/0:

1. Appendix S1: historical and recent records of *Buxbaumia viridis* in the Czech Republic.

References

1. Miko L, Hošek M, editors. State of nature and landscape in the Czech Republic. Prague: AOPK ČR; 2009.
2. Hallingbäck T, Hodgetts N, editors. Mosses, liverworts, and hornworts: status survey and conservation action plan for bryophytes. Gland: IUCN/SSC Bryophyte Specialist Group; 2000.
3. Andersson LI, Hytteborn H. Bryophytes and decaying wood – a comparison between managed and natural forest. Ecography. 1991;14(2):121–130.
4. Vellak K, Paal J. Diversity of bryophyte vegetation in some forest types in Estonia: a comparison of old unmanaged and managed forests. Biodivers Conserv. 1999;8(12):1595–1620.
5. Riffell S, Verschuyl J, Miller D, Wigley TB. Biofuel harvests, coarse woody debris, and biodiversity – a meta-analysis. For Ecol Manage. 2011;261(4):878–887.
6. Madžule L, Brūmelis G, Tjarve D. Structures determining bryophyte species richness in a managed forest landscape in boreo-nemoral Europe. Biodivers Conserv. 2012;21(2):437–450.
7. Söderström L. Conservation biology of bryophytes. Lindbergia. 2006;3(1–2):24–32.
8. CECC. Directive 92/43/EEC of 21 May 1992 on the conservation of natural habitats and of wild fauna and flora [Internet]. 1992 [cited 2013 Nov 25].
9. Wiklund K. Phosphorus concentration and pH in decaying wood affect establishment of the red-listed moss *Buxbaumia viridis*. Can J Bot. 2003;81(6):541–549.
10. Wiklund K. Substratum preference, spore output and temporal variation in sporophyte production of the epixylic moss *Buxbaumia viridis*. J Bryol. 2002;24(3):187–195.
11. Wiklund K, Rydin H. Ecophysiological constraints on spore establishment in bryophytes. Func Ecol. 2004;18(6):907–913.
12. Dierssen K. Distribution, ecological amplitude and phytosociological characterization of European bryophytes. Berlin: Cramer in der Gebrüder Borntraeger Verlagsbuchhandlung; 2001. (Bryophytorum bibliotheca; vol 56).
13. Plášek V. The moss *Buxbaumia viridis* (Bryopsida, Buxbaumiaceae) in the Czech part of the Western Carpathians – distribution and ecology. In: Stebel A, Ochyra R, editors. Bryological studies in the Western Carpathians. Poznań: Sorus; 2004. p. 37–44.
14. Vondráček M. Prodromus der Moose des Böhmerwaldes (Bryopsida) I. (Sphagnaceae–Bryaceae). Plzeň: Západočeské muzeum; 1990. (Folia musei rerum naturalium Bohemiae occidentalis: botanica; vol 31).
15. Soldán Z. Buxbaumia viridis – a candidate of red lists of bryophytes. Bryonora. 1992;9:40–44.
16. Söderström L. Sequence of bryophytes and lichens in relation to substrate variables of decaying coniferous wood in Northern Sweden. Nord J Bot. 1988;8(1):89–97.
17. Kučera J, Váňa J, Hradílek Z. Bryophyte flora of the Czech Republic: updated checklist and red list and a brief analysis. Preslia. 2012;84:813–850.
18. Ódór P, van Hees AFM. Preferences of dead wood inhabiting bryophytes for decay stage, log size and habitat types in Hungarian beech forests. J Bryol. 2004;26(2):79–95.
19. Jansová I, Soldán Z. The habitat factors that affect the composition of bryophyte and lichen communities on fallen logs. Preslia. 2006;78(1):67–86.
20. AOPK ČR. Habitat mapping layer – electronic georeferenced database. Prague: AOPK ČR; 2012.

21. Chytrý M, Kučera T, Kočí M, Grulach V, Luetyk P, editors. Katalog biotopů České republiky. Prague: AOPK ČR; 2010.
22. Lustyk P, Guth J. Methodology of habitat mapping in the Czech Republic. Prague: AOPK ČR; 2012.
23. Ellenberg H, Haeupler H, Hamann U. Arbeitsanleitung für die Kartierung der Flora Mitteleuropas (Ausgabe für die Bundesrepublik Deutschland). Mitt Flor Soz Arbeitsgem Niedersachs. 1968;13:284–296.
24. IUCN. IUCN red list categories and criteria. Version 3.1. Gland: IUCN Species Survival Commission; 2001.
25. Evans D, Arvela M. Assessment and reporting under Article 17 of the habitats directive. Explanatory notes & guidelines for the period 2007–2012 [Internet]. 2011 [cited 2013 Nov 25].
26. ArcGIS 9. Using ArcGIS Desktop. Redlands, CA: ESRI; 2006.
27. European Environment Agency. Corine land cover 2000 (CLC2000) seamless vector database [Internet]. 2000 [cited 2013 Nov 25].
28. R Development Core Team. R: a language and environment for statistical computing. Vienna: R Foundation for Statistical Computing; 2010.
29. Moose-Deutschland. Internetplattform der Zentralstelle Deutschland, Dienstleistungszentrum für die Darstellung biologischer Daten im Internet [Internet]. 2012 [cited 2013 Nov 25].
30. Eionet. Article 17 – reporting under habitats directive [Internet]. 2009 [cited 2013 Nov 25].
31. Šubová V, Ambróz J, editors. The atlas of species of European interest of Natura 2000 sites in Slovakia. Český Těšín: FINIDS; 2011.
32. Pawlaczyk P, Kepel A, Jaros R, Dzięciołowski R, Wylegała P, Szubert A, et al. Natura 2000 shadow list in Poland [Internet]. Warsaw: WWF; 2004 [cited 2014 Mar 25].
33. Hajek B. Rozmieszczenie, wymagania środowiskowe oraz fenologia rzadkiego mchu *Buxbaumia viridis* (Moug. ex Lam. & DC.) Brid. ex Moug. & Nestl. w Trójmiejskim Parku Krajobrazowym. Acta Bot Cassub. 2010;7–9:161–175.
34. Cykowska B, Vončina G. Recent occurrence of moss *Buxbaumia viridis* (Bryophyta, Buxbaumiaceae) in the Kłodzko region (Central and Eastern Sudetes, SW Poland). Čas Slez Zemského Muzea A. 2011;60(1):85–89.
35. Kozik J, Voncina G. Odkrycie bezlistu okrywowego *Buxbaumia viridis* (Bryophyta, Buxbaumiaceae) w Beskidzie Niskim (Karpaty Zachodnie). Rocz Bieszczadzkie. 2012;20:378–383.
36. Paciorek T. New locality of moss *Buxbaumia viridis* (Moug. ex Lam. & DC.) Brid. ex Moug. & Nestl. in relation to its distribution in the Beskid Sądecki range (Western Carpathians, Poland). Čas Slez Zemského Muzea A. 2013;61(3):285–288.
37. Šafář J. Olomoucko. In: Mackovčin P, Sedláček M, editors. Chráněná území ČR, svazek VI. Prague: AOPK ČR; 2003. p. 290–309.
38. Tolasz R, Míková T, Valeriánová A, Voženílek V, editors. Atlas podnebí Česka. Prague: ČHMÚ, UPOL; 2007.
39. Hallingbäck T, Lönnell N, Weibull H, Hedenäs L, von Knorring P. Nationalyckeln till Sveriges flora och fauna. Bladmossor: Sköldmossor. Bryophyta: *Buxbaumia–Lecobryum*. Uppsala: ArtDatabanken, SLU; 2006.
40. Chytrý M, Rafajová M. Czech national phytosociological database: basic statistics of the available vegetation-plot data. Preslia. 2003;75:1–15.

18

Alisma plantago-aquatica L.: the cytoskeleton of the suspensor development

Joanna Świerczyńska*, Jerzy Bohdanowicz
Department of Plant Cytology and Embryology, University of Gdańsk, Wita Stwosza 59, 80-308 Gdańsk, Poland

Abstract

The actin and the tubulin cytoskeleton organization during the differentiation of the embryo-suspensor in *Alisma plantago-aquatica* was studied in comparison with the development of embryo, using immunofluorescence detection and rhodamine-phalloidin assay. At the early stage of the suspensor basal cell development (from 2- to ~10-celled embryos) microfilaments form an abundant network in the cytoplasm of the basal cell, while the microtubules form a delicate network. At the mature stage of development (from a dozen to several dozen-celled embryos), in the suspensor basal cell, the microfilaments and microtubules were localized from micropylar to chalazal pole of the cell. At the micropylar end of the basal cell a high amount of actin and tubulin material was observed. The microfilaments were mainly arranged parallel whereas numerous bundles of microtubules distributed longitudinally or transversally to the long axis of the cell. At this stage of basal cell functioning, some bundles of microtubules appeared to pass close to the nucleus surface. Microtubules were also observed distributed at the chalazal pole of the basal cell. At the senescence stage of the suspensor basal cell (>100-celled embryos) the actin and tubulin filaments disorganize, some disrupted microfilaments and microtubules were observed in the cytoplasm of the basal cell. At all stages of the suspensor basal cell development in the embryo cells an extensive actin and tubulin network was observed.

Keywords: microfilaments; microtubules; immunofluorescence; *Alisma plantago-aquatica*; embryo-suspensor

Introduction

Embryogenesis is an important and complex period during the early stages of flowering plants development. The embryo of many species is differentiated into two parts: embryo-suspensor and embryo-proper. Suspensor is a terminally differentiated, short-lasted organ that keeps the embryo inside the embryo sac [1]. Numerous studies upon cytochemistry [2,3], ultrastructure [4], and synthetic activity [5,6] confirm that suspensor is involved in the absorption, synthesis, and movement of nutrients from surrounding tissues into the embryo, thereby enabling a proper course of the embryo differentiation [7,8]. In addition, the role of suspensor in transport of nutrients to the embryo is also confirmed by ultrastructure of the organ, especially development of walls typical for so-called transfer cells that are specialized in short-distance, active movement of substances by plasmalemma [9,10]. Presence of walls of the transfer character was described, among others, for suspensor in *Phaseolus* [4], or *Alisma* [2]. A few literature data on the cytoskeleton during formation, development, and differentiation of polyploid cells of embryo-suspensor have been reported [11]. During the karyological differentiation, polyploidization occurs in suspensor cells [12]. Endoreduplication is the most common polyploidization mechanism within the suspensor, which consists in cyclically repeating DNA synthesis – with no mitosis – which leads to multiplied ploidy level in suspensor cells [13]. At many plant species, endoreduplication leads to the high ploidy level during suspensor differentiation, e.g. 8192C in *Phaseolus coccineus* [14], 4096n in *Phaseolus hysterinus* and 2048C in *Tropaeolum majus* [15], and 1024n in *Alisma plantago-aquatica* [16]. It seems that polyploidization is widespread in seed plants and is especially prevalent in tissues of a secretory and trophic character [12]. On the other hand, up-to-date observations of actin and tubulin cytoskeleton organization during differentiation of highly polyploid suspensor cells, are limited to a few species such as *Gagea lutea* [17,18], *Sedum acre* [11], or *Phaseolus coccineus* [19]. In our previous preliminary studies in *Alisma plantago-aquatica* extremely abundant cytoskeleton in suspensor cells was recorded. The *A. plantago-aquatica* suspensor, in particular its basal cell, appears to be a convenient object to study the changes in the organization of cytoskeleton during differentiation of endopolyploid cells.

The aim of this study was to analyze changes in the organization of actin and tubulin skeleton during the suspensor

* Corresponding author. Email: joanna.swierczynska@biol.ug.edu.pl
Handling Editor: Elżbieta Bednarska-Kozakiewicz

basal cell differentiation in *Alisma plantago-aquatica* in comparison with the development of embryo-proper using immunofluorescence and rhodamine-phalloidin assay.

Material and methods

Plant materials

Plants of *Alisma plantago-aquatica* were obtained from natural habitats of Lipce near Gdańsk in northern Poland. Flowers in various developmental stages were collected in summer months (July and August).

Fluorescence assay of F-Actin

The ovules were isolated from ovaries and incubated for 15–60 min in 400-μM m-maleimidobenzoic acid N-hydroxysucinimide ester [20,21] in piperazine buffer containing 5% dimethyl sulfoxide (DMSO) to permeate the cells [22]. After that the ovules were fixed in 4% formaldehyde freshly prepared from paraformaldehyde in piperazine buffer containing 5% DMSO for 4 h at room temperature. The ovules were washed in piperazine buffer and stained with 0.33 μM rhodamine-phalloidin in piperazine buffer containing 5% DMSO for 1.5 h. Next, after several washes in piperazine buffer, the suspensor and embryo proper were isolated manually from ovules under a stereoscopic microscope and placed on a microscope slide.

Immunolocalization of microfilaments and microtubules

To localize of microfilaments, ovules were fixed as described above and then treated according to the procedure [23]. For the localization of microtubules, ovules were fixed in 4% formaldehyde (freshly prepared from paraformaldehyde) and 0.25% glutaraldehyde in piperazine buffer for 4 h at room temperature. Then the ovules were prepared using the procedure described previously [24]. Next, after several washes in piperazine buffer, they were dehydrated in an ethanol series, each containing 10 mM dithiothreitol [25] to minimize background of the cytoplasm. Then, the plant material was embedded with Steedman's wax [26]. The 5 μm thick sections were made and placed on microscope slides coated with Mayer's egg albumen. The sections were dried overnight, dewaxed in ethanol, re-hydrated in an ethanol-PBS series and rinsed in PBS. Tissue samples were pre-incubated in PBS containing 0.1% bovine serum albumin for 45 min at room temperature. For the localization of microfilaments and microtubules slides were incubated overnight at 4°C with mouse anti-actin monoclonal antibody (clone C4, ICN; diluted 1:1000) and mouse anti-β-tubulin monoclonal antibody (Amersham N357, diluted 1:200), accordingly. Then, the preparations were rinsed in PBS and incubated for 4 h in a secondary Alexa 488-conjugated anti-mouse (Molecular Probes; diluted 1:800) antibody. Sections were rinsed in PBS, nuclei stained by 4',6'-diamidino-2-phenylindole dihydrochloride (DAPI; 1 μg/ml, Sigma). Next slides were treated with 0.01% toluidine blue to diminish the autofluorescence of cell walls and mounted in an antifading solution (Citifluor, Agar). In control experiments, conducted in a similar manner the primary antibody was omitted, no actin and tubulin staining was detected. Fluorescence was observed with a Nikon Eclipse E 800 epifluorescence microscope equipped with a CCD cooled camera.

Results

In *Alisma plantago-aquatica*, the first zygote division is transverse, larger basal cell and smaller apical cell is formed, that divides mitotically and initiates the embryo and chalazal suspensor. The basal cell does not divide further; instead it rapidly grows in size and functions as a suspensor basal cell. Nucleus of the basal cell undergoes endoreduplication and then it becomes polytene and reaches its maximum ploidy level 1024n. Changes in the configuration of actin and tubulin cytoskeleton were observed during differentiation of the suspensor basal cell in *A. plantago-aquatica* as compared to the developmental stages of the embryo.

The early stage of the suspensor basal cell development (from 2- to ~10-celled embryos)

The two-celled embryo is differentiated into the large basal cell and a smaller apical cell. The basal cell occupies the micropylar end of the embryo sac. Oval, enlarged nucleus occupies a central place in the basal cell. A single, significantly expanded nucleolus is visible in the nucleus (Fig. 1a,b). At this stage of development, the actin skeleton has been examined both using antibodies (Fig. 1a) and rhodamine-phalloidin (Fig. 1b), and achieved results were comparable. The actin material revealed due to immunofluorescence technique is present in cytoplasm of the basal cell as a network of fine microfilaments (Fig. 1a). Bundles of microfilaments get denser at the micropylar end of the basal cell (Fig. 1b). In the apical cell, actin is visible as a few faint dots (Fig. 1a,b). After staining with rhodamine-phalloidin, bundles of microfilaments form a dense network filling the cytoplasm of the basal cell (Fig. 1c,d). Numerous of F-actin filaments are arranged in circles (Fig. 1c), while others are focused to form intensely fluorescent areas (Fig. 1d). Later in this stage of development, the suspensor basal cell significantly increases in size and the nucleus is also enlarged (Fig. 1e,f). The tubulin skeleton occurs as an irregular network of microtubules in the basal cell cytoplasm (Fig. 1e,f). Microtubules intensely fluoresce and take the form of cortical network in embryo-proper cells (Fig. 1e). Around embryo sac cavity a strong fluorescence of cutin is visible (Fig. 1e).

The mature stage of the suspensor basal cell development (from a dozen to several dozen-celled embryos)

At this development stage, the suspensor basal cell further grows in size, and its micropylar end is characteristically "elongated". The nucleus is enormous and moves toward the chalazal end of the basal cell (Fig. 2a). Actin filaments localized in the suspensor basal cell are mainly arranged along the micropylar-chalazal axis of the cell. Actin skeleton forms a dense network of microfilaments (Fig. 2a,b). At this stage of development, actin filaments are visible in large quantities at the micropylar end of a cell, where actin material forms denser arrangements. Actin material is also present in the vicinity of nuclear envelope, where it is visible as intensely fluorescent regions (Fig. 2a). The

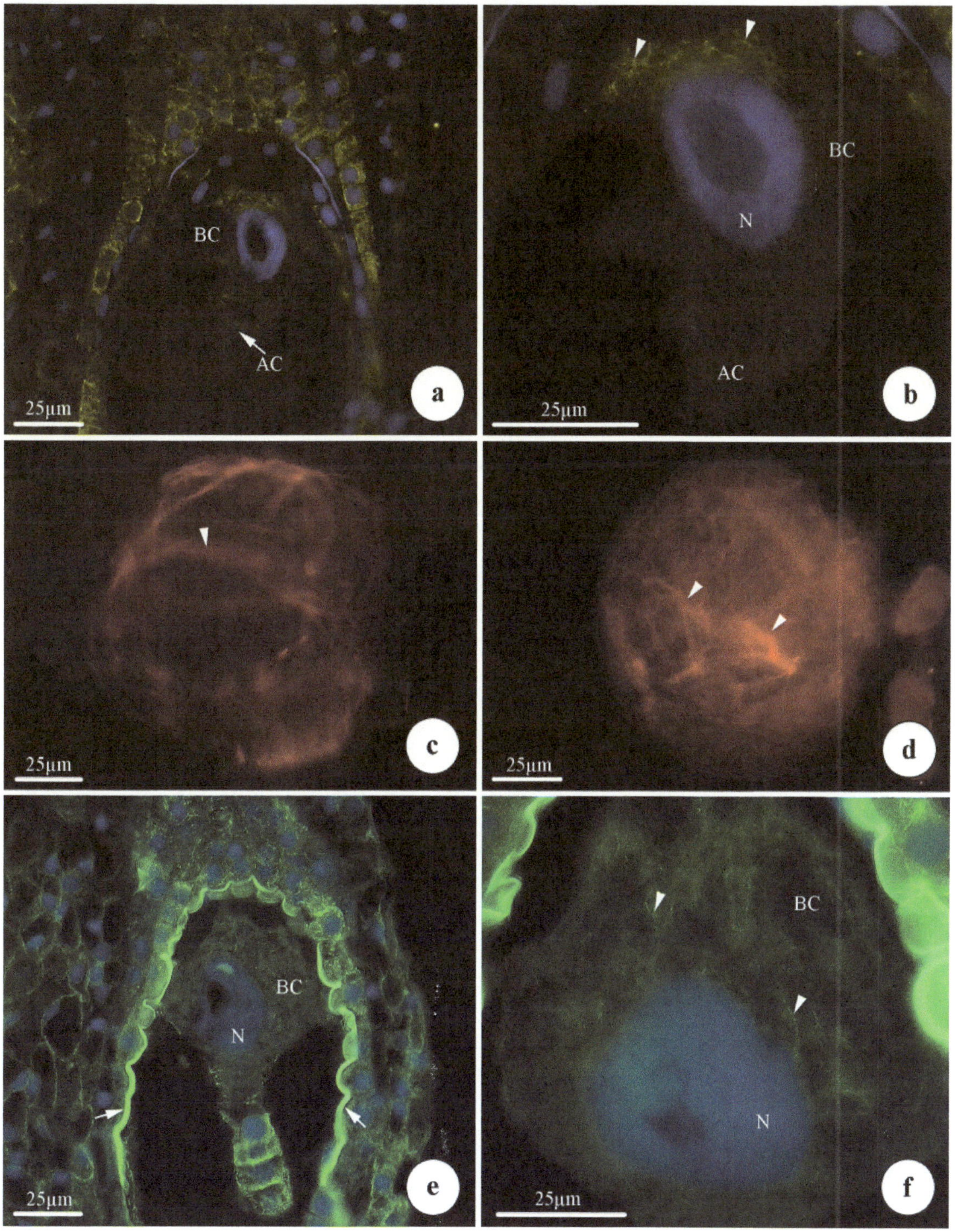

Fig. 1 The initial stage of the suspensor development of *Alisma plantago-aquatica*. **a,b** Images of immunofluorescence assay of microfilaments. **a** Two-celled embryo, visible basal cell and apical cell. **b** Microfilament network (arrows head) visible in the basal cell; enlarged nucleus and densities of actin filaments at the micropylar end of the cell. Few clusters of actin material in apical cell. **c,d** Images of F-actin visualized by rhodamine-phalloidin. **c** F-actin filaments form a dense network in the suspensor basal cell; numerous actin filaments are circularly arranged (arrow head). **d** Some microfilaments focus and form intensively fluorescent areas in the basal cell cytoplasm (arrows head). **e,f** Images of immunofluorescence assay of microtubules. **e** Five-celled embryo. Fine microtubule network visible in the basal cell cytoplasm and intensively fluorescent tubulin skeleton in embryo cells. A strong fluorescence of cutin around sac cavity is present (arrows). **f** Microtubule network (arrows head) in suspensor basal cell; changed basal cell shape and enlarging nucleus. AC – apical cell; BC – basal cell; N – nucleus.

actin skeleton in embryo cells takes the form that is typical for normal somatic cells; the cortical, near-wall networks dominate (Fig. 2a,b). Elements of tubulin skeleton are the most abundantly arranged at the micropylar, transfer wall of the suspensor basal cell. Densities of tubulin material are also present near the huge, polytene basal cell nucleus (Fig. 2c). In older suspensor basal cells, microtubules are parallel to the long axis of the cell in the direct vicinity of a contact surface, yet arranged in different directions in the distal part of the wall ingrowths (Fig. 2d,e). Strongly fluorescent microtubules are present at the nucleus (Fig. 2c,e). At this development stage, concentrations of tubulin material can be seen at the chalazal end of a cell (Fig. 2d). Tubulin skeleton in embryo-proper cells takes the configurations typical for intensively mitotically dividing cells (Fig. 2c). Close to the embryo-proper a strong fluorescence of tubulin densities are visible in endosperm (Fig. 2c).

The senescence stage of the suspensor basal cell (>100-celled embryos)

In the early stage of suspensor basal cell senescence, actin filaments are disorganized. Short, fragmented, and irregularly scattered microfilaments are present in the cytoplasm of the suspensor basal cell (Fig. 3a). There are a number of degraded F-actin filaments, which are concentrated at the micropylar end of a cell (Fig. 3b). Abundant, well-developed actin skeleton is visible in the cells of embryo (Fig. 3a,b). Microtubules are also disorganized and fragmented; disruption of parallel orientation of microtubules in the micropylar part of the suspensor basal cell can be found. Network of microtubules filling the basal cell cytoplasm is loosened (Fig. 3c). An increased cytoplasmic vacuolization is also observed in the suspensor basal cell (Fig. 3a,c).

Discussion

It is well known that during embryogenesis in the early stages of embryo development, two parts can be distinguished in it: the embryo-proper and suspensor that is a fast-growing, short-living organ being degraded before the development of a mature seed [27]. Suspensors are widespread among angiosperms and they stand out with diverse embryological morphology from other structures. Their presence, however, is not a universal feature, as there are species with embryos devoid of suspensor (*Lycopsis*, *Viola*), suspensors can be formed from a single cell (*Phaius tankervilliae*), several cells (*Arabidopsis*, *Capsella*), or a few hundred cells (*Phaseolus*) [7,28]. Our previous studies indicate that the suspensor in *Alisma plantago-aquatica* is composed of an enormous suspensor basal cell containing polytene nucleus and several cells of chalazal suspensor [2,16]. In many flowering plants, suspensor cell differentiation is accompanied by endoreduplication. It seems that endopolyploid cells are particularly important for tissues and organs that actively function for only a short period of development [29]. Endoreduplication leads to an increased size of the nucleus and multiplication of the genomic DNA quantity, which can significantly affect the increase in RNA transcription and thus accelerate the metabolism of a cell. Probably, endopolyploid cells are better adapted to uptake the nutrients from surrounding tissues [13,30,31]. It is known that cytoplasmic skeleton in plant cells plays an important role during growth and differentiation. So far, the organization of cytoskeletal components has been studied mainly during megasporogenesis and megagametogenesis [32–37]. On the other hand, little information relates to the cytoskeleton during differentiation of highly polyploid cells of embryonic suspensor. In this study, we present fluorescence and immunofluorescence detection of actin and tubulin skeleton in endopolyploid suspensor basal cell in *Alisma plantago-aquatica* L. At the initial stage of the suspensor development in *A. plantago-aquatica*, microtubular skeleton forms a delicate network in the cytoplasm of the basal cell, while numerous clusters of actin material are found on the micropylar end of the cell. The presence of such images of actin and tubulin material has been previously reported during the early stages of suspensor differentiation in *Phaius tankervilliae* [38], *Gagea lutea* [18], *Sedum acre* [11], in syncytium in *Utricularia* [39], or in the endosperm haustorium in *Rhinanthus serotinus* [40]. In the next, mature, development stage of suspensor in *Alisma*, actin and tubulin skeleton rebuilds and forms a dense and intricate network of microfilaments and microtubules in the cytoplasm of the basal cell. In our opinion, probably the changes in cytoskeleton organization are associated with the change in shape and enlargement of the basal cell. This has been earlier observed in magnifying suspensor in *Phaius tankervilliae* [38] or in the growing suspensor basal cell in *Sedum acre* [11]. In *A. plantago-aquatica*, the micropylar end of the basal cell, that grows deep into the embryonic tissue, is also characteristically elongated. At the same time, the microfilaments and microtubules density significantly increases at the micropylar end. Probably, the functional importance of cytoskeletal elements density at the micropylar end of the basal cell is related to the transfer nature of the wall. Previous cytochemical and ultrastructural studies revealed the presence of numerous wall ingrowths characteristic for the transfer cells. The presence of transfer walls suggests the possibility of an intensive exchange of nutrients, which occurs between the inner integument and suspensor basal cell, and then their transfer to the developing embryo [2]. Microtubules and microfilaments densities and accompanying numerous mitochondria were also observed in the vicinity of the wall ingrowths in the endosperm in *Vicia faba*, *Zea mays*, *Triticum aestivum* [41], the endosperm-placenta-syncytium in *Utricularia* [39,42], in endosperm haustorium in *Rhinanthus serotinus* [40], or in suspensor in *Sedum acre* [11] and *Gagea lutea* [18]. It is widely recognized that the presence of transfer walls significantly increases the plasmalemma surface area ensuring more efficient uptake of macromolecules from embryonic tissues [9,10,43]. In addition, endopolyploidization occurring in the suspensor basal cell in *Alisma* may indicate the specialization and high metabolic activity of this cell. It is also possible that both microfilaments and microtubules, besides intracellular transport [44,45], may be also involved in the transfer of intranuclear products to the cytoplasm [46–48]. It seems to be likely, since density of microtubules was observed close to the endopolyploid nucleus of the basal cell in *A. plantago-aquatica*. Presence of microtubules concentrations in the

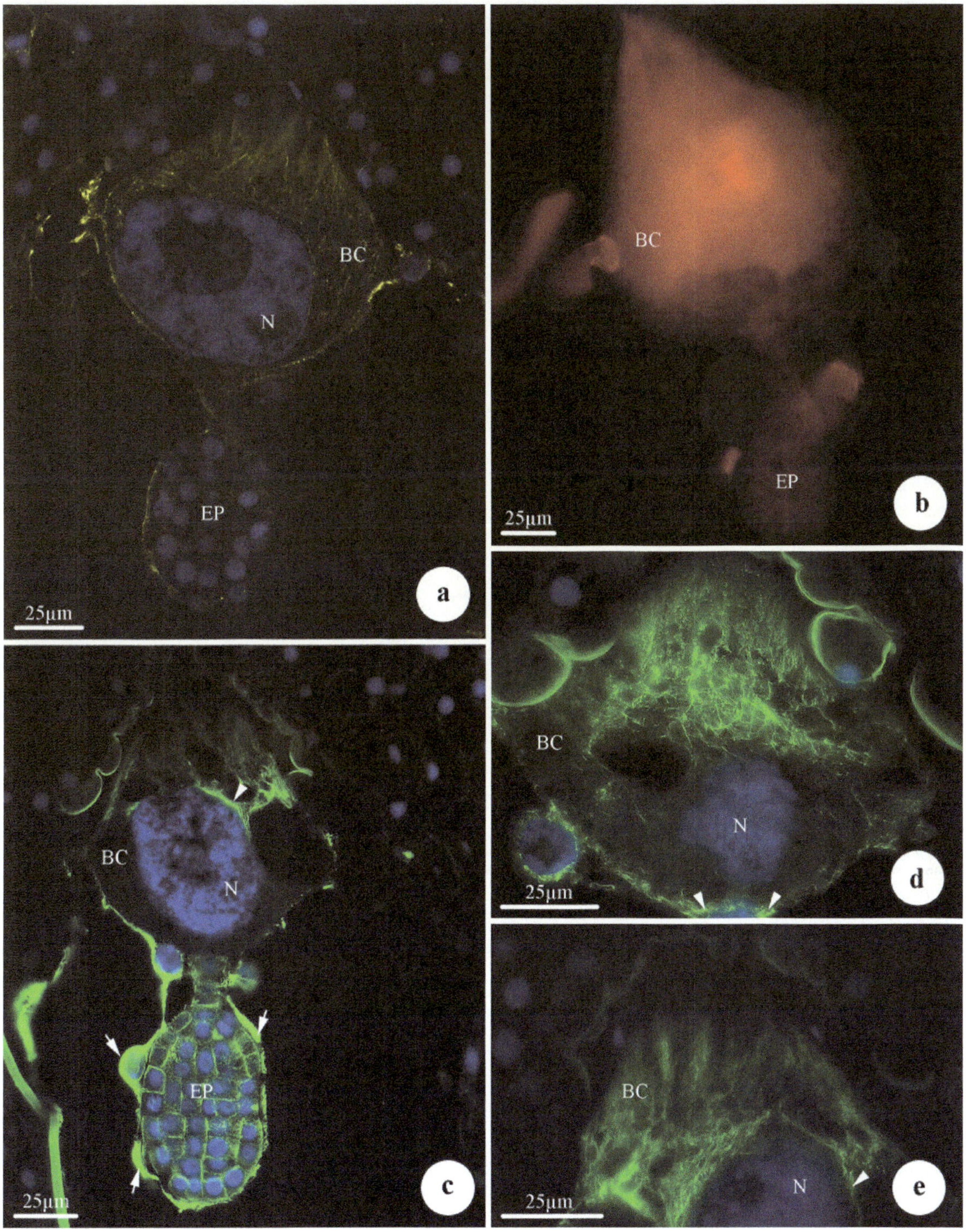

Fig. 2 The mature stage of the suspensor basal cell development of *Alisma plantago-aquatica*. **a** Image of immunofluorescence assay of microfilaments. Abundant network of microfilaments in the basal cell. Large amount of actin material occurs at the micropylar end of the cell; enormous polytene nucleus. The microfilament network visible in cells of embryo-proper. **b** Image of F-actin visualized by rhodamine-phalloidin. F-actin filaments form a dense network in the suspensor basal cell and cells of embryo-proper. **c–e** Images of immunofluorescence assay of microtubules. **c** Numerous microtubules visible in the basal cell cytoplasm, especially at the micropylar end. Intensively glowing microtubule clusters near nucleus (arrow). Intensively fluorescent tubulin skeleton visible in embryo-proper cells. Arrows point to the strong fluorescence of tubulin material present close to the embryo proper. **d** Bundles of microfilaments form an intricate network at the micropylar end of the basal cell, the tubulin densities are also visible at the chalazal end of the cell (arrows head). **e** Fragment of the basal cell, the microtubule network visible at the micropylar end. Microtubule concentrations visible near the nucleus (arrow head). BC – basal cell; E – embryo ; N – nucleus.

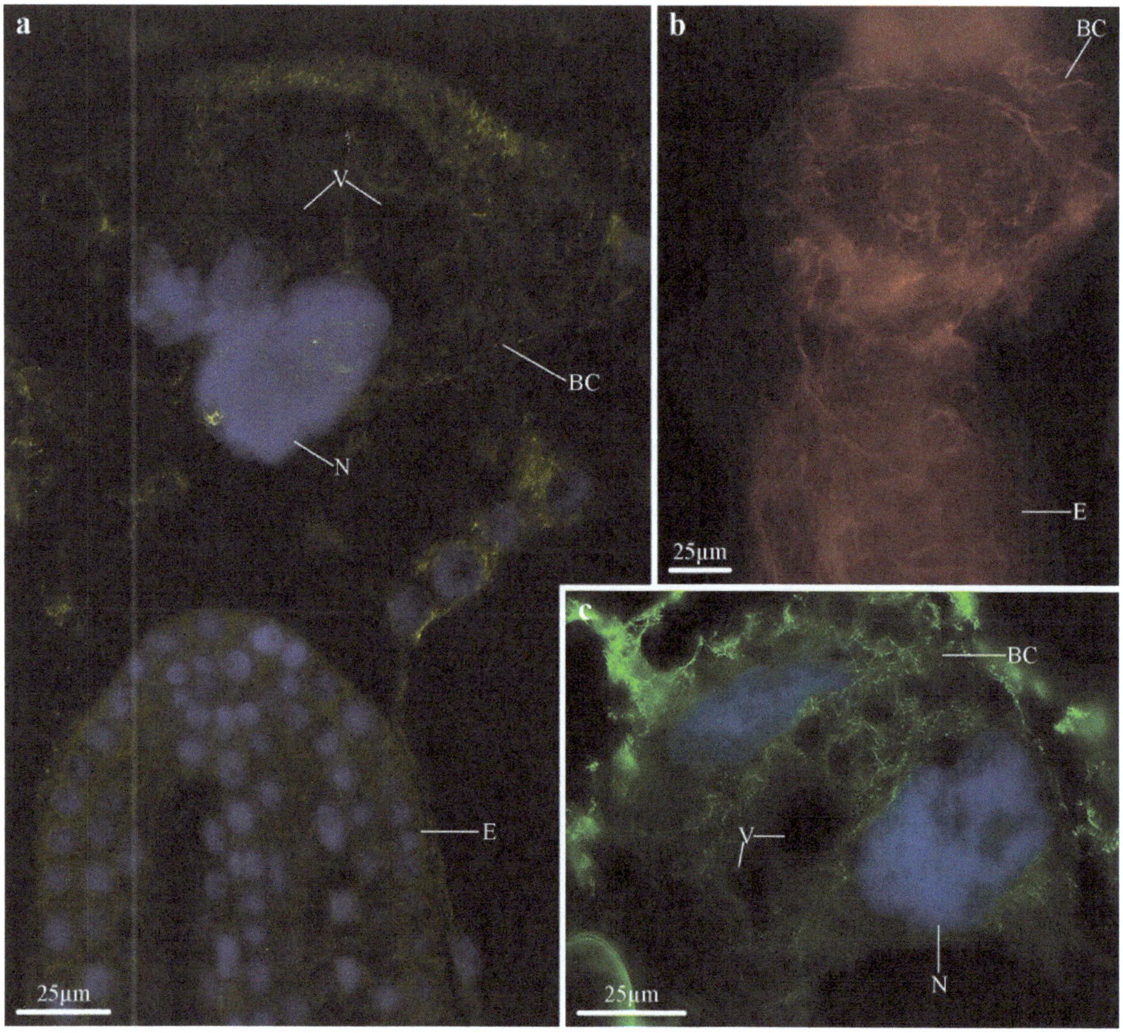

Fig. 3 The aging stage of suspensor development of *Alisma plantago-aquatica*. **a** Image of immunofluorescence assay of microfilaments. Short, fragmented, irregularly distributed microfilaments visible in the basal cell cytoplasm; pleated nucleus and numerous vacuoles. **b** Image of F-actin visualized by rhodamine-phalloidin. Numerous degraded F-actin filaments concentrating at the micropylar end of the basal cell. The microfilament network visible in embryo-proper cells. **c** Image of immunofluorescence assay of microtubules. Loosened microtubule network fulfilling the cellular cytoplasm visible in the basal cell; numerous vacuoles. BC - basal cell; E - embryo; N - nucleus; V - vacuoles.

vicinity of nuclear envelope was also reported in the chalazal endosperm haustorium and endosperm-placenta syncytia in *Utricularia* [42], the endosperm haustorium [40], as well as in polyploid cells of the suspensor [11,18]. It should not be excluded that microtubules may be involved in the positioning of the huge endopolyploid basal cell nucleus in *Alisma*. It is very reasonable, because nucleus is centrally located in the young basal cell, while in the old basal cell; nucleus changes its position and moves closer to the chalazal end of a cell. Similar observations were also reported during the differentiation of highly polyploid cell of endosperm haustorium in *Rhinanthus serotinus* [40]. In the aging suspensor basal cells in *Alisma* some disturbances in microfilaments and microtubule organization and increased cell vacuolization can be observed. These alterations are correlated with a progressive development of the embryo and endosperm. Probably, basal cell after fulfilling its functions with respect to the embryo begins to degrade; the literature data suggest that this process occurs by means of so-called programmed cell death [49–51]. In conclusion, the presence of abundant actin and tubulin cytoskeleton in endopolyploid suspensor basal cell in *Alisma plantago-aquatica* is probably associated with the positioning of the nucleus, movement of organelles, and transport of nutrients within the physiologically active cell, which appears to be essential for the proper development of the embryo.

Acknowledgments

This study was supported by the grants from the University of Gdańsk (BW 1410-5-0111-7).

Authors' contributions

The following declarations about authors' contributions to the research have been made: idea of the study: JŚ, JB; immunofluorescence and fluorescence assay: JŚ; microscopical analysis: JŚ, JB; writing of the manuscript: JŚ, JB; photographs: JŚ, JB.

References

1. Raghavan V. Embryogenesis in Angiosperms. A development and experimental study. Cambrige: Cambrige University Press; 1986.
2. Bohdanowicz J. *Alisma* embryogenesis: the development and ultrastructure of the suspensor. Protoplasma. 1987;137(2–3):71–83.
3. Kozieradzka-Kiszkurno M, Bohdanowicz J. Development and cytochemistry of the embryo suspensor in *Sedum*. Acta Biol Cracov Ser Bot. 2006;48(2):67–72.
4. Yeung EC, Clutter ME. Embryogeny of *Phaseolus coccineus*: the ultrastructure and development of the suspensor. Can J Bot. 1979;57(2):120–136.
5. Cremonini R, Cionini PG. Extra DNA synthesis in embryo suspensor cells of *Phaseolus coccineus*. Protoplasma. 1977;91(3):303–313.
6. Singh MB, Bhalla PL, Malik CP. Activity of some hydrolytic enzymes in autolysis of the embryo suspensor in *Tropaeolum majus* L. Ann Bot. 1980;45(5):523–527.
7. Yeung EC, Meinke DW. Embryogenesis in Angiosperms: development of the suspensor. Plant Cell. 1993;5(10):1371.
8. Schwartz BW, Vernon DM, Meinke DW. Development of the suspensor: differentiation, communication, and programmed cell death during plant embryogenesis. In: Larkins BA, Vasil IK, editors. Cellular and molecular biology of plant seed development. Dordrecht: Springer Netherlands; 1997. p. 53–72.
9. Gunning BES, Pate JS. Transfer cells. In: Robards AW, editor. Dynamic aspects of plant ultrastructure. London: McGraw-Hill; 1974. p. 441–480.
10. Talbot MJ, Offler CE, McCurdy DW. Transfer cell wall architecture: a contribution towards understanding localized wall deposition. Protoplasma. 2002;219(3–4):197–209.
11. Kozieradzka-Kiszkurno M, Świerczyńska J, Bohdanowicz J. Embryogenesis in *Sedum acre* L.: structural and immunocytochemical aspects of suspensor development. Protoplasma. 2011;248(4):775–784.
12. Nagl W. Endopolipoidy and polyteny in differentiation and evolution. Amsterdam: Oxford North-Holland Publishing Company; 1978.
13. D'Amato F. Polyploidy in cell differentiation. Caryologia. 1989;42(3–4): 183–211.
14. Brady T. Feulgen cytophotometric determination of the DNA content of the embryo proper and suspensor cells of *Phaseolus coccineus*. Cell Diff. 1973;2(2):65–75.
15. Nagl W. Early embryogenesis in *Tropaeolum majus* L.: ultrastructure of the embryo-suspensor. Bioch Physiol Pflanz. 1976;170:253–260.
16. Bohdanowicz J. Karyological anatomy of the suspensor in *Alisma* L. I. *Alisma plantago-aquatica* L. Acta Biol Cracov Ser Bot. 1973;16:235–248.
17. Świerczyńska J, Bohdanowicz J. The cytoskeleton of the embryo-suspensor in *Gagea lutea* (L.) Ker Gawl. Acta Biol Cracov Ser Bot. 2008;50(1 suppl):73.
18. Świerczyńska J, Bohdanowicz J. The immunocytochemical studies of the embryo-suspensor in *Gagea lutea* (L.) Ker-Gawl. Acta Biol Crac Ser Bot. 2010;52(1 suppl):41.
19. Świerczyńska J, Bohdanowicz J. The cytoskeleton of the embryo-suspensor in *Phaseolus coccineus* L. Acta Biol Crac Ser Bot. 2012;52(1 suppl):83.
20. Sonobe S, Shibaoka H. Cortical fine actin filaments in higher plant cells visualized by rhodamine-phalloidin after pretreatment with m-maleimidobenzoyl N-hydroxysuccinimide ester. Protoplasma. 1989;148(2–3):80–86.
21. Huang BQ, Russell SD. Fertilization in *Nicotiana tabacum*: cytoskeletal modifications in the embryo sac during synergid degeneration: a hypothesis for short-distance transport of sperm cells prior to gamete fusion. Planta. 1994;194(2):200–214.
22. Traas JA, Doonan JH, Rawlins DJ, Shaw PJ, Watts J, Lloyd CW. An actin network is present in the cytoplasm throughout the cell cycle of carrot cells and associates with the dividing nucleus. J Cell Biol. 1987;105(1):387–395.
23. Świerczyńska J, Bohdanowicz J. Microfilament cytoskeleton of endosperm chalazal haustorium of *Rhinanthus serotinus* (Scrophulariaceae). Acta Biol Crac Ser Bot. 2003;45(1):143–148.
24. Bohdanowicz J, Szczuka E, Świerczyńska J, Sobieska J, Kościńska-Pająk M. The distribution of microtubules during regular and disturbed microsporogenesis and pollen grain development in *Gagea lutea* (L.) Ker.-Gaw. Acta Biol Crac Ser Bot. 2005;47:89–96.
25. Brown RC, Lemmon BE, Mullinax JB. Immunofluorescent staining of microtubules in plant tissues: improved embedding and sectioning techniques using polyethylene glycol (PEG) and Steedman's wax. Bot Acta. 1989;102(1):54–61.
26. Vitha S, Baluška F, Jasik J, Volkmann D, Barlow PW. Steedman's wax for F-actin visualization. In: Staiger CJ, Baluška F, Volkmann D, Barlow PW, editors. Actin: a dynamic framework for multiple plant cell functions. Dordrecht: Springer Netherlands; 2000. p. 619–636.
27. Maheshwari P. An introduction to the embryology of angiosperms. New York, NY: McGrraw-Hill; 1950.
28. Kozieradzka-Kiszkurno M, Płachno BJ, Bohdanowicz J. New data about the suspensor of succulent angiosperms: ultrastructure and cytochemical study of the embryo-suspensor of *Sempervivum arachnoideum* L. and *Jovibarba sobolifera* (Sims) Opiz. Protoplasma. 2012;249(3):613–624.
29. Brodsky VY, Uryvaeva IV. Genome multiplication in growth and development. Biology of polyploid and polytene cells. London: Cambridge University Press; 1985.
30. Nagl W. Polyploidy in differentiation and evolution. Inter J Cell Clon. 1990;8(4):216–223.
31. Joubès J, Chevalier C. Endoreduplication in higher plants. Plant Mol Biol. 2000;43(5–6):735–745.
32. Willemse MTM, van Lammeren AAM. Structure and function of the microtubular cytoskeleton during megasporogenesis and embryo sac development in *Gasteria verrucosa* (Mill.) H. Duval. Sex Plant Reprod. 1988;82:631–634.
33. Bednara J, Willemse MTM, van Lammeren AAM. Organization of the actin cytoskeleton during megasporogenesis in *Gasteria verrucosa* visualized with fluorescent-labelled phalloidin. Acta Bot Neerl. 1990;39:43–48.
34. Zee SY, Ye XL. Changes in the pattern of organization of the microtubular cytoskeleton during megasporogenesis in *Cymbidium sinense*. Protoplasma. 1995;185(3–4):170–177.
35. Huang BQ, Ye. XL, Yeung E, Zee SY. Embryology of *Cymbidium sinense*: the microtubule organization of early embryos. Ann Bot. 1998;81(6):741–750.
36. Tung SH, Ye XL, Yeung EC, Zee SY. Ultrastructural aspects of megasporogenesis in *Cymbidium sinensie* (Orchidaceae). Lindleyana. 1999;14:178–192.
37. Tung SH, Ye XL, Zee SY, Yeung EC. The microtubular cytoskeleton during megasporogenesis in the nun orchid, *Phaius tankervillae*. New Phytol. 2000;146(3):503–513.
38. Ye XL, Zee SY, Yeung EC. Suspensor development in the nun orchid *Phaius tankervilliae*. Inter J Plant Sci. 1997;162:1053–1063.

39. Płachno BJ, Świątek P, Kozieradzka-Kiszkurno M. The F-actin cytoskeleton in syncytia from non-clonal progenitor cells. Protoplasma. 2011;248(3):623–629.

40. Świerczyńska J, Kozieradzka-Kiszkurno M, Bohdanowicz J. *Rhinanthus serotinus* (Schönheit) Oborny (Scrophulariaceae): immunohistochemical and ultrastructural studies of endosperm chalazal haustorium development. Protoplasma. 2013;250(6):1369–1380.

41. Thompson RD, Hueros G, Becker HA, Maitz M. Development and functions of seed transfer cells. Plant Sci. 2001;160(5):775–783.

42. Płachno BJ, Świątek P, Sas-Nowosielska H, Kozieradzka-Kiszkurno M. Organisation of the endosperm and endosperm–placenta syncytia in bladderworts (*Utricularia*, Lentibulariaceae) with emphasis on the microtubule arrangement. Protoplasma. 2013;250(4):863–873.

43. Offler CE, McCurdy DW, Patrick JW, Talbot MJ. Transfer cells: cells specialized for a special purpose. Ann Rev Plant Biol. 2003;54(1):431–454.

44. Kamal A, Goldstein LSB. Connecting vesicle transport to the cytoskeleton. Curr Opin Cell Biol. 2000;12(4):503–508.

45. Verchot-Lubicz J, Goldstein RE. Cytoplasmic streaming enables the distribution of molecules and vesicles in large plant cells. Protoplasma. 2010;240(1–4):99–107.

46. Muench DG, Park NI. Messages on the move: the role of the cytoskeleton in mRNA localization and translation in plant cells. Can J Bot. 2006;84(4):572–580.

47. Guo F, Yu L, Watkins S, Han Y. Orientation of microtubules suggests a role in mRNA transportation in fertilized eggs of Chinese pine (*Pinus tabulaeformis*). Protoplasma. 2007;231(3–4):239–243.

48. Zarnack K, Feldbrugge M. Microtubule-dependent mRNA transport in fungi. Eukaryot Cell. 2010;9(7):982–990.

49. Giuliani C, Consonni G, Gavazzi G, Colombo M, Dolfini S. Programmed cell death during embryogenesis in maize. Ann Bot. 2002;90(2):287–292.

50. Bozhkov PV, Filonova LH, Suarez MF. Programmed cell death in plant embryogenesis. Curr Top Dev Biol. 2005;67:135–179.

51. Lombardi L, Ceccarelli N, Picciarelli P, Lorenzi R. DNA degradation during programmed cell death in *Phaseolus coccineus* suspensor. Plant Physiol Biochem. 2007;45(3–4):221–227.

19

Influence of abiotic stress during soybean germination followed by recovery on the phenolic compounds of radicles and their antioxidant capacity

Sylwia Swigonska[1], Ryszard Amarowicz[2], Angelika Król[1]*, Agnieszka Mostek[1], Anna Badowiec[1], Stanisław Weidner[1]

[1] Department of Biochemistry, Faculty of Biology and Biotechnology, University of Warmia and Mazury in Olsztyn, M. Oczapowskiego 1A, 10-957 Olsztyn-Kortowo, Poland

[2] Division of Food Science, Institute of Animal Reproduction and Food Research of Polish Academy of Sciences, Tuwima 10, Box 55, 10-718 Olsztyn 5, Poland

Abstract

Abiotic stress factors are among the major causes of lower crop yields. It is known, that in response to cold and/or osmotic stress, crops activate various defense mechanisms, including morphological, physiological and metabolic adaptations. Secondary metabolism, especially phenolic compounds, seem to be an important factor of stress-induced metabolic re-engineering as their levels are alternated by abiotic stress in plants. Despite the fact, that the nature and function of phenolic compounds was already studied in various plant species, it is important to define tissue-specific changes induced by two most potent abiotic stressors – low temperature and decreased water potential. Moreover, in fields, the appearance of single stress is rather rare. Usually two or more factors are acting in parallel, which may potentially result in different effects. Therefore, the aim of this study was to analyze selected elements of secondary metabolism in roots of germinating soybean seeds under cold stress, osmotic stress and both stresses combined. In addition the effects of constant and persistent stress were compared to those induced by sudden and brief stress appearance, as well as after the post-stress recovery process. In the presented study standard methods for identification and quantification of phenolic acids and isoflavones were used and the antioxidant capacity of the radicle extracts was measured. The phenolic metabolism in plants was greatly intensified in response to cold and osmotic stress and remained at high level during the post-stress recovery. The amount and composition of both phenolic acids and identified isoflavones also changed in stress- and duration-dependent manner. This proves an important role of phenolic compounds in abiotic stress response of germinating soybean seeds and opens up new perspectives for further investigations.

Keywords: environmental stress; antioxidant capacity; phenolic compounds; phenolic acids; isoflavones

Introduction

It is believed that interactions of plants with their environment comprise biosynthesis of various secondary metabolites. Recent studies, have demonstrated that secondary metabolites, particularly phenolic compounds, perform important signaling and protective functions in plants. Environmental stresses often lead to intensive biosynthesis of phenolic compounds in plants and, as a consequence, they affect humans' and animals wellbeing [1–3]. It is noteworthy that the enhancement of plants' tolerance to adverse effects of environmental conditions would result in much higher yields obtained from the currently cultivated fields. Therefore, studies on basic mechanisms enabling plants to cope with environmental stresses are among high priority research areas in many countries [4].

Previously published results suggest that plants respond to a variety of stress factors by increasing their capacity to scavenge reactive oxygen species [4–6]. Environmental stresses induce accumulation of phenolic compounds to protect tissues from damage caused by free radical-induced oxidative stress or to incorporate them into the cell wall as suberin or lignin [6–8]. Therefore analyzing both quantities and composition of phenolic compounds synthesized in response to stress factors is important for understanding the mechanism of stress response as well as for possible future engineering of crops to overproduce antioxidant phenolics. The unfavorable effect of low temperature and water potential on germination and development of industrial crops has been frequently discussed in scientific literature. However, there are fewer reports on different concomitant stresses of different durations, or research dealing with the post-stress recovery process.

Apart from inducing the general phenylpropanoid pathway, we hypothesize that abiotic stress must influence the composition of polyphenols, which may also reflect

* Corresponding author. Email: angelika.krol@uwm.edu.pl

Handling Editor: Grzegorz Jackowski

stress-induced metabolic re-engineering. Therefore the aim of this study was to examine soybean radicles in terms of quantitative and qualitative changes occurring in selected elements of the secondary metabolism as a result of cold stress, osmotic stress or a combination of both stresses in different durations. Another question examined was the effect of post-stress recovery on changes in the secondary metabolism of roots.

Material and methods

Plant material and growth conditions

The plant material used in this experiment consisted of cultivar Aldana common soybean seeds (*Glycine max* L.), purchased in Strzelce (Poland). The analyzed material was composed of roots isolated from germinating soybean seeds. Seeds were surface sterilized in 0.5% sodium hypochlorite for 20 min, washed with sterilized water and placed on Petri dishes. The diagram of the experimental design is presented in Fig. 1. Control samples were germinated for 72 hours in +25°C/water (C72). To simulate conditions of cold and osmotic stress, seeds were divided into groups, germinating under following conditions: cold stress – +10°C/water (CS72), osmotic stress – +25°C/–0.2 MPa (OS72) and cold stress combined with osmotic stress – +10°C/–0.2 MPa (COS72). Apart from constant and prolonged stresses, sudden and short-lasting stresses were applied. For this purpose, after germinating another batch of seeds for 48 h under optimal conditions (+25°C/water – C48) seeds were transferred for 24 h to the following conditions: cold stress – +10°C/water (CS24), osmotic stress – +25°C/–0.2 MPa (OS24) and cold stress combined with osmotic stress – +10°C/–0.2 MPa (COS24). Additionally, seeds incubated under cold stress (CS24), osmotic stress (OS24) and combined cold and osmotic stress (COS24) underwent 24-h recovery in optimal conditions (+25°C/water), as a result of which three more samples were obtained: CR, OR and COR (Fig. 1). In order to simulate osmotic stress of −0.2 MPa, polyethylene glycol was used (molecular weight 8000; Sigma) at a concentration of 119.46 g/l during germination at temp. +25°C and 99.68 g/l at +10°C. Once the set germination time was over, roots of germinated seeds were collected, immediately frozen in liquid nitrogen and kept in −80°C for further analysis.

Extract preparation

Prior to the extraction frozen roots were homogenized and precisely weighed amount (10–20 mg) of homogenate was used as starting material. Phenolic compounds were extracted using 80% (v/v) methanol at a solids to solvent ratio of 1:10 (m/v) at 50°C for 30 min [9]. The extraction was carried out in Erlenmayer flasks using a shaking water bath (Elpan 357, Wrocław, Poland). The extraction was repeated three times, the supernatants were filtrated and pooled, and the organic solvent was evaporated under vacuum at 40°C in a Büchi rotary evaporator; the remaining aqueous solution was lyophilized. Prepared extracts were stored at −20°C until analyzed.

Content of total phenolics

The content of total phenolics in the crude extracts was estimated using Folin and Ciocalteu's phenol reagent [10]. (+)-Catechin was used as a standard.

HPLC-PAD analysis of phenolic acids

Phenolic acids (free and liberated from soluble glycosides) were isolated from the extracts according to the method previously described by Weidner et al. [11]. An aqueous suspension of the methanolic extract (200 mg/20 ml) was adjusted to pH 2 with 6 M HCl, and free phenolic acids were

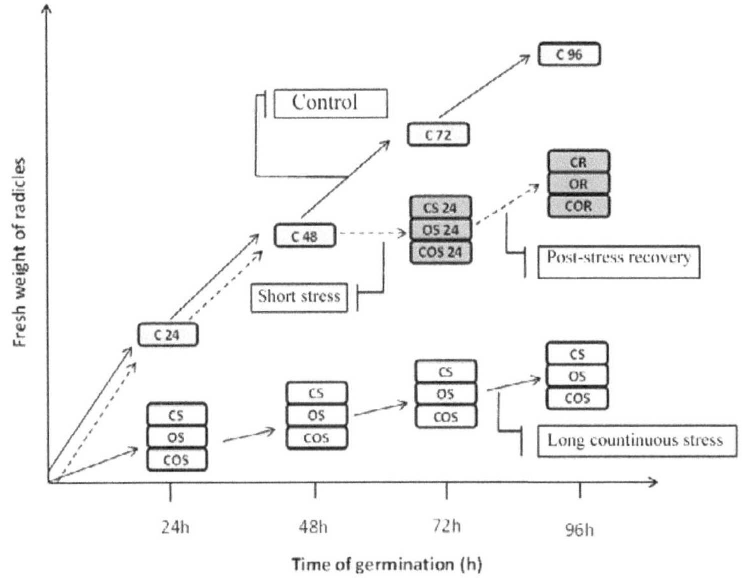

Fig. 1 Diagram of experimental design. Control samples were germinated for 72 hours in +25°C/water (C72). The study included three subsequent experiments: examining the effect of long and continuous stress (CS72 – cold stress, OS72 – osmotic stress, COS72 – combined stresses), short and rapid stress (CS24 – cold stress, OS24 – osmotic stress, COS24 – combined stresses) and post-stress recovery (CR – recovery after cold stress, OR – recovery after osmotic stress, COR – recovery after combined stresses). Detailed description in the text.

extracted five times into 20 ml of diethyl ether using a separatory funnel. The ether extract was evaporated to dryness under vacuum at room temperature. The water solution was neutralized and then lyophilized. The residue was dissolved in 20 ml of 2 M NaOH and hydrolyzed for 4 h under nitrogen atmosphere at room temperature. After acidification to pH 2 using 6 M HCl, phenolic acids released from soluble esters were extracted from the hydrolysate five times into 30 ml of diethyl ether. Nine milliliters of 6 M HCl were added to the water solution and the solution was placed in nitrogen atmosphere and hydrolyzed for 1 h a boiling water bath. Phenolic acids released from soluble glycosides were separated from the hydrolysate five times into 45 ml of diethyl ether. After ether evaporation, the dry residue was dissolved in 2 ml of methanol and filtered through a 0.45 μm nylon filter. The sample was injected onto an HPLC column. The same Shimadzu HPLC system was employed. The conditions of the separations were as follows: prepacked LUNA C18 column (5 μm, 4.6 × 250 mm; Phenomenex); mobile phase water-acetonitrile-acetic acid (88:10:2, v/v/v) [11] flow rate of 1 ml/min; injection volume of 20 μl; the detector was set at 280 and 320 nm; oven temperature was 20°C.

Scavenging of the DPPH radical

The scavenging effect of phenolics from the extracts was monitored as described by Amarowicz et al. [12]. A 0.1 ml methanolic solution containing between 0.02–0.10 mg of extract was mixed with 2 ml of deionized water and then added to a methanolic solution of DPPH• (2,2-diphenyl-1-picrylhydrazyl; 1 mM, 0.25 ml). The mixture was vortexed for 1 min, left to stand at room temperature for 20 min, and absorbance of the solution was then measured at 517 nm with the spectrophotometer.

Total antioxidant activity (TAA)

The total antioxidant activity of the extracts was determined according to the Trolox equivalent antioxidant activity (TEAC) assay described by Re et al. [13]. TAA was expressed as mmol Trolox equivalent/g of extract.

Qualitative and quantitative analysis of isoflavones

Determinations of the content of isoflavones in tissues of roots of germinating soybean seeds were made according to the modified method described by Amarowicz et al. [14].

Isoflavones were extracted from roots with 80% methanol (in a 1:8 ratio), by shaking for 15 minutes at temp. +75°C. Cooled supernatants were filtered while the solid fraction underwent three more runs of extraction. The filtrates were then mixed and methanol evaporated in a vacuum rotary evaporator at temp. +40°C. The extracts were lyophilized to remove all water and stored until analyses at temp −20°C. Reverse phase high performance liquid chromatography (RP-HPLC) was used for separation of isoflavones from the analyzed extracts. Extracts of phenolic compounds were placed on a chromatographer's column filled with RP-18 gel (2–43 μm; Merck) and then identification was performed with the HPLC method using a Shimadzu chromatographic system equipped with a C18 column (5 μm, 4.6 × 250 mm; Merck). The water-acetonitrile-acetic acid mixture (in a 88:10:2 ration, v/v) was used as the stationary phase [15]. Each time, a sample of 20 μl of an extract was placed on the column and the flow rate was 1 cm^3/min. Detection was performed at a wavelength of 254 nm. Concentration of compounds in samples was determined with reference to standards obtained commercially (Extrasynthese, Genay Cedex, France).

Statistical evaluation

All experiments were conducted in three biological replicates (radicles isolated from 100 seeds) and two technical replicates for each sample. The presented values are shown as means ± standard error of means (*SEM*). Statistically significant differences in the mean values were tested by student's *t*-test.

Results

Changes in the total content of phenolic compounds

The content of phenolic compounds in roots increased under constant and prolonged stresses, the total content of phenolic compounds in all examined samples significantly increased (Fig. 2a). It appeared that combination of cold and osmotic stress had the highest impact on the accumulation of phenolic compounds in roots.

Similar tendencies were observed in the subsequent experiment examining the effect of abrupt and short-lasting osmotic stresses (Fig. 2b). The total content of phenolic compounds increased significantly ($P < 0.05$) in all examined samples (Fig. 2b).

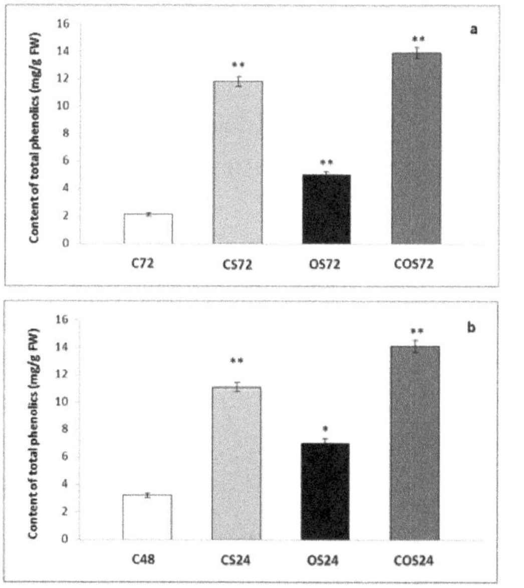

Fig. 2 Changes in the content of total phenolic compounds in soybean seeds germinating in optimal conditions and under: long-term continuous stress (**a**; C72 – control sample, CS72 – cold stress, OS72 – osmotic stress, COS72 – combined stresses), short-term rapid stress (**b**; C48 – control sample, CS24 – cold stress, OS24 – osmotic stress, COS24 – combined stresses). Results are presented as means ±*SEM*. Asterisks represent statistically significant differences at: $P < 0.05$ (*) or $P < 0.01$ (**).

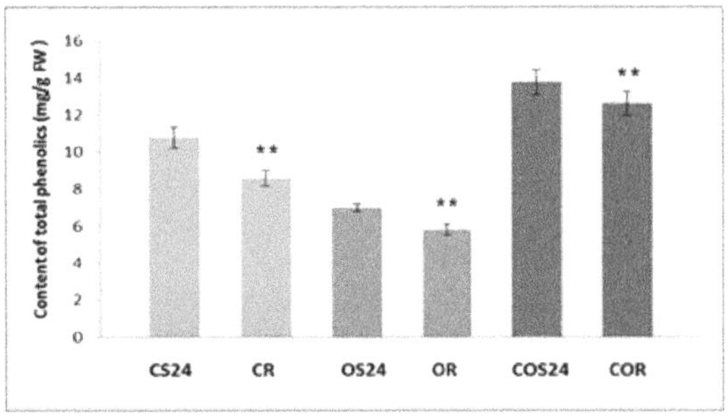

Fig. 3 Changes in the content of total phenolic compounds in soybean seeds germinating primarily for 48 h in optimal conditions, then for 24 h under short-term rapid stress conditions (CS24 – cold stress, OS24 – osmotic stress, COS24 - combined stresses) and finally transferred for 24 h to recovery: after cold stress (CR), after osmotic stress (OR), after cold stress combined with osmotic stress (COR). Results are presented as means ±*SEM*. Asterisks represent statistically significant differences at: $P < 0.05$ (*) or $P < 0.01$ (**).

Compared to the content of phenolic compounds in roots of seeds exposed to sudden and short-lasting stresses (serving as controls), the content of phenolic compounds after post-stress recovery decreased (Fig. 3).

Changes in the composition of phenolic acids

Five major phenolic acids were found in roots of germinating soybean seeds: vanillic, syringic, p-coumaric, ferulic and sinapinic acid. Despite the fact that the method applied in the experiment enables extraction of free phenolic acids from plant material, our analysis revealed only ester- or glycoside-bound phenolic acids. Glycosided phenolic acids were detected only in roots subjected to short-term stress (CS24, OS24, COS24) and after post-short-term stress recovery (CR, OR, COR).

Root extracts from the control sample (C72) contained a dominant share of syringic and vanillic acid (Fig. 4). The content of sinapinic, p-coumaric and ferulic acids in the control was much lower. The prolonged cold stress (CS72) caused a decrease in the content of syringic, p-coumaric and ferulic acids whereas the amount of sinapinic acid increased and the content of vanillic acid did not change significantly (Fig. 4a). Under osmotic stress (OS72), a rapid increase in vanillic and sinapinic acids in roots was noticed. Amounts of syringic, p-coumaric and ferulic acids in roots decreased versus the

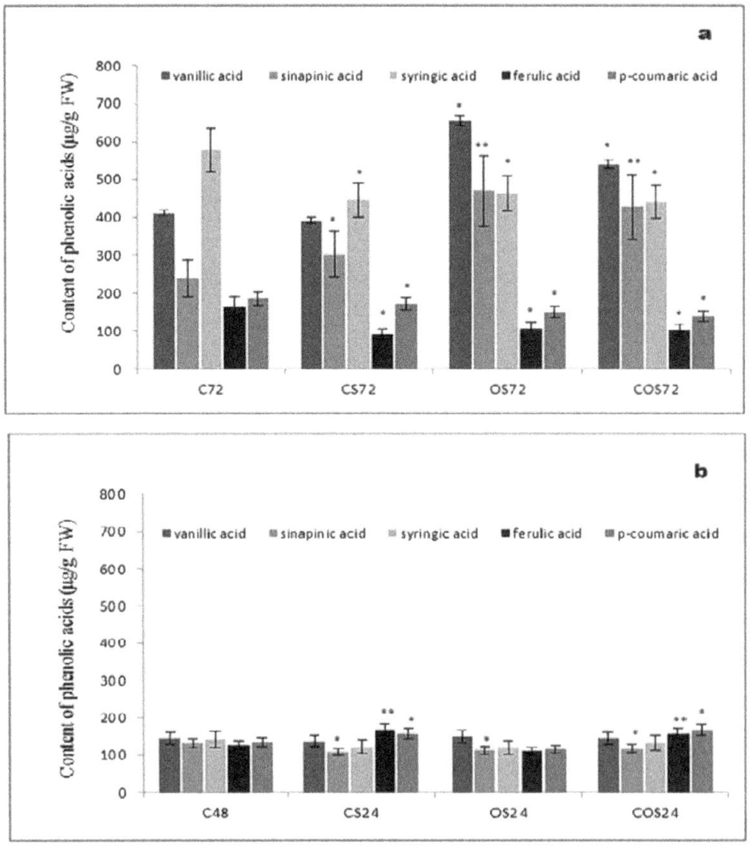

Fig. 4 Changes in the content of esterified phenolic acids in roots of soybean seeds germinating in optimal conditions (C72, C48) and subjected to: long-term continuous stress (**a**; CS72 – cold stress, OS72 – osmotic stress, COS72 – combined stresses) and short-term rapid stress (**b**; CS24 – cold stress, OS24 – osmotic stress, COS24 – combined stresses). Results are presented as means ±*SEM*. Asterisks represent statistically significant differences at: $P < 0.05$ (*) or $P < 0.01$ (**).

control sample. Under the combined stresses (COS72), the pool of phenolic acids isolated from roots contained higher amounts of vanillic and sinapinic acids (Fig. 4a).

After short-lasting cold stress (CS24), the content of ferulic and p-coumaric acids increased significantly relative to the control (C48), yet the amount of sinapinic acid was smaller and the content of vanillic and syringic acids did not change significantly (Fig. 4b). Short-lasting osmotic stress (OS24) resulted in lower amount of sinapinic acid. The content of vanillic, syringic, ferulic and p-coumaric acids did not differ significantly from the one found in the control sample (Fig. 4b). Roots of seeds exposed to short-lasting combined stress (COS24) were found to contain elevated amounts of ferulic and p-coumaric acids in the pool of the analyzed acids (Fig. 4b) in the content of sinapinic acid in roots however significantly decreased. Changes in the content vanillic and syringic acids were not statistically significant but demonstrated a decreasing tendency (Fig. 4b).

After post-cold recovery (CR) a strong increase in the content of vanillic sinapinic and syringic acids was observed, while the amounts of ferulic and p-coumaric acid in roots decreased (Fig. 5a). Recovery after osmotic stress (OR) also resulted in higher concentrations of vanillic, sinapinic and syringic acids. The amounts of ferulic and p-coumaric acids did not differ significantly (Fig. 5b). Similarly, allowing seeds to recover after the combined stresses (COR), their roots contained higher amounts of vanillic, syringic and sinapinic acids. The content of ferulic acid did not differ significantly compared to control. Interestingly, the content of p-coumaric acid in roots decreased. (Fig. 5c).

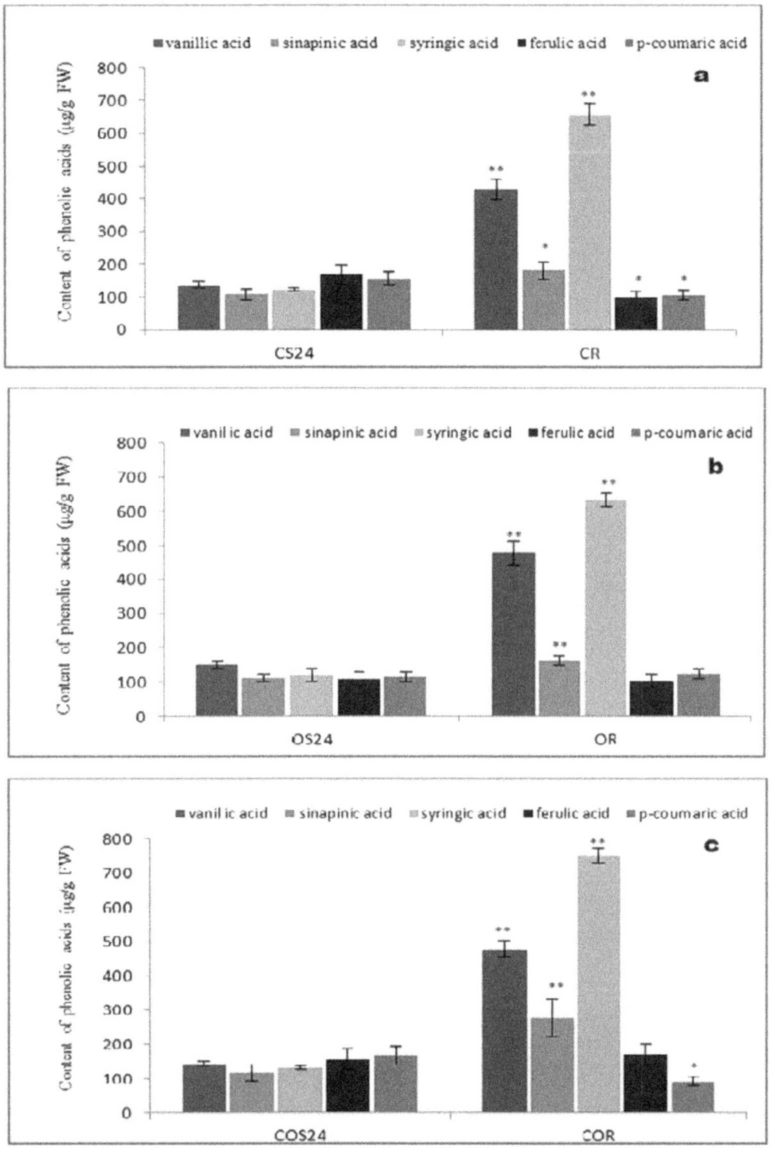

Fig. 5 Changes in the content of glycosided phenolic acids in roots of soybean seeds germinating primarily for 48 h in optimal conditions, then for 24 h under short-term rapid stress (CS24 – cold stress, OS24 – osmotic stress, COS24 – combined stresses) and finally transferred for 24 h to recovery after: cold stress (CR; **a**), osmotic stress (OR; **b**), and combined stresses (COR; **c**). Results are presented as means ±*SEM*. Asterisks represent statistically significant differences at: $P < 0.05$ (*) or $P < 0.01$ (**).

Tab. 1 Changes in isoflavone content in roots of soybean seeds subjected to stress and post-stress recovery.

Sample	Isoflavone content (mg/g FW) ±*SEM*			
	Daidzein	Genistein	Daidzin	Genistin
Long-term stress				
C72	1.86 ±0.09	1.94 ±0.03	0.80 ±0.01	0.20 ±0.04
CS72	6.58 ±0.1**	4.60 ±0.05**	1.38 ±0.05*	0.66 ±0.04*
OS72	5.35 ±0.12**	2.53 ±0.04*	1.22 ±0.08*	0.79 ±0.07*
COS72	10.94 ±0.25**	5.12 ±0.11**	1.67 ±0.09*	1.22 ±0.1*
Short-term stress				
C48	1.83 ±0.07	1.86 ±0.06	1.12 ±0.06	0.27 ±0.01
CS24	4.98 ±0.18**	4.52 ±0.13**	1.14 ±0.29	0.58 ±0.04*
OS24	3.95 ±0.06**	3.51 ±0.08**	1.22 ±0.33	0.55 ±0.07*
COS24	5.84 ±0.11**	4.77 ±0.17**	1.37 ±0.09*	0.73 ±0.03**
Post-stress recovery				
CR	1.31 ±0.02**	2.55 ±0.13**	0.85 ±0.03*	0.24 ±0.01*
OR	1.50 ±0.02**	2.85 ±0.15**	0.70 ±0.04*	0.19 ±0.01*
COR	1.59 ±0.03**	1.37 ±0.03**	0.93 ±0.01*	0.22 ±0.02*

Control samples were germinated for 72 hours in +25°C/water (C72). The study included three subsequent experiments: examining the effect of long and continuous stress (CS72 – cold stress, OS72 – osmotic stress, COS72 – combined stresses), short and rapid stress (CS24 – cold stress, OS24 – osmotic stress, COS24 – combined stresses) and post-stress recovery (CR – recovery after cold stress, OR – recovery after osmotic stress, COR – recovery after combined stresses). Results are presented as means ±*SEM*. Asterisks represent statistically significant differences at: $P < 0.05$ (*) or $P < 0.01$ (**).

Changes in the composition of isoflavones in seed roots

Two isoflavones: genistein and daidzein and their glycoside derivatives: genistein and daidzin were identified in roots of soybean seeds (Tab. 1, Fig. 6). In the control sample (C72), contribution of daidzein and genistein was bigger that daidzin and genistin. After long-term cold stress (CS72), a significant increase in the content of all the analyzed isoflavones was observed. Similarly prolonged osmotic stress (OS72) and the combined stresses (COS72) affected the accumulation of isoflavones (Tab. 1). The rapid increase in the quantity of daidzein under the combined stresses is especially interesting.

Under the short-lasting cold stress (CS24), a statistically significant increase was observed in the content of daidzein, genistin and genistein. The concentration of daidzin in response to cold stress also tended to increase, but the differences were not statistically significant. Similar results were achieved for seeds subjected to osmotic stress (OS24) Under the short-lasting impact of both stresses, roots were observed to contain higher concentrations of all examined isoflavones (Tab. 1).

Post-stress recovery also induced changes in the root content of the analyzed isoflavones. As a result of the recovery after cold stress (CR), concentrations of all the examined isoflavones decreased significantly. During the recovery after osmotic stress (OR), quantities of all the isoflavones considerably declined. Concentrations of the analyzed isoflavones in samples after regeneration after combined stresses (COR) decreased, analogously to the previously discussed stress and recovery conditions (Tab. 1).

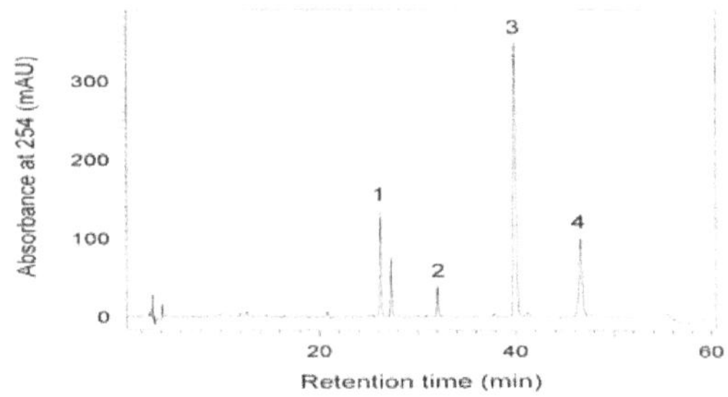

Fig. 6 HPLC chromatogram of isoflavones extracted from in soybean radicles: 1 – deidzin, 2 – genistin, 3 – deidzein, 4 – genistein.

Changes in antioxidant properties of root extracts

It has been shown that the biggest differences in absorbance between the control and samples submitted to stresses, resulting in the ability to scavenge the DPPH• free radical, occurred within the extract concentration range from 1 to 4 mg/100 µl of the sample (Tab. 2). It is noteworthy that a depressed value of absorbance correlates with an improved capacity of extracts to reduce the DPPH• free radical. By returning germinating seeds to optimal conditions their antioxidant capacity measured as the ability to scavenge DPPH• radical raised compared to controls. In all the samples differences were statistically significant at the concentration of 4 mg/100 µl (Tab. 2).

Tab. 2 Capability of soybean root extracts to scavenge DPPH˙ free radical after stress and post-stress recovery.

Extract concentration (mg/100 ul sample)	Absorbance at λ = 517 nm ±*SEM*										
	Prolonged stress				Short stress				Post-stress recovery		
	C72	**CS72**	**OS72**	**COS72**	**C48**	**CS24**	**OS24**	**COS24**	**CR**	**OR**	**COR**
0.5	1.70 ±0.02	1.69 ±0.03	1.71 ±0.04	1.70 ±0.02	1.77 ±0.06	1.70 ±0.03	1.70 ±0.07	1.69 ±0.13	1.70 ±0.15	1.71 ±0.21	1.70 ±0.14
1.0	1.54 ±0.09	1.44 ±0.07*	1.62 ±0.11	1.53 ±0.02*	1.59 ±0.05	1.47 ±0.06	1.50 ±0.09	1.50 ±0.08**	1.37 ±0.11	1.49 ±0.09	1.40 ±0.08
2.0	1.50 ±0.04	1.35 ±0.05*	1.49 ±0.09	1.38 ±0.09*	1.44 ±0.09	1.30 ±0.03	1.41 ±0.05	1.37 ±0.05*	1.15 ±0.03*	1.32 ±0.11	1.19 ±0.08
2.5	1.34 ±0.03	1.20 ±0.05*	1.32 ±0.04*	1.21 ±0.07*	1.38 ±0.04	1.10 ±0.01*	1.25 ±0.02*	1.19 ±0.02*	0.91 ±0.05	1.24 ±0.09	1.01 ±0.05
3.0	1.01 ±0.07	0.98 ±0.03	1.07 ±0.06	1.00 ±0.04	1.21 ±0.02	0.95 ±0.09*	1.15 ±0.03*	1.00 ±0.03*	0.80 ±0.09	1.14 ±0.06*	0.98 ±0.03*
3.5	0.92 ±0.05	0.89 ±0.05	0.90 ±0.05	0.84 ±0.07*	1.06 ±0.03	0.84 ±0.08	0.93 ±0.07	0.82 ±0.02**	0.54 ±0.04**	0.84 ±0.04**	0.76 ±0.03*
4.0	0.88 ±0.02	0.70 ±0.06*	0.77 ±0.05*	0.77 ±0.07*	0.99 ±0.01	0.80 ±0.02**	0.77 ±0.01**	0.70 ±0.02**	0.48 ±0.05**	0.67 ±0.03*	0.59 ±0.04**

Control samples were germinated for 72 hours in +25°C/water (C72). The study included three subsequent experiments: examining the effect of long and continuous stress (CS72 – cold stress, OS72 – osmotic stress, COS72 – combined stresses), short and rapid stress (CS24 – cold stress, OS24 – osmotic stress, COS24 – combined stresses) and post-stress recovery (CR – recovery after cold stress, OR – recovery after osmotic stress, COR – recovery after combined stresses). Results of absorbance at λ = 517 nm are presented as means ±*SEM*. Asterisks represent statistically significant differences at: $P < 0.05$ (*) or $P < 0.01$ (**).

The antioxidant capacity of extracts was also analyzed in terms of the ability to scavenge the $ABTS^{+\bullet}$ free radical (Fig. 7, Fig. 8). The free radical scavenging capacity was expressed in mmol of Trolox equivalent per gram of the analyzed extract. In samples exposed to prolonged stresses the antioxidant capacity was significantly higher than in the control (Fig. 7a), Similar results were obtained by analyzing the ability to scavenge $ABTS^{+\bullet}$ in extracts from roots of seeds germinating under short-lasting stresses (Fig. 7b). No statistically significant differences were detected in the scavenging capacity of extracts from roots of seeds submitted to recovery after stresses (Fig. 8).

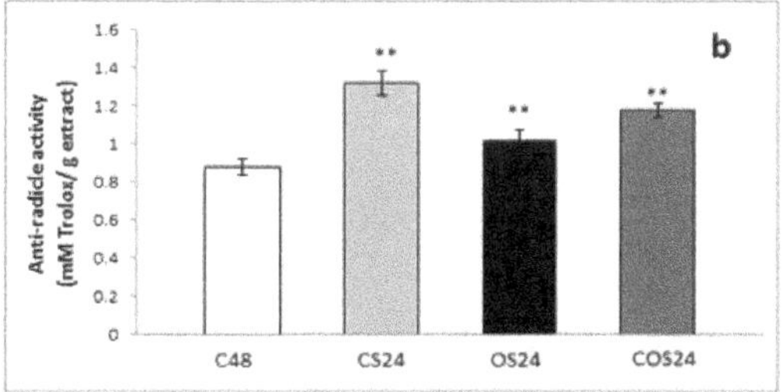

Fig. 7 Antioxidant activity of root extracts from soybean seeds germinating in optimal conditions (C72, C48) and long-term continuous stress (CS72 – cold stress, OS72 – osmotic stress, COS72 – combined stresses; **a**) and short-term rapid stress (CS24 – cold stress, OS24 – osmotic stress, COS24 – combined stresses; **b**). Results are presented as means ±*SEM*. Asterisks represent statistically significant differences at: $P < 0.05$ (*) or $P < 0.01$ (**).

Discussion

Plants are constantly exposed to stresses in their natural habitats. Low temperatures and water deficit are the main limiting factors affecting crop yields worldwide. In this study, germinating soybean seeds were exposed to cold stress, osmotic stress and concomitant cold and osmotic stress, which in their strength were analogous to stresses naturally occurring in the moderate climate. These stresses, both brief and prolonged, cause various morphological, physiological and metabolic changes in tissues of germinating seeds, but the actual response to a given stress factor can differ depending on the duration of stress [16]. Chilling temperatures and osmotic stress affect the germination and growth of plants by disturbing most of cellular processes [17]. In our study both long- or short-lasting stress inhibited of the growth of soybean radicles (data not presented). Similar results were obtained in the earlier research [18–23]. During post-stress recovery accelerated growth processes and more raid accumulation of root fresh matter were observed, which is analogous to the data reported by other authors [16,24,25].

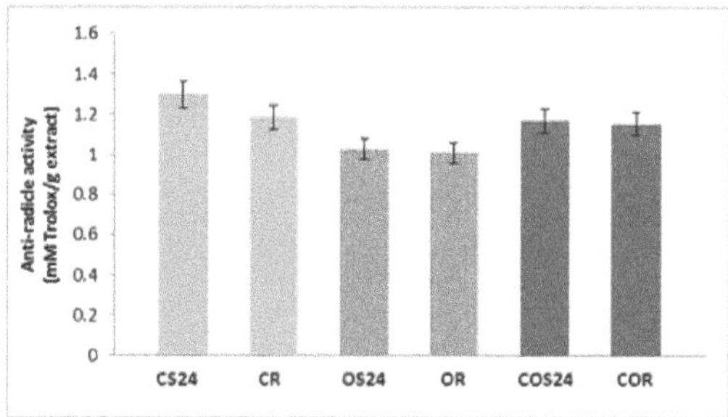

Fig. 8 Antioxidant activity of root extracts from soybean seeds germinating primarily for 48 h in optimal conditions, then for 24 h under short-term stress (CS24 – cold stress, OS24 – osmotic stress, COS24 – combined stresses) and finally transferred for 24 h to recovery after: cold stress (CR), osmotic stress (OR), cold stress combined with osmotic stress (COR). Results are presented as means ±*SEM*.

Our analysis of the total content of phenolics in roots demonstrated that the examined stresses (regardless of their duration) induced accumulation of phenolic compounds. These results are supported by the reports from other studies [6,19,26,27]. It is worth noticing that once the germinating seeds were transferred to optimal conditions, the total content of phenolics in roots declined, which suggests return to metabolic balance.

Plants produce a large variety of secondary products that contain a phenol group. Polyphenols are the most numerous group of natural antioxidants, highly varied in terms of structure, molecular weight or physical and chemical properties. In plants, they occur mostly as glycosides or esters [5]. Phenolic compounds in plants are said to play a key role as defense compounds counteracting negative effects of free radicals produced in stress conditions [5,6,28]. Thus, identification of the protective role of phenolic compounds, especially towards particular metabolites, requires monitoring.

Phenolic acids are an important link in the secondary metabolism of plants. They are a numerous group of derivatives of cinnamic acid (e.g. p-coumaric, ferulic, sinapinic acids) or benzoic acid (e.g. protocatechuic, gallic, vanillic acids). Their main function in plant defense mechanisms is antioxidant action [29], but they are also indicated (particularly vanillic and ferulic acids) as involved in the inhibition of cell elongation in growing organs [30]. Our results revealed five phenolic acids in soybean radicles, occurring only in the bound form, which stays in agreement with relevant references [31]. Vanillic and syringic acids dominated among the phenolic acids found in radicles under long and continuous stress. Intensive growth in the amount of vanillic and sinapinic acids in response to stress can be explained by their relatively high antioxidant capacity. Szwajgier et al. [31] demonstrated that these two acids are characterized by the highest free radical scavenging potential, and hence expected to protect roots from the adverse effects produced by free radicals [26]. Taken together this indicates that the stress-induced metabolic re-engineering involves not only the antioxidant activity but also a deceleration of cell growth.

Exposure of germinating seeds to sudden and brief stresses did not cause accumulation of vanillic and sinapinic acids, although the content of ferulic and p-coumaric acids in roots increased. Such modification in the composition of phenolic acids may be caused by differences in the plants' response to continuous and persistent stresses versus sudden and transient ones. Ferulic and p-coumaric acids are the key links in the biosynthesis of lignin, whose elevated accumulation is observed under stress conditions [32]. Despite the post-stress recovery period, the content of vanillic, synaptic and syringic acids in soybean radicles were higher than in stress-treated samples. This may be explained by hypothesis, that accumulation of phenolic acids during post-stress recovery could be associated with the occurrence of a secondary oxidative stress after a period of low temperature impact [26,33].

Our results confirmed close correlation between the content of phenolics and the antioxidant capacity of radicle extracts reported in previous studies [2,34,35]. It can be assumed that accumulation of phenolic compounds in roots of germinating seeds is used to minimize acute effects of secondary oxidative stress triggered by cold and osmotic stress. Increased antioxidant activity was also observed in extracts after post-stress recovery, which may be a result of the activation of the damage-repair system in tissues exposed to abiotic stress [36].

The content of all identified isoflavones increased in all stress-treated samples, which confirms previously published data [26,33,37,38]. Wu [39] demonstrated that soybean isoflavones show free radical scavenging potential, with aglycones (daidzein and genistein) being stronger scavengers than their glycoside conjugates (daizdin and genistin). On the other hand, the ability of isoflavones to reduce free radicals is much weaker than other phenolic compounds [40]. We hypothesize that accumulation of isoflavones is more likely due to synthesis of phytoalexins, hence daidzein and genistein serve as their precursors [41]. Enzymes of the isoflavone synthesis pathway may combine into multicomplexes with enzymes of the phenylpropanoid pathway [32,42–44], occurring in a cell loosely attached to the endoplasmic reticulum and connected to transmembrane proteins. Such complexes with enzymes mediating lignin biosynthesis have also been detected in soybean [45] and in *Arabidopsis* [41]. All this data confirms that the response of soybean radicles both to long and short lasting cold and osmotic stress, involves activation of phenylopropanoid pathways. However, mutual relationships between secondary metabolic pathways as well as mechanisms regulating stress defense require further investigation.

Acknowledgments

The presented research was funded by the Department of Biology and Biotechnology (No. 528-0209.806), University of Warmia and Mazury in Olsztyn.

Authors' contributions

The following declarations about authors' contributions to the research have been made: general concept of the study: SW; experimental work: SS, AK, AM, RA; data collection and processing: AB; design of experiments and manuscript writting: SS; manuscript revision: RA, SW.

References

1. França SC, Roberto PG, Marins MA, Puga RD, Rodrigues A, Pereira JO. Biosynthesis of secondary metabolites in sugarcane. Genet Mol Biol. 2001;24(1–4):243–250.
2. Oh MM, Carey EE, Rajashekar CB. Environmental stresses induce health-promoting phytochemicals in lettuce. Plant Physiol Biochem. 2009;47(7):578–583.
3. Zielińska-Dawidziak M, Siger A. Effect of elevated accumulation of iron in ferritin on the antioxidants content in soybean sprouts. Eur Food Res Technol. 2012;234(6):1005–1012.
4. Sharma P, Jha AB, Dubey RS, Pessarakli M. Reactive oxygen species, oxidative damage, and antioxidative defense mechanism in plants under stressful conditions. J Bot. 2012;2012:1–26.
5. Amarowicz R, Weidner S. Biological activity of grapevine phenolic compounds. In: Roubelakis-Angelakis KA, editor. Grapevine molecular physiology and biotechnology. Dordrecht: Springer Netherlands; 2009. p. 389–405.

6. Cheynier V, Sarni-Manchado P, Quidean S, editors. Recent advances in polyphenol research. Oxford: Wiley-Blackwell; 2012. (vol 3).
7. Chalker-Scott L, Fuchigami LH. The role of phenolic compounds in plant stress responses. In: Li PH, editor. Low temperature stress physiology in crops. Boca Raton, FL: CRC Press; 1989. p. 27–40.
8. Ippolito A, Nigro F, Lima G. Mechanism of resistance to *Botrytis cinerea* in wound of cured kiwifruits. Acta Hortic. 1997;444:719–724.
9. Amarowicz R, Raabe B. Antioxidative activity of leguminous seed extracts evaluated by chemiluminescence methods. Z Naturforsch C. 1997;52:709–712.
10. Amarowicz R, Pegg RB, Rahimi-Moghaddam P, Barl B, Weil JA. Free-radical scavenging capacity and antioxidant activity of selected plant species from the Canadian prairies. Food Chem. 2004;84(4):551–562.
11. Weidner A, Amarowicz R, Karamać M, Frączek E. Changes in endogenous phenolic acids during development of *Secale cereale* caryopses and after dehydration treatment of unripe rye grains. Plant Physiol Biochem. 2000;38(7–8):595–602.
12. Amarowicz R, Karamac M, Weidner S, Abe S, Shahidi F. Antioxidant activity of wheat caryopses and embryos extracts. J Food Lipids. 2002;9(3):201–210.
13. Re R, Pellegrini N, Proteggente A, Pannala A, Yang M, Rice-Evans C. Antioxidant activity applying an improved ABTS radical cation decolorization assay. Free Radic Biol Med. 1999;26(9–10):1231–1237.
14. Amarowicz R, Piskuła M, Honke J, Rudnicka B, Troszyńska A, Kozłowska H. Extraction of phenolic compounds from lentil seeds (*Lens culinaris*) with various solvents. Pol J Food Nutr Sci. 1995;4(3):53–62.
15. Amarowicz R, Weidner S. Content of phenolic acids in rye caryopses determined using DAD-HPLC method. Czech J Food Sci. 2001;19(6):201–205.
16. Uma S, Prasad TG, Kumar MU. Genetic variability in recovery growth and synthesis of stress proteins in response to polyethylene glycol and salt stress in finger millet. Ann Bot. 1995;76(1):43–49.
17. Raven PH. Biology of plants. 7th ed. New York, NY: W.H. Freeman; 2005.
18. Kosowska M, Frączek E, Amarowicz R, Karama M, Abe S, Weidner S. Water-deficit-induced changes in cytoskeleton-bound and other polysomal populations in embryonic tissue during triticale caryopsis germination. Acta Physiol Plant. 2004;26(1):67–74.
19. Wróbel M, Karama M, Amarowicz R, Fr czek E, Weidner S. Metabolism of phenolic compounds in Vitis riparia seeds during stratification and during germination under optimal and low temperature stress conditions. Acta Physiol Plant. 2005;27(3):313–320.
20. Costa França MG, Pham Thi AT, Pimentel C, Pereyra Rossiello RO, Zuily-Fodil Y, Laffray D. Differences in growth and water relations among *Phaseolus vulgaris* cultivars in response to induced drought stress. Env Exp Bot. 2000;43(3):227–237.
21. Caruso A, Morabito D, Delmotte F, Kahlem G, Carpin S. Dehydrin induction during drought and osmotic stress in *Populus*. Plant Physiol Biochem. 2002;40(12):1033–1042.
22. Ismail AM, Hall AE. Variation in traits associated with chilling tolerance during emergence in cowpea germplasm. Field Crops Res. 2002;77(2–3):99–113.
23. Munns R. Comparative physiology of salt and water stress. Plant Cell Env. 2002;25(2):239–250.
24. Duncan RF, Hershey JW. Protein synthesis and protein phosphorylation during heat stress, recovery, and adaptation. J Cell Biol. 1989;109(4 pt 1):1467–1481.
25. Weidner S, Kordala E, Brosowska-Arendt W, Karamać M, Kosińska A, Amarowicz R. Phenolic compounds and properties of antioxidants in grapevine roots (*Vitis vinifera* L.) under low-temperature stress followed by recovery. Acta Soc Bot Pol. 2009;78(4):279–286.
26. Posmyk MM, Bailly C, Szafrańska K, Janas KM, Corbineau F. Antioxidant enzymes and isoflavonoids in chilled soybean [*Glycine max* (L.) Merr.] seedlings. J Plant Physiol. 2005;162(4):403–412.
27. Chung IM, Kim JJ, Lim JD, Yu CY, Kim SH, Hahn SJ. Comparison of resveratrol, SOD activity, phenolic compounds and free amino acids in *Rehmannia glutinosa* under temperature and water stress. Env Exp Bot. 2006;56(1):44–53.
28. Solecka D. Role of phenylpropanoid compounds in plant responses to different stress factors. Acta Physiol Plant. 1997;19(3):257–268.
29. Karamać M, Kosińska A, Pegg RB. Comparison of radical-scavenging activities for selected phenolic acids. Pol J Food Nutr Sci. 2005;14(2):165–169.
30. Dixon RA, Paiva NL. Stress-induced phenylpropanoid metabolism. Plant Cell. 1995;7(7):1085–1097.
31. Szwajgier D, Pielecki J, Targoński Z. Antioxidant activities of cinnamic and benzoic acid derivatives. Acta Sci Pol Technol Aliment. 2005;4(2):129–142.
32. Dixon RA, Achnine L, Kota P, Liu CJ, Reddy MSS, Wang L. The phenylpropanoid pathway and plant defence – a genomics perspective. Mol Plant Pathol. 2002;3(5):371–390.
33. Janas KM, Cvikrová M, Pałagiewicz A, Szafranska K, Posmyk MM. Constitutive elevated accumulation of phenylpropanoids in soybean roots at low temperature. Plant Sci. 2002;163(2):369–373.
34. Naczk M, Shahidi F. Phenolics in cereals, fruits and vegetables: occurrence, extraction and analysis. J Pharm Biomed Anal. 2006;41(5):1523–1542.
35. López-Amorós ML, Hernández T, Estrella I. Effect of germination on legume phenolic compounds and their antioxidant activity. J Food Compos Anal. 2006;19(4):277–283.
36. Prasad TK. Mechanisms of chilling-induced oxidative stress injury and tolerance in developing maize seedlings: changes in antioxidant system, oxidation of proteins and lipids, and protease activities. Plant J. 1996;10(6):1017–1026.
37. Graham TL, Graham MY. Signaling in soybean phenylpropanoid responses (dissection of primary, secondary, and conditioning effects of light, wounding, and elicitor treatments). Plant Physiol. 1996;110(4):1123–1133.
38. Piślewska M, Bednarek P, Stobiecki M, Zielińska M, Wojtaszek P. Cell wall-associated isoflavonoids and β-glucosidase activity in *Lupinus albus* plants responding to environmental stimuli. Plant Cell Env. 2002;25(1):20–40.
39. Wu Q. Purification and antioxidant activities of soybean isoflavones [Master thesis]. Baton Rouge, LA: Departament of Food Science, Louisiana State University; 2003.
40. Lee CH, Yang L, Xu JZ, Yeung SYV, Huang Y, Chen ZY. Relative antioxidant activity of soybean isoflavones and their glycosides. Food Chem. 2005;90(4):735–741.
41. Yu O, Jung W, Shi J, Croes RA, Fader GM, McGonigle B, et al. Production of the isoflavones genistein and daidzein in non-legume dicot and monocot tissues. Plant Physiol. 2000;124(2):781–794.
42. Dixon RA, Harrison MJ, Paiva NL. The isoflavonoid phytoalexin pathway: from enzymes to genes to transcription factors. Physiol Plant. 1995;93(2):385–392.
43. Dixon RA, Summer LW. Legume natural products: understanding and manipulating complex pathways for human and animal health. Plant Physiol. 2003;131(3):878–885.

44. Winkel-Shirley B. Flavonoid biosynthesis. a colorful model for genetics, biochemistry, cell biology, and biotechnology. Plant Physiol. 2001;126(2):485–493.

45. Rawel HM, Czajka D, Rohn S, Kroll J. Interactions of different phenolic acids and flavonoids with soy proteins. Int J Biol Macromol. 2002;30(3–4):137–150.

20

The growth and saponin production of *Platycodon grandiflorum* (Jacq.) A. DC. (Chinese bellflower) hairy roots cultures maintained in shake flasks and mist bioreactor

Natalia Urbańska[1], Joanna Giebułtowicz[2], Olga Olszowska[1], Wojciech J. Szypuła[1]*

[1] Department of Pharmaceutical Biology and Medicinal Plants Biotechnology, Medical University of Warsaw, Banacha 1, 02-097 Warsaw, Poland

[2] Department of Bioanalysis and Drug Analysis, Medical University of Warsaw, Banacha 1, 02-097 Warsaw, Poland

Abstract

The growth and saponin accumulation were measured in two lines of transgenic hairy roots of *Platycodon grandiflorum*, Pl 6 and Pl 17, cultured for 8 weeks in 250-ml shake flasks containing 50 ml of hormone-free woody plant medium supplemented with 40 g/l sucrose and in the Pl 17 line cultured for 12 weeks in a 5-l mist bioreactor containing 1.5 l of the same medium. With both methods, the growth of transgenic hairy roots was assessed as both fresh and dry weight and the biomass growth was correlated with the conductivity and sucrose uptake. The accumulation of saponins was measured and compared with that in roots derived from the field cultivation. The saponin concentrations were significantly higher in the two hairy root lines cultured in shake flasks [6.92 g/100 g d.w. (g%) and 5.82 g% in Pl 6 and Pl 17, respectively] and the line cultured in the bioreactor (5.93 g%) than in the roots derived from the field cultivation (4.02 g%). The results suggest that cultures of *P. grandiflorum* hairy roots may be a valuable source for obtaining saponins.

Keywords: *Platycodon grandiflorum*; Chinese bellflower; hairy roots; saponins; mist bioreactor

Introduction

Hairy, or transgenic roots are characterized by a rapid and stable growth, which importantly can be initiated and maintained in a hormone-free medium. They often synthetize approximately the same and in some cases even greater amounts of secondary metabolites compared to their mother plants [1,2] and the rate of production remains stable. This makes hairy roots of medicinal plants a good source of valuable compounds and obviously their large-scale culture in bioreactors holds immense potential for the pharmaceutical industry [3].

Platycodon grandiflorum (Jacq.) A. DC. (Campanulaceae) is an ornamental plant growing wild in northern Asia, in China, Korea, Japan and east Siberia. For culture it needs light, medium moisture, organically rich, well-drained loams and full sun to part shade.

Platycodi radix, the root of *P. grandiflorum*, has been used as food and in traditional oriental medicine to treat bronchitis, asthma and other pulmonary diseases. Triterpene saponins, called platycosides, are responsible for the medicinal value of this plant material. They have demonstrated a broad spectrum of therapeutic effects, such as antitumor [4–7], anti-inflammatory [8,9], hepatoprotective [10], antiatherosclerotic [11] and antinociceptive [12]. Additionally, these compounds were found to be a potent adjuvant of specific cellular and humoral immune responses, with the potential use in vaccine production [13–15]. Finally, saponins from *Platycodi radix* stimulated osteoblast differentiation via *RUNX2*, a principal osteogenic master gene for bone formation [16].

Although the extensive pharmacological studies of these valuable compounds were undertaken, to date the sum of saponins, not the chosen, in hairy root cultures of *P. grandiflorum* was determined only by our research team [17]. Besides, this is the first report of the growth of *P. grandiflorum* transgenic roots cultured in a mist bioreactor and their yield of saponins. The method of *Platycodi radix* saponins determination published in "Chinese pharmacopeia" [18] exhibited as no suitable for quantitative analysis of saponins from *P. grandiflorum* transgenic roots. Therefore, the authors elaborated their own method [17].

The compounds mentioned above are the main chemicals found in *Platycodi radix*. They are typical composed of oleaene backbones with two side chains: one is a 3-O-glucose linked by glycosidic bond, and the other is a 28-O-arabinose-rhamnose-xylose-apiose linked by an ester bond [19]. To date, more than 55 triterpenoid saponins have been isolated from *Platycodi radix* [20]. Examples are platycodins A–I and polygalacins D and D2. According to WHO monographs

* Corresponding author. Email: wszypula@wum.edu.pl

Handling Editor: Jan Rybczyński

standards [21] the content of saponins should exceed 2%. An effective HPLC method to analyze 18 platycosides from three-year-old Chinese bellflower roots was developed. Amongst them platycoside E showed the highest content (2.00 mg/g), followed by polygalacin D2 (1.77 mg/g) and 3″-O-acetylplatyconic acid A (1.35 mg/g) [22]. Considerably smaller amounts of 7 platycosides in the wild and hairy roots of *P. grandiflorum* received Kim et al. [23]. It could be due to the other origin or age of the plants. No information is given about that in the article. Authors developed hairy roots of *P. grandiflorum* with inserted gene of 3-hydroxy-3-methylglutaryl-coenzyme A reductase. This enzyme catalyzes the synthesis of triterpenoids. The highest level of total platycosides (1.60 ±0.2 mg/g) was detected in the T7 line, what was 2.5-fold higher than that of the controls.

Chemical investigation on *Platycodi radix* revealed that, apart from saponins, in this plant material phytosterols Β-amyrin and α-spinasterol [23], polycaetylene lobetyolin (0.019%) [24], phenylpropanoid esters [25], a mixed compound of n-tetracosanoic acid (lignoceric acid), n-hexacosanoic acid (cerotic acid), n-octacosanoic acid and α-monopalmitin [26], polysaccharides named PGPN, PGPA1, PGPA2 and PGPA3 [27], arabinogalactan PGAW1 [28], homogalacturonan PGA4-3b [linear poly-(1→4)-α-D-galactopyranosyluronic acid] [29] and polyunsaturated fatty acids: linoleic acid (20%), palmitic acid (7%), stearic acid (<1%) and oleic acid (<1%) [30] are present.

Aerial parts of *P. grandiflorum* contain no saponins but phenolics, particularly luteolin-7-O-glucoside and apigenin-7-O-glucoside [31]. From seeds flavoplatycoside, grandoside and four flavonoids: (2R,3R)-taxifolin, quercetin 7-O-glucoside, luteolin-7-O-glucoside and quercetin 7-O-rutinoside were isolated [32]. Flowers of *P. grandiflorum* comprise the following constituents: apigenin, apigenin-7-O-Β-D-glucopyranoside, apigenin-7-O-(6″-O-acetyl)-Β-D-glucopyranoside, luteolin, luteolin-7-O-Β-D-glucopyranoside, luteolin-7-O-(6″-O-acetyl)-Β-D-glucopyranoside, isorhamnetin-3-O-neohesperoside, 4-O-caffeoylquinic acid, chlorogenic acid methyl ester, 4-O-Β-D-glucopyranosylcaffeic acid, lobetyolin, cordifolioidyne C, isomultiflorenyl acetate, Β-sitosterol glucoside and α-spinosterol [33].

In vitro *P. grandiflorum* plants regeneration was elaborated by some scientists, including Urbańska et al. [34]. The aim was achieved by two ways: by somatic embryogenesis [35] or by two-step action: repeated shoot-section and following rooting of the developed new shoots [34,36,37]. In both methods developed plantlets, which grew to maturity in the soil.

Somatic embryogenesis was conducted as follows: mature zygotic embryos formed embryogenic calluses on Murashige and Skoog (MS) medium [38] supplemented with 2,4-dichlorophenoxyacetic acid (2,4-D; 4.52 μM). From calli cell suspension culture was established using the same, but liquid, medium and the same phytohormone at the same dose. Following transfer to solid MS basal medium, cells gave rise to somatic embryos.

The MS medium was chosen as well for proliferation of shoots. For good proliferation rate medium should be supplemented with benzyladenine (BA; 4.4 μM) [36], with 2-isopenthenyladenine (2iP; 4.92 μM) and indole-3-acetic acid (IAA; 2.85 μM) [37], or with kinetine (KIN; 0.01 mg/l) and α-naphtaleneacetic acid (NAA; 0.01 mg/l) [34]. The highest rate of rooting of obtained shoots was induced on MS medium containing 4.90 μM or 9.80 μM indole-3-butyric acid (IBA) [37], 0.49 μM IBA [36] or on half-strength MS medium with IBA (0.05 mg/l) and saccharose (40 g/l) [34]. The medium enrichment on saccharose at a dose 40 g/l profitably influenced on the plantlets development, both shoots and roots [34]. The root, shoot and leaf thickness was found to be increased, while the length of shoots was decreased.

Urbańska elaborated the best conditions for maintenance the culture of non-transformed roots of *P. grandiflorum* (data not published). As the explants the plantlets growing by three passages on half-strength MS (MS/2) medium with IBA (1 mg/l) and saccharose (40 g/l) were used. The roots were cut from the shoots and their growth was tested on the following liquid media: MS/2, Gamborg medium (B5) [39] and woody plant medium (WPM) [40] full or half-strength media all with NAA (1 mg/l) and saccharose (40 g/l), the last mentioned also with IBA (1 mg/l) instead of NAA. The best for growing proved to be WPM medium with IBA (1 mg/l) and saccharose (40 g/l).

Besides *P. grandiflorum*, the other plants, which accumulate triterpene saponins in their roots, occur in the nature. They belong to genus *Panax* and *Glycyrrhiza*. For both genuses in vitro culture of different cells and tissues have been established successfully.

For many years the ability to increase secondary metabolites production by colchicine is used in the practice. For this purpose, the seeds of *Glycyrrhiza glabra* var. *glandulifera* were treated with colchicine. The amount of glycyrrhizic acid, the main biologically active saponin among that produced by licorice (*Glycyrrhiza*), in callus formed from the plantlets grown from seeds treated with colchicines increased [41]. The other way to enhance the amount of glycyrrhize acid is the employment of elicitors. Methyl jasmonate was most efficient in increment glycyrrhize acid production in *G. inflata* hairy roots. What is more, sucrose (6%) was optimal for growth and glycyrrhizic acid accumulation [42]. The ammonia/nitrate ratio and phosphate concentration influence glycyrrhizic acid accumulation as well. The content of glycyrrhize acid in adventitious roots of *G. uralensis* reached the optimum (0.47 mg/g) at an ammonia/nitrate ratio of 15:15 and when 0.625 μM phosphate was applicated [43]. The method of elicitation to improve glycyrrhizic acid productiveness was applied in the roots of 65-day old plants of *G. glabra*. Treatment of plantlets with 0.1–2 mM of methyl jasmonate or with 0.1 and 1 mM salicylic acid enhanced the production of glycyrrhizic acid, as compared to the controls [44]. There are the other ways for increasing the amount of glycyrrhizic acid. In the hairy roots of *G. uralensis* with inserted gene of squalene syntase, the enzyme, which plays a regulatory role in the biosynthesis of triterpene saponins, the highest glycyrrhizic acid content was obtained [45]. Finally, it worth to add that the scale-up of the suspension culture of *G. inflata* was established successfully in a bioreactor with a low-shear impeller [46]. The balloon-type bubble bioreactor can be utilized as well to provide a suspension culture [47].

Elicitation as the method for increase *Panax ginseng* saponins, known as ginsenosides, was used in hairy root culture of this species [48]. Jasmonic acid in the range 4.8–23.8 μM strongly improved total ginsenosides production. The same elicitor at 2.0 mg/l increased significantly ginsenoside yield in adventitious root culture of *P. ginseng* in flasks and balloon type airlift bioreactor [49]. The other elicitor, vanadate at a dose 50 μM, was used successfully to improve ginsenoside production in suspenson cultures of *P. ginseng* [50].

Initial sucrose concentration and inoculum size influence ginseng saponin production by suspension culture of *P. ginseng*. The maximum saponin production of 275 mg/l was achieved at 6 g/l of inoculum size and 60 g/l initial medium sucrose [51]. Temperature and light quality influence ginsenoside production by hairy root culture of *P. ginseng* in the bubble bioreactor. Ginsenoside production was optimal under 20°C/13°C day (12 h)/night (8 h) cycle. Ginsenoside accumulation was optimum in the cultures grown under fluorescent light [52]. Comprehensive review of influence of various factors on ginsenoside production published Wu and Zhong [53].

Material and methods

Hairy root culture induction

Hairy roots of *Platycodon grandiflorum* were established using *Agrobacterium rhizogenes* ATCC 15834 [17]. The bacteria were grown on yeast mannitol broth (YMB) solid medium [54] for 120 h at 25°C, in the dark. Next, single colonies were inoculated into yeast extract broth (YEB) liquid medium [55] with acetosyringone (0.2 mg/l) 72 h at 25°C, in the dark, with continuous shaking. The bacterial cultures were diluted (1:4) with YEB liquid medium before transformation. The leaves and stems of 50 micropropagated plants developed from the seeds and next growing for three passages on the MS/2 medium with IBA (1 mg/l) and saccharose (40 g/l) were directly wounded with a sterile needle containing bacterial suspension. The explants were cultured on phytohormone-free MS solid medium with sucrose (40 g/l), for 96 h at 25°C, in the dark, and then transferred to the same medium with the addition of Claforan® (500 mg/l). The roots emerged at the wounding site after 15 days of incubation under 40 μmol m^{-2} s^{-1} fluorescent lights for 18 h per day (Fig. 1).

To prove the transformed nature of hairy roots PCR analysis was done. Total genomic DNA was extracted from hairy roots using DNeasy Plant Mini Kit (Qiagen, USA). Primers for detecting *rolA* gene (5′-CGT TGT CGG AAT GGC CCA GAC C-3′ and 5′-CGT AGG TCT GAA TAT TCC GGT CC-3′), *rolB* gene (5′-CGT TGT CGG AAT GGC CCA GAC C-3′ and 5′-CGT AGG TCT GAA TAT TCC GGT CC-3′) and *rolC* gene (5′-TGT GAC AAG CAG CGA TGA GC-3′ and 5′-GAT TGC AAA CTT GCA CTC GC-3′) were used in amplification process. Primers amplified a 248-bp fragment of *rolA* coding region, 652-bp fragment of *rolB* coding region and a 490-bp fragment of *rolC* coding region respectively. Primers for detecting the *virC* gene (5′-ATCATTTGTAGC GACT-3′ and 5′-AGCTCAAACCTGCTTC-3′) outside the T-DNA of Ri plasmid were also used to eliminate the possibility of *A. rhizogenes* contamination of the hairy roots lines. The negative controls were DNA from non-transformed roots, clone number 26 and positive control was *A. rhizogenes*, strain ATCC 15834 respectively. The amplification conditions for *rolA* and *rolC* were as follow: 2 min melting at 94°C followed by 35 cycles of a 1 min melting at 94°C, a 1 minute annealing at 55°C and 1 min elongation at 72°C and final elongation for 5 min at 72°C. The amplification conditions for *rolC* were as follow: 3 min melting at 94°C followed by 30 cycles of a 1 min melting at 94°C, a 1 minute annealing at 55°C and 1 min elongation at 72°C and final elongation for 5 min at 72°C. PCR products were visualized by electrophoretic separation on 1% (w/v) agarose gels in 1× TBE buffer and staining with ethidium bromide.

Hairy root culture in shake flasks

Two hairy root lines (Pl 6, Pl 17) showing rapid growth were maintained in 250-ml Erlenmyer flasks with 50 ml of hormone-free WPM liquid medium with sucrose (40 g/l), which proved to be the best for optimal growth. The pH value was adjusted to 5.7 before autoclaving. The culture was incubated in the dark at 25°C on a rotary shaker (100 rpm) and subcultured every six weeks. The average inoculum size was 0.4 ±0.1 g fresh weight, which corresponded to 0.04 g

Fig. 1 **a** *Platycodon grandiflorum* hairy roots growing from inoculated site of plantlet, week 2 after inoculation. **b** Hairy roots on hormone-free WPM medium with sucrose (40 g/l). Culture week 2.

of dry biomass. After 14 passages, the transgenic line Pl 17, whose growth was superior to that the Pl 6 line, was used for studies in a bioreactor.

Hairy root culture in a bioreactor

The mist bioreactor used in this study measured 400 mm in height and 200 mm in diameter. It consisted of one 5 l-glass vessel fitted with a stainless steel mesh placed at the level of 150 mm from the bottom. The air inlet and the air outlet with filters were situated on the top of the column. The initial volume of the medium was 1.5 l. A peristaltic pump (MasterFlex® L/S, Cole Pomer, USA) maintained recirculation and dosing of the hormone-free WPM medium supplemented with 40 g/l sucrose through a polypropylene dispersal nozzle situated at the level of 70 mm from the bottom of the vessel. The operating time of the pump was 60 s with the medium supplied at 10 s intervals. The air flow rate was 0.4 l/min. The bioreactor was inoculated with four fragments of hairy roots. The inoculum size was 18 g fresh weight, which was proportional to the inoculum size 0.4 g in 50 ml of medium in the shake flasks. The roots grew at 23°C, in the dark, for 12 weeks. After 8 weeks of culture the medium was exchanged and 2 weeks later 1 l of the nutrient was added to the medium.

Measurement of root growth

The growth of transgenic roots was assessed as fresh and dry weight and by correlating the biomass growth with conductivity and sucrose uptake. For the fresh weight determination, the roots were gently pressed on filter paper to remove excess water, and weighed. The growth rate (*GR*) was calculated using the formula [56]: *GR* = final weight (g) – initial weight (g)/initial weight (g) × 100%.

The dry weight was determined after lyophilization. The conductivity (expressed in miliSiemens, mS) was determined using a conductivity meter (model CC-317, Elmetron, Poland; sensitivitity 0.5%). The sucrose concentration (expressed as percentage) was measured using a saccharimetr (model RR 10, PZO, Poland; sensitivity 0.1%). The pH was determined using a pH-meter (model N5170, ELWRO, Poland; sensitivity 0.01).

The contents of three flasks of the transgenic hairy root line Pl 17 were harvested and analyzed for biomass, conductivity and sucrose concentration at 7 days intervals during the 8-week experiment. Samples of the medium from bioreactor obtained with a sampler were tested for conductivity and sucrose uptake from the beginning of week 8.

The convenient formula to describe the inverse relationship between biomass and conductivity during the shake flask culture was $\Delta m = a\,\Delta k + b$ where Δm (g) was the increase of dry biomass of hairy roots and Δk (mS/cm) was the decrease in conductivity. The *a* and *b* values were found to be 0.20 (0.03) and –0.08 (0.07) respectively. The correlation coefficient was 0.95 ($P = 0.0001$).

The relationship between biomass and sucrose level was described by the formula $\Delta m = a\,\Delta s + b$ where Δm (g) was the increase of dry biomass of hairy roots and Δs (%) was the decrease of sucrose level. The *a* and *b* values were found to be 0.30 (0.03) and –0.23 (0.06), respectively. The correlation coefficient was 0.98 ($P = 0.0001$).

Determination of saponins accumulation

The saponin content was determined in the hairy roots lines Pl 17 and Pl 6 cultivated in flasks and in the line Pl 17 cultured in the bioreactor and compared with the accumulation of saponins in 6-year-old roots from the field cultivation. Mother plant used as a control was derived from in vitro culture. Plantlets before transfer to the soil were micropropagated as mentioned above [34]. The 6-year-old roots were chosen for the experiment because the saponin content increases with the age of the plant.

Dried, pulverized roots (2 g) were put in a Soxhlet apparatus, soaked in methanol (150 ml), allowed to macerate for 17 hours and next refluxed for 6 hours, left overnight, and finally filtered. The filtrate was evaporated in vacuum to afford a dark brown residue, which was dissolved in methanol (50 ml). The methanol solution was poured into water (1/1, v/v) and then methanol was removed in vacuum. The aqueous residue was successively extracted 10 times with diethyl ether (4/1, v/v) and subsequently 5 times with butanol saturated with water (4/1, v/v). The butanol layers were combined and evaporated in vacuum. The residue was dissolved in methanol (10 ml). The methanol solution was poured into diethyl ether (35 ml). The resulting precipitates of the crude saponins were collected by filtration and dried at 105°C to constant weight.

The mean value and standard deviation represent the replicate of six determinations. Normal distribution and homogeneity of variances of the data were checked with the Shapiro–Wilk test and the Leaven's test respectively. Since in any of the analyzed groups there was not enough evidence to reject null hypothesis ($P > 0.05$), subsequent statistical evaluation of significance was performed by ANOVA test. Post-hoc Dunnett's test was used for individual comparison of groups with the control group. Otherwise LSD test (least significant differences) was used. Calculations were performed using the STATISTICA version 9.0 software.

Results

Induction of hairy roots

Infections of *Platycodon grandiflorum* leaf explants with *Agrobacterium rhizogenes* ATCC 15834 resulted in formation of hairy roots (Fig. 1). An analysis of amplification products obtained in PCR confirmed the presence of the *rolA*, *rolB* and *rolC* genes in genomic DNA of hairy roots. PCR analysis result for the *P. grandiflorum* are shown in Fig. 2.

Studies in shake flask cultures

Platycodon grandiflorum hairy roots grew in liquid WPM medium without growth regulators and demonstrated a typical transformed phenotype. Hairy roots tended to proliferate root hairs, which appeared to be white and healthy (Fig. 3). During the culture of the transgenic line Pl 17 the phases of growth typical of hairy roots were observed: lag phase, exponential phase and stationary phase. The exponential phase started from week 2 of growth and lasted until week 7. The growth rate after this phase was 3770%.

During this phase the considerable biomass increase was correlated with a marked decrease in the sucrose level in the

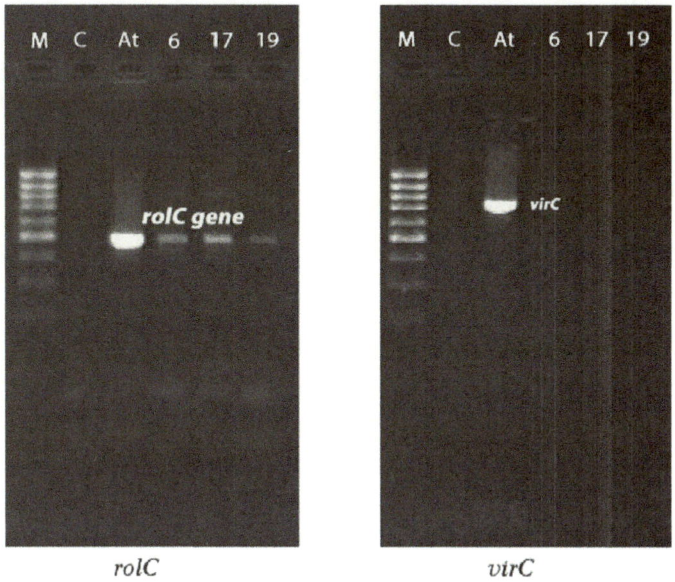

Fig. 2 Confirmation of transformation in hairy root lines by PCR reaction. The pictures shows agarose gel electrophoresis of PCR-amplified products performed with *rolC* primers and *virC* primers. Non transformed roots of *Platycodon grandiflorum*, clone No. 26 were used as negative control (C), while template for the positive (At) control was *Agrobacterium rhizogenes*, strain 15834. 100 bp DNA ladder was used for the estimation of DNA fragment size. No. 6, 17 and 19 corresponds to the transformed roots of *P. grandiflorum*.

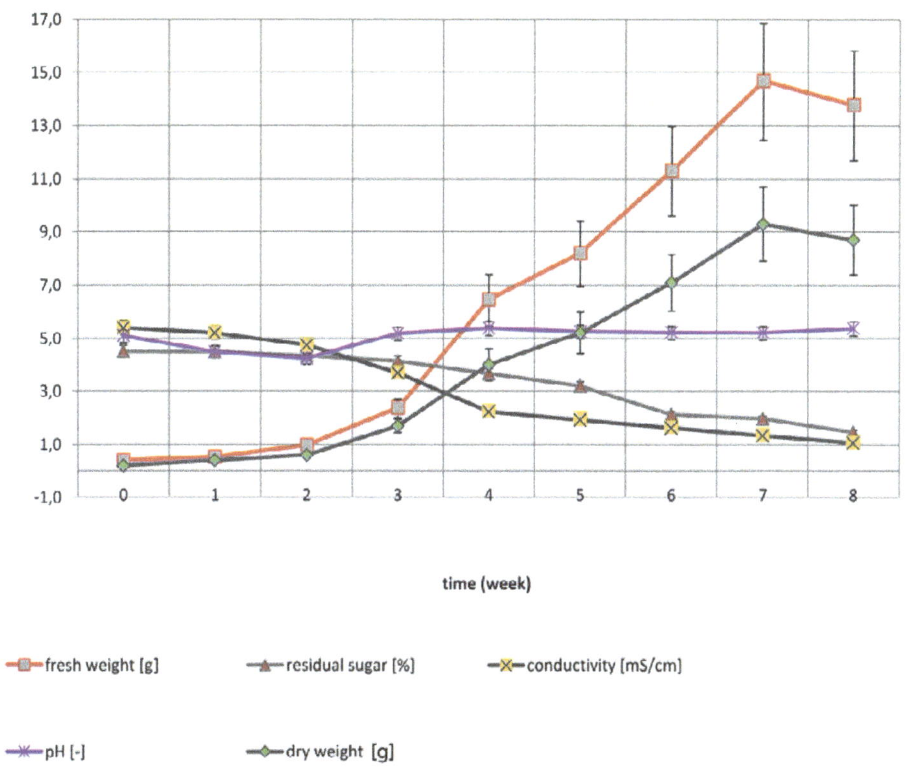

Fig. 3 Changes in medium conductivity and sucrose concentration in the medium and the biomass decrease during the culture of *Platycodon grandiflorum* hairy root line Pl 17 in shake flasks. Each point is the mean value of three estimates (three Erlenmeyer flasks from one passage).

medium (from 4.0 to 1.7%) and conductivity (from 5.30 to 1.24 mS/cm). The changes of hairy root biomass, conductivity, pH and sucrose level is shown in Fig. 3.

Post-hoc Dunett's test revealed that saponin concentration was significantly higher in both hairy roots lines ($P <<$ 0.0001 for line Pl 6, $P = 0.0005$ for line Pl 17) compared to normal roots derived from the field cultivation. The highest concentration was described for line Pl 6 ($0.0264 << P <<$ 0.0001). The mean values [g/100 g d.w. (g%)] of saponins concentrations for line Pl 6, line Pl 17 and field roots were 6.92 (0.88), 5.82 (0.84) and 4.02 (0.51) respectively.

Bioreactor studies

In our study, *P. grandiflorum* hairy roots grew well in a mist bioreactor. The roots were supported on a steel nest and sprinkled with nutrient WPM supplemented with 40 g/l sucrose. The roots occupied the whole space available and their growth was demonstrated by enhanced root length and lateral root induction (Fig. 4). After 8 weeks of culture the medium was replaced. The kinetic parameters from the medium poured out from the bioreactor provided evidence for considerable consumption of the mineral components and sucrose. The conductivity was decreased from 5.25 to 2.11 mS cm^{-1} and sucrose concentration decreased from 4.8 to 2.5%. In week 8 of culture a very rapid growth of *P. grandiflorum* hairy roots was observed which was correlated with fast exhaustion of the medium components. The conductivity decreased from 5.25 to 1.74 mS/cm and the sucrose concentration fell from 4.4 to 3.4%. The differences in conductivity and sucrose concentrations in the medium observed over the entire period of *P. grandiflorum* hairy root line Pl 17 culture in the bioreactor are shown in Fig. 5.

Fig. 4 *Platycodon grandiflorum* hairy root culture of Pl 17 line in 5-l mist bioreactor with liquid WPM medium supplemented with 40 g/l sucrose, cultivated for 12 weeks. After 8 weeks of culture medium was exchanged and 2 weeks later medium was refilled by 1 l of nutrient.

After culture day 71, when 1 l of medium was added, the conductivity and sucrose concentration did not change as much as described above. It was due to the necrosis of the hairy roots occupying the top of the mesh, which were not sufficiently sprinkled with the medium. The hairy roots from the bottom of the mesh continued to demonstrate enhanced length and lateral roots were induced.

After the culture of *P. grandiflorum* hairy roots in the bioreactor 591 g of fresh biomass was obtained, which corresponded to 36.6 g of dry weight (Fig. 6). The growth rate of the hairy roots was 3180%, comparable to the growth rate of hairy roots of the same line, when the exponential phase of growth in the shake flasks was completed. Moreover, comparable saponin contents were observed during both cultures of the hairy roots line 17 ($P = 0.80$). The mean saponin content of the hairy roots grown in the bioreactor was 5.93 g% (0.48).

Discussion

For the experiment WPM medium with increased up to 40 g/l amount of saccharose was chosen. Our previous study [34] showed that plantlets growing on MS medium with saccharose at a dose 40 g/l developed thicker roots compared to those cultivated on the MS medium with saccharose at a dose 30 g/l.

Hairy root growth can be easily monitored by measurement of the conductivity. An inverse relationship between biomass and conductivity was observed not only by our research team, but by Suresh et al. [57,58] as well. Authors [57] described this relationship occurred in hairy root culture of *Tagetes patula* by the same formula as used in *P. grandiflorum* hairy root shake flask cultures. The decrease in conductivity can be attributed to the intake of nutrients by growing hairy roots leading to a decrease in the number of ions.

Besides our experiment, an inverse relationship between biomass and sucrose level was observed by Sykłowska-Baranek et al. [59]. However, our results indicate that this correlation is weaker than that between biomass increase and conductivity decrease.

A nutrient mist bioreactor, used in this study, offers some advantages for growing hairy roots. It provides an environment with a low shear stress, where hairy roots receive rapid replenishment of nutrients and removal of toxic metabolites [60]. What is more, the dispersal of liquid medium in the gas phase could also facilitate better oxygen transfer due to its higher solubility in air than in aqueous medium [61,62]. A very similar type of bioreactor was used by Kuźma et al. [63]. In their study, *Salvia sclarea* hairy roots elicited by methyl jasmonate in bioreactor produced 9 and 3.8 times as much aethiopinone and salvipisone respectively, as elicited roots cultured in shake flasks. The nutrient sprinkle bioreactor

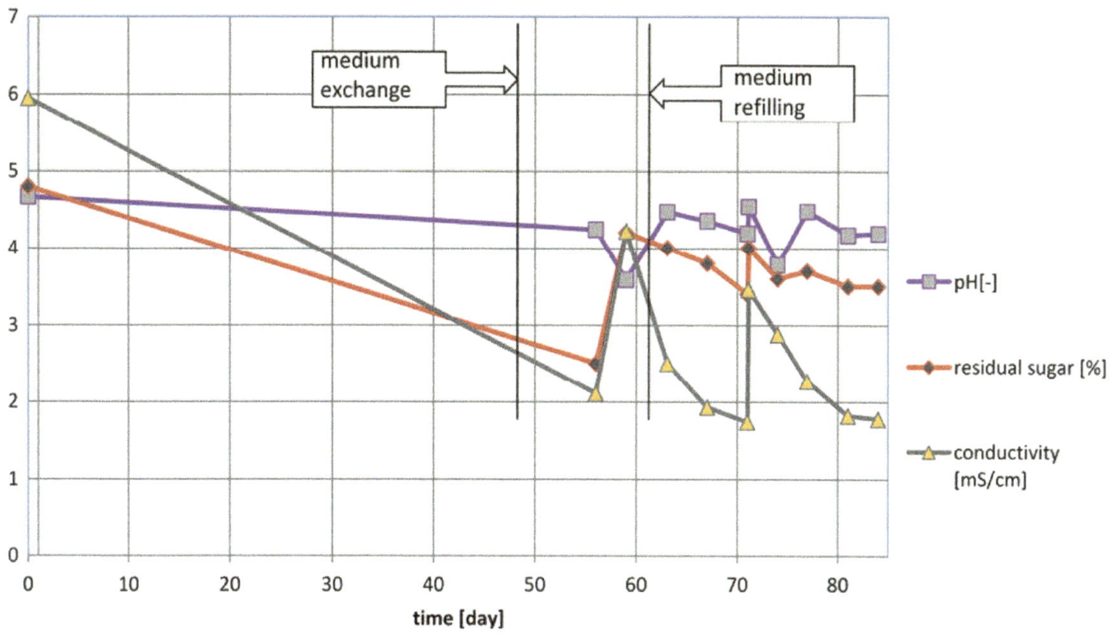

Fig. 5 Changes in medium conductivity and sucrose concentration in the medium during the culture of *Platycodon grandiflorum* hairy root line Pl 17 in bioreactor.

was used as well by Kochan et al. [64] to provide the culture of *Panax quinquefolium* hairy roots. The contents of six examined ginsenoside showed twofold higher level in hairy roots cultivated in bioreactor than in roots growing in the shaken flasks.

The medium exchange during *Platycodon grandiflorum* hairy root culture in the bioreactor considerably accelerated sucrose and nutrient consumption. These results were consistent with the experiment by Pawłowska and Chmiel [65]. They cultured *Paulownia tomentosa* hairy roots in a similar bioreactor using the same medium. After the entire medium exchange they observed a greater biomass increase and culture productivity than with the medium refilled with phosphate or ammonium nitrate only. The similar observations were done as well by Palazon et al. [66]. The medium exchange in the bioreactor was the requirement of the increased productivity of ginsenosides.

In the latter part of the *P. grandiflorum* hairy root culture in the bioreactor cell lysis in a population of roots put on the top of the mesh occurred, possibly because the culture medium did not remain in suitable contact with the cultured roots. Thus a modification of this bioreactor construction is required. Liu et al. [60] fitted a concentric draught tube with holes into their mist bioreactor which ensured better mist distribution compared to a standard mist bioreactor. The alternative can be the Wave bioreactor employment [66], in which the medium reach to any cell. The satisfactory results were received by Jeong and Park as well [67]. The most effective mass production of *Panax ginseng* hairy roots was achieved in several differently sized air bubble bioreactors compared to all other bioreactor types.

Fig. 6 *Platycodon grandiflorum* hairy root biomass of transgenic line Pl 17 after the culture period of 12 weeks in 5-l bioreactor with liquid WPM medium supplemented with 40 g/l sucrose.

Conclusions

Saponins obtained from the roots of *Platycodon grandiflorum* are valuable compounds showing a broad spectrum of pharmacological activities. It is noteworthy that the hairy root culture of *P. grandiflorum* presents weak haemolytic activity what makes this plant material safe in use [17]. Our results suggest that the hairy root culture is a good alternative to the field cultivation of the plant.

Acknowledgments

This study was supported by the Polish Ministry of Science and Higher Education, grant No. NN 405 219 539.
We thank Dr. D. Ruszkowski from Laboratory Of Environmental Contamination Analysis, University of Warsaw, for his help during chemical analysis of saponin.

Authors' contributions

The following declarations about authors' contributions to the research have been made: concept of the study: NU, WJS; laboratory research and data analyses: NU, JG, WJS; writing of the manuscript: NU, JG, OO, WJS.

References

1. Kim Y, Wyslouzil BE, Weathers PJ. Secondary metabolism of hairy root cultures in bioreactors. In Vitro Cell Dev Biol Plant. 2002;38(1):1–10.
2. Guillon S, Trémouillaux-Guiller J, Pati PK, Rideau M, Gantet P. Hairy root research: recent scenario and exciting prospects. Curr Opin Plant Biol. 2006;9(3):341–346.
3. Liu C. Bioreactor: a power tool for pharmaceutical production from plant hairy root cultures. J Biotech. 2008;136:S496.
4. Ahn KS, Hahn BS, Kwack KB, Lee EB, Kim YS. Platycodin D-induced apoptosis through nuclear factor-κB activation in immortalized keratinocytes. Eur J Pharmacol. 2006;537(1–3):1–11.
5. Zhang L, Liu ZH, Tian JK. Cytotoxic triterpenoid saponins from the roots of *Platycodon grandiflorum*. Molecules. 2007;12(4):832–841.
6. Kim J Y, Park KW, Moon KD, Lee MK, Choi J, Yee ST, et al. Induction of apoptosis in HT-29 colon cancer cells by crude saponin from *Platycodi radix*. Food Chem Toxic. 2008;46(12):3753–3758.
7. Lee KJ, Hwang SJ, Choi JH, Jeong HG. Saponins derived from the roots of Platycodon grandiflorum inhibit HT-1080 cell invasion and MMPs activities: Regulation of NF-κB activation via ROS signal pathway. Cancer Lett. 2008;268(2):233–243.
8. Kim YP, Lee EB, Kim SY, Li D, Ban HS, Lim SS, et al. Inhibition of prostaglandin E2 production by platycodin D isolated from the root of *Platycodon grandiflorum*. Planta Med. 2001;67(4):362–364.
9. Kim JY, Hwang YP, Kim DH, Han EH, Chung YC, Roh SH, et al. Inhibitory effect of the saponins derived from roots of *Platycodon grandiflorum* on carrageenan-induced inflammation. Biosci Biotechnol Biochem. 2006;70(4):858–864.
10. Lee K. Protective effect of saponins derived from roots of *Platycodon grandiflorum* on tert-butyl hydroperoxide-induced oxidative hepatotoxicity. Toxicol Lett. 2004;147(3):271–282.
11. Zhao HL, Cho KH, Ha YW, Jeong TS, Lee WS, Kim YS. Cholesterol-lowering effect of platycodin D in hypercholesterolemic ICR mice. Eur J Pharmacol. 2006;537(1–3):166–173.
12. Choi SS, Han EJ, Lee TH, Han KJ, Lee HK, Suh HW. Antinociceptive profiles of platycodin D in the mouse. Am J Chin Med. 2004;32(02):257–268.
13. Xie Y, Pan H, Sun H, Li D. A promising balanced Th1 and Th2 directing immunological adjuvant, saponins from the root of *Platycodon grandiflorum*. Vaccine. 2008;26(31):3937–3945.
14. Xie Y, Sun HX, Li D. Platycodin D is a potent adjuvant of specific cellular and humoral immune responses against recombinant hepatitis B antigen. Vaccine. 2009;27(5):757–764.
15. Xie Y, Ye YP, Sun HX, Li D. Contribution of the glycidic moieties to the haemolytic and adjuvant activity of platycodigenin-type saponins from the root of *Platycodon grandiflorum*. Vaccine. 2008;26(27–28):3452–3460.
16. Jeong HM, Han EH, Jin YH, Hwang YP, Kim HG, Park BH, et al. Saponins from the roots of *Platycodon grandiflorum* stimulate osteoblast differentiation via p38 MAPK- and ERK-dependent *RUNX2* activation. Food Chem Toxic. 2010;48(12):3362–3368.
17. Urbańska N, Nartowska J, Skorupska A, Ruszkowski D, Giebułtowicz J, Olszowska O. Determination and haemolytic activity of saponins in hairy root culture of *Platycodon grandiforum* A. DC. Herba Pol. 2009;55(3):103–108.
18. Pharmacopoeia Commission of PRC. Pharmacopoeia of the People's Republic of China (1988 English edition). Beijing: People's Medical Publishing House; 1988.
19. Ha IJ, Kang M, Na YC, Park Y, Kim YS. Preparative separation of minor saponins from *Platycodi radix* by high-speed counter-current chromatography. J Sep Sci. 2011;34(19):2559–2565.
20. Zhan Q, Zhang F, Sun L, Wu Z, Chen W. Two new oleanane-type triterpenoids from *Platycodi radix* and anti-proliferative activity in HSC-T6 cells. Molecules. 2012;17(12):14899–14907.
21. WHO. WHO monographs on selected medicinal plants. Geneva: WHO; 1999. (vol 1).
22. Yoo DS, Choi YH, Cha MR, Lee BH, Kim SJ, Yon GH, et al. HPLC-ELSD analysis of 18 platycosides from balloon flower roots (*Platycodi radix*) sourced from various regions in Korea and geographical clustering of the cultivation areas. Food Chem. 2011;129(2):645–651.
23. Kim YK, Kim JK, Kim YB, Lee S, Kim SU, Park SU. Enhanced accumulation of phytosterol and triterpene in hairy root cultures of Platycodon grandiflorum by overexpression of *Panax ginseng* 3-hydroxy-3-methylglutaryl-coenzyme A reductase. J Agric Food Chem. 2013;61(8):1928–1934.
24. Tada H, Shimomura K, Ishimaru K. Polyacetylenes in *Platycodon grandiflorum* hairy root and Campanulaceous plants. J Plant Physiol. 1995;145(1–2):7–10.
25. Lee JY, Yoon JW, Kim CT, Lim ST. Antioxidant activity of phenylpropanoid esters isolated and identified from *Platycodon grandiflorum* A. DC. Phytochemistry. 2004;65(22):3033–3039.
26. Fu WW, Dou DQ, Shimizu N, Takeda T, Pei YH, Chen YJ. Studies on the chemical constituents from the roots of *Platycodon grandiflorum*. J Nat Med. 2006;60(1):68–72.
27. Liu W, Liu H, Han M. Polysaccharides from *Platycodon grandiflorum*. Chem Nat Compd. 2013;48(6):927–929.
28. Xu Y, Dong Q, Qiu H, Cong R, Ding K. Structural characterization of an arabinogalactan from *Platycodon grandiflorum* roots and antiangiogenic activity of its sulfated derivative. Biomacromolecules. 2010;11(10):2558–2566.
29. Xu Y, Dong Q, Qiu H, Ma CW, Ding K. A homogalacturonan from the radix of *Platycodon grandiflorum* and the anti-angiogenesis activity of poly-/oligogalacturonic acids derived therefrom. Carbohydr Res. 2011;346(13):1930–1936.
30. Reininger EA, Bauer R. Prostaglandin-H-synthase (PGHS)-1 and -2 microtiter assays for the testing of herbal drugs and in vitro inhibition of PGHS-isoenzyms by polyunsaturated fatty acids from *Platycodi radix*. Phytomedicine. 2006;13(3):164–169.
31. Jeong CH, Choi GN, Kim JH, Kwak JH, Kim DO, Kim YJ, et al. Antioxidant activities from the aerial parts of *Platycodon grandiflorum*. Food Chem. 2010;118(2):278–282.
32. Inada A, Murata H, Somekawa M, Nakanishi T. Phytochemical studies of seeds of medicinal plants. II. A new dihydroflavonol glycoside and a new 3-methyl-1-butanol glycoside from seeds of *Platycodon grandiflorum* A. de Candolle. Chem Pharm Bull Tokyo. 1992;40(11):3081–3083.
33. Jang DS, Lee YM, Jeong IH, Kim JS. Constituents of the flowers of

Platycodon grandiflorum with inhibitory activity on advanced glycation end products and rat lens aldose reductase in vitro. Arch Pharm Res. 2010;33(6):875–880.

34. Micropropagation of *Platycodon grandiflorum*. In: Conference proceedings of the 53. conference of Polish Botanical Society: Polish nature in the natural heritage of Europe. 6–11 September 2004, Toruń, Bydgoszcz, Poland. Toruń: Polish Botanical Society; 2004. p. 120.
35. Kim SW, Liu JR. Somatic embryogenesis and plant regeneration in zygotic embryo cultures of balloon flower. Plant Cell Tissue Organ Cult. 1999;58(3):227–230.
36. Hosoki T, Mochida M. Mass propagation of balloon flower (*Platycodon grandiflorum* A. DC.) by repeated shoot-sectioning and separation of axillary shoot. Seibutsu Kankyo Chosetsu. 1995;33(3):213–216.
37. Iapichino G, Airò M. Micropropagation of *Platycodon grandiflorus*. Italus Hortus. 2009;16(2):128–131.
38. Murashige T, Skoog F. A revised medium for rapid growth and bio assays with tobacco tissue cultures. Physiol Plant. 1962;15(3):473–497.
39. Gamborg OL, Miller RA, Ojima K. Nutrient requirements of suspension cultures of soybean root cells. Exp Cell Res. 1968;50(1):151–158.
40. Lloyd G, McCown B. Commercially-feasible micropropagation of mountain laurel, *Kalmia latifolia*, by use of shoot-tip culture. Comb Proc Int Plant Propagators Soc. 1980;30:421–427.
41. Bernard F, Moghbel N, Hassannejad S. Treatment of licorice seeds with colchicine: changes in seedling DNA levels and anthocyanin and glycyrrhizic acid contents of derived callus cultures. Nat Prod Commun. 2012;7(11):1457–1460.
42. Wongwicha W, Tanaka H, Shoyama Y, Putalun W. Methyl jasmonate elicitation enhances glycyrrhizin production in *Glycyrrhiza inflata* hairy roots cultures. Z Naturforsch C. 2011;66(7–8):423–428.
43. Yin S, Zhang Y, Gao W, Wang J, Man S, Liu H. Effects of nitrogen source and phosphate concentration on biomass and metabolites accumulation in adventitious root culture of *Glycyrrhiza uralensis* Fisch. Acta Physiol Plant. 2014;36(4):915–921.
44. Shabani L, Ehsanpour AA, Asghari G, Emami J. Glycyrrhizin production by in vitro cultured *Glycyrrhiza glabra* elicited by methyl Jasmonate and salicylic acid. Russ J Plant Physiol. 2009;56(5):621–626.
45. Lu HY, Liu JM, Zhang HC, Yin T, Gao SL. Ri-mediated transformation of *Glycyrrhiza uralensis* with a squalene synthase gene (*GuSQS1*) for production of glycyrrhizin. Plant Mol Biol Rep. 2008;26(1):1–11.
46. Li YL, Yang Y, Fu CH, Yu LJ. Production of glycyrrhizin in cell suspension of *Glycyrrhiza inflata* Batalin cultured in bioreactor. Biotechnol Biotechnol Equip. 2012;26(5):3231–3235.
47. Wang J, Zhang J, Gao W, Wang Q, Yin S, Liu H, et al. Identification of triterpenoids and flavonoids, step-wise aeration treatment as well as antioxidant capacity of *Glycyrrhiza uralensis* Fisch. cell. Ind Crops Prod. 2013;49:675–681.
48. Yu KW, Gao WY, Son SH, Paek KY. Improvement of ginsenoside production by jasmonic acid and some other elicitors in hairy root culture of ginseng (*Panax ginseng* C.A. Meyer). In Vitro Cell Dev Biol Plant. 2000;36(5):424–428.
49. Yu KW, Gao W, Hahn EJ, Paek KY. Jasmonic acid improves ginsenoside accumulation in adventitious root culture of *Panax ginseng* C.A. Meyer. Biochem Eng J. 2002;11(2–3):211–215.
50. Huang C, Zhong JJ. Elicitation of ginsenoside biosynthesis in cell cultures of *Panax ginseng* by vanadate. Process Biochem. 2013;48(8):1227–1234.
51. Akalezi CO, Liu S, Li QS, Yu JT, Zhong JJ. Combined effects of initial sucrose concentration and inoculum size on cell growth and ginseng saponin production by suspension cultures of *Panax ginseng*. Process Biochem. 1999;34(6–7):639–642.
52. Yu KW, Murthy HN, Hahn EJ, Paek KY. Ginsenoside production by hairy root cultures of *Panax ginseng*: influence of temperature and light quality. Biochem Eng J. 2005;23(1):53–56.
53. Wu J, Zhong J. Production of ginseng and its bioactive components in plant cell culture: current technological and applied aspects. J Biotech. 1999;68(2–3):89–99.
54. Hooykaas PJJ, Klapwijk PM, Nuti MP, Schilperoort RA, Rorsch A. Transfer of the *Agrobacterium tumefaciens* TI plasmid to avirulent Agrobacteria and to Rhizobium ex planta. J Gen Microbiol. 1977;98(2):477–484.
55. Vervliet G, Holsters M, Teuchy H, van Montagu M, Schell J. Characterization of different plaque-forming and defective temperate phages in *Agrobacterium* strains. J Gen Virol. 1975;26(1):33–48.
56. Street HE, Henshaw GG. Introduction and employed in plant tissue culture. In: Willmer EN, editor. Cells and tissue culture. London: Academic Press; 1966. p. 459–532. (vol 3).
57. Suresh B, Rajasekaran T, Rao SR, Raghavarao KSMS, Ravishankar GA. Studies on osmolarity, conductivity and mass transfer for selection of a bioreactor for *Tagetes patula* L. hairy roots. Process Biochem. 2001;36(10):987–993.
58. Suresh B, Bais HP, Raghavarao KSMS, Ravishankar GA, Ghildyal NP. Comparative evaluation of bioreactor design using *Tagetes patula* L. hairy roots as a model system. Process Biochem. 2005;40(5):1509–1515.
59. Sykłowska-Baranek K, Pietrosiuk A, Gawron A, Kawiak A, Łojkowska E, Jeziorek M, et al. Enhanced production of antitumour naphthoquinones in transgenic hairy root lines of *Lithospermum canescens*. Plant Cell Tissue Organ Cult. 2012;108(2):213–219.
60. Liu CZ, Wang YC, Zhao B, Guo C, Ouyang F, Ye HC, et al. Development of a nutrient mist bioreactor for growth of hairy roots. In Vitro Cell Dev Biol Plant. 1999;35(3):271–274.
61. Srivastava S, Srivastava AK. In vitro azadirachtin production by hairy root cultivation of *Azadirachta indica* in nutrient mist bioreactor. Appl Biochem Biotechnol. 2012;166(2):365–378.
62. Weathers PJ, Wyslouzil BE, Wobbe KK, Kim YJ, Yigit E. The biological response of hairy roots to O_2 levels in bioreactors. In Vitro Cell Dev Biol Plant. 1999;35(4):286–289.
63. Kuźma Ł, Bruchajzer E, Wysokińska H. Methyl jasmonate effect on diterpenoid accumulation in *Salvia sclarea* hairy root culture in shake flasks and sprinkle bioreactor. Enzyme Microb Technol. 2009;44(6–7):406–410.
64. Kochan E, Królicka A, Chmiel A. Growth and ginsenoside production in *Panax quinquefolium* hairy roots cultivated in flasks and nutrient sprinkle bioreactor. Acta Physiol Plant. 2012;34(4):1513–1518. http://
65. Pawłowska B, Chmiel A. Scaling up the *Paulownia tomentosa* hairy roots culture. BioTechnologia. 2003;4(63):142–153.
66. Palazón J, Mallol A, Eibl R, Lettenbauer C, Cusidó RM, Piñol MT. Growth and ginsenoside production in hairy root cultures of *Panax ginseng* using a novel bioreactor. Planta Med. 2003;69(4):344–349.
67. Jeong GT, Park DH. Comparative evaluation of modified bioreactors for enhancement of growth and secondary metabolite biosynthesis using *Panax ginseng* hairy roots. Biotechnol Bioprocess Eng. 2005;10(6):528–534.

21

How a river course influences the species richness and ecological requirements on two opposite riverbanks in a forest area

Bożenna Czarnecka[1]*, Anna Rysiak[1], Łukasz Chabudziński[2]

[1] Department of Ecology, Faculty of Biology and Biotechnology, Maria Curie-Skłodowska University, Akademicka 19, 20-033 Lublin, Poland

[2] Geoinformation Laboratory, Faculty of Earth Sciences and Spatial Management, Maria Curie-Skłodowska University, Kraśnicka 2c,d, 20-718 Lublin, Poland

Abstract

The goal of the present research was to find correlations between the topographic attributes of a river valley and local ground-floor vegetation and its habitat requirements expressed by ecological indicator values (EIV), using the geographical information systems (GIS), digital elevation model (DEM), and multivariate statistical analysis. We paid special attention to the river course, which determines the differentiation in slope aspects and the amount of solar radiation reaching the ground surface. The model object was an almost latitudinal, ca. 4-km-long break section of the Sopot river, crossing the escarpment zone of the Central Roztocze Highlands, southeastern Poland. The main material comprised species lists (with estimated abundance) for each ca. 200-m-long section, according to the river valley course, separately for the left and right riverbanks, 40 sections altogether, ca. 15 000 vegetation records, and physical and chemical soil measurements. A 3-meter resolution DEM was derived from a 1:10 000 topographic map. We calculated the correlations between the topographic attributes of the valley, species richness, and the EIVs for all the species recognized in each section of the valley. We found 241 herb plant species in the ground-floor vegetation of the study area. We did not find significant differences between the two riversides (61 ±13 species per one section for the left and 63 ±17 for the right side). Thus, the parallel course of the river valley does not change the species richness on a more "sunny" and more "shiny" riverbank. However, this factor "cooperating" with other topographic attributes of the valley significantly differentiates the shape of species showing various requirements for basic habitat resources: light, moisture, soil trophy, reaction, dispersion, and organic-matter content.

Keywords: species diversity; ground-floor vegetation; riparian landscape; ecological indicator values; solar radiation; wetness index; Digital Elevation Model; multivariate analysis

Introduction

Habitat complexes of river valleys are characterized by considerable heterogeneity defined as variability of spatial and temporal patterns and processes [1]. Environmental heterogeneity of the riparian landscape is a result of interfaces between terrestrial and aquatic ecosystems, i.e. river fluvial activity (erosion, transport, retention of inorganic sediments and nutrients), organic-matter dynamics, climatic factors, and hydrological relationships between the abiotic and biotic elements of the ecosystem [2–8]. The riverine landscapes are usually characterized by specific zonal toposequence of plant communities as well as great biodiversity, despite the rather small proportion of the total watershed area in a given region [9–14].

For the last decades, application of geographical information systems (GIS) and digital elevation models (DEM) has provided very useful tools for studying the biodiversity indicators and relationships between topographic and ecological characters of different landscapes, including river systems (e.g. [13–18]). The GIS facilitates classification of natural systems based on linking a variety of spatial components or observation results using multivariate canonical analyses [8,14,19–23]. Regardless of the scale of the study, some spatial elements remain unchanged. The basis of the analytical procedures is two groups of topographic attributes calculated from the DEM [24–26]: primary topographic attributes – slope, aspect, planar (or contour), vertical (or profile), and total curvature; and secondary topographic attributes – solar radiation (SRAD), and the topographic wetness index (TWI).

The approach based on "ecological indicator values" (EIV) of plants as predictors of the environment quality has widely developed since the first attempt made by Ellenberg [27], who defined EIVs, which reflect the realized optima for species of Central Europe expressed as ordinal numbers. The EIV system provides a very valuable tool for habitat calibration and recently it has been applied for modeling

* Corresponding author. Email: bozenna.czarnecka@poczta.umcs.lublin.pl

Handling Editor: Zygmunt Dajdok

plant distribution at various spatial scales. However, investigations conducted at the regional and landscape levels [15,22,28–32] are still much more common than studies on local sites [14,17,21,33].

Our studies were focused on a break section of a small-scale river valley characterized by an almost latitudinal course of the entire analyzed fragment, which is the main factor determining the slope aspect and, in consequence, the disproportion between the amounts of energy supplied to the right and left banks of the river valley. We assumed that solar radiation reaching the Earth's surface is the most important environmental factor determining species richness and diversity on the two riverbanks. The goal of the present research was to find correlations between the terrain characters of the river valley and local vascular flora and its ecological requirements expressed by the EIVs in the entire valley and on its right and left sides separately, using the GIS, DEM, and multivariate statistical analysis. Finally, we have discussed the usefulness of the GIS and DEM for studying the relationships between the topographic (abiotic) and ecological (biotic) attributes of the riparian vegetation.

Material and methods

Study area and material collection

The study area was a break section (i.e. between the 14th and 18th km of the river course) of the Sopot river (IV order river), the largest right-side tributary of the upper Tanew river, crossing the escarpment zone of the Central Roztocze Highlands, southeastern Poland (Fig. 1a). The Sopot river is a 24-km-long stream with a catchment area of approximately 122 km^2 and discharge varying from 913 to 1638 dm^3 s^{-1}. A characteristic feature of the "strict" zone of the river break (over the first 1.1 km of the study section) is a large slope of the riverbed (19.1‰); over the length of ca. 4 km – 5.3‰, on average; a large incision of the valley bottom and presence of numerous rock steps and springs, all of which give the valley a mountainous character. In the upper part of the study area the valley was built of Miocene formations and in the lower part – Quaternary formations, on which different types of soils have been evolved [11,34,35]. The valley section to be analyzed (4 km long; 80.5 ha) is wooded in 99.5% and only slightly transformed by human activity (map at the 1:5000

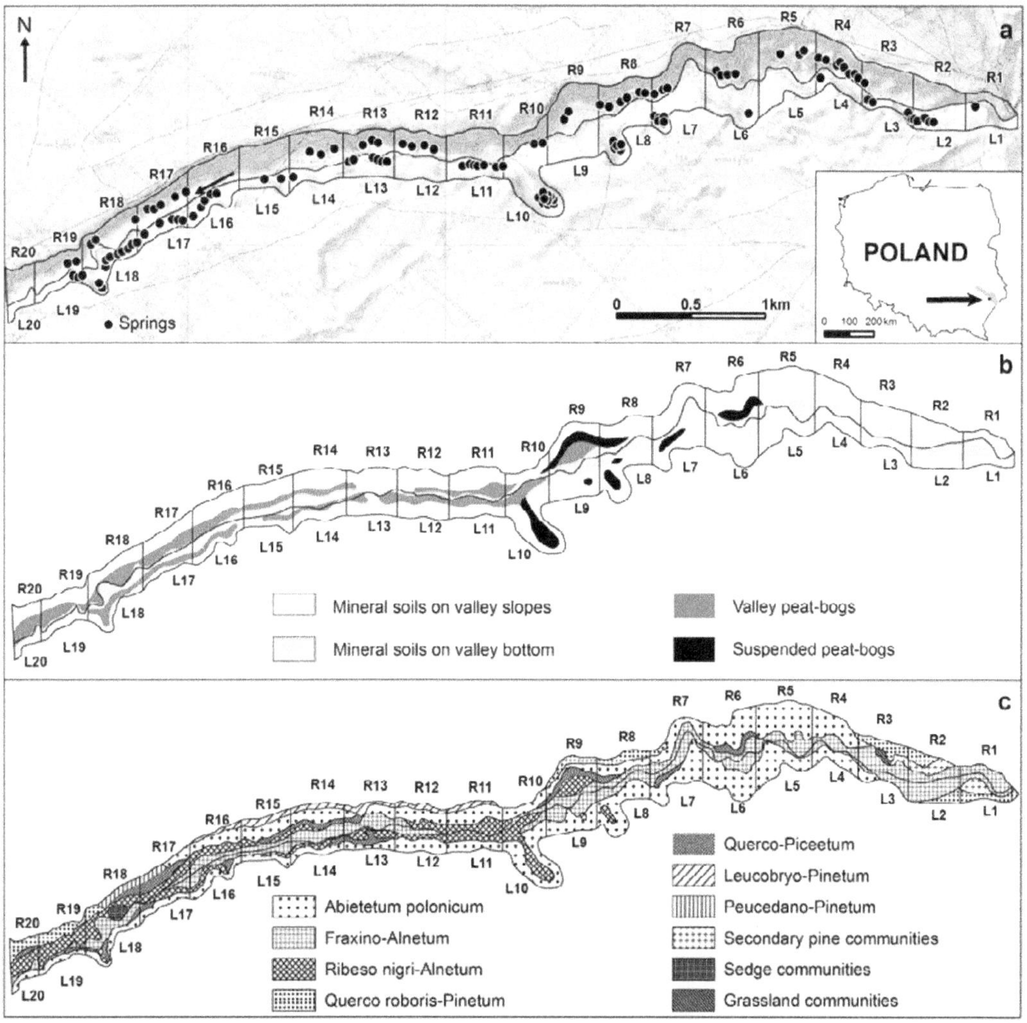

Fig. 1 The study section of the Sopot river valley. **a** Distribution of springs. **b** Distribution of soil complexes. **c** Distribution of plant communities. L1–L20 – left-side sections of the valley. R1–R20 – right-side sections of the valley. Vertical exaggeration = 3.

scale from 1998; Czarnecka, unpblished data). The main character of the vegetation landscape of the steep slopes is created by upland mixed fir forest *Abietetum polonicum*, a community of European importance [36]. The remaining area (0.5%) is covered by sedge and meadow vegetation (Tab. 1). The study valley is characterized by zonal arrangement of soil complexes and plant communities (Fig. 1b,c). The most valuable break part of the valley with forests on the slopes and river terrace has been preserved since 1958 in a landscape reserve called "Czartowe Pole" ("The Devil's Field"). The study object was also included in the Natura 2000 network within both types of protected sites: Special Protected Area (PLB 060012) and Special Area of Conservation (PLH 060018).

The main material comprised species lists (with estimated abundance) for each ca. 200-m-long section, according to the river valley course, separately for the left and right riverbanks, 40 sections altogether (Fig. 1). We have established the frequency and abundance of each vascular plant species in the ground-floor vegetation, using a simplified, combined scale, where: 1 – sporadic species; 2 – rare and not-abundant species (covering <10% of the section area); 3 – frequent and abundant species (covering 10–50% of the section area); 4 – common and very abundant species (covering >50% of the section area). To estimate the real habitat conditions in the whole study area and each section and the ecological scale of particular plant species, we also used other field materials: phytosociological relevés in different types of plant communities (60 in total), and soil pits (27 in total) distributed proportionally to the area, and the diversity of the identified communities. Using commonly accepted methods [37,38], the following properties were determined for 140 soil samples: the content of organic matter and/or organic carbon, active acidity, calcium carbonate and basic ions (Ca, K, Na, Ma, Fe, P, and N in the form of ammonia and nitrate).

Analysis of terrain attributes

Spatial data were obtained from a topographic map at the 1:10 000 scale by successive digitization of contour lines, elevation points, valley edges and their height. For calculation of the terrain attributes, we used tools available in the ArcToolbox of the ArcGIS 10 software and the Spatial Analyst extension. Using the Topo to Raster tool, we generated the DEM and its derivatives with a resolution of 3 m [39]. Based on the DEM and the derivatives, the following commonly used topographic attributes were calculated [24–26]: primary – slope, aspect, and planar, vertical, and total curvature; and secondary – the SRAD and TWI. Apart from the above-listed attributes, we also took into account additional primary terrain characters: denivelation (elevation gradient), total area, flat area (i.e. ≤2° of terrain slope), and upslope area (>2° of terrain slope) of a given section. Subsequently, each attribute was analyzed for each valley section of the river using the Zonal Statistics tool, with which a statistic was calculated for each zone defined by a zone dataset (in our case, these were particular sections of the valley), based on values from the other datasets (DEM, slope aspect, planar, vertical and total curvature, SRAD, and TWI). A single output value was computed for every zone in the input zone dataset. For detailed description, see Czarnecka and Chabudziński [14].

Statistical analysis

In the next step, the mean value of specific EIVs in each valley section was calculated using a modified formula for the weighted average [14]:

$$W_A = \frac{\sum_{i=1}^{n}(A_i^2 I_i)}{\sum_{i=1}^{n} A_i^2}$$

where: W_A – weighted average, A_i – abundance of the cover of the *i*-th species in a given section of the valley, I_i – ecological indicator value for the *i*-th species, n – number of species in the section.

We calculated the correlations of topographic attributes of the valley, species richness, and the EIVs for all the species in each section. According to the suggestions of some authors [21,28,40], we analyzed the EIVs only for one vegetation layer, i.e. ground-floor vegetation (pteridophytes

Tab. 1 Characteristics of the study area of the Sopot river valley ([34,35] and Czarnecka unpblished data).

Plant community	Area (%)	Soil types	Active acidity (pH)	Organic matter/ humus content (%)
Upland mixed fir forest *Abietetum polonicum*	46.1	Podzolic soil	3.64–5.32	0.11–0.65
		Rusty-podzolic soil	4.07–4.40	0.23–2.13
Riverside ash-alder forest *Fraxino-Alnetum*	21.2	Brown pararendzina	7.57–7.98	1.20–2.77
		Alluvial soil	5.53–7.73	0.21–21.61
		Gley podzolic soil	4.06–8.15	0.27–13.19
		Low peat soil	4.46–8.18	39.53–80.79
Bog alder forest *Ribeso nigri-Alnetum*	14.2	Low peat soil	5.34–7.75	18.15–85.76
Oak-pine mixed forest *Querco roboris-Pinetum*	7.0	Rusty-podzolic soil	4.56–6.37	0.81–1.09
Moist mixed coniferous forest *Querco-Piceetum*	5.7	Low peat soil	3.51–5.56	35.98–90.99
		Gley podzolic soil	4.24–5.49	6.49–56.18
Suboceanic pine forest *Leucobryo-Pinetum* Subcontinental pine forest *Peucedano-Pinetum*	3.2	Podzolic soil	4.74–6.68	0.10–0.61
Secondary pine communities	0.8	Podzolic soil	4.00–4.53	0.14–0.93
Sedge communities	0.3	Low peat soil	5.10–7.11	28.26–77.20
Grassland communities	0.2	Leached brown soil	5.32–6.42	2.07

and spermatophytes, except for the shrub and tree species). The full data set comprises almost 15 000 vegetation records taken during the field survey. We took into account the following 6 indicator values describing the most typical habitat conditions of the species according to Zarzycki et al. [41]: light (L), soil moisture (W), trophy (Tr), soil/water acidity (R), soil granulometry = dispersion (D), and organic matter content (H). The EIVs used in the analyzed valley were calibrated by adjustment to the range of habitat conditions found during the field and laboratory studies. The continentality (C) and temperature (T) values were purposely not taken into account, as their value is constant in a small study area.

To establish whether particular topographic variables have or do not have normal distribution, we used the Shapiro–Wilk test. Because a majority of the values of the analyzed attributes did not show a normal distribution, Spearman's rank correlation coefficients (*r*) were calculated between the number of species and the mean value for a specific EIV in each section and all topographic attributes of the valley. The mean and sum value of the primary and secondary topographic attributes were taken into account. The correlation analysis was performed for the entire valley section and separately for both riverbanks, due to the varied slope aspect and terrace asymmetry (cf. Fig. 1). All statistics were calculated with the use of Statistica PL.

To analyze the relationships between weighted averages for the particular EIV of local vascular flora and topographic attributes of the valley, we also used multivariate ordination methods in Canoco version 4.5 [19,20]. According to the length of the gradient from preliminary detrended correspondence analysis (DCA), a linear model was used – redundancy analysis (RDA). A manual procedure with 499 Monte Carlo significance permutation tests was used to find the minimum number of statistically significant variables. The eigenvalues and percentages of floristic and topographic variance explained by the first four axes were calculated. Finally, the pattern obtained from the classification was transferred onto a graph with sample groups marked in the RDA. The first two axes were shown on the graph.

Results

Correlations between species richness and topographic attributes

In total, we found 241 herb plant species (spermatophytes and pteridophytes) in the ground-floor vegetation of the study area. The average number of species per one section was variable (mean ±standard deviation = 62 ±15) and ranged from 32 (section R1) to 99 (R3). We did not notice significant differences between the two riversides (61 ±13 for the left, 63 ±17 for the right). Species richness increased significantly with the rise of the total area ($r = 0.52$) and the flat area ($r = 0.67$) of a given section; the correlation with the upslope area was weaker for the total study area ($r = 0.34$), but stronger for the left riverside ($r = 0.64$; Tab. 2). The number of species was not correlated with terrain denivelation but it decreased significantly with the mean slope ($r = -0.49$), CP_mean ($r = -0.36$), CT_mean ($-0.33 < r < -0.48$) and CT_sum ($-0.50 < r < -0.61$). In turn, species richness was positively correlated with CV_mean ($0.49 < r < 0.56$) and CV_sum ($0.57 < r < 0.58$).

The lowest number of species (≤50 species) was recorded in sections, in which at least two of the following three factors were combined: average slope >10°, denivelation >20 m, proportion of flat area <10% (sections: L1–3, L19–20,

Tab. 2 Statistically significant correlation coefficients between topographic attributes and ecological indicator values of ground-floor vegetation in the Sopot river valley and its opposite riverbanks.

Topographic attributes	Number of species			Ecological indicator values												
				L		W		Tr			R		D		H	
	All	Left	Right	All	Right	All	Right	All	Left	Right	All	Right	All	Right	All	Right
TA	0.52***								0.48*							
FA	0.67***						0.51*	0.38*			0.36*		0.33*	0.56**		0.53*
UA	0.34*	0.64**			−0.51*		−0.65**		0.48*					−0.51*		−0.54*
DN				−0.54***	−0.80***	−0.43**	−0.70***			−0.53*		−0.53*	−0.45**	−0.59**	−0.46**	−0.59**
SL_mean	−0.49**			−0.49**	−0.58**		−0.60**	−0.46**			−0.49**		−0.37*	−0.52*		−0.56*
CP_mean	−0.36*							−0.33*	−0.46*		−0.32*					
CP_sum								−0.35*			−0.34*					
CV_mean	0.56***	0.49*	0.56**													
CV_sum	0.58***	0.58*	0.57**		0.46*											
CT_mean	−0.33*		−0.48*		−0.63**		−0.84***							−0.51**		−0.82***
CT_sum	−0.50*	−0.54*	−0.61**		−0.47*		−0.72***							−0.58**		−0.75***
SRAD_mean		0.48*	−0.56*		−0.48*		−0.64**									−0.65**
SRAD_sum	0.46**	0.69***							0.47*							
TWI_mean	0.69***	0.72***	0.63**	0.37*	0.48*	0.48**	0.66**	0.42*			0.43**		0.47**	0.59	0.46**	0.66**
TWI_sum	0.57***	0.71***							0.47*							

Explanations: All – all sections of the valley; Left – left sections of the valley; Right – right sections of the valley; TA – total area; FA – flat area; UA – upslope area; DN – denivelation; SL_mean – mean slope; CP_mean – mean planar curvature; CP_sum – sum planar curvature; CV_mean – mean vertical curvature; CV_sum – sum vertical curvature; CT_mean – mean total curvature; CT_sum – sum total curvature; SRAD_mean – mean solar radiation; SRAD_sum – sum solar radiation; TWI_mean – mean topographic wetnes index; TWI_sum – sum topographic wetnes index. Significance level: * 0.01 < *P* ≤0.05; ** 0.0001 < *P* ≤ 0.01; *** *P* ≤ 0.0001.

R1–2, R4). The highest species richness (≥75 species) was characteristic for sections with the largest proportion of the flat area (>20%, in some sections even >40%); in this case, denivelation and the mean slope are less important (L7, L9–10, L16, R9, R18–20). Section R3 is an exception (99 species; mean slope >10°; flat area = 16.7%), as it comprises the only afforested fragment of the valley slope overgrown with grassland community with the share of some thermophilous elements (cf. Fig. 1c).

The vascular ground flora displayed a great number of high correlations between species richness and the secondary topographic attributes for the entire valley or at least for the individual banks. A different effect of the SRAD_mean on the species richness was observed for both riverbanks: positive for the left ($r = 0.48$) and negative for the right one ($r = -0.56$), i.e. for the bank with domination of the southern and northern aspects and their derivatives, respectively. This is correlated with the disproportion between the amounts of energy reaching the surface of ground-floor layer during vegetation season (April–October) in the right- and left-side sections of the river valley, 150 kWh m^{-2} on average, which is similar along the entire valley (Fig. 2). The SRAD_sum and TWI_sum were not significant for the right-side sections of the valley, while the mean TWI had a highly positive impact on the number of the species for all the sections studied ($r = 0.69$), and the left and right banks separately ($r = 0.72$ and $r = 0.63$, respectively).

Correlations between species diversity and topographic attributes

The topographic attributes exert a different effect on the ecological diversity of flora expressed by the EIVs (Tab. 2). The relationships differed between the valley banks; they were much stronger and more significant for the right side (for all indicators except the Tr value). The denivelation, which was not significant for the species richness, had a negative impact on the share of species with higher requirements for light ($-0.54 < r < -0.80$), soil humidity ($-0.43 < r < -0.70$), trophy ($r = -0.53$), acidity ($r = -0.53$), dispersion ($-0.45 < r < -0.59$), and humus content ($-0.46 < r < -0.59$), especially on the right bank. The mean slope plays a similar role for species diversity. The upslope area affects negatively the L, W, D, and H values ($-0.51 < r < -0.65$), but positively the Tr value ($r = 0.48$) on the left riverside.

The higher the mean total curvature of the right valley slope, the significantly lower the number of species with higher values of the L ($r = -0.63$), W ($r = -0.84$), D ($r = -0.51$), and H ($r = -0.82$) indicators. In general, slightly lower correlations were found for the CT_sum: L ($r = -0.47$), W ($r = -0.72$), D ($r = -0.58$), and H ($r = -0.75$). The SRAD_mean was negatively correlated with the value of L ($r = -0.48$), W ($r = -0.64$), and H ($r = -0.65$) for the right riverbank while the SRAD_sum influenced significantly only the Tr value on the left bank ($r = 0.47$), i.e. the bank with the northern aspect and its derivatives. A similar situation was observed for the TWI_sum only for the Tr value in the left-side sections ($r = 0.47$). In contrast, the correlations for the TWI_mean were more frequent and higher: L ($0.37 < r < 0.48$), W ($0.48 < r < 0.66$), Tr ($r = 0.42$), R ($r = 0.43$), D ($r = 0.47$), and H ($0.46 < r < 0.66$).

The RDA analysis revealed topographic attributes of the valley that were the most significant for species richness and diversity (Tab. 2, Fig. 3). The analysis confirmed the positive correlation between the number of species in the sector and its total area and, in particular, with the flat area and the sum and mean TWI (correlation with axis 1). These factors were equally significant, which was evidenced by the length of their vectors. There was a considerably lower correlation between the same axis and the CT_sum. The D value was highly correlated with the TWI_mean and CV_sum. The values of the other EVIs (L, H, W, R, and Tr) were interrelated and positively correlated with axis 2. Denivelation and upslope area had a reverse vector, and the vectors of the CP_sum and SL_mean were shorter. All the factors were negatively correlated with axis 1 (Fig. 3). The CP_mean, CT_mean, and CT_sum were negatively correlated with axis 2. The alignment of the particular sections of the valley along the axes and vectors of the investigated attributes indicates which attributes exert an impact on the species number and their ecological diversity in each section.

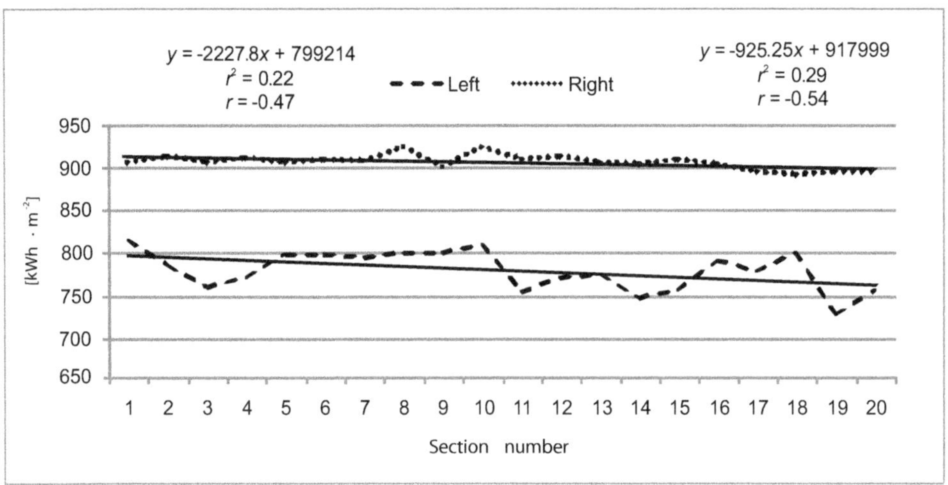

Fig. 2 Amounts of solar energy supplied for the left and right riverbanks and trend line equations for mean SRAD along the Sopot river valley.

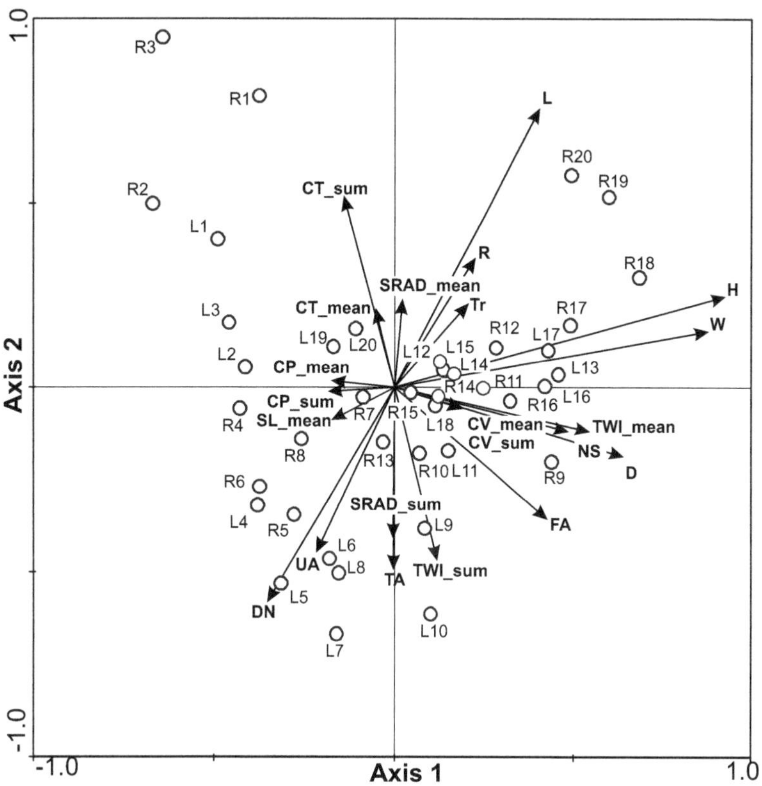

Fig. 3 Ordination diagram showing the result of the RDA for species richness, ecological indicator values and topographic attributes for all sections of the Sopot river valley. Eigenvalues: Axis 1 – 19.10, Axis 2 – 11.22, Axis 3 – 7.13, Axis 4 – 4.61. NS – number of species. Remaining abbreviations as in Tab. 2.

The cumulative percentage of ground-floor species and topographic variance explained by the first two RDA axes was 57.4%. For all the variables analyzed for 40 valley sections, the sum values of the Monte Carlo test and forward selection of species–topographic relationships explained 73.2% of the variety; for statistically significant variables see Tab. 3. There was a statistically significant relationship between the occurrence of species and the gradients of both RDA canonical axes ($P < 0.01$). F ratio for the first canonical axis was even higher (4.25; $P = 0.002$).

Tab. 3 Results of the Monte Carlo permutation tests and forword selection for relations between the ground-floor vegetation, primary and secondary topographic attributes and ecological indicator values in the Sopot river valley.

	All sections		Left bank		Right bank	
Variable	**Lambda A**	**F ratio**	**Lambda A**	**F ratio**	**Lambda A**	**F ratio**
No. of species	0.08	4.19**	0.15	1.98**	0.10	2.99**
Flat area			0.05	1.51*		
SRAD_mean	0.03	1.64**				
CT_sum	0.02	1.49*				
CT_mean	0.02	1.40*				
H	0.17	7.97**	0.20	4.60**	0.22	5.07**
L	0.09	4.26**	0.15	3.86**	0.14	3.87**
W	0.04	2.78**	0.07	1.84**	0.07	2.12**
Tr	0.05	2.41**	0.05	1.78**	0.06	2.02**
R	0.03	1.43*	0.005	1.78**		
D	0.02	1.62*				

Significance level: * $0.01 < P \leq 0.05$; ** $0.0001 < P \leq 0.01$. Explanations as in Tab. 2.

Ground-floor vegetation on two riverbanks

The ordination diagrams for both riversides showed varied significance of the topographic attributes analyzed in the individual sections of the valley. For the left riverbank (northern aspect and its derivatives), the following variables are positively correlated with axis 1 (Tab. 3, Fig. 4): CT_sum, W, H, L; a significantly weaker correlation was found for the CP_sum and CT_mean. This is characteristic for sections L11–L15, L17, and L18. The SL_mean and CP_mean are negatively correlated with axis 1. These factors are important for sections L19 and L20, and less important for L1–L3. The following variables are positively correlated with axis 2: the number of species, total and flat area, TWI_sum, TWI_mean, and, more weakly, CV_mean. These factors are the most important in the case of sections L9 and L10. The vectors for the Tr and R have a similar length, whereas that for D is considerably longer and similar to vectors W and H correlated with axis 1. With its position, section 16 is related to the group of sectors correlated with axis 1, i.e. L17 and L13 in particular, but section L16 exhibits a significant share of species associated with heavier soil formations, e.g. loams, clayey sands, and clays. The greatest importance for the ground-floor vegetation in sections L6–L8 is attributed to the SRAD_sum, upslope area, denivelation, and SRAD_mean, which are negatively correlated with axis 2. The aforementioned factors play a lesser role in sections L4 and L5.

For the left bank, the cumulative percentage of ground-floor species and topographic variance explained by the first two RDA axes was 54.9%. The Monte Carlo test of significance of the first canonical axis showed the value 33.8%. There were 7 statistically significant variables shown by the forward selection of species–topographic relationships (Tab. 3).

On the right side of the valley (southern aspect and its derivatives), the highest positive correlation with axis 1 was observed for the following attributes: denivelation, CT_sum, CT_mean, and SRAD_mean; a weaker correlation was found in the case of the upslope area, SL_mean, CP_mean and CP_sum, SRAD_sum, and total area. These factors have the most substantial role in sections R4–R8 and R13. A negative correlation with axis 1 is exhibited by the D value and TWI_sum. The latter factor seems to have an impact on the ground-floor vegetation in sections R9–R12 and R14–R16. Three EIVs, i.e. L, W, and H have the strongest correlation with axis 2 and are statistically significant for species diversity. The share of species with high requirements for light, moisture, and humus content exhibits a strong correlation with the TWI_mean, CV_mean, and CV_sum, a slightly weaker relationship with the flat area (section R17). This is particularly important in sections R18–R20. The position of sections R1–R3 in the graph indicates their distinctness from the others. The Tr and R values are important for the flora of these sections (Tab. 3, Fig. 5).

The cumulative percentage of ground flora and topographic attributes for the right riverside (southern aspect and its derivatives) explained by the two first RDA axes was 58.8%. The Monte Carlo test of significance of the first canonical axis showed the value 24.9%. Forward selection of the relationships between the ecological variables revealed 5 statistically significant variables.

Discussion

Out of the primary topographic attributes [24–26], the following ones seem to be of great ecological relevance for ground-floor vegetation: elevation, slope (surface gradient), aspect (surface orientation), and planar, vertical, and total curvatures. In the present study, we also found other terrain characters to be important for species richness and diversity: the total area of a given section of the river valley, share of the flat area (i.e. floodplain terrace), and share of the upslope area in the sections. Despite the specific zonal toposequence of soil complexes and forest communities in the valley, we found significant differences between the ground-floor vegetation on the opposite riverbanks.

Elevation gradients (denivelations) determine soil variables, species distribution, vegetation production levels, and patterns of disturbance [3,17,18,42,43]. Surprisingly, the denivelation of the Sopot river valley (incision into the substratum: mean = 20 m, max = 27 m) does not alone influence significantly the species richness but shapes the values of all ecological indicators for ground-floor vegetation: L, W, Tr, R, D and H, particularly for the more "sunny" right bank. That terrain attribute was not significant for species richness and ecological variety of flora in another river valley, the Szum river, crossing the escarpment zone of the same highland region, similar in its mountainous character but incised in the bedrock for only 10 m on average; max = 15 m [14].

We found other primary attributes to be significantly important for both richness and diversity of the ground-floor vegetation. The species richness was positively correlated with the total area of a given section and first of all with the proportion of the flat area for the entire study area. The width of the floodplain terrace seems to be of great importance for the species richness because of the presence of the richest hygrophilous and nitrophilous vegetation, i.e. ash-alder forests *Fraxino-Alnetum* and bog alder forests *Ribeso nigri-Alnetum*. Their occurrence is promoted, from the riverbed side, by sedimentation of heavier and hence more nutrient-rich mineral formations: loams, clayey sands, heavy and silty clays [34,35]. This is also evidenced by significant correlations between the flat area and some EIVs: Tr, D, and H; the intercorrelations are noticeably higher for the right riverbank.

Significant negative correlations were found for the mean slope in the individual valley sections and all of the study EIVs: L, W, Tr, R, D, and H. The correlations were usually higher for the right riverbank, which is characterized by southern aspect and its derivatives. The upslope area also influences the decrease in the share of species with higher requirements for light, humidity, heavier soil formations, and humus content. Greater slopes, higher insolation, and lower humidity offer less favorable conditions for accumulation and humification of organic matter. Soils in such conditions are more acidic and poorer podzolics overgrown with different coniferous forests, first of all upland mixed fir forest *Abietetum polonicum* [11,12,35].

On the right-side bank, a positive influence of vertical curvatures on the species number and the increase in species with higher requirements for light was noted. In contrast,

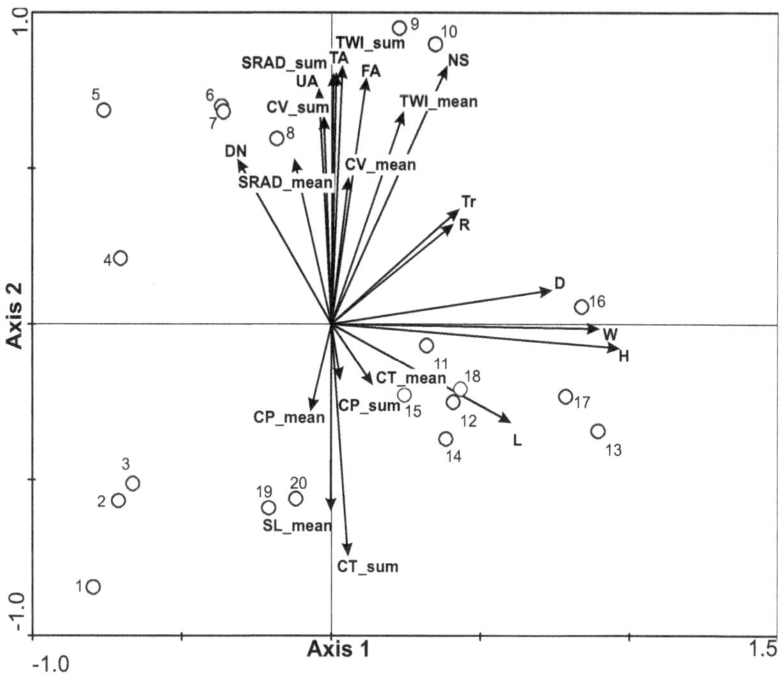

Fig. 4 Ordination diagram showing the result of the RDA for species richness, ecological indicator values and topographic attributes for the left-side sections of the Sopot river valley. Eigenvalues: Axis 1 – 21.32, Axis 2 – 16.93, Axis 3 – 9.50, Axis 4 – 7.10. Abbreviations as in Tab. 2 and Fig. 3.

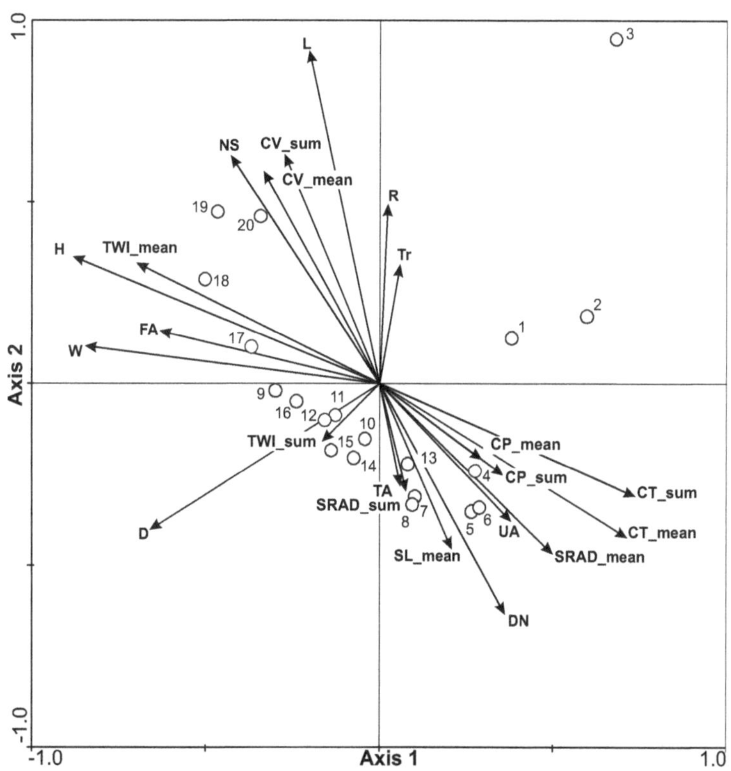

Fig. 5 Ordination diagram showing the result of the RDA for species richness, ecological indicator values and topographic attributes for the right-side sections of the Sopot river valley. Eigenvalues: Axis 1 – 24.92, Axis 2 – 15.11, Axis 3 – 10.70, Axis 4 – 8.11. Abbreviations as in Tab. 2 and Fig. 3.

the total curvatures contributed significantly to the decrease in the share of species with higher L, H, D, and H values. This indicates that the slopes of the valley are dominated by convex rather than concave forms. Simultaneously, we also found negative correlations between the planar curvature and the species number, and the Tr and R values. This implies that the planar curvature of the contour lines has more "ridges" than "valleys", which promotes divergence of flowing water over the convergence of water [24–26]. This situation, locally together with the narrow terrace and water velocity, reduces the repository role of the river, thereby leading to reduction of the nutrient content and reaction of soil [4,5,7].

The SRAD values are significantly correlated with the number of species and diversity of ground-floor vegetation in the study river valley. This secondary topographic attribute is highly variable from place to place due to changing slope and aspect [17,18,44]. In the present study, the decisive condition shaping the energy amounts delivered for both riverbanks was the almost parallel course of the valley in a greater part of the analyzed fragment. The disproportion is on average 150 kWh m^{-2}, in favor of the right-side section. That is approximately one fifth of the total amounts of solar energy reaching the ground-floor layer during the vegetation season on the left-site bank. On the right riverside, the mean SRAD has a negative impact on the species number and share of species with higher light, humidity, and organic matter content requirements. In turn, on the more "shiny" left riverside, the sum SRAD promotes highly the rise of species richness and the trophy level of soils.

Since exact measurement of the water regime is very difficult, the TWI derived from DEM is commonly used in analyses of relationships between abiotic and biotic environmental characters. The TWI correlates well with soil attributes such as horizon depth, groundwater table, silt and organic matter contents, and thereby serves as a good indicator of habitat productivity [18,23,42,45,46]. In our research, the TWI, particularly its mean values, is significantly positively correlated with the species abundance and all the EIVs studied. As a rule, the TWI was more important for the ground flora on the right (more "sunny") than the left (more "shiny") riverbank, with the exception of the Tr value.

Among the EIVs, the R value seems to be most frequent subject of studies. However, some authors found during field measurements that pH values for some ecological groups of species were different than those expected from Ellenberg's scale, which strongly limits the use of the R values and requires re-calibration [28–30,33,47]. The well-known influence of soil acidity and calcium content on species occurrence [33] may also apply to the study area due to the presence of the Ca-rich formations of various origins and age (2.5–89% of $CaCO_3$ content). The diverse water-bearing horizons of these formations result in approximately a ninefold differentiation of water mineralization [11,35].

In conclusion, we have to stress that the GIS, DEM, and multivariate ordination methods have become a useful tool for detection of species richness and distribution of their ecological groups in correlation with primary and secondary topographic attributes of a given terrain, including riparian landscapes. We have also proved that the EIV system, which has been widely applied for modeling plant distribution at large spatial scales, i.e. on the global or regional levels, may be used on a local level, e.g. in a minor river valley and its particular banks. In the light of our study, the course of the river valley which determines the differentiation in slope aspects and, in consequence, between the amount of solar radiation reaching the ground-floor vegetation, does not change the species richness on a more "sunny" and more "shiny" riverbank. However, this factor "cooperating" with other topographic attributes may significantly differentiate the shape of herb species showing various requirements for basic habitat resources: light, moisture, soil trophy, reaction, dispersion, and organic matter content.

Acknowledgments

The authors are grateful to two anonymous reviewers for their criticism and valuable comments on the manuscript and to Prof. Józef Superson, for discussion on geomorphology problems. We would also like to thank Mrs. Anna Zoń, M.A., who has improved our English. This research was supported financially by the Faculty of Biology and Biotechnology, and Faculty of Earth Sciences and Spatial Management, Maria Curie-Skłodowska University in Lublin, Poland.

Authors' contributions

The following declarations about authors' contributions to the research have been made: concept of the study: BC; field and laboratory work: BC; statistical analyses: AR, ŁC; GIS and DEM: ŁC; writing the manuscript: BC.

Competing interests

No competing interests have been declared.

References

1. Kolasa J, Rolle CD. Introduction. The heterogeneity of heterogeneity: a glossary. In: Kolasa J, Pickett STA, editors. Ecological heterogeneity. New York, NY: Springer-Verlag; 1999. p. 1–23.
2. Gregory SV, Swanson FJ, McKee WA, Cummins KW. An ecosystem perspective of riparian zones: focus on links between land and water. Bioscience. 1991;41(8):540–551.
3. Lyon J, Sagers CL. Structure of herbaceous plant assemblages in a forested riparian landscape. Plant Ecol. 1998;138:1–16.
4. Werner KJ, Zedler JB. How sedge meadows, soils, microtopography, and vegetation respond to sedimentation. Wetlands. 2002;22(3):451–466.
5. Francis RA, Tibaldeschi P, McDougall L. Fluvially-deposited large wood and riparian plant diversity. Wetlands Ecol Manage. 2008;16(5):371–382.
6. Duru M, Ansquer P, Jouany C, Theau JP, Cruz P. Comparison of methods for assessing the impact of different disturbances and nutrient conditions upon functional characteristics of grassland communities. Ann Bot. 2010;106(5):823–831.
7. Jolley RL, Lockaby BG, Cavalcanti GG. Changes in riparian forest composition along a sedimentation rate gradient. Plant Ecol. 2010;210(2):317–330.
8. Angiolini C, Nucci A, Frignani F, Landi M. Using multivariate analyses to assess effects of fluvial type on plant species distribution in a Mediterranean river. Wetlands. 2011;31(1):167–177.
9. Tabacchi E, Correll DL, Hauer R, Pinay G, Planty-Tabacci AM, Wissmar RC. Development, maintenance and role of riparian vegetation in the river landscape. Freshw Biol. 1998;40(3):497–516.

10. Decocq G. Patterns of plant species and community diversity at different organization levels in a forested riparian landscape. J Veg Sci. 2002;13(1):91–106.

11. Czarnecka B, Janiec B. Przełomy rzeczne Roztocza jako modelowe obiekty w edukacji ekologicznej. Lublin: Maria Curie-Skłodowska University Press; 2002.

12. Czarnecka B, Janiec B. Krajobrazy roślinne jako wyraz naturalnych i antropogenicznych przemian środowiska małych dolin rzecznych Roztocza. Problemy Ekologii Krajobrazu. 2006;16:171–184.

13. Dunn WC, Milne BT, Mantilla R, Gupta VK. Scaling relation between riparian vegetation and stream order in the Whitewater River Network, Kansas, USA. Landsc Ecol. 2011;26(7):983–977.

14. Czarnecka B, Chabudziński Ł. Assessment of flora diversity in a minor river valley using ecological indicator values, geographical information systems and digital elevation models. Centr Eur J Biol. 2014;9(2):220–231.

15. Muller E. Mapping riparian vegetation along rivers: old concepts and new methods. Special issue: geographic information systems and remote sensing in aquatic botany. Aquatic Bot. 1997;58(3–4): 411–437.

16. Tappeiner U, Tasser E, Tappeiner G. Modelling vegetation patterns using natural and anthropogenic influence factors: preliminary experience with a GIS based model applied to an Alpine area. Ecol Modell. 1998;113(1–3):225–237.

17. Pfeffer K, Pebesma EJ, Burrough PA. Mapping alpine vegetation using vegetation observations and topographic attributes. Landsc Ecol. 2003;18(8):759–776.

18. Fitterer JL, Nelson TA, Coops NC, Wulder MA. Modelling the ecosystem indicators of British Columbia using Earth observation data and terrain indices. Ecol Indic. 2012;20:151–162.

19. Ter Braak CJF, Šmilauer P. CANOCO reference manual and CanoDraw for Windows. User's guide: software for canonical community ordination, Version 4.5. Ithaca, NY: Microcomputer Power; 2002.

20. Lepš J, Šmilauer P. Multivariate analysis of ecological data using Canoco. Cambridge: Cambridge University Press; 2003.

21. Seidling W. Ground floor vegetation assessment within the intensive (level II) monitoring of forest ecosystems in Germany: chances and challenges. Eur J For Res. 2005;124(4):301–312.

22. Petřik P, Wild J. Environmental correlates of the patterns of plant distribution at the meso-scale: a case study from Northern Bohemia (Czech Republic). Preslia. 2006;78:211–234.

23. Kopecký M, Čižkova Š. Using topographic wetness index in vegetation ecology: does the algorithm matter? Appl Veg Sci. 2010;13(4):450–459.

24. Willson JP, Gallant JC, editors. Terrain analysis. Principles and applications, New York, NY: John Wiley and Sons; 2000.

25. Kraak MJ, Ormeling F. Cartography. Visualisation of geospatial data. Glasgow: Bell and Brain Ltd.; 2003.

26. Urbański J. GIS w badaniach przyrodniczych [Internet]. 2012 [cited 2015 Mar 16].

27. Ellenberg H. Zeigerwerte der Gefäßpflanzen Mitteleuropas. Scr Geobot. 1974;9:9–160.

28. Ertsen ACD, Alkemade JRM, Wassen MJ. Calibrating Ellenberg indicator values for moisture, acidity, nutrient availability and salinity in the Netherlands. Plant Ecol. 1998;135:113–124.

29. Wamelink GWW, Joosten V, van Dobben HF, Berendse F. Validity of Ellenberg indicator values judged from physico-chemical field measurements. J Veg Sci. 2002;13(2):269–278.

30. Lawesson JE, Fosaa AM, Olsen E. Calibration of Ellenberg indicator values for the Faroe Islands. Appl Veg Sci. 2003;6(1):53–62.

31. Crosti R, de Nicola C, Fanelli G, Testi A. Ecological classification of beech woodlands in the Central Apennine through frequency distribution of Ellenberg indicators [Internet]. Ann Bot (Roma). 2010 [cited 2015 Mar 16].

32. Balkovič J, Kollár J, Šimonovič V. Experience with using Ellenberg's R indicator values in Slovakia: oligotrophic and mesotrophic submontane broad-leaved forests. Biologia. 2012;67(3):474–482.

33. Schaffers AP, Sýkora KV. Reliability of Ellenberg indicator values for moisture, nitrogen and soil reaction: a comparison with field measurements. J Veg Sci. 2000;11(2):225–244.

34. Czarnecka B, Janiec B. Factors affecting the distribution and properties of forest soils in river breaks of Roztocze. Acta Agrophys. 2001;50:81–93.

35. Janiec B, Czarnecka B. The "Czartowe Pole" landscape reserve in Roztocze (SE Poland) in the light of interdisciplinary research. Ekologia (Bratislava). 2001;20(4 suppl):222–232.

36. European Commission 2007. Interpretation manual of European Union habitats. Eur 27. DG Environment.

37. Dobrzański B, Uziak S, Klimowicz Z, Melke J. Badanie gleb w laboratorium i w polu. Lublin: Maria Curie-Skłodowska University Press; 1992.

38. Sapek A, Sapek B. Metody analizy chemicznej gleb organicznych. Falenty: IMUZ; 1997.

39. Hengl T. Finding the right pixel size. Comput Geosci. 2006;32(9):1283–1298.

40. Bergès L, Gègout JC, Franc A. Can understory vegetation accurately predict site index? A comparative study using floristic and abiotic indices in sessile oak (*Quercus petraea* Liebl.) stands in northern France. Ann For Sci. 2006;63(1):31–42.

41. Zarzycki K, Trzcińska-Tacik H, Różański W, Szeląg Z. Wołek J, Korzeniak U. Ecological indicator values of vascular plants of poland. Biodiversity of Poland 2. Kraków: W. Szafer Institute of Botany, Polish Academy of Sciences; 2002.

42. Dorner B, Lertzman K, Fall J. Landscape pattern in topographically complex landscapes: issues and techniques for analysis. Landsc Ecol. 2002:17(8):729–743.

43. Kumar L, Skidmore AK, Knowles E. Modelling topographic variation in solar radiation in a GIS environment. Int J Geogr Inf Sci. 1997;11(5):475–497.

44. Moore DM, Lees BG, Davey SM. A new method for predicting vegetation distributions using decision tree analysis in a geographic information system. Environ Manage. 1991;15(1):59–71.

45. Sørensen R, Zinko U, Seibert J. On the calculations of the topographic wetness index evaluation of different methods based on field observation. Hydrol Earth Syst Sci. 2006;10(1):101–112.

46. Franklin J. Predictive vegetation mapping: geographic modelling of biospatial patterns in relation to environmental gradients. Prog Phys Geogr. 1995;19(4):474–499.

47. Feldmeyer-Christe E, Ecker K, Küchler M, Graf U, Waser L. Improving predictive mapping in Swiss mire ecosystems through re-calibration of indicator values. Appl Veg Sci. 2007;10(2):183–192.

22

Pre-Linnaean herbaria viva of Helwing in the collections of the National Library of Poland and the University of Warsaw

Krzysztof Spalik*

Department of Plants Systematics and Geography, Faculty of Biology, University of Warsaw, Aleje Ujazdowskie 4, 00-478 Warsaw, Poland

Abstract

Georg Andreas Helwing (1666–1748) was the author of two important early accounts on the flora of former East Prussia: "Flora qusimodogenita" and "Supplementum florae prussicae". Along with his son-in-law Matthias Ernst Boretius, he prepared several herbaria viva. Four of these herbaria survived until WWII; however, their whereabouts since WWII have been generally unknown. In this paper, two of these herbaria are described: one preserved in the collections of the National Library of Poland and another in the herbarium of the Faculty of Biology of the University of Warsaw. Both were formerly in the possession of the Königsberg city library. These herbaria document not only Helwing's studies on the native flora of Prussia but also his experiments with acclimation of exotic species in his garden in Stullichen (Stulichy, Poland). They are also an important source of vernacular Polish and German names of plants.

Keywords: botanic history; Boretius; ethnobotany; Helwing; herbarium vivum; Prussia

Introduction

Pre-Linnaean botanical collections are invaluable for the interpretation of early botanical accounts and species names and particularly for tracing the cultivation of exotic species. In this paper, I describe two extant pre-Linnaean herbaria preserved in the Polish National Library and the herbarium of the Faculty of Biology, University of Warsaw that are linked to the Prussian botanist Georg Andreas Helwing (1666–1748).

Although during his lifetime he was revered as "Prussian Pliny" or "Prussian Tournefort" [1], Georg Andreas Helwing has been almost completely forgotten. He was born in Angerburg (at present Węgorzewo, Poland) to a family of a Lutheran pastor. Continuing family tradition, he started theological studies at Alberts-Universität or Albertina in Königsberg (at present Kaliningrad, Russia). Refraining from obtaining a master degree at Albertina, he commenced a "Bildungsreise" continuing his studies in Wittenberg, Leipzig and Jena. In Jena, he also studied natural sciences, particularly botany, with Georg Wolfgang Wedel, a physician and botanist at the university. Having graduated from the university in 1688, he carried on his botanical studies staying a year with Herman Boerhaave in Leiden. Summoned by his father, he returned to Angerburg in 1691 to serve as his adjunct in the Lutheran parish and after the death of his father in 1705 he replaced him as the provost [1–3]. Although busy with his church duties, he did not renounce his botanical interests. He investigated the flora of the vicinity of Angerburg. He also acquired a patch of land in a neighboring village Stullichen (Stulichy, Poland), where he established an acclimation garden sowing seeds of exotic plants obtained from Johann Philipp Breyn, Christoph Ritter and Herman Boerhaave [4].

His major botanical work was "Flora qusimodogenita" [5], in which he enumerated indigenous plants of Prussia, not only identifying species listed in earlier accounts but also describing new ones. A continuation of this work was "Supplementum florae prussicae" [6]. Two important works of Helwing remained in manuscript: "Tournefortius Borussicus" and "Index plantarum latino-polonicus" [1]. The former may be identified with "Tournefortius prussicus, sive enumeratio plantarum in Prussia indigenarum, partim a scriptoribus Prussicis recensitarum, partim noviter detectarum, ad methodum Josephi Pittonis Tournefortii adornata" described by Georg Christoph Pisanski ([7], p. 549). Pisanski, a grandson of Helwing, wrote that the manuscript was known from several copies and the plants were arranged following the system of Tournefort allowing for identification based on diagnostic characters. The plant index with Latin and Polish names was kept in the City Library of Königsberg and was a major source for Rostafiński's [8] treatise on early vernacular Polish names of plants.

* Email: spalik@biol.uw.edu.pl

Handling Editor: Beata Zagórska-Marek

Extant herbarium materials associated with Helwing

Sherardian Herbarium (OXF)

In the biographical note of Helwing for the Taxonomic Literature, Stafleu and Cowan [9] indicated that some of his herbarium material is kept at OXF. Indeed, several specimens that may be associated with Helwing are within the Sherardian Herbarium collection; the majority bear annotations by William Sherard (Serena Marner, Oxford, personal communication). A specimen of *Quercus robur* associated with Helwing but actually coming from Herman Boerhaave's suburban garden explains the provenance of this material. Helwing spent almost a year with Boerhaave in Leiden and having returned to Angerburg he kept in touch with the Dutch botanist and regularly exchanged plants and seeds with him. Boerhaave distributed this material to other botanists; based on his correspondence with Sherard, he also acted as an intermediary between the British botanist and his Prussian colleague [10].

Herbaria viva

Among the works of the Angerburg pastor, his biographers included several herbaria viva, some prepared with the assistance of his son-in-law Matthias Ernst Boretius [1,7,11]. According to Pisanski [7], the first herbarium initially consisted of two volumes in folio but with successive supplements it grew to six volumes. It was kept in the Castle Library of Königsberg. Other six-volume sets were sent to the Electorial Library in Dresden and to St. Petersburg. A copy was in the possession of Jacob Theodor Klein, a botanist from Danzig (Gdańsk, Poland). In the City Library of Königsberg there was a five-volume set described as "Matth. Ern. Boretii Herbarium vivum, plantarum et florum in Porussia nascentium methodo Tournefortiana, in classes divisum; adscriptis nominibus Plantarum Latinis, Germanicis, Polonicis, cum Indice. Vol. V" [12]. However, Pisanski [7] claimed that Helwing in fact prepared this herbarium while Boretius merely wrote plant names. A third Königsberg copy belonged to Carl Gottfried Hagen; the plants were arranged according to the system of Tournefort and annotated with Latin, German and Polish names ([11], p. 495). This copy was examined by Caspary [13], who described it as a set of five volumes in folio bound in pigskin.

The Dresden copy was burnt during the riots of 1848 [13]; the whereabouts of the St. Petersburg copy are unknown; at least, it has not been preserved in the collections of the Komarov Botanical Institute (personal inquiry). Jacob Klein sold his copy to Friedrich, margrave of Brandenburg-Bayreuth; it was kept in the collections of the University of Erlangen until the end of the 19th century when Hugo Conwentz acquired it by exchange for the collection of the Westpreussiches Landesmuseum in Danzig. After the Second World War, surviving botanical collections from this museum were transferred to the Nicolaus Copernicus University in Toruń but Helwing's herbarium was not among them [4]. The fact that one of Helwing's herbaria viva has been preserved in the collection of the National Library in Warsaw has been known to Polish historians of botany [14]. The information on the preservation of another copy in the herbarium of the University of Warsaw has not been hitherto published.

Herbarium vivum in the Polish National Library

The herbarium set kept in the National Library in Warsaw includes two volumes; a third volume figuring in the catalogue has been obviously mistakenly associated with Helwing as it includes several loose herbarium sheets with plants named with Linnaean binomials. Both volumes are covered with lavishly decorated leather. Each volume also includes a watercolor illustrating the goddess Chloris (or Flora). In the first picture, she is assisting a botanist arranging plants in a herbarium book; a faun holding a cornucopia with flowers stands beside her (Fig. 1). The caption praises herbarium books ("Donec erunt herba, donec florentia serta herbarum, vivat florida fama libro"). The watercolor in the second volume illustrates Chloris in the clouds pointing to a female gardener carrying a basket of flowers. The caption expresses the goddess's surprise that out of the many plants so few are cultivated ("Hesperidum ingentes quorsum miraberis hortos, cum tot opes unus hortulus iste ferat?").

The first volume bears the title "Herbarium Vivum Plantarum sponte in Prussia nascentium, Herculeo labore comparatum M. Georgii Andreae Helwingii Pastoris Angerburgensium. et Reg. Societ. Scient. Brand. Membrii". The information about the membership of the Royal Brandenburg

Fig. 1 The goddess Chloris assisting a botanist. Illustration from the first volume of Georg Andreas Helwing's herbarium vivum from the National Library of Poland.

Scientific Society is written in different ink. Helwing was elected to the membership of the Berliner Akademie der Wissenschaften in 1709 [11]; the herbarium, therefore, must have been prepared before that date. The volume includes 449 pages with dried plants numbered 1–452; numbers 41, 155 and 403 are missing. The arrangement of plants does not follow any system; however, some natural groups (legumes, umbellifers, grasses etc.) are recognized. Each plant is annotated with a Latin phrasename without an authorship. Some specimens are also annotated with German and Polish names. As suggested by the title, this volume includes indigenous and naturalized species.

The title page of the second volume bears the sentence "Florae Napaeisque Sacrum! Civis, amice, advena, qui herbarii amoenitate cupis oblectarier, modestiam et continentiam comites habeas florae custodes, quod rogat, monet, expetit M. Georgius Andreas Helwing, pastor Angeburgensis, operis huius faber." It is a paraphrase of an inscription from public botanic and ornamental gardens summoning visitors to behave properly and to refrain from damaging the plants. Such warnings are known from the summerhouse of the botanic garden of the university of Altdorf ([15], p. 469) and from the ornamental garden of count Leonardo Valmarana in Vicenza ([16], p. 79–80). The sentence in Helwing's herbarium is almost an exact repetition of that in Altdorf and this connection may have had a special meaning. The botanical garden of Altdorf was the place where the ideas of the modern botanical garden and of investigating local flora were born, mostly due to the work of Ludwig Jungermann and Caspar Hoffman [17]. Opened in 1626, the garden constituted a "hortus publicus" and also served for the acclimation of exotic plants. In 1635, Jungermann published the catalogue of plants that survived German winters. In this catalogue, Jungermann placed both exotic and indigenous species. As Cooper [17] points out, for the German botanist, exotic plants in gardens and their local counterparts in surrounding fields were clearly at opposite poles of a rhetorical opposition, but they were at the same time linked by this very opposition. Similar parallelism of the local and the exotic may be seen in the work of Helwing. He investigated the flora of the vicinity of Angerburg while at the same time experimented with cultivation of exotic plants in his garden in Stullichen. The second volume of his herbarium vivum is a testimony to these efforts as it includes cultivated plants. The page following the watercolor with Chloris expressing her dissatisfaction with gardeners' efforts starts with an exclamation: "Chloris prussica fave!" meaning "Smile, Prussian Chloris", because numerous exotic plants are now being grown in "blessed land under the Prussian sky".

The pages with plants are numbered 1-195. Numbering is not continuous: numbers 126–159 are missing and there is one page without a number inserted after page 37. The majority of specimens in the second volume are botanical species; however, some ornamental varieties, for instance with double flowers, are also represented. Horticultural varieties include the members of *Primula* (p. 19–20, 181), *Dianthus* (p. 55), *Delphinium* (p. 110–111, 227), *Aquilegia* (p. 168–172) and *Tulipa* (p. 173–176, 222).

Both volumes bear a stamp of the City Library of Königsberg. However, this herbarium does not match the set that was kept in the City Library as described by Bernoulli and Pisanski but rather those two original volumes of the set from the Castle Library of Königsberg [7,12]. The herbarium is accessible on-line at the digital library Polona (http://www.polona.pl). The website is in Polish, a direct link to the first volume is http://polona.pl/item/7971108/0/ while the second volume is available at http://polona.pl/item/7971109/0/.

Herbarium vivum at the University of Warsaw (WA)

The set preserved in the herbarium of the Faculty of Biology originally comprised five volumes but only four have been preserved: volume 3 is missing. The volumes bear a stamp of the City Library of Königsberg. The covers are paperboard lined with parchment. The bindings of all volumes are damaged by fire; many pages are also partly burnt. Each volume has a title page with the number of the volume and the ordinal numbers of classes in the classification system of Tournefort. There is no title page to the entire herbarium and its author is not indicated. Pages with mounted plants are numbered and volume 5 includes Latin, German and Polish indices with reference to the volume, class and page number. Plants are not numbered continuously: in each class, the last number is omitted. Moreover, in volume 5 there are no plants on pages 31–67. Volume 1 comprises classes 1–4 (272 pages), vol. 2: 5–6 (274 pp.), vol. 4: 13–16 (272 pp.), vol. 5: 17–22 (175 pp.).

The herbarium includes mostly native species; there are no exotics as in the previously described set. However, some presumed garden escapes are present including tobacco (*Nicotiana tabacum* L.), cucumber (*Cucumis sativus* L.), ornamental hollyhocks (*Alcea rosea* L.), bladder cherry (*Physalis alkekengi* L.), lavender (*Lavandula angustifolia* Mill.) and others. Moreover, several varieties of native species are included, particularly those differing in color of flowers or with double flowers ("flore pleno"). They may comprise either rarely occurring natural forms of the species or their cultivated ornamental varieties. The plants are annotated with Latin phrasenames and, sometimes, German and Polish names. Some phrasenames are referred to Caspar Bauhin, Johann Bauhin, Johannes Loesel, Carolus Clusius, John Ray, John Parkinson, Joseph Pitton de Tournefort, Jacobus Theodorus Tabernaemontanus and others.

This set matches two Königsberg herbaria viva: the one that was initially kept in the City Library and the one that was in the possession of Hagen. Both comprised five volumes and were arranged according to the system of Tournefort, probably referring to the aforementioned Helwing's manuscript "Tournefortius prussicus". The set belonging to Hagen was examined by Caspary [13]. In his paper on *Senecio vernalis* Waldst. & Kit., he described a specimen mounted on page 66 of volume 4 and annotated by Helwing or Boretius as "*Jacobaea Senecionis folio incano perennis* Raji hist. 285". The specimen was characterized as measuring 27 cm and having a rosette of ten leaves and two flower stalks with a single flowerhead preserved. The volume, the page number and the phrasename correspond to the herbarium vivum preserved in the collection of the University of Warsaw. However, the specimen does not match Caspary's description as it constitutes a flowering stem 25.5 cm long without a basal rosette, with five leaves and three head stalks; only two capitula are preserved. It seems, therefore, that this herbarium is not the copy belonging initially to Hagen but rather that attributed to Boretius and kept in the Stadtbibliotek Königsberg [7,11].

Helwing's herbarium vivum at WA has been digitalized. It has not been made accessible on-line yet but photos are available at request to the curator of the herbarium. Latin, German and Polish indices are accessible from the web page http://www.zielnik.biol.uw.edu.pl/?page_id=2941.

Apart from those four finished volumes, there are three additional volumes containing loose plants without any arrangement but labeled and apparently prepared for successive mounting. The first volume is bound in parchment with "Plantar[um] Exotic[um] tom. I." written on the spine. The two remaining volumes have cardboard covers. These volumes contain mostly exotic plants.

The significance of Helwing's herbaria viva for taxonomy, botanical history and ethnobotany

Similarly to many other pre-Linnaean publications, the works of Georg Andreas Helwing are mostly of historical interest. Although his two major floristic treatises, "Flora quasimodogenita" and "Supplementum florae prussicae", were known to Linnaeus and cited in the "Hortus Cliffortianus" [18], they do not seem to be a source of types for Linnaean names (Charles E. Jarvis, personal communication). Recently, Helwing's works were mostly cited in the context of history of botany in the former East Prussia [2–4]. However, they also constitute invaluable source documenting the spread of certain cultivars and naturalization of alien species [19–21], the more that Helwing's accounts may be now verified against his herbarium materials. For instance, Wein [19] discussed the spread of Jerusalem artichoke (*Helianthus tuberosus* L.) in Europe and noticed that Helwing listed this species in his "Flora quasimodogenita" as indigenous because "it grows in many Angerburg gardens without cultivation". This observation is documented in Helwing's herbarium with specimens of this species of sunflower. For Polish ethnobotanists, particularly valuable are glossas with Polish vernacular names. They allow for an interpretation of early ethnobotanical accounts.

Those two herbaria also illustrate the change that occurred in perceiving diversity of plants at that time. In the herbarium kept at the National Library, plants are shown as the gifts of Nature that should be studied for their potential use. The herbarium preserved in the collections of the University of Warsaw underlines that plants might be classified based on their internal features marking therefore the beginning of modern taxonomy.

Acknowledgments

I am particularly grateful to Friederike Neuffer for her contribution in identifying the herbarium vivum at WA, I also thank Janina Andrearczyk, Maja Graniszewska, Łukasz Kozak, Charlie Jarvis, Serena Marner, Jarosław Sak, Maks Szot and Ewa Bartnik for their help during the research; this study was supported with the statutory funds of the Faculty of Biology, University of Warsaw.

References

1. Dunkel JGW. Historisch-critische Nachrichten von verstorbenen Gelehrten und deren Schrifften. Köthen: Cörner; 1757.
2. Sukertowa-Biedrawina E. Helwing – "Pliniusz pruski z Węgorzewa". Przegląd Zach. 1950;6:364–369.
3. Augusiewicz S, Jasiński J, Oracki T. Wybitni Polacy w Królewcu. Olsztyn: Wydawnictwo Littera; 2005.
4. Schwarz Z. Prywatne ogrody botaniczne a rozwój nauk przyrodniczych w ośrodku gdańskim w XVI–XVII wiekach. Kwart Hist Nauki Tech. 1986;31:411–444.
5. Helwing GA, Breyn JP. Flora quasimodogenita, sive enumeratio aliquot plantarum indigenarum in Prussia. Gdańsk: Joannes Daniel Stollius; 1712.
6. Helwing GA. Supplementum florae prussicae. Gdańsk: Thomas Johannes Schreiber; 1726.
7. Pisanski GC. Entwurf einer preußischen Literärgeschichte. Königsberg: Hartung; 1886.
8. Rostafiński J. Prowincyonalne, polskie nazwy roślin XVIII w., z Prus Książęcych głównie z rękopisu Andrzeja Helwinga. Rozpr Akad Umiejętności Wydz Filol. 1905;25:209–238.
9. Stafleu FA, Cowan RS. Taxonomic literature: a selective guide to botanical publications and collections with dates, commentaries and types. Utrecht: Bohn, Scheltema and Holkema; 1979. (vol 2).
10. Lindeboom GA. Boerhaave's correspondence. Leiden: E.J. Brill; 1962.
11. Hagen KG. Bruchstücke zur Lebensbeschreibung des Probst Helwing zu Angerburg. Beitr. Zur Kunde Preußen. 1818;1:435–506.
12. Bernoulli J. Reisen durch Brandenburg, Pommern, Preußen, Curland, Rußland und Pohlen. Leipzig: Caspar Fritsch; 1779.
13. Caspary R. Senecio vernalis W. et K. schon um 1717 in Ostpreussen gefunden. Schriften Phys-Ökon Ges Zu Königsb. 1886;27:104–108.
14. Drobnik J. Zielnik i zielnikoznawstwo. Warsaw: Polish Scientific Publishers PWN; 2007.
15. Skippon P. An account of a journey made thro' part of the Low-Countries, Germany, Italy, and France. In: Churchill A, Churchill J, editors. Collection of voyages and travels, some now printed from original manuscripts, others now first published in English. London: John Walthoe; 1732. p. 359–736.
16. Schott F, Capugnano H. Itinerarii Italiae rerumque Romanarum libri tres. Antwerp: Officina Plantiniana; 1625.
17. Cooper A. Inventing the indigenous: local knowledge and natural history in early modern Europe. Cambridge: Cambridge University Press; 2007.
18. Heller JL. Linnaeus's Hortus Cliffortianus. Taxon. 1968;17(6):663–719.
19. Wein K. Die Einführungsgeschichte von *Helianthus tuberosus* L. Kulturpflanze. 1963;11(1):41–91.
20. Wein K. Die Geschichte der Entstehung und Verbreitung von *Rubus idaeus* var. *leucocarpus*. Kulturpflanze. 1962;10(1):195–211.
21. Wein K. Die Geschichte des Rettichs und des Radieschens. Kulturpflanze. 1964;12(1):31–74.

Analysis and glycosyl composition of the exopolysaccharide isolated from submerged fermentation of *Ganoderma lucidum* CG144

Rosália Rubel[1], Leandro Freire dos Santos[1]*, Herta Stutz Dalla Santa[1], Luciana Porto de Souza Vandenberghe[1], Adenise Lorenci Woiciechowski[1], Júlio Cesar de Carvalho[1], Sascha Habu[1], Francisco Menino Destéfanis Vítola[1], Christopher Augur[2], Carlos Ricardo Soccol[1]

[1] Bioprocess Engineering and Biotechnology Division, Federal University of Paraná (UFPR), Jardim das Américas, CEP 81531-990, Curitiba, Paraná (PR), Brazil

[2] Institute of Research for Development, Unite Biotrans – IMEP, Boite 441, Paul Cézanne University, Av. Escadrille Normandie-Niemen, F-13391 Marseille Cedex 20, France

Abstract

The evaluation of glycosyl composition is an essential step to guide future research designs applied in bioactivity. In the same way, the unexplored potential bioactivity of exopolysaccharide from *Ganoderma lucidum* is huge. Therefore, this study investigated the glycosyl composition of the exopolysaccharide isolated from submerged fermentation of *G. lucidum* to serve as guide for future studies on bioactivity. Glycosyl content and composition were evaluated by combined GC/MS of the TMS derivatives of the monosaccharide methyl glycosides produced from the sample by acidic methanolysis. Glycosyl composition analysis showed that the dominant carbohydrate component in all samples of exopolysaccharide isolated from submerged fermentation of *G. lucidum* CG 144 was glucose (58.1%), mannose (26.6%) and galactose (12.5%) which can be referred to as heteroglycan. These results suggest that this *Ganoderma* exopolysaccharide may be a new immunomodulatory agent.

Keywords: glycosyl composition; exopolysaccharide; *Ganoderma lucidum*; submerged fermentation; heteroglycan

Introduction

Ganoderma lucidum (Curtis) P. Karst is a famous fungus (mushroom) in Chinese herbal medicine. Recently, polysaccharides from *G. lucidum* have been identified as active ingredients responsible for its biological activities. For instance, polysaccharides from *G. lucidum* fruiting bodies have been shown to have antimicrobial property [1]. In the same way, *G. lucidum* mycelium have been demonstrated to possess (1,3;1,6)-beta-D-glucans which are recognized for their many health claims [2]. Finally, after the biological evaluation of intracellular polysaccharides (polysaccharides from fruiting bodies and mycelium), studies also have reported biological activities such as antioxidant properties of the exopolysaccharides (extracellular polysaccharides) from *G. lucidum* cultured in submerged fermentation [3]. In addition, there are only a few reports available on biological activities of the exopolysaccharides from *G. lucidum* and new biological activities cannot be carried out randomly.

Similarly, the glycosyl composition is a key factor to understand the underlying biological activities and possible metabolic mechanisms of fungal polysaccharides. For example, studies have been reported the immunostimulatory and antitumor activities of glucans and mannans [4]. Galactose may prevent cognitive deficits associated with glucose hypometabolism in Alzheimer's disease [5]. Recent studies have also focused on the effect of glucans on redox balance [6]. Acetylglucosamine may modulate function of the skin fibroblasts and it also has anti-inflammatory and anti-senescent activity in mesothelial cells [7]. Thus, the glycosyl composition is an essential step to guide future research designs applied in bioactivity. Moreover, a great lack of knowledge exists concerning the glycosyl composition of the exopolysaccharides from *G. lucidum*.

Considering the glycosyl composition as an essential step to guide future researches applied in bioactivity and the unexplored bioactivity of exopolysaccharide from *Ganoderma lucidum*, here we investigated the glycosyl composition of the exopolysaccharide isolated from submerged fermentation of *Ganoderma lucidum* to serve as guide for future studies on bioactivity.

Material and methods

Mushroom strain

The strain of *G. lucidum* CG 144 (botanical family Polyporaceae) was maintained by the standard stock of

* Corresponding author. Email: leandrofreire@onda.com.br

Handling Editor: Łukasz Łuczaj

Bioprocess Engineering and Biotechnology Division at Federal University of Paraná (UFPR), Curitiba – PR, Brazil. It was maintained on potato-dextrose agar (PDA) medium, incubated at 30°C for 10 days and subsequently stored at 4°C. Sub-cultures were made every 2 months to maintain the strain active.

Inoculums preparation

For preparation of inoculums, the *G. lucidum* CG 144 was transferred from agar PDA slants into 125 ml Erlenmeyer flasks containing 20 ml of potate dextrose broth and incubated in a shaker at 30°C, agitated at 120 rpm for 7 days.

Flask culture

The culture medium was composed (in grams per liter) of: glucose 35, peptone 5, yeast extract 2.5, KH_2PO_4 0.88 and $MgSO_4{\cdot}7H_2O$ 0.5; the initial pH of the medium was adjusted to 5.5. Erlenmeyer flasks (250 ml) containing 120 ml medium were inoculated at 30°C in a rotatory shaker, at 120 rpm, for 10 days. A 4% (v/v) of inoculum was added to the flask culture. Cells were removed from the culture medium by centrifugation at 12 000 g for 15 min at 4°C [8].

Isolation and purification of exopolysaccharide

Exopolysaccharide was obtained by diluting the fermented culture medium with 3 volumes of cold 95% ethanol and leaving for 2 h at 4°C for precipitation to be complete, followed by centrifuging (7500 *g* for 10 min at 5°C) and dialysis through a membrane with 10 kDa cutoff. The exopolysaccharide was isolated as a lyophilized material [9].

Glycosyl content and composition

Glycosyl composition analysis was performed by combined gas chromatography/mass spectrometry (GC/MS) of the per-O-trimethylsilyl (TMS) derivatives of the monosaccharide methyl glycosides produced from the sample by acidic methanolysis. An aliquot (10 µg) was taken from the sample and added to separate screw-cap tubes with 40 µg and 60 µg of inositol as the internal standard. The samples were lyophilized. Methyl glycosides were then prepared from the dry sample (0.3 mg) following the mild acid treatment by methanolysis in 1 M HCl in methanol (1 M) at 80°C (16 h), followed by re-N-acetylation with pyridine and acetic anhydride in methanol. The sample was then per-O-trimethylsilylated by treatment with Tri-Sil (Pierce) at 80°C (0.5 h). GC/MS analysis of the TMS methyl glycosides was performed on an AT 6890N GC interfaced to a 5975B MSD (mass selective detector), using a Supelco EC-1 fused silica capillary column (30 m × 0.25 mm ID) [10–12].

Results and discussion

The main objective of this study was to evaluate the glycosyl composition of the exopolysaccharide from *G. lucidum* CG144. Certainly, the glycosyl composition is an essential step to guide future researches applied in bioactivity and must be studied in promising bioactive molecules such as exopolysaccharides, especially from *G. lucidum*.

Glycosyl content and composition were evaluated by combined GC/MS of the TMS derivatives of the monosaccharide methyl glycosides produced from the sample by acidic methanolysis. This technique has been used extensively to evaluate the glycosyl composition of polysaccharides. Detection limits of a few nanograms can be achieved with this methodology [10–12].

Glycosyl composition analysis showed that the dominant carbohydrate component in all samples of exopolysaccharide isolated from submerged fermentation of *G. lucidum* CG 144 was glucose (58.1%), mannose (26.6%) and galactose (12.5%). Traces of fucose (<3%) were detected (Tab. 1). For this reason, this exopolysaccharide may be referred to as heteroglycan.

Tab. 1 Glycosyl content and composition of exopolysaccharide isolated from submerged fermentation of *Ganoderma lucidum* CG 144.

Glycosyl residue	Mass	Mole
Arabinose (Ara)	n.d.	...
Rhamnose (Rha)	n.d.	...
Fucose (Fuc)	0.5 mg	2.8%
Xylose (Xyl)	n.d.	...
Glucuronic acid (GlcA)	n.d.	...
Galacturonic acid (GalA)	n.d.	...
Mannose (Man)	5.2 mg	26.6%
Galactose (Gal)	2.4 mg	12.5%
Glucose (Glc)	11.3 mg	58.1%
N-Acetyl Galactosamine (GalNAc)	n.d.	...
N-Acetyl Glucosamine (GlcNAc)	n.d.	...
N-Acetyl Mannosamine (ManNAc)	n.d.	...
Σ=	19.4 mg	
Total carbohydrate by weight	**3.9%**	

n.d. – none determinated.

Interestingly, the carbohydrate composition of the exopolysaccharide resembles the carbohydrate compositions in other fungal systems (other heteroglycans), which are known about their immunostimulatory activity and antioxidant properties [13,14]. Thus, research focused on exopolysaccharide from *G. lucidum* and all aspects of the immune system is certainly promising and should guide future studies. Exopolysaccharide lyophilization yielded 0.45 $g\,l^{-1}$ of submerged fermentation. Cell dry weight reached a maximum value of 7.22 $g\,l^{-1}$.

Conclusion

In summary, these experiments have shown that the exopolysaccharide from *G. lucidum* CG 144 was composed of glucose (58.1%), mannose (26.6%) and galactose (12.5%), which can be referred to as heteroglycan. Further studies should be made in order to evaluate their immunostimulatory activity.

Acknowledgments

The authors would like to thank the National Research Council (CNPq), and the Paraná's Institute Technology (TECPAR), Brazil, for providing the financial and technical support. We thank Dr. Arailde Fontes Urben (Brazilian Agricultural Research Corporation – EMBRAPA) who kindly provided the *G. lucidum* CG 144 strain.

Authors' contribution

The following declarations about authors' contributions to the research have been made: conceived and designed the experiments: RR, LPSV, ALW, JCC, CA, CRS; performed the experiments: RR, HSDS, SH, FMDV; performed the analysis: RR; wrote the paper: RR, LFS; revised the paper: RR, LFS, CRS.

References

1. Skalicka-Woźniak K, Szypowski J, Łoś R, Siwulski M, Sobieralski K, Głowniak K, et al. Evaluation of polysaccharides content in fruit bodies and their antimicrobial activity of four *Ganoderma lucidum* (W Curt.: Fr.) P. Karst. strains cultivated on different wood type substrates. Acta Soc Bot Pol. 2012;81(1):17–21.
2. Wang CH, Hsieh SC, Wang HJ, Chen ML, Lin BF, Chiang BH, et al. Concentration variation and molecular characteristics of soluble (1,3;1,6)-β-d-glucans in submerged cultivation products of *Ganoderma lucidum* mycelium. J Agric Food Chem. 2014;62(3):634–641.
3. Yuan B, Zhang W, Yu Z, Zhang R. In vitro evaluation of antioxidant property of the exopolysaccharides peptides from *Ganoderma lucidum* CAU5501 in submerged culture. J Food Agric Env. 2012;10(1):97–101.
4. Batbayar S, Lee DH, Kim HW. Immunomodulation of fungal β-glucan in host defense signaling by dectin-1. Biomol Ther Seoul. 2012;20(5):433–445.
5. Salkovic-Petrisic M, Osmanovic-Barilar J, Knezovic A, Hoyer S, Mosetter K, Reutter W. Long-term oral galactose treatment prevents cognitive deficits in male Wistar rats treated intracerebroventricularly with streptozotocin. Neuropharmacology. 2014;77:68–80.
6. Mihailović M, Arambašić J, Uskoković A, Dinić S, Grdović N, Marković J, et al. Β-glucan administration to diabetic rats reestablishes redox balance and stimulates cellular pro-survival mechanisms. J Funct Foods. 2013;5(1):267–278.
7. Połubinska A, Cwalinski J, Baum E, Bręborowicz A. N-acetylglucosamine modulates function of the skin fibroblasts. Int J Cosmet Sci. 2013;35(5):472–476.
8. Xu P, Ding ZY, Qian Z, Zhao CX, Zhang KC. Improved production of mycelial biomass and ganoderic acid by submerged culture of *Ganoderma lucidum* SB97 using complex media. Enzyme Microb Technol. 2008;42(4):325–331.
9. Mandal AK, Yadav KK, Sen IK, Kumar A, Chakraborti S, Islam SS, et al. Partial characterization and flocculating behavior of an exopolysaccharide produced in nutrient-poor medium by a facultative oligotroph *Klebsiella* sp. PB12. J Biosci Bioeng. 2013;115(1):76–81.
10. Haas A, Wild C. Composition analysis of organic matter released by cosmopolitan coral reef-associated green algae. Aquat Biol. 2010;10(2):131–138.
11. Lee HI, Donati AJ, Hahn D, Tisa LS, Chang WS. Alteration of the exopolysaccharide production and the transcriptional profile of free-living *Frankia* strain CcI3 under nitrogen-fixing conditions. Appl Microbiol Biotechnol. 2013;97(24):10499–10509.
12. Fisher ML, Allen R, Luo Y, Curtiss R. Export of extracellular polysaccharides modulates adherence of the cyanobacterium synechocystis. PLoS ONE. 2013;8(9):e74514.
13. Sen IK, Maji PK, Behera B, Mallick P, Maiti TK, Sikdar SR, et al. Structural characterization of an immunoenhancing heteroglycan of a hybrid mushroom (pfls1h) of *Pleurotus florida* and *Lentinus squarrosulus* (Mont.) Singer. Carbohydr Res. 2013;371:45–51.
14. Patra S, Patra P, Maity KK, Mandal S, Bhunia SK, Dey B, et al. A heteroglycan from the mycelia of *Pleurotus ostreatus*: structure determination and study of antioxidant properties. Carbohydr Res. 2013;368:16–21.

24

Molecular evidence for the hybrid origin of *Potamogeton ×subrufus* Hagstr. (Potamogetonaceae)

Joanna Zalewska-Gałosz*, Dagmara Kwolek
Institute of Botany, Jagiellonian University, Kopernika 31, 31-501 Cracow, Poland

Abstract

Potamogeton ×subrufus Hagstr. was described as a hybrid between *P. lucens* L. and *P. nodosus* Poir.; however, the taxon had not been widely accepted and regarded as conspecific with *P. ×fluitans* Roth, the hybrid between *P. lucens* and P. *natans* L. The origin of *P. ×subrufus* had been obscured till 2010, when, based on morpho-anatomical treatment, it was shown that *P. ×subrufus* displays several characters consistently different from those of *P. ×fluitans*. Here we report a successful amplification and sequencing of nuclear ribosomal ITS1 region from a 115 year-old herbarium specimen of *P. ×subrufus*, collected in locus classicus by J. Baagöe and preserved in the Herbarium of the Institute of Botany, Jagiellonian University (KRA). Based on the additive polymorphism pattern expressed in the ITS1 sequences of *P. ×subrufus*, we demonstrate that one of the parents of this hybrid was *P. nodosus*, as was claimed by Hagström.

Keywords: aquatic plants; cloning; hybridization; ITS1; molecular identification; *Potamogeton*; sequencing; taxonomy

Introduction

Interspecific hybridization is one of the important factors shaping *Potamogeton* diversity [1–3]. Although *Potamogeton* hybrids are mostly sterile, due to vegetative propagation they often create stable populations of high ecological significance [3]. Moreover, the hybrids are able to persist at a locality for even hundreds or thousands of years [4–6], what makes them an important component of local floras.

Potamogeton ×subrufus Hagstr. was described as a hybrid between *P. lucens* L. and *P. nodosus* Poir. [1]. Subsequently, *Potamogeton* taxonomists had not generally accepted this taxon and usually regarded *P. ×subrufus* as conspecific with fairly common and morphologically similar hybrid *P. lucens* × *P. natans*, named *P. ×fluitans* Roth [2,7]. Only recently, based on morpho-anatomical treatment of the original materials of *P. ×subrufus*, it was shown that this taxon displays several, mainly anatomical characters different from those of *P. ×fluitans* [6]. Although anatomical characters are highly informative for identification of some *Potamogeton* hybrids [8–10], modern taxonomy of the genus focuses in particular on the molecular assessment and unequivocal confirmation of the hybrid origin by means of the molecular tools [11–15]. Bearing in mind the controversy on the taxonomic status of *P. ×subrufus*, a molecular evidence for the hybrid origin of this taxon is highly desirable. Therefore, the aim of this study is to analyze the molecular characters of the herbarium specimen of *P. ×subrufus* collected in locus classicus, together with European broad-leaved *Potamogeton* species, which could have potentially served as its parental species, and to resolve the origin of the hybrid.

Material and methods

Plant material used for molecular analysis

DNA was extracted from one leaf taken from the specimen of *P. ×subrufus*, collected by J. Baagöe in Jutland, in the river Gudenå (Denmark), on 1 August 1899 and preserved in the herbarium of the Institute of Botany, Jagiellonian University (KRA). Aliquot of DNA extracted from the specimen of *P. ×subrufus* (KRA 91746) was deposited in the Institute of Botany, Jagiellonian University. As a reference for the analysis of the putative hybrid genotype, a comprehensive set of broad-leaved *Potamogeton* species representing potential parental taxa was included in the molecular analyses. Species were mostly represented by one to three samples collected from distant populations based on earlier observations that intraspecific sequence polymorphism was observed only exceptionally in the analyzed DNA regions of *Potamogeton* [12,16]. Fresh material was collected in the field and stored in small plastic tubes filled with silica gel. The origin of the sequenced *Potamogeton* samples is provided in our previous studies [12,17]. Taxonomic delimitation of species and taxa names followed Wiegleb and Kaplan [2], unless a more recent reference is cited.

---

* Corresponding author. Email: joanna.zalewska-galosz@uj.edu.pl
Handling Editor: Krzysztof Spalik

Tab. 1 Polymorphism in the ITS1 sequences from the hybrid individual of *P.* ×*subrufus* and the putative parental *Potamogeton* species, obtained by direct sequencing and cloning.

Taxon*	GB number	22	23	25	37	38–47	48	51	53	62	71	80	87	91–92	99	103	117	149	151	171	173	179	180	191	202	203	204	211	216	218	219	220	221	223	230	231	233	234
P. lucens	FJ883592	T	A	C	A	AACACGTGA	A	G	A	G	T	A	C	--	T	C	G	T	C	G	A	G	G	G	A	T	A	T	G	A	C	T	A	T	C	T	-	G
P. praelongus	JX012091	.	.	.	.	---------	.	.	.	.	.	.	.	--	.	.	.	.	.	.	.	.	.	.	.	.	.	.	.	.	.	.	.	.	.	.	-	.
P. gramineus	FJ883587	.	.	.	.	---------	.	.	.	.	.	.	.	--	.	.	.	.	.	.	.	.	T	.	.	.	.	.	.	.	.	.	.	.	.	.	-	.
P. nodosus	FJ883594	.	.	T	.	---------	.	C	.	T	.	.	.	--	C	.	.	.	.	T	.	.	.	.	T	.	.	.	.	.	.	.	.	.	T	.	T	.
P. perfoliatus	FJ883583	.	.	T	.	---------	T	C	.	T	.	T	T	--	C	.	.	.	T	T	.	.	.	.	.	.	.	.	.	.	.	.	.	.	T	.	-	.
P. natans	JX012090	.	.	.	.	---------	.	.	G	.	.	.	.	--	.	.	.	.	.	.	.	.	.	.	.	.	G	.	T	.	.	.	.	.	.	.	-	.
P. crispus	FJ883590	C	.	.	T	---------	.	.	.	.	.	.	.	--	.	.	.	.	.	A	.	.	.	C	.	.	.	.	T	T	T	A	.	.	T	.	-	.
P. alpinus	FJ883580	.	.	.	.	---------	.	A	C	.	.	.	.	--	C	.	.	.	.	.	.	.	.	.	.	A	.	.	.	.	.	.	.	.	T	.	T	T
P. polygonifolius	JX012087	.	C	.	.	--CCTCTCG	.	.	.	.	C	.	.	AG	.	T	A	G	.	.	C	T	.	A	T	.	.	C	.	.	.	.	T	A	T	C	-	.
P. ×*subrufus* (direct seq.)	KJ734980	.	.	Y	.	---------	.	S	.	K	.	.	.	--	Y	.	.	.	.	K	.	.	.	.	W	.	.	.	.	.	.	.	.	.	.	.	(T)	R
P. ×*subrufus* (clone 3-3)	KJ734974	.	.	.	.	---------	.	C	.	T	.	.	.	--	C	.	.	.	.	T	.	.	.	.	T	.	.	.	.	.	.	.	.	.	T	.	T	.
P. ×*subrufus* (clone 2-1)	KJ734975	.	.	.	.	---------	.	.	.	.	.	.	.	--	.	.	.	.	.	.	.	.	.	.	.	.	.	.	.	.	.	.	.	.	.	.	-	.
P. ×*subrufus* (clone 2-3)	KJ734972		.	T	.	---------	.	C	.	T	.	.	.	--	C	.	.	.	.	.	.	.	.	.	.	.	.	.	.	.	.	.	.	.	.	.	-	.
P. ×*subrufus* (clone 2-8)	KJ734973	.	.	T	.	---------	.	C	.	T	.	.	.	--	C	.	.	.	.	.	.	.	.	.	.	.	.	.	.	.	.	.	.	.	T	.	-	.
P. ×*subrufus* (clone 2-5)	KJ734976	.	.	.	.	---------	.	.	.	.	.	.	.	--	.	.	.	.	.	.	.	.	.	.	.	.	.	.	.	.	.	.	.	.	T	.	-	.
P. ×*subrufus* (clone 3-5)	KJ734977	.	.	.	.	---------	.	.	.	.	.	.	.	--	C	.	.	.	.	.	.	.	.	.	.	.	.	.	.	.	.	.	.	.	T	.	-	.
P. ×*subrufus* (clone 3-10)	KJ734978	.	.	.	.	---------	.	.	.	.	.	.	.	--	C	.	.	.	.	.	.	.	.	.	.	.	.	.	.	.	.	.	.	.	T	.	-	.
P. ×*subrufus* (clone 3-1)	KJ734979	.	.	.	.	---------	.	.	.	.	.	.	.	--	C	.	.	.	.	.	.	.	.	.	.	.	.	.	.	.	.	.	.	.	T	.	-	.

Polymorphic nucleotide sites are coded using the IUPAC code. Symbol in brackets indicates the presence of a base only in one of the ITS copies in the hybrid genome. * Origin of all samples except *P.* ×*subrufus* is provided in Zalewska-Gałosz et al. [12] and Zalewska-Gałosz and Ronikier [17].

DNA isolation, PCR amplification, sequencing and cloning

DNA isolation, PCR amplification, direct sequencing of ITS1 and the sequence editing were done as described by Zalewska-Gałosz et al. [12]. ITS1 region was amplified using the primers ITS A and ITS C [18]. Additionally, the ITS1 region obtained from the type specimen of *P. ×subrufus* was cloned to verify its additivity pattern, following the method described by Zalewska-Gałosz et al. [19]. The ITS1 sequences of *P. ×subrufus* were submitted to GenBank under the numbers KJ734972–80.

Phylogenetic analysis

Phylogenetic relationships were estimated using MEGA version 5.0 [20]. The maximum likelihood (ML) method was performed with Jukes–Cantor substitution model [21], which was selected based on BIC (Bayesian information criterion). The tree with the highest log likelihood (−598.87) is shown. Bootstrap support values from 1000 replicates are shown next to the branches. Initial trees for the heuristic searches were obtained with neighbor-joining algorithm and a matrix of pairwise distances estimated using the maximum composite likelihood (MCL) approach. The tree was drawn to scale, with branch lengths estimated based on the number of substitutions per site using Jukes–Cantor model. The analysis involved 17 nucleotide sequences. There were a total of 234 positions in the final dataset.

Results

The DNA sample extracted from the herbarium specimen of *Potamogeton ×subrufus* showed signs of degradation. Judging from the appearance of the total DNA in the agarose gel it contained mainly fragments shorter than 300 bp. It is not surprising therefore that all attempts to amplify the entire ITS region (ITS1, 5.8S, ITS2) were unsuccessful. Similarly, we were not able to obtain shorter fragments of ITS using the primers suggested by Blatter [18] as well as the sequence of cpDNA *rpl32-trn*L intergenic spacer despite using internal primers. The only sequence we have succeeded to amplify was ITS1. The fragment obtained by direct sequencing was 479 base pair long and covered entire ITS1 region and the beginning of the gene 5.8S. The sequence displayed an additive polymorphism at eight positions. Examination of the polymorphism pattern in these positions allowed indicating *P. lucens*/*P. praelongus* and *P. nodosus* as the parental taxa of *P. ×subrufus* (Tab. 1). The chromatogram had a little baseline noise, what was probably caused by poor quality of DNA template (A_{260}/A_{280} ratio was 1.66) and could lead to misinterpretation; therefore, cloning of the ITS1 sequence was performed. Sequences of eight clones were obtained, each 320 base pair long. Clone 3-3 was identical with ITS1 sequence of *P. nodosus* while clone 2-1 was identical with ITS1 sequence of *P. lucens* and *P. praelongus* (Tab. 1, Fig. 1). None of the other clones of *P. ×subrufus* was identical with any sequence obtained from the *Potamogeton* species. Clones 2-3 and 2-8 had a chimeric character. The first part of these sequences, to the position 99 at the alignment, was identical with *P. nodosus* while the second part resembled *P. lucens*/*P. praelongus*, except the position 230 in the clone 2-8 (Tab. 1). The clone 2-5 was almost identical with *P. lucens*/*P. praelongus* sequence except the position 230, where it had T instead of C (Tab. 1). The clones 3-1, 3-5 and 3-10 were similar to the ITS1 sequences of *P. lucens*/*P. praelongus* except two positions: 99, where the clones had C, and 230, where clones had T (Tab. 1). Because there is no differences in the ITS1 sequence between *P. lucens* and *P. praelongus*, it is impossible to determine which of these species served as an ITS ribotype donor for *P. ×subrufus*.

Discussion

Potamogeton ×subrufus was described based on the specimens collected from two separate localities in Europe and North America over than a century ago [1]. European population of *P. ×subrufus* in the Gudenå River in Denmark probably disappeared [22]. American locality was enigmatically described and difficult to locate. Therefore, despite that DNA extracted from herbarium materials is often poor-quality, its examination was the only possibility for checking the molecular characteristics of this taxon.

The aim of the present study was to resolve the controversy whether *P. ×subrufus* originated from the cross between *P. lucens* and *P. natans* or *P. lucens* and *P. nodosus*. We have found that one of the parents of *P. ×subrufus* is *P. nodosus* and definitely not *P. natans* (Tab. 1, Fig. 1). Although the second parent could not be univocally identified based on the ITS1 sequences, the parentage of *P. lucens* is well supported by morphology of *P. ×subrufus*. *Potamogeton lucens* is the only pondweed species that develops stipules with winged ridges. Narrowly winged ridges of stipules are also characteristic for *P. ×subrufus* [6]. Also, *P. ×subrufus* alike *P. lucens* and *P. nododsus* has petiolate submersed leaves while *P. praelongus* and all of its hybrids develop sessile submersed leaves [1–3].

The present results, although limited due to degraded DNA extracted from an old herbarium specimen, supplement the previous morpho-anatomical treatment of *P. ×subrufus* [6]. Finally, based on all available evidence, it can be stated that *P. ×subrufus* is a hybrid of *P. lucens* and *P. nodosus*.

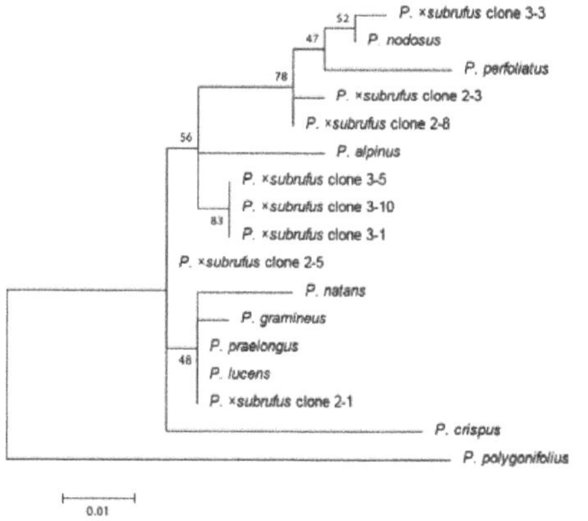

Fig. 1 Identification of the hybrid origin based on cloned ITS1 sequences inferred using maximum likelihood method with MEGA5. Bootstrap support values from 1000 replicates are shown next to the branches. Cloned hybrid sequences are numbered and their accession numbers are provided in Tab. 1.

Acknowledgments

We would like to thank Dr. Wacław Bartoszek, curator of the Herbarium of the Institute of Botany, Jagiellonian University, for his permission for DNA extraction from the herbarium specimen. The research was financially supported by grant No. N N303 564439 from the Polish Ministry of Science and Higher Education.

Authors' contributions

The following declarations about authors' contributions to the research have been made: idea of study, research design, interpretation of the results, writing the manuscript: JZG; molecular analysis, results consulting, editing the manuscript: DK.

References

1. Hagström JO. Critical researches on the *Potamogeton*. K Sven Vetenskapsakademiens Handl. 1916;55:1–281.
2. Wiegleb G, Kaplan Z. An account of the species of *Potamogeton* L. (Potamogetonaceae). Folia Geobot. 1998;33:241–316.
3. Preston CD. Pondweeds of Great Britain and Ireland. London: Botanical Society of the British Isles; 1995. (BSBI Handbook No. 8).
4. Hollingsworth PM, Preston CD, Gornall RJ. Isozyme evidence for the parentage and multiple origins of *Potamogeton* ×*suecicus* (*P. pectinatus* × *P. filiformis*, Potamogetonaceae). Plant Syst Evol. 1996;202(3–4):219–232.
5. King RA, Gornall RJ, Preston CD, Croft JM. Molecular confirmation of *Potamogeton* ×*bottnicus* (*P. pectinatus* × *P. vaginatus*, Potamogetonaceae) in Britain. Bot J Linn Soc. 2001;135(1):67–70.
6. Zalewska-Gałosz J. *Potamogeton* ×*subrufus* Hagstr.: a neglected *Potamogeton* hybrid. Ann Bot Fenn. 2010;47(4):257–260.
7. Dandy JE, Taylor G. Studies of British *Potamogetons* – XVI ×*Potamogeton olivaceus* (*P. alpinus* × *P. crispus*). J Bot Br Foreign Lond. 1942;80:117–120.
8. Wiegleb G. The importance of stem anatomical characters for the systematics of the genus *Potamogeton* L. Flora. 1990;184(3):197–208.
9. Kaplan Z. *Potamogeton* ×*fluitans* (*P. natans* × *P. lucens*) in the Czech Republic. I. Morphology and anatomy. Preslia. 2001;73(4):333–340.
10. Zalewska-Gałosz J. *Potamogeton* ×*jutlandicus*, a binominal for the hybrid between *P. lucens* and *P. praelongus* (Potamogetonaceae). Nord J Bot. 2011;29(4):473–476.
11. Kaplan Z, Fehrer J, Hellquist CB. New hybrid combinations revealed by molecular analysis: the unknown side of north american pondweed diversity (*Potamogeton*). Syst Bot. 2009;34(4):625–642.
12. Zalewska-Gałosz J, Ronikier M, Kaplan Z. The first European record of *Potamogeton* ×*subobtusus* identified using ITS and cpDNA sequence data. Preslia. 2009;81:281–292.
13. Zalewska-Gałosz J, Ronikier M. Are linear-leaved *Potamogeton* hybrids really so rare? Molecular evidence for multiple hybridizations between *P. acutifolius* and *P. compressus* in central Europe. Nord J Bot. 2010;28(3):257–261.
14. Zalewska-Gałosz J, Ronikier M. *Potamogeton* ×*maëmetsiae* – a new hybrid of linear-leaved pondweeds from central Europe. Preslia. 2011;83(3):259–273.
15. Bobrov AA, Zalewska-Gałosz J, Chemeris EV. *Potamogeton* ×*clandestinus* (*P. crispus* × *P. natans*, Potamogetonaceae), a new natural pondweed hybrid discovered in Europe. Phytotaxa. 2013;149(1):31.
16. Kaplan Z, Fehrer J. Erroneous identities of *Potamogeton* hybrids corrected by molecular analysis of plants from type clones. Taxon. 2011;60(3):758–766.
17. Zalewska-Gałosz J, Ronikier M. Molecular evidence for two rare *Potamogeton natans* hybrids with reassessment of *Potamogeton* hybrid diversity in Poland. Aquat Bot. 2012;103:15–22.
18. Blattner FR. Direct amplification of the entire ITS region from poorly preserved plant material using recombinant PCR. BioTechniques. 1999;27(6):1180–1186.
19. Zalewska-Gałosz J, Jopek M, Ilnicki T. Hybridization in *Batrachium* group: controversial delimitation between heterophyllous *Ranunculus penicillatus* and the hybrid *Ranunculus fluitans* × *R. peltatus*. Aquat Bot. 2014 (in press).
20. Tamura K, Peterson D, Peterson N, Stecher G, Nei M, Kumar S. MEGA5: molecular evolutionary genetics analysis using maximum likelihood, evolutionary distance, and maximum parsimony methods. Mol Biol Evol. 2011;28(10):2731–2739.
21. Jukes TH, Cantor CR. Evolution of protein molecules. In: Munro HN, editor. Mammalian protein metabolism. New York, NY: Academic Press; 1969. p. 21–132.
22. Kaplan Z, Fehrer J. Molecular identification of hybrids from a former hot spot of *Potamogeton* hybrid diversity. Aquat Bot. 2013;105:34–40.

25

Wild food plants used on the Dubrovnik coast (south-eastern Croatia)

Katija Dolina[1], Łukasz Łuczaj[2]*

[1] Institute for Marine and Coastal Research, Botanical Garden on the island Lokrum, University of Dubrovnik, Kneza Damjana Jude 12, P.O. Box 83, 20000 Dubrovnik, Croatia

[2] Department of Botany, Institute of Applied Biotechnology and Basic Sciences, University of Rzeszów, Werynia 502, 36-100 Kolbuszowa, Poland

Abstract

Croatia's versatile plant use traditions are still not sufficiently documented. The aim of this study was to record local traditions of wild food plant use on the Dubrovnik coast. We interviewed 40 inhabitants of 23 villages, mainly knowledgeable informants. On average 19 species were listed, which in total produced an inventory of 95 food plant species (including species whose leaves or inflorescences are used as recreational teas).

The most commonly collected are: *Sonchus oleraceus* and *S. asper*, *Asparagus acutifolius*, *Dioscorea communis*, *Cichorium intybus*, *Crepis zacintha*, *Allium ampeloprasum*, *Picris echioides* and *Foeniculum vulgare* (all of them used as vegetables), the fruits of *Rubus ulmifolius* (mainly eaten raw), the fruits of roses (*Rosa sempervirens* and *R. canina*) and the leaves of *Salvia officinalis* (both roses and salvia are used for making recreational teas). A particular feature of the local gastronomy is the collection of young *Ruscus aculeatus* shoots.

Keywords: ethnobotany; wild vegetables; wild edible plants; foraging

Introduction

The documentation of traditionally used wild food plants has been carried out in many European countries over the last several years [1–6], and reviews of older ethnographic literature concerning this issue have been published [7–13]. Some results of field ethnobotanical studies have also been published in the countries of the former Yugoslavia (e.g. [14–19]), though only a few of them are solely focused on wild foods [19–21]. The Adriatic coast has a particularly rich tradition of using wild foods in nutrition. This probably stems from a combination of two factors: the low productivity of rocky soils, which made people utilize every possible food resource, and Mediterranean cultural influences (the widespread appreciation of wild greens in the Mediterranean diet). The proper documentation of wild food ethnobotany in this region has only recently begun. Wild greens sold in Dalmatian markets were documented [20], and the traditions of using wild foods in a selected site in northern Dalmatia, near Lake Vrana, were recorded [21]. Additional information on wild food plant use can be found in older ethnographic sources, wild food guides and ethnolinguistic publications [22–26]. Unfortunately no ethnobotanical studies had previously been carried out at the very southern edge of the Croatian coast – in the Dubrovnik area. As traditional ecological knowledge is eroding fast in most of Europe, the aim of our study was to document the wild food plants used now or in the past around the city of Dubrovnik.

Material and methods

Study site

The study site is located around Dubrovnik, in the Croatian section of the Adriatic coast between the areas of the coast belonging to Bosnia-Herzegovina and Montenegro (Fig. 1). The investigated area was part of the Republic of Ragusa (Dubrovnik Republic) in Medieval times, and later formed part of the Habsburg Monarchy (Austro-Hungarian Monarchy), then Yugoslavia and finally Croatia. A Dalmatian/Venetian (Romanesque) language elite lived in the city of Dubrovnik for centuries, while the lower classes spoke a dialect of Croatian. Gradually the Croatian language took over, sealed by the expulsion of the last remaining Italians after World War II [27]. The Dubrovnik coast is now inhabited almost exclusively by Croatian speaking Roman-Catholics.

The climate is Mediterranean; summers are hot with short periods of drought, while the remaining seasons are characterized by abundant rainfall and moderate temperatures. According to the Meteorological and Hydrological Service of Croatia, for the period 1970–2005 the average annual mean temperature of Dubrovnik was 16.4°C, the coldest months of the year were January and February, with an average temperature of 9.3°C, and the warmest was August with an average temperature of 24.9°C. The average total annual rainfall in the same period was 1064.7 mm. The lowest amounts of

* Corresponding author. Email: lukasz.luczaj@interia.pl

Handling Editor: Beata Zagórska-Marek

rainfall were recorded in July, an average of 30.1 mm, and the maximum in November, an average of 141.8 mm [28].

From the phytogeographical point of view the study area is situated in the Mediterranean Region, Eastern Mediterranean Subregion, Adriatic Province, and Epiro-Dalmatian Sector [29]. According to Trinajstić [30], the area belongs to the Eumediterranean vegetational zone of evergreen holm-oak forests [the *Quercion ilicis* Br.-Bl. (1931) 1936 alliance], while the warmest and driest parts of the study area belong to the Stenomediterranean vegetational zone of wild olive and carob woodlands (alliance *Oleo Ceratonion* Br.-Bl. 1931). Vegetation in the Dubrovnik hinterland belongs to the Submediterranean vegetational zone, or the Mediterranean-montane vegetational belt of a deciduous vegetational zone of hop hornbeam and oriental hornbeam forests [alliance *Ostryo-Carpinion orientalis* Horvat (1954) 1959].

Climazonal forest vegetation can be found only in a small part of this area because, due to intensive anthropogenic activities, it has been degraded to a greater or lesser extent. Such degradation caused the formation of very significant, permanent vegetative forms like shrublands (macchia), garrigue, dry grasslands and rocky pastures.

Field study

We interviewed people (all of Croatian nationality) from 23 villages: Buići, Doli, Dubravka, Dunave, Gabrili, Gromača, Gruda, Ljuta, Majkovi, Mlini, Mokošica, Mravinjac, Mrčevo, Orašac, Osojnik, Palje Brdo, Petrača, Popolica, Riđica, Smokovljani, Šumet, Topolo and Visočani (Fig. 1). A relatively large area was chosen for the study, as the knowledge and use of wild edible plants has strongly decreased and more effort was required to find interviewees than in the case of a similar study in northern Dalmatia in the Zadar area [21]. Also for this reason, we decided to present all the species mentioned even by one respondent providing we were sure that they were identified correctly, with the awareness that these are dying out relic uses, which should be recorded.

The study was carried out throughout 2013, mainly in the spring months (March–May). The first author has spent her life in the area and has family ties in a few of the studied villages, which helped in gathering extra information on local plant uses and in choosing the informants.

Fig. 1 The study area and the distribution of the studied villages.

The research was carried out following the American Anthropological Association Code of Ethics [31] and the International Society of Ethnobiology Code of Ethics [32]. Data were collected mainly by free listing methods. Interviews were carried out in Croatian.

Participants were selected either as the most knowledgeable people in the village, based on recommendation, or simply approached outside during their farm or garden work. Altogether, 40 free listing questionnaires were obtained, from middle-aged and elderly inhabitants (27 women, 10 men and three couples). The mean age of informants was 62 and the median 64, with the youngest interviewee aged 40 and the oldest 81.

The participants were asked their age, place of origin and present habitation. Then they were asked the following questions: what wild vegetables have you collected? What wild roots have you collected? What wild fruits have you collected? What wild mushrooms have you collected? Do you drink any herbal teas on an everyday basis?

Walks were made with some of the participants, which enabled the gathering of plants and identification of folk taxa. Voucher specimens were collected and deposited in the herbarium of the Faculty of Biology of Warsaw University (WA).

Results

Altogether, 95 taxa of wild food plants (and 3 taxa of fungi) were recorded in the 76 species of plants that were used or remembered by at least two informants (Tab. 1). If species used only for infusions ("recreational teas") or to spice alcohol were excluded, the total list of plants would include 78 species. Leaves and other above-ground vegetative organs are the largest category (58 species), followed by fruits (23 species), flowers (19) and underground organs (3). The leaves or shoots of 43 species are used for boiled or raw salads, 18 species of fruits are eaten raw or in the form of preserves, 14 species of plants are used to spice rakija (including the herbal rakija called "travarica") and sweet liqueurs, and 7 species are used as culinary spices. Various organs of 27 species are used as recreational teas.

On average 19.2 species (median = 16.5) were listed by each informant, including 10.0 species of wild vegetables, 4.8 species of fruits, 2.5 species used as spices, 0.1 species with edible roots, and 0.2 species of fungi.

Wild vegetables are still widely used in the Dubrovnik area, though only a few species are used on a regular basis. These are usually weeds growing in old fields, on roadsides and in hedgerows. They are mainly collected in spring (February–April). Unfortunately wild food plants are much less popular among young people, most people who use wild foods are elderly, and a large percentage of the population have stopped using them completely.

Most wild vegetables are boiled in a mix, which is called "pazija", [a Turkish name for beet (*Beta*) [33]] or "pakoleć" (dominant species is *Bunias erucago* [34]). They are usually boiled for 20–30 min., strained and sprinkled with olive oil, then served as a side dish. Usually only a few of the most abundant species are collected and used (sow thistle *Sonchus* spp., wild leek *Allium ampeloprasum*, bristly ox-tongue *Picris echioides*, fennel *Foeniculum vulgare*, red poppy *Papaver rhoeas*, *Urospermum picroides* and *Bunias erucago*). Other species are used for pazija only rarely.

Asparagus shoots are boiled separately and served in a similar way. They can also be fried or added to soup, often with boiled eggs. Black bryony *Dioscoraea communis* (syn. *Tamus communis*), and more rarely butcher's broom *Ruscus aculeatus* (and *Smilax aspera* mentioned only by one person), are commonly used "like asparagus" ("kao šparoge"), as the shoots have a similar shape. Some Asteraceae species are also eaten raw or blanched, and dressed in olive oil (*Cichorium*, *Crepis*, *Taraxacum* etc.).

Several aromatic wild plants are used in cooking. Fennel (*Foeniculum vulgare*) is used both as a vegetable and a spice, added to pazija, and other dishes. Another common spice is bay leaf (*Laurus nobilis*), which is used to flavor meat dishes and preserves.

A large number of taxa are used as recreational teas, i.e. infusions, drunk without a particular health target. The most common plants used are rose hips (*Rosa* spp.) and sage (*Salvia officinalis*) leaves, but a number of other Lamiaceae species, both cultivated, semi-wild, and wild are used (see Tab. 1).

Wild fruits are now rarely collected, apart from snacking on blackberries or collecting *Rosa* fruits for tea. Wild mushrooms are practically unknown (only two respondents mentioned collecting parasol mushrooms *Macrolepiota procera*, and single informants mentioned collecting porcini (*Boletus* sp.) and chanterelles (*Cantarellus cibarius*).

Discussion

The presented list of wild foods is similar to the plants collected in the Vrana area, and also typical for other Mediterranean areas. Most of the wild vegetables and fruits collected in the Dubrovnik area are also widely used in Spain, Italy and western Turkey [1–5,35]. On the other hand, worthy of attention is the use of *Ruscus aculeatus*, used similarly to asparagus. This use has only been observed in some parts Italy [36,37] and in the Istrian part of the Dalmatian coast [38]. It is not known in other parts of Dalmatia [21]. We also found that it is the only place in Dalmatia that *Tordylium apulum*, a common plant of roadside verges in southern Croatia, is still occasionally eaten in the wild vegetable mix [this species was incorrectly provisionally identified in our previous article as "*Pimpinella peregrina* (?)" [20]].

Out of the 36 wild food plant taxa listed by Bakić and Popović [22] as the most frequently used on the ex-Yugoslav coast, as many as 26 species (or their close relatives from the same genus) are used or remembered in the Dubrovnik area.

In spite of the fact that many of the studied households are located on the coast, the only truly seaside plant collected is *Crithmum maritimum*. The plants used for everyday infusions in the Dubrovnik area are species commonly used around the Mediterranean or in other Slavic countries (e.g. [39]).

The erosion of local plant knowledge in the Dubrovnik area seems to be more extensive than on some other parts of the Croatian coast, for instance in the Vrana area. In Vrana less species were recorded per interview [21], but most people gather wild vegetables and were willing to be interviewed. On the Dubrovnik coast only a few elderly people collect wild foods now and it is very difficult to find people with knowledge of wild foods, hence this study was focused on selected key informants. The difference between Vrana and Dubrovnik can probably largely be explained by the fact that the Dubrovnik area is a region with a higher income than other parts of Dalmatia (see e.g., http://www.dzs.hr/Hrv_Eng/publication/2013/12-01-02_01_2013.htm).

It should however be borne in mind that in the Dubrovnik area a much larger area was studied than in the study on the wild food plants of the lake Vrana area.

Local differences in the list of the species still collected for vegetable mixes should be noted. For example, in the Dubrovnik area neither *Tragopogon* nor *Scolymus hispanicus* were recorded by us, although their leaves are used as food in northern Dalmatia. And vice versa – *Tordylium apulum*, which is used (but rarely) in Dubrovnik area is not known as edible in northern Dalmatia, although it also occurs there. It is difficult to establish whether these differences occur only due to differences in the abundance of species (we assume that if a species is more abundant it may be remembered better and used more frequently), or if knowledge about the edibility of some lesser used species has died out, similarly to the knowledge of wild edible greens in Eastern Europe (compare [6,8,9,13,40]). Unfortunately it is likely that the latter is true, which should urge us to extend culinary ethnobotanical research to other parts of the Balkans where this knowledge is disappearing as well.

Tab. 1 Wild food plants and mushrooms used in the area.

Scientific name (plant names follow the Plant List [41])	Local name	Part used	Use	Frequency $N = 40$	Voucher specimen No. starting from WA0000370
Achillea millefolium L. (Asteraceae)	hajdučka trava, sporiš	FL	tea	1	
Allium ampeloprasum L. (Amaryllidaceae)	poriluk, divlji luk, lukovac	L	in pazija*	30	35
Allium roseum L. (Amaryllidaceae)	divlji luk, lukovac	L	in pazija	3	62
Allium subhirsutum L. (Amaryllidaceae)	divlji luk, lukavac	L	in pazija	2	
Althaea officinalis L. (Malvaceae)	bijeli sljez	L	tea	12	74
Amaranthus cruentus L. (Amaranthaceae)	trator	L	boiled	1	
Amaranthus spp. (Amaranthaceae)	štir	L	cooked	3	44

Tab. 1 *(continued)*

Scientific name (plant names follow the Plant List [41])	**Local name**	**Part used**	**Use**	**Frequency** $N = 40$	**Voucher specimen No. starting from WA0000370**
Arbutus unedo L. (Ericaceae)	maginja	F	raw, rakija**	20	87
Arctium lappa L. (Asteraceae)	čičak	S	young stems eaten by children	1	
Artemisia absinthium L. (Asteraceae)	osjenač	L	rakija, tea	4	
Asparagus acutifolius L. (Asparagaceae)	šparoge	S	boiled and served with eggs, fried with (or without) eggs, soup	38	51
Asphodeline lutea (L.) Rchb. (Xanthorrhoeaceae)	pečak	FL, S	young stem and inflorescences, harvested before bloom, boiled like asparagus or baked in oven (very sweet taste)	1	66
Asplenium ceterach L. (Aspleniaceae)	paprat, zlatna paprat	W	tea	7	54
Beta vulgaris L. (Amaranthaceae)	divlja blitva	L	in pazija	11	19
Boletus edulis Bull. s.l. (Boletaceae)	vrganj	M	grilled, fried (with eggs), stewed (pasta sauce)	2	
Bunias erucago L. (Brassicaceae)	pakoleć	L	in pazija	18	37
Cantharellus cibarius Fr. (Cantharellaceae)	lisičarka	M	grilled, fried (with eggs), stewed (pasta sauce)	1	
Capparis orientalis Veill (Capparaceae)	kapare	FL	preserved in vinegar, used as spice/vegetable	8	49
Capsella bursa-pastoris (L.) Medik. (Brassicaceae)	šurlin	L	in pazija	1	33
Celtis australis L. (Cannabaceae)	košćela	F	children snack	9	57
Ceratonia siliqua L. (Fabaceae)	rogač	F	raw, rakija, cakes	11	52
Chamomilla recutita (L.) Rauschert (Asteraceae)	kamomila, kamilica	FL	tea	3	
Chenopodium album L. (Amaranthaceae)	loboda	L	cooked	1	45
Chondrilla juncea L. (Asteraceae)	zvečak	L	in pazija	3	
Cichorium intybus L. (Asteraceae)	žučenica	L	raw or shortly boiled	35 (not distinguished from *Crepis* spp.)	
Citrus aurantium L. (Rutaceae)	ljuta naranča	F	roasted meat, tea, rakija, juice, marmelade	2	
Crepis zacintha (L.) Babc. and possibly other related species from the genus (Asteraceae)	žučenica	L	raw or shortly boiled	35 (not distinguished from wild *Cichorium*)	27, 85
Cornus mas L. (Cornaceae)	drenjina, drjen	F	raw, jam, syrup	7	61
Crataegus monogyna Jacq. (Rosaceae)	glog, gloginja	F	raw, tea	5	41
Crithmum maritimum L. (Apiaceae)	motar, mota	L	bloom, boiled like asparagus or bake in oven (very sweet taste)	8	47
Daucus carota L. (Apiaceae)	divlja mrkva	L, R	in pazija	2	20
Dioscorea communis (L.) Caddick & Wilkin (Dioscoreaceae)	kuke	S	boiled and served with eggs, fried with (or without) eggs	35	21
Diplotaxis tenuifolia (L.) DC. (Brassicaeceae)	divlja rikula	L	raw salad (usually mixed with Sonchus) or boiled	6	63
Foeniculum vulgare Mill. (Apiaceae)	morač	L	in pazija or used as spice for other dishes, rakija, preservation of olives	31	29
Fragaria vesca L. (Rosaceae)	divlja jagoda, fragula	F	raw	2	
Fraxinus ornus L. (Rutaceae)	jasen	L	tea	1	
Helianthus tuberosus L. (Asteraceae)	čičoka	R	emergency food	1	

Tab. 1 *(continued)*

Scientific name (plant names follow the Plant List [41])	**Local name**	**Part used**	**Use**	**Frequency $N = 40$**	**Voucher specimen No. starting from WA0000370**
Helichrysum italicum (Roth) G.Don (Asteraceae)	smilj	FL, L	tea	4	25
Hypericum perforatum L. (Hypericaceae)	gospina trava	FL	tea	7	60
Juniperus oxycedrus L. (Cupressaceae)	smrča, smrekina	F	rakija	10	26
Lactuca serriola L. (Asteraceae)	[no name]	L	in pazija	2	78
Laurus nobilis L. (Lauraceae)	lovorika	L	spice and preservative	16	42
Lavandula angustifolia Mill. (Lamiaceae)	lavanda	FL, L	tea	4	82
Macrolepiota cf *procera* (Scop.) Singer (Agaricaceae)	sunčanice, sunčica	M	grilled, fried (with eggs), stewed (pasta sauce)	2	
Malva sylvestris L. (Malvaceae)	crni sljez	L, FL, R	tea	10	68
Melissa officinalis L. (Lamiaceae)	matičnjak, pčelinja ljubica, melisa	L	tea	4	80
Mentha spp. including *M. piperita* L. and *M. spicata* L. (Lamiaceae)	metvica	L	tea	6	71, 77
Morus alba L., *M. nigra* L. (Moraceae)	bijela murva (for *M. alba*), crna murva (for *M. nigra*)	F	raw	9	67
Myrtus communis L. (Myrtaceae)	mrča	F	liqueur, rakija	18	69
Opuntia ficus-indica (L.) Mill. (Cactaceae)	divlja smokva, indijska smokva, kaktus	F	raw	3	
Origanum majorana L. (Lamiaceae)	mažuran	FL, L	spice	2	70
Origanum vulgare L. (Lamiaceae)	origano	FL, L	spice	5	
Paliurus spina-christi Mill. (Rhamnaceae)	crna drača	F	tea and children's snack	15	50
Papaver rhoeas L. (Papaveraceae)	mak	L	in pazija	15	48
Picris echioides L. (Asteraceae)	hrastej, rastej, pazija	L	in pazija	25	
Phlomis fruticosa L. (Lamiaceae)	[no name]	FL	nectar as children snack	1	73
Pimpinella anisum L. (Apiaceae)	aniš	F	tea	2	
Pinus pinea L. (Pinaceae)	pinj	F	raw	1	
Plantago lanceolata L. (Plantaginaceae)	muška bokvica	L	tea (mainly), in pazija	6	58
Plantago major L. (Plantaginaceae)	ženska bokvica	L	tea (mainly), in pazija	1	
Portulaca oleracea L. (Portulacaceae)	tušt, tušac , štir , medenica	L	boiled with onion or raw salad	4	65
Prunus cerasus L. (Rosaceae)	višnja	F	raw, liqueur	1	
Prunus spinosa L. (Rosaceae)	trnina	F	raw	2	
Punica granatum L. (Lythraceae)	divlji šipak, ljuti šipak	F	raw, juice	11	55
Pyrus amygdaliformis Vill. (Rosaceae)	trnovača	F	raw	9	84
Quercus pubescens Willd. (Fagaceae)	žir (about acorns), hrast (the tree)	F	emergency food, baked, remembered only from stories	3	39
Ranunculus neapolitanus Ten. (Ranunculaceae)	[no name]	L	in pazija	1	23
Reichardia picroides (L.) Roth (Asteraceae)	sladić	L	in pazija	4	79
Rosa sempervirens L. and *R. canina* L. (Rosaceae)	šipurak, šipurika, šipak, divlja ruža, srbi guzica	F	tea, jam	23	34
Rosa ×*centifolia* L. = *R. gallica* var. *centifolia* (L.) Regel (Rosaceae)	vonjača, čentifolia, ruža	FL	liqueur (rozulin)	3	
Rosmarinus officinalis L. (Lamiaceae)	rusmarin, ružmarin	L	spice for marinades, rakija	15	16
Rubus ulmifolius Schott (Rosaceae)	kumpiena, kupjena, kupina	F	raw, jams, syrup	33	53
Rumex pulcher L. (Polygonaceae)	štavolj, štavelj	L	in pazija	2	56
Ruscus aculeatus L. (Asparagaceae)	koštrava, koštrika	S	boiled, eaten like asparagus, nowadays rarely	15	38
Salvia officinalis L. (Lamiaceae)	pelin, kadulja	L, FL	tea, syrup from flowers, rakija	23	72
Sambucus nigra L. (Adoxaceae)	zovina	FL	tea, syrup	7	64

Tab. 1 *(continued)*

Scientific name (plant names follow the Plant List [41])	Local name	Part used	Use	Frequency $N = 40$	Voucher specimen No. starting from WA0000370
Satureja montana L. (Lamiaceae)	konjski vrjesak	FL, L	tea, rakija	5	24
Scorzonera laciniata L. (Asteraceae)	kozja brada	L	in pazija	1	30
Silene vulgaris (Moench) Garcke (Caryophyllaceae)	pucalica	L	in pazija	2	59
Silybum marianum (L.) Gaertn. (Asteraceae)	sjekavica	L	in pazija	1	
Sisymbrium officinale (L.) Scop. (Brassicaceae)	[no name]	L	in pazija	1	
Smilax aspera L. (Smilacaceae)	tetivika	S	eaten like asparagus	1	46
Sonchus oleraceus L. and *Sonchus asper* (L.) Hill (Asteraceae)	čevčeg, kostriječ, kostrješ	L	main component of pazija	38	17
Smyrnium olusatrum L. (Asteraceae)	vrati muž, lesandra, divlja selen	L	in pazija	0 (but found sold in the market in 2013)	
Sorbus domestica L. (Rosaceae)	oskoruša	F	raw, jam	8	83
Taraxacum officinale Weber (Asteraceae)	maslačak	L	raw or shortly boiled salad	11	
Teucrium montanum L. (Lamiaceae)	ivan trava, dubačak	FL, L	tea, rakija	1	40
Thymus sp. (Lamiaceae)	popovac, poponac, popunac, majčina dušica	FL	tea, rakija, spice	9	36
Tilia spp. (Tiliaceae)	lipa	FL	tea	3	86
Tordylium apulum L. (Apiaceae)	vrati muž, mužobrad	L	in pazija	2	31
Tropaeolum majus L. (Trapaeolaceae)	ljubidrag, dragušac	FL	like capers	1	
Urospermum picroides (L.) Scop. ex F.W. Schmidt (Asteraceae)	pleština, plešti guzica, pakoleć, pazija	L	in pazija	8	22, 32
Urtica dioica L. (Urticaceae)	kopriva	L	in pazija, tea	12	43
unidentified Apiaceae	divlji petrusin	L	in pazija	1	
unidentified Asteraceae (*Cirsium* sp.? *Carduus* sp.?)	badelj	L	in pazija	2	

Acknowledgments

We are grateful to the vice-Rector of the University of Rzeszów, Prof. Sylwester Czopek for granting additional financial support to this study.

Authors' contributions

The following declarations about authors' contributions to the research have been made: wrote the text and led all the interviews: KD; wrote the text and took part in half of the interviews: ŁŁ.

Competing interests

The following declarations about authors' competing interests have been made: ŁŁ: I have no competing interests apart from the fact that I am a section editor in the Acta Societatis Botanicorum Poloniae journal; KD: no competing interests.

References

1. Hadjichambis AC, Paraskeva-Hadjichambi D, Della A, Giusti ME, De Pasquale C, Lenzarini C, et al. Wild and semi-domesticated food plant consumption in seven circum-Mediterranean areas. Int J Food Sci Nutr. 2008;59(5):383–414.
2. Ghirardini M, Carli M, del Vecchio N, Rovati A, Cova O, Valigi F, et al. The importance of a taste. A comparative study on wild food plant consumption in twenty-one local communities in Italy. J Ethnobiol Ethnomed. 2007;3(1):22.
3. Tardio J, Pardo-De-Santayana M, Morales R. Ethnobotanical review of wild edible plants in Spain. Bot J Linn Soc. 2006;152(1):27–71.
4. Bonet MÀ, Vallès J. Use of non-crop food vascular plants in Montseny biosphere reserve (Catalonia, Iberian Peninsula). Int J Food Sci Nutr. 2002;53(3):225–248.
5. Leonti M, Nebel S, Rivera D, Heinrich M. Wild gathered food plants in the European Mediterranean: a comparative analysis. Econ Bot. 2006;60(2):130–142.
6. Łuczaj Ł, Pieroni A, Tardío J, Pardo-de-Santayana M, Sõukand R, Svanberg I, et al. Wild food plant use in 21st century Europe: the disappearance of old traditions and the search for new cuisines involving wild edibles. Acta Soc Bot Pol. 2012;81(4):359–370.
7. Łuczaj Ł. Dziko rosnące rośliny jadalne użytkowane w Polsce od połowy XIX w. do czasów współczesnych. Etnobiologia Pol. 2011;1:57–125.
8. Kalle R, Sõukand R. Historical ethnobotanical review of wild edible plants of Estonia (1770s–1960s). Acta Soc Bot Pol. 2012;81(4):271–281.
9. Dénes A, Papp N, Babai D, Czúcz B, Molnár Z. Wild plants used for food by Hungarian ethnic groups living in the Carpathian Basin. Acta Soc Bot Pol. 2012;81(4):381–396.
10. Łuczaj Ł. Ethnobotanical review of wild edible plants of Slovakia.

Acta Soc Bot Pol. 2012;81(4):245–255.

11. Svanberg I. The use of wild plants as food in pre-industrial Sweden. Acta Soc Bot Pol. 2012;81(4):317–327.
12. Redžić S. Wild edible plants and their traditional use in the human nutrition in Bosnia-Herzegovina. Ecol Food Nutr. 2006;45(3):189–232.
13. Łuczaj Ł, Köhler P, Pirożnikow E, Graniszewska M, Pieroni A, Gervasi T. Wild edible plants of Belarus: from Rostafiński's questionnaire of 1883 to the present. J Ethnobiol Ethnomed. 2013;9(1):21.
14. Pieroni A, Giusti ME. The remedies of the folk medicine of the Croatians living in Ćićarija, northern Istria. Coll Antropol. 2008;32(2):623–627.
15. Šavikin K, Zdunić G, Menković N, Živković J, Ćujić N, Tereščenko M, et al. Ethnobotanical study on traditional use of medicinal plants in South-Western Serbia, Zlatibor district. J Ethnopharmacol. 2013;146(3):803–810.
16. Rexhepi B, Mustafa B, Hajdari A, Rushidi-Rexhepi J, Quave CL, Pieroni A. Traditional medicinal plant knowledge among Albanians, Macedonians and Gorani in the Sharr Mountains (Republic of Macedonia). Genet Resour Crop Evol. 2013;60(7):2055–2080.
17. Zlatković BK, Bogosavljević SS, Radivojević AR, Pavlović MA. Traditional use of the native medicinal plant resource of Mt. Rtanj (eastern Serbia): ethnobotanical evaluation and comparison. J Ethnopharmacol. 2014;151(1):704–713.
18. Pieroni A, Rexhepi B, Nedelcheva A, Hajdari A, Mustafa B, Kolosova V, et al. One century later: the folk botanical knowledge of the last remaining Albanians of the upper Reka Valley, Mount Korab, Western Macedonia. J Ethnobiol Ethnomed. 2013;9(1):1–19.
19. Redzić S. Use of wild and semi-wild edible plants in nutrition and survival of people in 1430 days of siege of Sarajevo during the war in Bosnia and Herzegovina (1992–1995). Coll Antropol. 2010;34(2):551–570.
20. Łuczaj Ł, ZovkoKončić M, Miličević T, Dolina K, Pandža M. Wild vegetable mixes sold in the markets of Dalmatia (southern Croatia). J Ethnobiol Ethnomed. 2013;9(1):2.
21. Łuczaj Ł, Fressel N, Perković S. Wild food plants used in the villages of the Lake Vrana Nature Park (northern Dalmatia, Croatia). Acta Soc Bot Pol. 2013;82(4):275–281.
22. Bakić J, Popović M. Nekonvencionalni izvori u ishrani na otocima i priobalju u toku NOR-a. Beograd. Beograd: Izd. Mornaričkog glasnika; 1983.
23. Grlić L. Enciklopedija samoniklog jestivog bilja. Rijeka: Ex Libris; 2005.
24. Grlić L. Enciklopedija samoniklog jestivog bilja. Zagreb: August Cesarec; 1986.
25. Sardelić S. Samoniklo jestivo bilje – mišanca, gruda, parapač... Wild edible herbs – Mišanca, Gruda, Parapač... Etnološka Istraživanja. 2008;1(12–13):387–396.
26. Šugar I. Hrvatski biljni imenoslov. Nomenclator botanicus Croaticus. Zagreb: Matica Hrvatska; 2008.
27. Harris R. Dubrovnik: a history. London: Saqi Books; 2006.
28. Dolina K. Forecasting of airborne pollen seasons in the Mediterranean region of Croatia [PhD thesis]. Zagreb: University of Zagreb, Faculty of science, Department of Biology; 2012.
29. Rivas-Martínez S, Penas A, Díaz TE. Biogeographic map of Europe. Cartographic service. León: University of León; 2004.
30. Trinajstić I. Plantgeographical division of forest vegetation of Croatia. Zagreb: Razred za prirodne znanosti, Hrvatska akademija znanosti i umjetnosti; 1995. (Annales Forestales; vol 20/2).
31. American Anthropological Association Code of Ethics [Internet]. 2013 [cited 2013 Sep 10].
32. International Society of Ethnobiology Code of Ethics (with 2008 additions) [Internet]. 2013 [cited 2013 Sep 10].
33. Iveković F, Broz I. Rječnik hrvatskoga jezika. Zagreb: Štamparija Karla Albrechta; 1901.
34. Kovačić S, Nikolić L, Ruščić M, Milović M, Stamenković V, Mihelj D, et al. Flora jadranske obale i otoka. Zagreb: Školska knjiga; 2008.
35. Dogan Y. Traditionally used wild edible greens in the Aegean Region of Turkey. Acta Soc Bot Pol. 2012;81(4):329–342.
36. Lentini F, Venza F. Wild food plants of popular use in Sicily. J Ethnobiol Ethnomed. 2007;3(1):15.
37. Dreon AL, Paoletti MG. The wild food (plants and insects) in Western Friuli local knowledge (Friuli-Venezia Giulia, North Eastern Italy). Contrib Nat Hist. 2009;12:461–488.
38. Pericin C. Erbe di campo nella cucina istriana e la saggezza dei proverbi. La Ricerca. 1999;25–26:13–16.
39. Sõukand R, Quave CL, Pieroni A, Pardo-de-Santayana M, Tardío J, Kalle R, et al. Plants used for making recreational tea in Europe: a review based on specific research sites. J Ethnobiol Ethnomed. 2013;9(1):58.
40. Łuczaj Ł. Changes in the utilization of wild green vegetables in Poland since the 19th century: a comparison of four ethnobotanical surveys. J Ethnopharmacol. 2010;128(2):395–404.
41. The Plant List: a working list of all plant species [Internet]. 2014 [cited 2014 Sep 10].

26

Orobanche elatior and *O. kochii* (Orobanchaceae) in Poland: distribution, taxonomy, plant communities and seed micromorphology

Renata Piwowarczyk[1]*, Łukasz Krajewski[2]

[1] Department of Botany, Institute of Biology, Jan Kochanowski University, Świętokrzyska 15, 25-406 Kielce, Poland

[2] Department of Nature Protection and Rural Landscape, Institute of Technology and Life Sciences, Hrabska 3, Falenty, 05-090 Raszyn, Poland

Abstract

Species of the genus *Orobanche* (Orobanchaceae), parasitic on *Centaurea* in Central Europe, were previously considered to belong to the *O. elatior* group. At present, the taxon is differentiated into two species, *O. elatior* Sutton and *O. kochii* F.W. Schultz. The paper presents for the first time the distribution of *O. elatior* and *O. kochii* in Poland based on a critical revision of herbarium and the literature data, as well as the results of field studies conducted between 1999 and 2014. The majority of the species' localities are in south Poland: Silesia-Cracow, Małopolska and the Lublin Uplands. The distribution of both species in Poland is mapped and chronologically organized, and is thus the most recent in Europe. The taxonomy, host preferences, and ecology are also discussed. Seeds of both species were also investigated using light and scanning electron microscopy, which resulted in the designation of diagnostic features. The new color form of *O. kochii* f. *citrina* is described and illustrated. An account of all revised herbarium specimens collected from Poland, deposited in Poland and neighboring countries, is presented.

Keywords: *Orobanche elatior*; *Orobanche kochii*; taxonomy; distribution; phytocoenoses; SEM

Introduction

Orobanche and *Phelipanche* (syn. *O.* sect. *Trionychon*), are the largest genera in the Orobanchaceae family, and comprise more than 200 species, that lack chlorophyll and are root holoparasites of other vascular plants. Broomrapes are subcosmopolitan, with largest number of species being found in the warmer parts of the Mediterranean, Northern Africa, North America and Western and Central Asia. Several representatives, like *P. ramosa*, *P. aegyptiaca*, and *O. crenata*, parasitize important crops, e.g. tomatoes, tobacco, carrots [1–3]. In Europe they usually grow in the warmest regions, mostly in the Mediterranean countries. On the contrary, in central and northern parts of Europe the genera comprise about 30 species (17 established and 2 ephemerophytes from Poland), which are mostly rare, endangered or declining [e.g. 2,4–20]. These parasitic genera consist of plants, which possess very reduced vegetative organs and are highly variable. They are widely known for notorious difficulties in identifying, especially in herbarium material. For these reasons, there are a lot of erroneous determinations of species.

Research into the species of the genus *Orobanche* parasitic on *Centaurea scabiosa*, previously considered as *Orobanche elatior* group, in Central Europe revealed the existence of two distinct and unrelated species. Their proper names proved to be *O. kochii* F.W. Schultz and *O. elatior* Sutton [21]. *Centaurea scabiosa* is the main host of *O. kochii* although single records have been reported on *C. jacea*, *C. tenuifolia*, *C. triumfettii* subsp. *axillaris*, *C. stoebe*, *C. sadleriana*, *C. montana*, and *C. ruthenica*; *O. elatior* is mainly parasitic on *C. scabiosa* and sporadically on *C. jacea* and *C. tenuifolia* [9,13,21–23]. It is interesting that two separate species parasitize the same host and some unrelated broomrapes have been found to parasitize the same host, e.g. *Phelipanche bohemica* (syn. *O. bohemica*), *P. arenaria* (syn. *O. arenaria*), *O. coerulescens*, or *O. artemsiae-campestris* on *Artemsia campestris*. Different species of the genus *Artemisia* are known as hosts of over 20 species of the genera *Orobanche* and *Phelipanche* [1].

Prior to studies conducted by Zázvorka [21], the name *Orobanche elatior* had been commonly used for over 100 years for an undifferentiated taxon of Eurasian distribution extending from Britain to China [1,2,24,25]. The distribution area of *O. elatior* (excluding *O. kochii*) is restricted to Western and Central Europe, also Nordic and Baltic countries. *Orobanche elatior* occurs mainly in England, the Netherlands, southern Sweden, southern Norway, Denmark, Switzerland, France, Italy and Germany, while eastwards its distribution

* Corresponding author. Email: renata.piwowarczyk@ujk.edu.pl

Handling Editor: Zygmunt Dajdok

area reaches the Baltic states, Poland, Slovakia, the Czech Republic, Slovenia, Romania [13,21,23] (also Piwowarczyk unpublished data, and http://www.nahuby.sk]. Its distribution eastwards is not yet known precisely. It is a rare species in the whole range.

Orobanche kochii is more common. It is Eurasian, continental species stretching from Central Europe to Central Asia, central China and India. Its distribution in Europe covers southeastern France, Italy, Austria, Hungary, the Czech Republic, Slovakia, Poland, Slovenia, Croatia, Serbia, Bosnia and Herzegovina, Romania, Bulgaria, Greece, Ukraine and European Russia. The species may occur in the Caucasus, Siberia, up to Central Asia and the Himalayas ([e.g. 13,21,23,26–30]). Distribution maps of *O. kochii* and *O. elatior* in Central Europe are included in the study by Zázvorka [21].

In Central Europe, the two species are primarily noted in xerothermic habitats, i.e. xerothermic grasslands of the class *Festuco-Brometea* and thermophilous fringes of the class *Trifolio-Geranietea sanguinei*, developed on limestone, marl, basalt, loess, at altitudes 200–750 m. *Orobanche elatior* is found in more semixerothermic and mesophilous communities mostly of the alliances *Arrhenatherion* and *Bromion erecti*, while *O. kochii* is encountered in more typical xeric steppe habitats [21]. The highest localities of *O. elatior* s.l. in Europe have been reported from the Swiss Alps at 1800 m and from Asia up to 2440 m [1,2].

The aim of this study was to identify the distribution of *Orobanche elatior* and *O. kochii* in Poland based on field investigations and verified herbarium and literature data. Prefered habitats, communities, hosts, taxonomic problems and threats to both species are also discussed. The new color form of *O. kochii* f. *citrina* is also described and illustrated. The objective of the present study is to investigate the micromorphology (SEM) of seeds of *O. elatior*, and *O. kochii* in order to establish if there are any micromorphologic characters diagnostic for the investigated taxa.

Material and methods

Study area

Field research was carried out in Poland in the xerothermic areas of the greatest concentrations of *Orobanche* s.l., i.e. in the Silesia-Cracow and Małopolska, and Lublin-Lviv, Volhynia Uplands, Polesie, Lower Oder and Vistula River, the Carpathian Mts and Carpathian Foothills, Sudetes Mts. The field investigations were conducted in 1999–2014, but mostly in 2006–2014. In parallel, the first author revised all the herbarium materials of *O. elatior* and *O. kochii* collected in Poland. The materials are deposited in Polish herbaria CHRZ, KRA, KRAM, KTC, KTU, LBL, LOD, POZ, WA, WRSL, the Czech Republic (PR), Hungary (BP), Germany (B, GLM), and in private herbaria. Herbarium acronyms are given after Mirek et al. [31] and Thiers [32].

Relevé data

The study is based on the analysis of phytosociological relevés with the occurrence *O. elatior* and *O. kochii*, according to Braun-Blanquet approach [33] and originating from Poland. We performed more than 100 phytosociological relevés, but in this paper we presented the 39 (20 for *O. elatior*, 19 for *O. kochii*) most representative or significant. The nomenclature of vascular plants follows Mirek et al. [34]. The nomenclature of syntaxa is based on Matuszkiewicz [35].

Data processing

The localities of the both species are listed alphabetically as ATPOL cartogram units (10 × 10 km) based on Zając [36]. Only localities recorded in our observations and identified or confirmed as well as verified herbarium data are listed. Published data not confirmed by us in the field or undocumented by herbarium materials are not reliable due to frequent determination errors and these incorrect/doubtful localities are listed separately ("List of localities of *Orobanche elatior* and *O. kochii* in Poland" section below) and marked with a special symbol on the map.

Localities are described as follows: ATPOL grid unit, geographic location, habitat description, abundance in brackets. The following information is also given for most localities: geographic coordinates and altitude (above sea level), and for revised exsiccata the collector and collection date, exsiccatum number and the herbarium acronym. Archival maps of western Poland and military maps were used for localities reported in old data (http://mapy.amzp.pl/maps.shtml, http://www.mapywig.org/).

Symbols and abbreviations: app. – approximately, ATPOL – database, DG – Dąbrowa Górnicza, distr. – district, ecol. site – ecological site, exp. – exposure, gras. – grassland, leg. – collected by, n. – near, phot. – photography, res. – reserve, US – University of Silesia database, v. – valley, xer. gras. – xerothermic grassland, vid. – seen by, ? – doubtful locality.

Host analysis

Host plants were observed by delicately exposing the soil with a gardening shovel. A total of 80 soil pits were observed in the field and in the analysis of herbarium materials containing an attached host.

SEM analysis of seeds

Seeds were extracted from 16 samples from 8 localities coming from Poland. The investigated material of *O. elatior* (4 localities: Baldram, Gipsowa Góra, Leszczany, Pińczów) and *O. kochii* (4 localities: Pęczelice, Wesołówka, Toporowice, Boria), has been collected in the field by the first author (see "List of localities of *Orobanche elatior* and *O. kochii* in Poland" section below); these specimens are deposited in KTC. Seeds were extracted from dried plants, mounted on stubs, coated with gold, and examined using a Joel JSM-7400F SEM, at various magnifications. Microscopic observations (LM) were carried out with a stereo microscope NIKON SMZ800 and a biological microscope NIKON Eclipse 50i.

Twelve characters (length, width, and outline of the seed and of the cells on their surface; height, width, and ornamentation of cell walls; presence of a median trough; form and diameter of seed wall perforation) were recorded. For each investigated taxon at least 30 seeds were examined. The terminology of seed surface is given after Barthlott [37,38], Stearn [39], Plaza et al. [40], Bojnansky and Fargasová [41], and Black et al. [42]. The collected data were subject to elementary statistical procedures [43], the mean, the standard deviation, and the lower and upper quartiles were calculated for each character of each of the three studied taxa. Statistical calculations were performed with Statistica 7.1.

Results

Taxonomic notes

Orobanche elatior Sutton

Orobanche elatior Sutton Transact. Linn. Soc. London 4: 178, 1798 [44]; Syn.: *O. stigmatodes* Wimmer Fl. Schles. 280, 1840 [45]; *O. centaureae scabiosae* Holandre Nouvelle Fl. Moselle, ed. 2, p. 520, 1842 [46]; *O. confusa* F.W. Schultz Fl. Pfalz 339, 1846 et Flora, Regensburg, 30: 66, 1847 [47], nom. inval., in syn. *O. stigmatodes*; *O. major* auct. non L. [48].

Lectotype (Foley 2001: 230): Norfolk, England, herb. Smith 1087.14 n. 1, LINN [49].

Orobanche kochii F.W. Schultz

Orobanche kochii F.W. Schultz Flora Regensburg, 30(5): 66, 1847 [47]; *O. borbasiana* Beck Bibl. Bot. (Monogr. Orobanche) 19: 173, 1890 [26]; *O. elatior* auct. (e.g. Beck in Halácsy et Braun, Nachträge zur Flora von Nieder-Oesterreich, p. 128, 1882) [50], non Sutton; *O. major* auct. non L. [48]; *O. major* f. *typica* Beck, f. *exigua* Beck Bibl. Bot. (Monogr. Orobanche) 19: 170, 1890 et f. *microphya* Beck l. c., p. 171 [26]; *O. major* f. *moravica* Beck Fedde Repert. 18: 36, 1922 [51]; *O. major* f. *remotiflora* Beck in Engler, Das Pflanzenreich (Monogr. Orobanchaceae) 96 (IV/261): 250, 1930 [27].

Lectotype: "*Orobanche amethystea* Thuill.? Auf mehreren Pflanzen des Gratzer Schlossberges, auch auf Anthericum ramosum !! – Blüht im Juli – August wenn an derselben Orten die *O. rubens* et *O. galii* ganz vertrocknet sind. 1840. Dr. Maly" (L, 0835152; cf. [21]).

The interesting history of *O. elatior* and *O. kochii* goes back to 1840 when Maly collected a specimen from Austria and included it into *O. elatior* Sutton [52] but later changed the identification to *O. amethystea*. Maly sent the specimen to Koch, who determined it as *O. stigmatodes* Wimmer in 1844. Koch forwarded the same material to Schultz, who described a new species, *O. kochii* F.W. Schultz, based on it in 1847 [47]. Schultz was the first to distinguish *O. stigmatodes* (=*O. elatior*) from *O. kochii* and also provided a key to their determination. Bohemian, Moravian and Silesian authors accepted this division and the two species were separated in subsequent studies (e.g. [53–55]).

Wimmer [45] described *O. stigmatodes* parasitic on *C. scabiosa* based on plants collected from Poland (Silesia) from the Góra Gipsowa hill in Dzierżysław near Kietrz. The species has occurred there until now [56]. The morphological description of *O. stigmatodes* and the herbarium material collected by Kelch (GLM, see list of localities) undoubtedly belongs to *O. elatior* Sutton s.str. Two species, *O. elatior* s.str. and *O. caryophyllacea*, have occurred on the Gipsowa Góra hill. A herbarium revision shows that the two species, especially *O. elatior* (e.g. [26,27,54]), have been mistaken with *O. lutea* and *O. alsatica*.

The confusion regarding the separation of *O. kochii* followed the publication of monographs by Beck [26,27], who included a non-homogenous group of taxa parasitic on *Centaurea*, *Echinops* or other Compositae, and even Ranunculaceae, into one species, *O. major* L. The division proposed by Beck was later followed by subsequent authors. Beck included eleven forms into *O. major*, five of which should have been included into *O. kochii* (after [21], see synonyms therein).

Some other forms of *O. major* distinguished by Beck were also later classified as separate species. They include f. *krylowii* (Beck) Beck, which was considered by Russian authors as a separate species, *O. krylowii* Beck, occurring in Russia and Central Asia. It is a parasite of the genus *Thalictrum* (Ranunculaceae) [28,29,57]. Surprising recent studies by Frajman et al. [58] reported *O. krylowii* in Albania, 2500 km of known localities. Based on morphological evidence, nuclear ribosomal ITS sequences, Frajman et al. have shown that the Albanian population belongs to *O. krylowii* and that its closest relative is the European *O. lycoctoni*, but not *O. elatior* as previously thought.

The form f. *ritro* (Grenier at Godron) Beck was considered as a separate species parasitic on *Echinops*, *O. ritro* Gren. & Godr. [59] as early as in 1853. Some botanists treated *O. ritro* as a synonym for *O. major* or *O. elatior* (e.g. [60,61]). Zázvorka [21] includes it as a synonym of *O. kochii*. Recently, Pujadas-Salvà [62] reinstated its species rank as *O. ritro*, and has been adopted by other authors [20]. Carlón et al. [63] used *O. loscosii* to name the Iberian plants identified as *O. ritro*, however, Pujadas-Salvà [64] considers, that this species should be included among the synonyms for *O. ritro*.

Orobanche icterica Pau from the Iberian Penisula, traditionally synonymized with *O. elatior* Sutton, was recently reinstated for Spain by Pujadas-Salvà et al. [65] as a separate species, parasitizing *C. aspera*, *C. fontanesii* and *Rhaponticoides linaresii*, changing the earlier name *O. leptantha* Pomel [63]. Main differences were presented between *O. ritro*, *O. elatior* and *O. icterica* [62], and *O. icterica* and *O. lepthanta* [64]. Further research, including molecular analysis, may be interesting to explain relationship between *O. elatior*, *O. ritro*, *O. kochii* and Iberian species *O. lepthantha* and *O. icterica*. Moreover, recently has been discovered the locality of *O. elatior* from Lérida in Spain [63].

The form *hypochoeridis* (Druce ex Beck) Beck, described from the Chanel Islands, is at present classified in *O. minor* subsp. *maritima* var. *hypochoeridis* (Beck) Rumsey, parasitic on *Hypochoeris radicata* [66,67].

Zázvorka [21] rightly renewed the division into *O. elatior* and *O. kochii* after nearly 100 years. During investigations into the genus *Orobanche* in Poland, the first author also noticed a lack of homogeneity of specimens considered as *O. elatior*. Lightly colored specimens with a dense inflorescence were initially treated as hypochromatic plants [68]. The above examples clearly show that many different taxa were placed in *O. elatior* and that this group of species requires further studies. A historical survey of *Orobanche* parasitizing *Centaurea* and the history and taxonomic contents of the related names are briefly discussed in a study by Zázvorka [21].

The most important characters distinguishing *Orobanche elatior* from *O. kochii*

Both species are quite easily distinguishable when fresh. The corolla and stem in *O. elatior* are yellowish to ochre to light-brown or dirty-pale-pink (very rarely to dark-purple, rusty black, e.g. locality in Podskale) while *O. kochii* has a characteristic carrot-red to whitish-rosaceous or raspberry coloration (Fig. 1). Other characters are helpful in determining herbarium materials. *Orobanche elatior* is usually a large plant, with dense flowers; the corolla is regularly curved throughout, leaves are long and markedly narrow, linear-lanceolate. *Orobanche kochii* is mainly medium-sized; the corolla back is straighter in the middle, lower leaves on the stem are ovate to triangular, the spike subcylindrical and below lax. The two species also differ by the flowering period. *Orobanche elatior* flowers briefly and early, usually in late June, while *O. kochii* has a long flowering period, from late June until late August. A list of distinguishing characters is given in Tab. 1.

A new form *Orobanche kochii*

Orobanche kochii f. *citrina* R. Piwowarczyk & Ł. Krajewski, f. nov. (Fig. 1).

Lemon-yellow forms of *O. kochii* are found in Boria and Jaworzno-Długoszyn. However, they are sometimes an extremely small part of a normally-colored population (Boria), or comprise 50% of the population (Jaworzno-Długoszyn). The yellow color of the stems in typically pigmented populations was observed during two summer seasons. This new taxon is very similar to typical *O. kochii*, and is compliant with morphometric characters in the description of the plant after Zázvorka [21], parasitizing *C. scabiosa*, but differs significantly by color; in typical *O. kochii* they are red or rosaceous, sometimes pale rosaceous-creamy, whereas in the form *citrina* whole plant are lemon-yellow with brownish leaves and bracts, drying out much earlier than the rest of the stem (Fig. 1). In addition, the citrina form individuals are slightly smaller – plants are shorter and sometimes have a shorter corolla. Interestingly, the creamy color corolla and stems are found in *O. ritro*, parasitizing *Echinops*, a taxon considered being a synonym of *O. kochii* [21] or a separate species by some authors [62] (see also "Taxonomic notes").

TYPE. Southern Poland, Boria (N of Ćmielów), xerothermic grassland on the slope of the Kamienna river valley, inclination SE, 50°58′15″ N, 21°32′53″ E, alt. 167 m, 27 June 2013, R. Piwowarczyk (holotype KTC!).

DESCRIPTION. Plants medium-sized, flowering stem simple, (5–)15–25(–40) cm high, lemon-yellow, glandular-pubescent, glands whitish to pale yellow; dried specimens light-brown, cream-brown or cinnamon (do not always retain the lighter color after drying). Leaves broad at the base, ovate-triangular. Inflorescence (5–)12–15(–20) cm long, subcylindrical, tapering to the apex, usually 15–30-flowered. Bracts as long as the corolla or shorter. During florescence the bracts dry out quickly and are light-brown, particularly on their ends, which is very clearly seen in contrast to the lemon-yellow corolla and stem. Calyx segments are free or occasionally connate at base, 12–16 mm long, asymmetrically ovate and unequally bidentate. Corolla lemon-yellow (rarely whitish-yellow, whitish-lemon), (15–)20–28 mm long, dorsal line curved at the base, than nearly straight, glandular-pubescent, glands pale lemon-yellow or whitish. Stamens inserted 4–6 mm above the corolla base. Style sparsely glandular, stigma lobes deep yellow and usually darker than the corolla. Flowering period (late June) July to August. Parasitizing *C. scabiosa* (in Poland).

DISTRIBUTION AND ECOLOGY. *Orobanche kochii* f. *citrina* is currently known only from S Poland – Małopolska and the Silesia-Cracow Uplands (Boria and Jaworzno-Długoszyn). It is found in xerothermic grasslands of the *Cirsio-Brachypodion pinnati* alliance, on gentle slopes of hills, on soils rich in calcium carbonate, southern inclination, altitude 160–280 m.

ETYMOLOGY. The form name originated from the lemon-yellow coloring of plants.

PARATYPES. Southern Poland, Jaworzno-Długoszyn E, hill with iron cross SE of church, secondary grassland *Cirsio-Brachypodion pinnati*, inclination SW, 50°14′43″ N, 19°15′53″ E, alt. 280 m, 4 August 2010, Ł. Krajewski (KTC!).

Seeds micromorphology

Orobanche elatior seeds are oblongoid, rare ovoid or subrectiangular, (300–)386(–450) × (190–)242(–320) µm. Epidermal seed coat cells are isodiametric to elongated; rare irregular and elongated, (40–)105(–160) × (30–)59(–100) µm. Perforation diameter ranged (4–)7.9(–13) µm, and its shape is elliptic to subcircular. The narrow trough is always present on the upper side of cell walls and is continuous. The wall width reaches (3–)6.1(–8) µm (Tab. 2, Fig. 2).

Orobanche kochii seeds are oblongoid, rare ovoid or subrectiangular, (200–)328(–410) × (130–)206(–300) µm. Epidermal seed coat cells are isodiametric to elongated; rare irregular and elongated, (55–)102(–170) × (35–)52(–90) µm. Perforation diameter ranged (2–)2.7(–4) µm, and its shape is subcircular. The narrow trough is always present on the upper side of cell walls and is continuous, sometimes with ornamentation. The wall width reaches (3–)4.2(–6) µm, pitted, sometimes with additional and less ornamentation (Tab. 2, Fig. 2).

Preliminary results of micromophological examinations of seeds of *O. kochii* and *O. elatior* (then treated as two forms of *O. elatior* s.l.) have revealed differences in seed ornamentation, especially of the periclinal walls [68]. Our detailed studies demonstrated significant differences of diagnostic importance between *O. elatior* and *O. kochii*. The most important factor is the diameter and shape of the wall perforation. In *O. elatior* the perforation diameter is much larger, 4–13 µm, and is usually elliptical, while in *O. kochii* the diameter is smaller, 2–4 µm, and subcircular. Other characters of seed size and their cells are less important, but the analysis of a larger sample clearly demonstrated that seeds of *O. elatior* are larger than those of *O. kochii* (Tab. 2, Fig. 2).

Determination problems and discriminating from other species

Orobanche elatior belongs to a diversified section *Orobanche* (syn.: sect. /grex/ *Curvatae* Beck) [27], containing about 15 species. However, *O. kochii* probably has a relationship to another section *Minores*, this is supported by morphological characteristics, and preliminary molecular

a
b
c
d
e
f
g
h
i
j
k
l

Fig. 1 General habit of the studied species. **a–d** *Orobanche elatior*. **a** Young bud, Turowiec, 03.06.2011, R. Piwowarczyk. **b** Young shoots, Dąbrowa Górnicza-Gołonóg, 10.06.2011, Ł. Krajewski. **c** Lax inflorescence, Baldram, 10.07.2010, R. Piwowarczyk. **d** Typical dense inflorescence, Dąbrowa Górnicza-Gołonóg, 10.06.2011, Ł. Krajewski. **e–h** *O. kochii*. **e** Young bud, Wesołówka, 30.06.2005, R. Piwowarczyk. **f** Inflorescence, Kaczkowice, 2008, R. Piwowarczyk. **g** Inflorescence, Wesołówka, 30.06.2005. **h** General habit of shoot, Gartatowice, 09.07.2013. **i–l** *O. kochii* f. *citrina*. **i** Young bud. **j** Inflorescensce, 27.06.2013, Boria, R. Piwowarczyk. **k** Inflorescence, Jaworzno-Długoszyn, 05.08.2011, Ł. Krajewski. **l** General habit of shoot of *O. kochii* f. *citrina*, left and typical *O. kochii*, right, Boria, 27.06.2013, R. Piwowarczyk.

Tab. 1 The most important characters distinguishing *Orobanche kochii* and *O. elatior* [21] (corrected after our observations).

Characters	*Orobanche kochii*	*Orobanche elatior*
Stem, height	Usually (5)20–30(60) cm	Usually (10)30–40(95) cm
Leaves, shape	Broad at base, ovate-triangular	Narrow, elongated, linear-lanceolate
Inflorescence	Subcylindrical (tapering to the apex) and below lax	Cylindrical and dense (often with an additional, narrow, acute top of unopened flower buds)
Corolla and stem, color (in fresh materials)	Carrot red to whitish-rosaceous, raspberry; Rarely also entire plants lemon-yellow; stem usually with stronger coloration, corollas lighter	Deep yellow to pale brown (ochre), pale rosaceous, very rare to dark purple violet (almost black); stem and corolla in the same color
Back of corolla	The middle of the back nearly straight	Regularly curved throughout
Dried specimens	Brownish-rusty, cinnamony, darker	Pale brown (ochre), paler
Flowering period	Lengthy, (late June) July-August (September)	Very short (only 2 weeks), mid-June
Host (Central Europe)	*Centaurea scabiosa*, occasionally *C. stoebe*	*Centaurea scabiosa*

Tab. 2 A comparison of the main characters distinguishing seeds of *Orobanche elatior* and *O. kochii*.

Characters	*Orobanche elatior*	*Orobanche kochii*
Seed length (μm)	(300–)386(–450)	(200–)328(–410)
Seed width (μm)	(190–)242(–320)	(130–)206(–300)
Cell length (μm)	(40–)105(–160)	(55–)102(–170)
Cell width (μm)	(30–)59(–100)	(35–)52(–90)
Perforation diameter (μm)	(4–)7.9(–13)	(2–)2.7(–4)
Wall width (μm)	(3–)6.1(–8)	(3–)4.2(–6)
Seed shape	Oblongoid, rare ovoid or subrectiangular	Oblongoid, rare ovoid or subrectiangular
Cell shape	Isodiametric to elongated; rare irregular	Isodiametric to elongated; rare irregular
Perforation shape	Elliptic, rarely subcircular	Subcircular
Wall depth	Medium	Medium
Wall ornamentation	Pitted or slightly pitted	Pitted, sometimes with additional and less ornamentation
Narrow trough	Evident around cells	Evident around cells, sometimes with ornamentation

evidence (Piwowarczyk et al. unpublished); it was also initially indicated in the work of Carlón et al. [63].

Typical *O. kochii* is clearly distinguished from *O. elatior* and they are easily differentiated. *Orobanche kochii* is a distinctive species and it does not have any morphologically similar taxa in Central Europe. However, *O. elatior* in Central Europe is similar in morphology to *O. alsatica* and *O. bartlingii*. Some specimens in herbarium materials lacking information on the host are difficult to diagnose. It is crucial to note the host of plants of the genus *Orobanche*, especially for *O. elatior*. *Orobanche elatior* is parasitic on *Centaurea* (mainly *C. scabiosa*), *O. alsatica* on *Peucedanum*

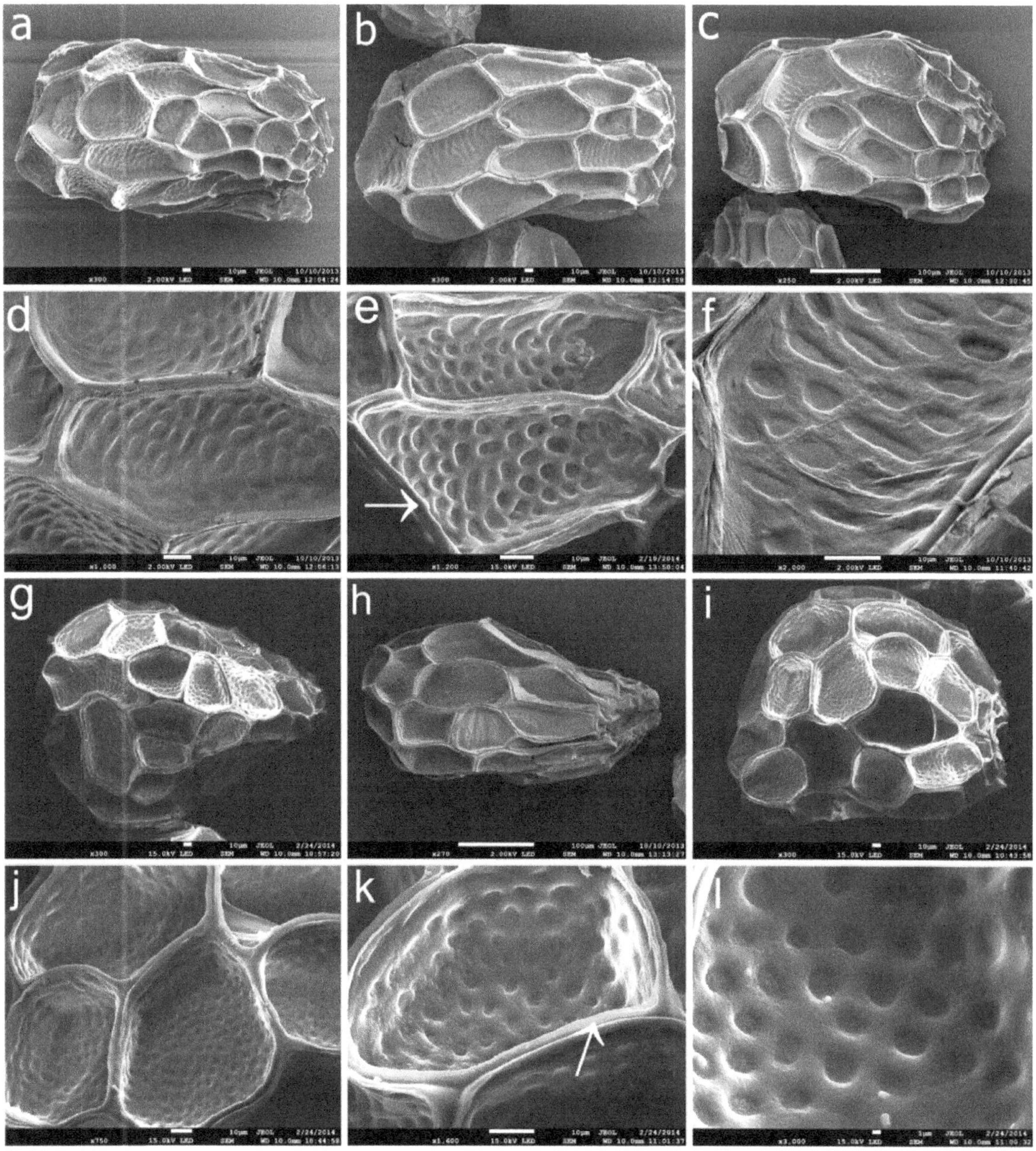

Fig. 2 Micromorphology of seeds of the studied species in SEM micrographs (surfaces). **a–f** *Orobanche elatior*. **g–l** *O. kochii*. **a–c,g–i** – general habit of seeds; **d–f,j–l** – variation of sculpture, perforation diameter and their shape; arrows indicate the narrow trough.

(chiefly *P. cervaria*, sporadically *P. alsaticum*), rarely on *Seseli osseum* and *S. austriacum*, while *O. bartlingii* on *Libanotis pyrenaica*. Even results of molecular studies have revealed close relationships between *O. elatior* and *O. alsatica* [69]. It seems that these relationships will be especially close between *O. bartlingii* and *O. flava*. Incorrect determinations can also be related to *O. elatior* and *O. flava*. However, *O. flava* is evidently distinguished from other taxa by different phytoceonoses, usually mountain hydric scrub on streams, and by parasitism on the genus *Petasites*. Lack of information regarding the host can also cause determination problems.

Monstrous shoots of *O. elatior*, branched into 2–4 inflorescences, with a tape-like stem, multiflorous inflorescence, with incorrectly placed flowers, are encountered occasionally (rarely), e.g. Leszczany in the Lublin region and DG-Błędów in the Dąbrowa Basin. This is probably related to herbicide on farmland near localities.

Distribution in Poland

The revision of and detailed field investigations into *O. elatior* and *O. kochii* in Poland have revealed considerable differences in the distribution, especially in relation to the previous distribution map of *O. elatior* s.l. [70]. The study also significantly changed and updated the previous distribution map for both species in central Europe prepared by Zázvorka [21].

At present *O. elatior* in Poland mainly occurs in the Silesia-Cracow, Małopolska and Lublin Uplands, rarely in the Lower Vistula Valley, Central Sudetes Mts and the Carpathian Foothills. *Orobanche kochii*, which is more common than *O. elatior*, has numerous localities but almost exclusively in the Polish uplands in a very compact range (Fig. 3, Fig. 4).

Only data verified in the field or confirmed by the herbarium material or photographic documentation is listed below. Due to frequent incorrect determinations, localities reported in the literature and unpublished localities that are not documented by herbarium material and not verified by us in the field are listed separately and marked with a special symbol on the map. Localities in the maps have been differentiated into three time periods, in order to illustrate the most current and the actual state distribution of both species (Fig. 3, Fig. 4).

Many literature reports may not be reliable due to the host they provide, e.g. *Medicago falcata*, where *O. elatior* s.l. was probably mistaken with *O. lutea* and others (e.g. [71–77]). Corrections are also needed for photographs of *O. kochii* [68,78–80] or *O. lutea* [81], captioned as *O. elatior.*

List of localities of *Orobanche elatior* and *O. kochii* in Poland

Orobanche elatior

AC: 30 - Bielinek, leg. R. Schulz 1914, 1925 (B) [82–85]; E of Bielinek, on a grassy SW slope of the Oder v.; leg. J. Mądalski, 21.09.1955 (KRAM, 494779); leg. Exc. Inst. Bot. U. P., 06.1952 (POZ); **BF: 24** - Słoszów, xer. gras. of the alliance *Bromion erecti* on the margin of an old limestone quarry W of Słoszów, (<10), 630 m [86]; Raczyn, overgrowing gras. on the forest margin, *Bromion erecti*, (<100), 700 m [86]; Zielone, thermophilous gras. of the alliance *Bromion erecti* (a few hundred specimens), 680–700 m [87–89]; between Zielone and Słoszów, a belt of dry gras. on the forest margin, (8), 650 m [86,87]; **36** - Śnieżnik Massif, Krowiarki, 0.3 km NW of the summit Kamiennik, SW slope xer. gras. of the alliance *Bromion erecti*, a narrow strip between the cultivated field and spruce forest, (7), 50°19′46″ N, 16°43′38″ E, 560 m, vid. M. Karakula, det. M. Smoczyk, 17.06.2014 (phot.); 0.3 km to the NW and SE from the top of the mountain Kamiennik, xer. gras. of the *Bromion erecti* and *Trifolion medii* aliances, (7), 550–560 m, vid. M. Karakula, M. Smoczyk, 01.07.2014 (phot.); **CE: 00** - Kaszowo, Milice distr., thermophilous meadow, NW exp. (40), vid. L. Matacz, 23.06.2014 (phot.), **CF: 46** - between Borzysławice and Karchów (Borislawitz u. Karchwitz) [54,55,90]; Silesia, Kędzierzyn-Koźle distr., Pawłowiczki; Borzysławice village, small forest in the direction of Karchów (Borislawitz im Waldchen gegen Karchwitz), leg. R. Uechtritz, 07.1867 (PR); **65** - Gipsowa Góra res., Dzierżysław n. Kietrz, Góra Gipsowa hill n. Kietrz, leg. Kelch, 18?? (GLM, 138802), leg.?, 18?? (GLM, 138799); leg. Wetschky, 06.1872 (WA, 30508); Silesia, distr. Głubczyce, Kietrz village, Góra Gipsowa nature res. [21,27,45,54–56,91]; xer. gras., leg. A. Sendek, 14.05.1972 (KRA, 0138561, 01385670); leg. J.J. Wójcicki, 29.06.1973 (KRAM, 405945); leg.?, 17.07.1964 (KRAM, 356938); sunny slope, leg. E. Kozioł, 15.06.1983, 15.06.1985 (WRSL, 93091, 93092, 93093); leg. M. Buchalik, 26.06.1999 (KTU, 99023); leg. A. Sendek, 14.06.1972 (KTU, 01055); meadow below the reserve, leg. K. Rostański, 04.07.1978 (KTU, 29273), 50°03′21.04″ N, 17°59′33.3″ E, 262 m, leg. R. Piwowarczyk 11.07.2010; 68 - Kozłówki (Kösling) [54]; Czernica (Czernitz zwischen Rybnik und Ratibor) [54,55]; **DB: 52** - Baldram, xer. gras., (10), 53°45′56.2″ N, 18°55′54.1″ E, 40 m, vid. R. Piwowarczyk, 10.07.2010; **DE: 11** - cfr., Burzenin, Sieradz distr., alfa alfa cultivation, leg. A. Janković, 28.07.1967 (LOD, 154399); **DF: 22** - between Orzech and Radzionków, W of railway, field baulk with *Solidago* and termophilous scrubs, SW exp., (20), 50°25′10″ N, 18°54′51,5″ E, 310 m, leg. Ł. Krajewski, 29.07.2011, (Bitoft 1979, US); **24** - Podskale NW, hill 317, gras. on field baulk, SE exp., (very rare, mixed with *O. lutea*), 50°25′20″ N, 19°13′09″ E, 310 m, leg. Ł. Krajewski, 27.06.2010 (KTC), vid. Ł. Krajewski 2009–2011, 2013; DG-Ujejsce E, W of the Recki Las forest, dry, sandy old fallow, overgrowing *Pinus* and *Calamagrostis*, NE exp., (40, 150, 30), 50°24′04″ N, 19°16′04″ E, 335 m, leg. Ł. Krajewski, 05.08.2010 (KTC), vid. Ł. Krajewski 2010–2011, 2013; DG-Gołonóg, Mała Góra Gołonoska (hill 334), xer. gras., overgrowing *Crataegus* and *Solidago*, SW, S exp., (25, 50, 20, 15, 10, 15, in upper part of site close to *O. kochii* and from 2011 also *O. lutea*), 50°20′05″ N, 19°14′22″ E, 324–332 m, vid. Ł. Krajewski 2009–2014, phot.; **26** - between Grabowa and Kromołowiec rocks, rendzina fallow in field roadside, under power line, N exp., (very rare), 50°23′50″ N, 19°26′58″ E, 415 m, leg. Ł. Krajewski, 02.10.2009 (KTC); cfr., Grabowa E, SE slopes of Wierzgóry (hill 423), above road to Piaski, old fallows, (very rare), 50°23′15″ N, 19°28′32″ E, 345 m, leg. Ł. Krajewski, 18.09.2009 (KTC), later not found; **35** - DG-Łęka NW, hill 361 (Łosy), calcareous gras. with *Thalictrum simplex*, strongly overgrowing *Prunus spinosa* and *Crataegus monogyna*, thermophilous scrubs, field margins, (30, 25, 3, 50, 20; from 2011 with *O. lutea, O. kochii* also occurred in 2013), 50°22′11″ N, 19°20′02″ E, 360 m, leg. Ł. Krajewski, 27.09.2009 (KTC), vid. Ł. Krajewski 2008–2011, 2013; Sławków, NE Tomanówka (hill 350, Góra Wielka massif), ecotone of fallows and calcareous/calamine gras. with *Libanotis*, overgrowing *Solidago* and *Calamagrostis*, NW, N, NE exp., (25, 50, 30; mixed with *O. lutea* in upper part, only in 2009 also present *O. picridis*), 50°19′29″ N, 19°21′17″ E, 50°19′28″ N, 19°21′14″ E, 335–340 m, leg. Ł. Krajewski, 09.08.2009, 10.08.2009 (KTC), vid. Ł. Krajewski 2009–2011, 2013; Sławków-Kozioł N, *Geranion sanguinei* with dominant *Libanotis*, (very rare, with *O. lutea*, close to *O. picridis*), 50°18′41″ N, 19°21′57″ E, 332 m, leg. Ł. Krajewski, 09.07.2011 (KTC); Sławków N, 650 m NNE of hill 345, S of field road, calcareous rocky gras./thermophilous scrubs, NE exp., (very rare, mixed with *O. lutea* and *O. bartlingii*), 50°19′06,3″ N, 19°23′05,4″ E, 310 m, leg. Ł. Krajewski, 23.07.2010 (KTC), vid. Ł. Krajewski 2010–2011, 2013; **36** - DG-Błędów, hill 341 m (Młyńska Mt or Kamieniec; entire ridge 1.5 km long), old fallow rendzinas, margins of ploughed fields, baulks, thermophilous scrubs, small limestone quarries, calcareous grasslands, meadows, N, NW, W, SE exp., (150), 50°21′13″ N, 19°27′01″ E, 50°21′12″ N, 19°26′54″ E, 50°20′54,9″ N, 19°27′46,3″ E, 50°20′57″ N, 19°27′51″ E (only here mixed with *O. kochii* in lower part), 50°20′53″ N, 19°28′09″ E, 320–335

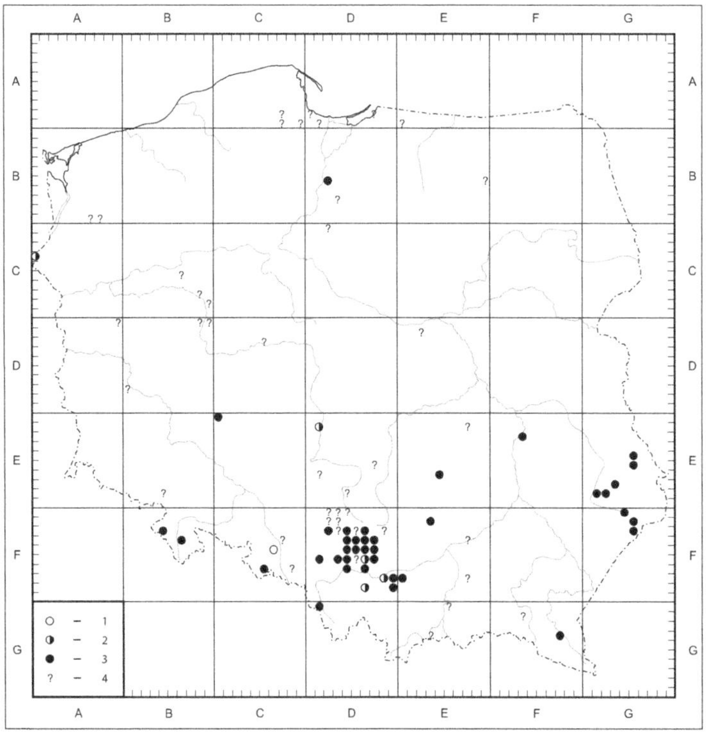

Fig. 3 Distribution of *Orobanche elatior* in Poland: 1 - before 1950, 2 - 1950–2000, 3 - after 2000, 4 - only literature data of *O. elatior* s.l., uncertain.

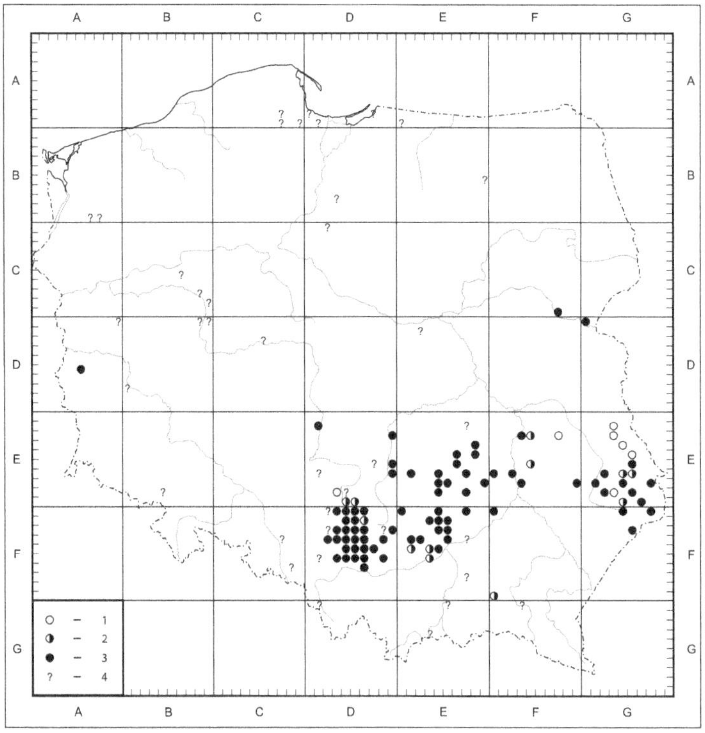

Fig. 4 Distribution of *Orobanche kochii* in Poland: 1 - before 1950, 2 - 1950–2000, 3 - after 2000, 4 - only literature data of *O. elatior* s.l., uncertain.

m, leg. Ł. Krajewski, 25.09.2009, 26.09.2009, 10.07.2011 (KTC, also monstrous shoots - stem forked to 2, 4 inflorescences and flat), vid. Ł. Krajewski 2009–2014; Chechło-Pustkowie, Buczna Góra hill, roadside, NW exp., (very rare, close to locality *O. lutea*), 50°21′45″ N, 19°32′21″ E, 375 m, leg. Ł. Krajewski, 26.09.2009 (KTC); **37** - Kolbark, near Wilcze Doły, xer. gras., fallows, baulks, (>5), 50°21′38,5″ N, 19°39′28,9″ E, 407 m, 50°21′34,5″ N, 19°39′33,1″ E, 379 m, vid. R. Piwowarczyk, 31.09.2011; Cieślin N, Organistówka hill, S slopes 100 m E of church, old rendzina fallow with xer. gras., (15, 25), 50°22′17″ N, 19°37′35″ E, 345 m, leg. Ł. Krajewski 10.09.2013, vid. Ł. Krajewski, 13.06.2014; Domaniewice S, 50 m S of road Bydlin-Wolbrom, balk of field road, fallow with xer. gras., (10), 50°22′51″ N, 19°40′19″ E, 375 m, vid. Ł. Krajewski 13.06.2014; **44** - Jaworzno-Niedzieliska, Warpie NW (Wał hill), N of allotments, *Thalictro-Salvietum pratensis* evolving to *Prunetalia spinosae*, E exp., (35, 25; mixed with less numerous *O. lutea*), 50°24′04″ N, 19°16′04″ E, 310 m, leg. Ł. Krajewski, 04.08.2010 (KTC), vid. Ł. Krajewski, 12.06.2011; **45** - Jaworzno-Ciężkowice SW, Góra Wielkanoc Mt (hill 331), calcareous gras., S, SW exp., (10, 20; mixed with *O. lutea*, *O. kochii* 100 m W), 50°12′39″ N, 19°20′40″ E, 327 m, leg. Ł. Krajewski, 08.08.2010 (KTC), vid. Ł. Krajewski 2010–2011; **46** - Żurada, S of Kolonia I, NE slopes of hill 439, calcareous gras. and scrubs, fallows, field baulks, N exp., (50), 50°14′58″ N, 19°32′22″ E, 425–430 m, leg. Ł. Krajewski, 13.09.2011 (KTC); **47** - Olkusz NE, hill 423, S slope, baulk between field road and new house, calcareous gras. (20), 50°17′10″ N, 19°34′44″ E, 415–420 m, leg. Ł. Krajewski, 15.09.2011 (KTC); Kosmolów N, S of Podłużne forest, close to Eagles' Nests red trail, NE slopes of hill 438, dry, sandy meadow, xer. gras. on fallow rendzinas, roadsides, NE, N exp., (60; 500 m E of *O. kochii*'s site), 50°16′48″ N, 19°39′29″ E, 50°16′52,5″ N, 19°39′23″ E, 50°16′54″ N, 19°39′29″ E, 400–420 m, leg. J. Drobnik, 06.07.2001 (KTU, 99021), leg. Ł. Krajewski, 10.07.2011 (KTC) [92]; **51** - Mikołów-Mokre, Fiołkowa Góra hill, (Nikolai: Mokrauer Kalkberg) [93]; dry meadow on N slope, leg. M. Dobrzańska, 28.08.1977 (KTU, 16211); leg. M. Kłączek, 01.08.1983 (KTU, 38299) (Babczyńska-Sendek 1998, Urbisz 2001, US); Mikołów, Sośnia Góra (hill 329), gras. areas of the botanical garden, (10), vid. W. Podstolski 06.2013, phot.; **53** - Lędziny SW, Góra Klimont Mt (hill 302), SW of St. Clement church, xer. gras., (very rare, mixed with *O. kochii*), 50°07′13″ N, 19°06′08″ E, 295 m, leg. Ł. Krajewski, 13.09.2010 (KTC); **54** - between Imielin-Pasieczki and Mysłowice-Dziećkowice, hill 310, calcareous xer. gras., S exp., (15, 250 m to W of *O. kochii* on summit mound), 50°09′55″ N, 19°12′11,5″ E, 287 m, leg. Ł. Krajewski, 04.09.2010 (KTC); **56** - between Dulowa and Młoszowa, n. Trzebinia, calcareous slope close to road, leg. A. & J. Kornasiowie, 08.08.1955 (KRAM, 014846); **57** - between Wola Filipowska and Krzeszowice, xer. gras. on the border of hamlets Pasternik and Podgórki, on a hillside Skałka, (32), 50°07.230′N, 19°35.893′E, 305 m, leg. B. Binkiewicz, 08.07.2014 (KTC); **64** - Chełm Śląski E, Smutna Góra Mt (hill 285 with cholera cemetery), xer. gras., old limestone quarries, (20), 50°06′30″ N, 19°12′44″ E, 50°06′26″ N, 19°12′50″ E, 275–280 m, leg. Ł. Krajewski, 04.09.2010 (KTC), vid. P. Grzegorzek, 07.2014 (phot.); **66** - between Wygiełzów and Zagórze, pass between Grodzisko Wielkie and Małe hills, in *C. scabiosa* population, very abundant, leg. P. Grzegorzek, 10.07.1989 (Grzegorzek private herbarium); **79** - Ochojno Świątniki Górne distr., n. Stronie estate, overgrowing *Festuco-Brometea* gras. with *C. scabiosa*, on a small hill n. houses in the estate, W exp., (Pogórze Wielickie, Western Carpathian Mts), (>40), leg. W. Bartoszek, 15.06.2002 (KRA, 0260657, 0260658, 0260579, 0260586); **78** - cfr., between Rzozów and Radziszów n. Skawina, NW slope of Lipie hill, leg. H. Trzcińska-Tacik, 1980 (KRA); **86** - Wadowice, between road bridge and railway bridge, leg. E. Markiewicz, 19.06.1992 (KTU, 45891); **89** - Siepraw (Sieprawski), S slope, gras., leg. K. Hajduk, 20.09.2009 (KRA); **DG: 01** - Tuł, calcareous hill between Goleszów and Leszna (Śląsk Cieszyński), leg. A. Żmuda, (KRAM, 061500); In Getriedefeldern am Tuł bei Ustroń (Schlesien), leg. F. Pax, 06.1888 (BP); Tuł bei Teschen, leg. Fiek, 16.07.1858 (WA, 30506), leg. Fiek, 18.07.1888 (BP), (Silesia, Cieszyn distr., Tuł hill SW of Ustroń town); Tuł [21,54,55,94–97]; vid. M. Śniegoń, 06.2000 (phot.), vid. T. Beczała, 2003 (has not occurred since 2008); **EE: 64** - Zachełmie n. Zagnańsk, xer. gras. on the edges of a closed quarry n. Chełmowa hill (<10), 50°58′10″ N, 20°41′28″ E, 354 m, vid. A. Przemyski, 09.2011 (phot.) [13]; **EF: 13** - Góry Pińczowskie Mts, NW of Pińczów, margin of a field and a fallow (3), 50°32′08,4″ N, 20°30′46″ E, 219 m , leg. R. Piwowarczyk, 09.06.2009 (KTC) [13,14]; Polana Polichno res. n. Młodzawy, xer. gras., oak-hornbeam communities (>8), 50°28′01,9″ N, 20°28′06,7″ E, 215 m, leg. R. Piwowarczyk, 27.06.2006, (KTC), vid. R. Piwowarczyk, 06.2009 [13]; **70** - Koźmice Wielkie n. Wieliczka, dry gras. *Festuco-Brometea*, at the foot of a mound (Pogórze Wielickie, Western Carpathian Mts), leg. W. Bartoszek, 12.06.2002 (KRA, 0260473, 0260474, 0260484, 0260583, 0260584, 0260585, 0260590) [79]; **EG: 05** - cfr., Znamirowice on lake Rożnowskie, W slope of Grochówki [98]; **FG: 37** - Myczkowce n. Olszanica, slope on the San river, vid. M. Szewczyk (phot.), 07.06.2012; **GE: 45** - Wólka Leszczańska, a narrow belt of xer. gras. between the "Żmudź" res. and the village (>5), 51°00′41″ N, 23°39′09″ E, leg. P. Chmielewski, R. Piwowarczyk, 08.06.2008 (KTC) [9]; Leszczany, a fallow field and an arable field adjacent to a former limestone excavation pit, by the Żmudź–Kumów Majoracki road, near village buildings (<50). Some specimens were deformed, probably following treatment with weed killers. 51°01′29″ N, 23°34′53″ E, leg. P. Chmielewski, R. Piwowarczyk, 09.06.2007 (KTC) [9]; **55** - Żmudź, xer. gras. and scrub in the "Żmudź" reserve and along its borders, ca. 0.5 km W of Kolonia Żmudź (>10), 51°00′29″ N, 23°40′27″ E, leg. R. Piwowarczyk, P. Chmielewski, 08.06.2007 (KTC) [9]; Podlesie, a xer. gras. and blackthorn scrub, ca. 1 km S of Wólka Leszczańska (>20), 51°00′03″ N, 23°39′03″ E, vid. P. Chmielewski, 2007 [9]; Turowiec, a vast complex of initial xer. gras. and juniper scrub, ca. 1.5 km N of the cemetery near Turowiec and Popówka villages (<5), 50°58′00″ N, 23°38′12″ E, vid. P. Chmielewski 2007, xer. gras. with *Juniperus communis*, (2), 50 57 59,5N, 23 38 11,9E, vid. R. Piwowarczyk 03.06.2011 [9]; **55/56** - Kolonia Teresin, roadsides, fallow fields, xer. gras. and sparse blackthorn scrub on a vast calcareous scrub ca. 0.5 km S of the village (>30), 50°59′41″ N, 23°40′54″ E; 50°58′43″ N, 23°41′57″ E; 50°59′19″ N, 23°41′29″ E; 50°58′49″ N, 23°42′09″ E; 50°59′01″ N, 23°41′11″ E; 50°58′46″ N, 23°41′25″E;

50°59'20" N, 23°42'00" E, leg. P. Chmielewski, R. Piwowarczyk, 08.06.2007 (KTC), [9], probably also *O. kochii*; **72** – Chomęciska Małe, a xer. gras. on a hillside, ca. 0.5 km SE of village buildings, at an intersection of field roads (<5), 50°48'26" N, 23°12'34" E, vid. P. Chmielewski, R. Piwowarczyk, 09.06.2007 [9]; **73** – Huszczka Duża, xer. gras. belonging to association *Inuletum ensifoliae* and in communities with *Brachypodium pinnatum*, on a vast calcareous slope (<15), 50°50'10" N, 23°19'11" E, leg. A. Cwener, 30.06.2006 (LBL) [9]; **81** – Niedzieliska, fringe communities on the edge of a juniper scrub complex, ca. 700 m S of village buildings (one specimen), 50°42'00" N, 23°04'53" E, vid. A. Cwener, P. Chmielewski, 28.07.2009 [9]; **82** – Kąty II, initial xer. gras. near a dirt road from the Wymysłówka village to Wychody, on a midfield hill known as Wieprzecka Góra (<10), 50°40'46" N, 23°07'35" E; vid. R. Piwowarczyk, W. Michalczuk & P. Chmielewski, 2006 [9]; Kąty n. Zamość, pine forest on humic chalky rendzina, leg. D. Fijałkowski, 05.06.1959 (LBL); sunny chalky slope, leg. D. Fijałkowski, 11.08.1965 (LBL); Kąty II, oak forest on chalk, leg. D. Fijałkowski, 29.06.1951 (LBL); Kąty II n. Zamość, *Inuletum ensifoliae*, leg. A. Brzeg, 04.08.1984 (POZ); vicinity of Zawada n. Zamość, on a baulk between arable fields, field, leg. D. Fijałkowski, 29.06.1951 (LBL); **GF: 04** – Przecinka, xer. gras. on a chalky hillside (Kamienna hill), between Przecinka and Justynówka villages (>20), 50°29'22" N, 23°29'37" E, vid. P. Chmielewski, 15.06.2008 [9]; **15** – Kolonia Korhynie, a fallow field on chalky rendzina, in the W part of the village. The locality was destroyed by ploughing in spring 2010. 50°25'41" N, 23°33'02" E, leg. P. Chmielewski, R. Piwowarczyk, 13.07.2008 (KTC) [9]; Kolonia Jarczów I, a fallow field in the W part of the village, (<10), 50°26'03" N, 23°33'46" E, vid. P. Chmielewski, 2008 [9]; Żurawce, an initial xer. gras. near the "Żurawce" ecol. site, on the slope of the Sołokija v. (>20), 50°23'33" N, 23°33'35" E, vid. P. Chmielewski, 2007 [9]; Przeorsk, przydroże, vid. P. Chmielewski, 30.06.2011; **25** – Machnów Stary, xer. gras. in "Machnowska Góra" res. (>10), 50°22'10" N, 23°35'21" E; vid. P. Chmielewski, 2007 [9].

Orobanche kochii

AD: 05(?) – cfr., Lubusz Land, Bielice, escarpment on a roadside ditch, leg. H. Ratyńska, M. Wojterska, 18.06.2007 (POZ); **DE: 11** – Łęg hamlet, between Burzenin and Szczawno, gras. on the left bank of the Warta river, 51°26'45" N, 18°48'47" E, 164 m, vid. P. Niedźwiedzki, 02.08.2011 (phot.); between Wielka Wieś n. Burzenin and Borki forest (n. Winnica res.), abandoned fallow, excavation pit, on the right bank of the Warta river, 51°25'58" N, 18°50'20" E, 155 m, vid. P. Niedźwiedzki, 02.08.2011 (phot.); **29** – E of Sulejów-Podklasztorze, on the right bank of the Radońka river (tributary of the Pilica river), on the side of an unfrequented road (overgrown with grass), by a grain field on one side and a low gras. on the other, (1 specimen), 51°21'55.3" N, 19° 53'26.3" E, vid. P. Niedźwiedzki, 29.06.2011 (phot.); **57** – between Przedbórz and Radomsko [99]; **59** – cfr. Brzostek, Przedbórz distr., hill by the forest near the Brzostek Policzko-Chojna road, thermophilous gras., leg. M. Kurzac, 07.07.2010 (LOD, 154948); Góry Suche village near Góry Mokre village, W part, NNE-facing xer. gras. and scrubs (>15), 51°01'45,57" N, 19°58'24" E, 279 m, leg. R. Piwowarczyk, 03.05.2006 (KTC) [13]; **59/69** – Murawy Dobromierskie res. n. Dobromierz, edge of a former excavation pit and baulks (>10), 51°00'34,9" N, 19°54'37,2" E, 273 m, leg. R. Piwowarczyk, 04.06.2008 (KTC) [13]; **69** – Stara Wieś n. Góry Mokre village, xer. gras. (>10), leg. A. Adamiec, 2006, leg. R. Piwowarczyk, 20.07.2007 (KTC) [13]; **83** – Częstochowa-Ostatni Grosz, fields, crops, grains, leg. F. Karo, 1865, 25.08.1874 (WA, 018913 = 28454; 018918 = 28440) [100]; **94** – Skrajnica, Dolny Ostrówek hill, Olsztyn distr., limestone quarry, xer. gras., leg. J. Hereźniak, 21.08.1986 (LOD, 154405), [101]; **95** – Przymiłowice-Kotysów, Olsztyn distr. n. Częstochowa, midfield scrub, leg. J. Kołodziejek, 16.08.1987 (LOD, 084145), [101]; **DF: 03** – between Woźniki and Gniazdów, NE slope of hill 359, near yellow route, steep field baulk (very rare), 50°35'14" N, 19°05'09" E, 350 m, leg. Ł. Krajewski, 28.07.2011 (KTC; Gniazdów; Babczyńska-Sendek 1995, US); between Wojsławice and Cynków, hill 345, W of road, secondary xer. gras., fallow rendzinas, S exp., (250), 50°34'11" N, 19°07'25" E, 345 m, leg. Ł. Krajewski, 28.07.2011 (KTC), Wojsławice Duże, S margin, steep slopes (Babczyńska-Sendek 1996, US); **04** – Rzeniszów N, in front of chapel near junction of Koziegłówka road and Katowice–Warszawa motorway, secondary xer. gras. (very rare), 50°34'36,45" N, 19°09'47,75" E, 317 m, leg. Ł. Krajewski, 28.07.2011 (KTC); Żarki E, Góra Parchowatka Mt (rocky hill 407), xer. gras., SW exp., (40), 50°37'28" N, 19°13'22" E, 390–395 m, leg. Ł. Krajewski, 24.09.2011 (KTC); **05** – Góra Włodowska SW (hill 410), NW slope, xer. gras. on field roadside, (10, close to *O. lutea*), 50°34'45" N, 19°25'04,5" E, 375 m, leg. Ł. Krajewski, 07.09.2011 (KTC); **06** – Kolonia Hucisko E, xer. gras., SE slope, (15), 50°35'50" N, 19°30'26" E, 394 m, vid. Ł. Krajewski, 08.09.2014 (phot.); **14** – Markowice SW, fallows, mown meadows, field baulks, S exp., (125), 50°33'17" N, 19°09'36" E, 355 m, leg. Ł. Krajewski, 28.07.2011 (KTC); Brudzowice W, hill 368, W slopes, xer. gras. on field baulk, (very rare), 50°30'15" N, 19°10'37" E, 340 m, leg. Ł. Krajewski, 23.07.2011 (KTC); **15** – Skałka N, hill Skałka, rendzina fallows, SW exp. (3, mixed with *O. lutea*), 50°32'02.4" N, 19°24'22.5" E, 370 m, vid. Ł. Krajewski 26.10.2014, phot.; **16** – Włodowice, dry calcareous hill, leg. M. Molenda, 13.08.1978 (KTU, 25638); **23** – Sączów W, hill 333, N slopes, fallows with dominant *C. scabiosa*, SE, W exp., (10), 50°26'26" N, 19°01'22" E, 330 m, leg. Ł. Krajewski, 29.07.2011 (KTC); Góra Siewierska S/Strzyżowice, Równa Góra Mt (hill 390), xer. gras. (>20), 50°23'42,3" N, 19°04'50,7" E, 370 m, leg. R. Piwowarczyk, 05.06.2008 (KTC), Strzyżowice (Babczyńska-Sendek 1998 – US, 102); Strzyżowice NW, Krzasek (hill 372), steep S slopes, old limestone quarries, calcareous gras. *Adonido-Brachypodietum pinnati*, *Thalictro-Salvietum pratensis*, S, SW exp. (20, mixed with *O. lutea*), 50°23'37,3" N, 19°04'20" E, 50°23'37,7" N, 19°04'16" E, 365–370 m, vid. Ł. Krajewski 10.09.2010, 06.08.2012 (phot.); Rogoźnik, Bobrowniki distr., xer. gras., leg. M. Bosek, 03.08.1985 (KTU, 41265); Siemonia NE, Góra Korzystna Mt (hill 383), *Cirsio-Brachypodion pinnati*, fallows, S, NE, NW exp., (80 in 3 subpopulations), 370–380 m, leg. Ł. Krajewski, 03.09.2010 (KTC), Łubianki n. Siemonia, Góra Korzystna Mt, (Babczyńska-Sendek 1998 – US; Nowak 1997 – US, as *O. lutea*); Nowa Wieś, Łubianki, field baulk, leg. B. Hanak, 17.08.1996 (KTU, 99028), between Łubianki and

Nowa Wieś, Góra Łubianki Mt (hill 398 m), *Cirsio-Brachypodion pinnati*, S exp., (25), 50°25′23″ N, 19°04′51″ E, 385 m, leg. Ł. Krajewski, 03.09.2010 (KTC); **24** – Przeczyce W, Bukowice (hill 314, 0.2 km W of dam), xer. gras. with numerous *Peucedanum oreoselinum* on old rendzina fallow, calcareous gras., N, S exp., (50), 50°26′18″ N, 19°10′05″ E, 312–314 m, leg. Ł. Krajewski, 04.09.2013; Toporowice NE, Ostra Góra (hill 351), xer. gras., margin of termophilous scrubs, (100), 50°25′37″ N, 19°09′06,2″ E, 334 m, leg. R. Piwowarczyk, 05.06.2008 (KTC), calcareous gras. under rocky steep slopes, fallow rendzinas on summit plateau, baulks, roadsides, W, NW, N, NE, E, SE exp., (208), 50°25′36″ N, 19°09′06″ E, 50°25′38″ N, 19°09′09″ E, 50°25′44,7″ N, 19°09′01″ E, 320–345 m, leg. Ł. Krajewski, 25.10.2009 (KTC; Babczyńska-Sendek 1988 – US); between Toporowice and Stara Wieś, 0.5 km E of Ostra Góra Mt (hill 351), rendzina fallows with numerous *C. scabiosa*, SE exp., (15), 50°25′36″ N, 19°09′31″ E, 305 m, vid. Ł. Krajewski 04.09.2013 (phot.); Siewierz-Chmielowskie E, calcareous sands with dominant *C. scabiosa*, SW exp., (very rare), 50°26′68,9″ N, 19°12′18,1″ E, 320 m, leg. Ł. Krajewski, 17.07.2011 (KTC); between Podwarpie and Podskale, hill 317, W slope, field baulk, (very rare, close to *O. lutea*, 200 m to *O. elatior* s.str.), 50°25′26″ N, 19°12′58″ E, 312 m, leg. Ł. Krajewski 04.09.2013; DG, Podwarpie SE, fallow rendzina, SW exp., (25), 50°25′16,8″ N, 19°14′16,2″ E, 315 m, leg. Ł. Krajewski, 17.07.2011 (KTC); DG-Trzebiesławice NW, 0.5 km W of cemetery, *Adonido-Brachypodietum pinnati*, (15), 50°26′21″ N, 19°14′22″ E, 330 m, leg. Ł. Krajewski, 17.07.2011 (KTC); DG, Ujejsce E, Góra Chendówka hill, calcareous gras., SW exp., (1, numerous *O. lutea*), 50°23′32″ N, 19°14′45″ E, 325 m, vid. Ł. Krajewski 17.07.2011, phot.; **25** – Chruszczobród S, Gawronów/DG, Tucznawa-Bugaj, field baulk with numerous *Peucedanum oreoselinum*, overgrowing *Prunus spinosa* and *Solidago*, SE exp., (10), 50°23′51″ N, 19°17′51″ E, 335 m, leg. Ł. Krajewski, 28.10.2009 (KTC), not found later; **26** – Zawiercie-Bzów, close to road to Podzamcze, fallows, (<10), 50°28′07″ N, 19°31′29″ E, 432 m, vid. R. Piwowarczyk 07.2013; Ogrodzieniec S, near Dół Zarazki, fallow with *Prunus spinosa* scrubs, leg. P. Kurek, 08.2009 (KTC); Grabowa NE, Wierzgóry (hill 423), calcareous gras., SW, E, SE exp., (50, rarely mixed with *O. lutea*), 50°23′46″ N, 19°28′04″ E; 50°23′34″ N, 19°27′44,5″ E; 50°23′28″ N, 19°28′11″ E, 390–410 m, leg. Ł. Krajewski, 18.09.2009, 02.10.2009 (KTC); **29** – Uniejów-Rędziny, fallow between the forest n. the Tunel railway station and the Biała Góra res., between the road to Kępa and the forest, fallow behind a single house, leg. M. Bogdański, 27.07.1988 (KRA, 0171466); Tunel n. Miechów, steppe slope, leg. A. Jasiewicz, 01.07.1956 (KRAM, 414025); NEE of Kępie, xer. gras. near railway line (1 shoot), 50°27′37″ N, 19°57′49″ E, 345 m, vid. B. Binkiewicz, 19.08.2005; **32** – Piekary Śląskie-Szarlej, NE of railway station, xer. grass on old calamine quarries and above cliff, S, SE exp., (30), 50°22′24″ N, 18°55′10″ E, 50°22′29″ N, 18°55′25″ E, 290–300 m, leg. Ł. Krajewski, 29.07.2011 (KTC), (=Scharley; as *O. lutea* in [55]); Bobrowniki-Namiarki, xer. gras., vid. Ł. Depa, P. Siwy, A. Szczepańczyk, 07.2009, phot.; between Bobrowniki and Namiarki, E of A1 motorway, xer. gras. on steep field baulk, S exp., (very rare), 50°22′39,4″ N, 18°58′12,7″ E, 295 m, leg. Ł. Krajewski, 07.07.2011 (KTC); Bobrowniki S, 0.2 km NWW of cemetery, W slopes N of field road, calcareous gras. by old limestone quarries, roadsides, SW, W exp., (25), 50°22′37″ N, 18°59′33″ E, 300–310 m, leg. Ł. Krajewski, 13.07.2011 (KTC); **33** – Rogoźnik, xer. gras., leg. M. Bosek 03.08.1985 (KTU, 41265); Wojkowice, Kolonia Maszyńska, xer. gras. on baulk n. former "Saturn" cement plant (Będzin, Kolonia Maszyńska), leg. A. Sendek, 06.09.1976 (KTU, 34397; as *O. minor* in [103]); Wojkowice, hill between Kolonia Maszyńska and Brynica v., secondary xer. gras., W, SW exp., (40), 50°21′38″ N, 19°01′14″ E, 275–280 m, leg. Ł. Krajewski, 01.07.2011 (KTC); Wojkowice-Żychcice/Bobrowniki NW, hill 329, summit and NE slopes, NE, SE, S, SW exp., (90), 50°22′36″ N, 19°00′28″ E, 322–328 m, leg. Ł. Krajewski, 13.07.2011 (KTC); Wojkowice Komorne, Żychcice, field baulk close to "Saturn" cement plant, leg. A. Sendek, 06.09.1976 (KTU, 34396); Wojkowice S, between "Jowisz" former coal mine and Brynica v., fallow rendzinas, SW, S, SE, E exp., (150), 50°21′08″ N, 19°02′53″ E, 285–295 m, leg. Ł. Krajewski, 01.07.2011 (KTC); Będzin, Grodziec W-Boleradz, overgrowing field baulks with numerous *P. oreoselinum* and *C. scabiosa*, W exp., (20), 50°21′05,6″ N, 19°04′02,9″ E, 302 m, leg. Ł. Krajewski, 13.07.2011 (KTC); Będzin, Grodziec NE, Góra Świętej Doroty (hill 381), oat field, leg. K. Zagórna, 08.1969 (LOD, 154406), NW slope, *Adonido-Brachypodietum pinnati*, (very rare), 50°21′09″ N, 19°05′48″ E, 350 m, leg. Ł. Krajewski, 13.07.2011 (KTC; Nowak 1997 – US, as *O. lutea*); Będzin-Małobądz, hill 305, W of iron radio tower, S of field road, calcareous xer. gras. overgrowing *Solidago*, W exp., (20; 12 in one clump), 50°18′27″ N, 19°07′17″ E, 300 m, leg. Ł. Krajewski, 17.07.2010 (KTC); **34** – Będzin, Warpie-Wschód, hill 301, secondary xer. gras. close to old limestone quarries, (15, mixed with *O. lutea*), 50°18′59″ N, 19°09′13″ E, 297 m, leg. Ł. Krajewski, 27.07.2009 (KTC), later not found; DG-Gołonóg, Mała Góra Gołonoska (hill 334), calcareous xer. gras. overgrowing scrub, W, SW, S exp., (2, 25, 10, mixed with *O. lutea*, close to locality *O. elatior* s.str.), 50°20′05″ N, 19°14′22″ E; 50°20′03,8″ N, 19°14′31,5″ E, 324–332 m, leg. Ł. Krajewski, 04.08.2009 (KTC), vid. Ł. Krajewski 2009–2013; DG-Ujejsce, xer. gras., leg. T. Nowak, 14.07.1994 (KTU, 99016); DG, Pańska Góra (hill 337), between Ujejsce and Ząbkowice, calcareous gras. on summit mound, (very rare in numerous *O. lutea*), 50°23′03,5″ N, 19°14′48,5″ E, 335 m, vid. Ł. Krajewski 2009, leg. Ł. Krajewski, 17.07.2011 (KTC); DG, NE of Pańska Góra hill, fallow rendzina, S exp., (very rare), 50°23′08,6″ N, 19°14′56,3″ E, 330 m, leg. Ł. Krajewski, 17.07.2011 (KTC); **35** – DG, Ząbkowice E, S of dolomite quarry, close to Gościniec Pasieka route, old rendzina fallows overgrowing *Crataegus* and *Solidago*, (10–15; mixed with *O. lutea*), 50°22′05,3″ N, 19°17′29″ E, 344 m, (5), 50°22′06″ N, 19°17′45″ E, 356 m, (25), 50°22′17,5″ N, 19°16′57″ E, 340 m, leg. Ł. Krajewski, 13.09.2009 (KTC), vid. Ł. Krajewski, 13.11.2012, 4.01.2014, vid. R. Piwowarczyk 2014, not found 2010–2011, DG-Ząbkowice, Góra Kamionka hill [104]; DG, Łęka NW, hill 361 (Łosy), xer. gras. overgrowing *Prunus spinosa*, (very rare, mixed with *O. elatior* s.str., close to *O. lutea*), S exp., 50°22′11″ N, 19°20′03,5″ E, 358 m, leg. Ł. Krajewski, 06.09.2013 (KTC), not observed 2008–2011; DG, Łosień N, W of medieval Trakt Siewierski route, 25 m S Las Bienia forest, old fallows, S exp. (very rare), 50°21′38″ N,

19°19'16" E, 370 m, leg. Ł. Krajewski, 27.09.2009 (KTC); DG-Łosień N, hill 390, NW slopes, field roadside with xer. gras., (very rare), 50°21'28" N, 19°19'40,25" E, 380 m, vid. Ł. Krajewski 06.08.2009, leg. Ł. Krajewski, 01.08.2010 (KTC); DG-Łosień, 1 km S, (10, mixed with numerous *O. lutea*), xer. calamine gras., SW exp., 50°20'29,5" N, 19°19'27,5" E, 355 m, leg. Ł. Krajewski, 06.08.2009 (KTC, later not found); Sławków W, Góra Kołdaczka Mt (hill 322), N slope, old rendzina fallow overgrowing *Libanotis*, (very rare, close to *O. lutea*), 50°18'05,4" N, 19°22'33" E, 318 m, leg. Ł. Krajewski, 15.09.2009 (KTC); Sławków E, Świnia Góra Mt (hill 312), xer. gras. overgrowing scrubs above old limestone quarry and road Sławków-Podlipie, (very rare, 100 m E of *O. lutea*), 50°17'48,8" N, 19°24'28,1" E, 308 m, leg. Ł. Krajewski, 16.09.2009 (KTC), later not found; **36** – DG, Błędów, Góra Kamieniec (Młyńska, hill 341), W side of asphalt road crossing N slope, secondary xer. gras. on fallow rendzinas, (20, mixed with *O. elatior* s.str. in upper part), 50°20'58" N, 19°27'51" E, 325–330 m, leg. Ł. Krajewski, 26.09.2009, 10.07.2011 (KTC), vid. Ł. Krajewski, 06.09.2013; **38** – Gołaczewy Stara Wieś, xer. gras. n. quarry, NE exp., (<5), 50°21'52,9" N, 19°42'52,3" E, 397 m, vid. R. Piwowarczyk, 31.09.2011; **44** – Jaworzno-Długoszyn E, hill with iron cross SE of church, secondary gras. *Cirsio-Brachypodion pinnati*, SW exp., (40, 30), 50°14'43" N, 19°15'53" E, 280 m, leg. Ł. Krajewski, 04.08.2010, 05.08.2011 (KTC, also f. *citrina* in 2010 and 2011); **45** – Jaworzno, Szczakowa-Gadlin, hill W of cemetery, horse pasture, small patch of xer. gras. sourrounded by *Solidago* and *Calamagrostis*, (20), 50°14'05" N, 19°17'48" E, 285 m, leg. Ł. Krajewski, 07.08.2010 (KTC); Jaworzno-Ciężkowice, arable fields – fallow, leg. J. Nowak, 24.07.1953 (CHRZ); Jaworzno-Ciężkowice, Góra Wielkanoc Mt (hill 331), *Thalictro-Salvietum pratensis*, S, SW exp., (25), 50°12'41" N, 19°20'32" E, 318–320 m, leg. Ł. Krajewski, 08.08.2010 (KTC); Jaworzno NE, hill 341 under power line, E of Góra Przygoń hill, mesophilous gras. *Thalictro-Salvietum pratensis* varying to *Origano-Brachypodietum pinnati*, NW, W exp., (45), 50°12'28" N, 19°23'18" E, 340 m, leg. Ł. Krajewski, 13.08.2010 (KTC); **46** – Podlipie S, hill 362 m, xer. gras. on old fallow rendzinas, (very rare in numerous *O. lutea*), 50°17'32" N, 19°25'45" E, 360 m, leg. Ł. Krajewski, 16.09.2009 (KTC); Krążek S, hill 335, poor calamine gras./ fallow rendzina, slightly sandy, overgrowing *Pinus*, SW, W exp., (very rare), 50°17'17" N, 19°27'18" E, 330 m, leg. Ł. Krajewski, 18.09.2011 (KTC); Olkusz NW, Góra Pomorska (hill 418), poor xer. gras. on fallow rendzinas, SW, W exp. (10), 50°17'34" N, 19°33'08" E, 405–415 m, leg. Ł. Krajewski, 15.09.2011 (KTC); **47** – Sikorka n. Olkusz, leg. A. Zalewski, 27.07.1893 (WA, 018895 = 28421, 018894 = 28450); Witeradów SE, Góra Kadzielnia (hill 417) and to E, xer. gras., baulks, fallows, W, S exp., (30), 50°14'30" N, 19°34'57" E, 410–420 m, leg. Ł. Krajewski, 13.09.2011 (KTC); Kosmolów N, S of Podłużne forest, closed Eagles' Nests red trail, dry meadow, xer. gras. on fallow rendzinas, (50), 50°16'56" N, 19°38'54" E, 50°16'52,4" N, 19°38'52" E, 400–420 m, leg. Ł. Krajewski, 10.07.2011 (KTC); **53** – Lędziny SW, Góra Klimont (hill 302), SW of St. Clemens' church, xer. gras., (20, mixed with *O. elatior* s.str.), 50°07'13" N, 19°06'08" E, 295 m, leg. Ł. Krajewski, 13.09.2010 (KTC); **54** – Dziećkowice (Grabina b. Dzieckowitz), as var. *kochii* [54,55,105]; between Imielin-Pasieczki and Mysłowice-Dziećkowice, hill 310, S of A4 motorway, calcareous xer. gras., S exp., (15, close to *O. elatior* s.str.), 50°09'50" N, 19°12'37" E, 300–307 m, leg. Ł. Krajewski, 04.09.2010 (KTC), Imielin, Pasieczki [55,106], Myslovitz [27]; **55** – Jaworzno-Śródmieście, above old limestone quarry NW of Góra Grodzisko hill, overgrowing xer. gras., SW exp., (20, mixed with numerous *O. lutea*), 50°11'20" N, 19°17'33" E, 315 m, leg. Ł. Krajewski, 04.08.2010 (KTC); Jaworzno NE, hill W of Trzebinia-Stara Maszyna, fallows with dominant *C. scabiosa*, S, SW exp., (10), 50°12'13" N, 19°23'33" E, 347 m, leg. Ł. Krajewski, 13.08.2010 (KTC), 2011 ploughed (=Siersza, Stara Maszyna, W of village; Chmiel 1993 – US); between Chrzanów and Balin, Wójtowa Góra hill on S side, baulk, leg. P. Grzegorzek, 21.09.1985 (CHRZ); **56** – Trzebinia-Podbuczyna, hill under power line between Podbuczyna and Podkrystynów, *Adonido-Brachypodietum pinnati*, SW, S exp., (15), 50°10'44" N, 19°25'59" E, 372 m, leg. Ł. Krajewski, 24.08.2010 (KTC); Trzebinia E, hill 375 m, NE of the ZWM district, *Thalictro-Salvietum pratensis*/*Prunetalia spinosae*, SW exp., (very rare, mixed with *O. lutea*), 50°09'42" N, 19°28'44" E, 355 m, leg. Ł. Krajewski, 24.08.2010 (KTC); **58** – Dolina Kobylańska valley [80] (phot.); **66** – Bolęcin, n. Skałka Triasowa, vid. P. Grzegorzek, 01.08.2012 (phot.); Grojec, Alwernia distr., xer. gras. on the hillside between the hamlets Brandyska and Wymiarki, 50°05.621'N, 19°33.080'E, 353 m, leg. B. Binkiewicz, 26.07.2014 (KTC); **EF: 00** – between Sędziszów and Sosnowiec, xer. gras. (>10), leg. B. Piwowarski, 28.07.2011 (KRA); **04** – Gartatowice, xer. gras., fallows, on gypsum (<100), 50°34'34" N, 20°37'33" E, 237 m, vid. R. Piwowarczyk 2012–2013, leg. M. Nobis, 2012 (KRA) [13]; between Chomentówek and Sędziejowice, xer. gras. (<10), 50°34'23" N, 20°39'44" E; vid. R. Piwowarczyk 2012–2013 [13]; Samostrzałów SSE, xer. gras., fallows, (<5), 50°35'16" N, 20°38'33" E, 248 m, vid. R. Piwowarczyk, 09.07.2013; **07** – Jabłonica, xer. gras., leg. A. Pierścińska, 14.05.2009 (KRA); Jabłonica, Jabłonickie Mts, dry meadow, leg. A. Pierścińska, 09.09.2009 (KRA) [13]; NW of Szydłów, xer. gras. on a hill known as Lisi Kamień (>10), 50°36'10,25" N, 20°59'35,27" E, 274 m, leg. M. Gwardjan, 2010 (KTC) [13]; Kotuszów, rye cultivation; leg. M. Faniczak, 19.07.1979 (KRA, 0233676); **13** – Góry Pińczowskie Mts, NW of Pińczów, wastelands, xer. gras. (>10), 50°32'08,4" N, 20°30'46" E, 219 m, leg. R. Piwowarczyk, 09.06.2009 (KTC) [13]; **14** – NW of the road from Pasturka village to the Las Pasturski forest, xer. gras., field margins (>5), 50°30'24,77" N, 20°34'04" E, 225 m, leg. R. Piwowarczyk, 14.07.2008 (KTC) [13]; Wełecz n. Busko Zdrój, S-facing xer. gras. on a hill by a hardened road from Wełecz towards Siesławice (>20), 50°28'14,92" N, 20°41'12" E, 261 m, leg. R. Piwowarczyk, 21.07.2006 (KTC) [13]; **15** – Łagiewniki n. Busko Zdrój, S and SW-facing xer. gras. (Zimne Wody res.; <10), 50°28'58,72" N, 20°44'00" E, 254 m, leg. R. Piwowarczyk, 21.07.2006 (KTC) [13]; **24** – Krzyżanowice res. n. Gacki, S part, xer. gras. and wasteland (>10), 50°27'11,65" N, 20°34'04,92" E, 219 m, vid. R. Piwowarczyk, 2007 [13]; Skotniki Górne res. n. Skotniki Górne village, xer. gras. between road and railway lines, S and SW-facing, sporadically N and NE (>50), leg. R. Piwowarczyk, 07.2007 (KTC) [13]; Skorocice res. n. Skorocice village, xer. gras. (>10), vid. R. Piwowarczyk, 2007 [13]; Wola

Zagojska, dry hill, leg. I. Latkowska, 24.07.2006 (KTC); **25** - Żerniki Górne, xer. gras. (>10), 50°27'29" N, 20°47'28" E, 260 m, leg. M. Nobis, 2009 (KRA) [13,107]; xer. gras. located on the slope of the road leading through Żerniki Górne, 50°27'29,45" N, 20°47'26,1" E, vid. I. Dembicz, Ł. Kozub, 02.07.2013, phot.; S of Ostra Góra ecol. site, S of Pęczelice, S-facing wastelands and field margins (>10), 50°26'34" N, 20°47'05" E, 239 m, leg. R. Piwowarczyk, 13.07.2010 [13,14]; A hill S of the Łatanice–Hołudza road, xer. gras. (>30), 50°24'54,8" N, 20°42'25,3" E, 197 m, leg. R. Piwowarczyk, 30.09.2009 (KTC) [13]; Przęślin res. n. Chotel Czerwony, xer. gras. (>5), 50°22'41,87" N, 20°42'59" E, 188 m, vid. R. Piwowarczyk, 2006 [13]; **31** - Opalonki res., xer. gras., leg. D. Wołkowycki, 04.08.2010 (3377, D. Wołkowycki private herbarium); **32** - Pałecznica, xer. gras., leg. K. Towpasz, 08.07.2003; vicinity of Działoszyce, gypsum sink area on a slope between Szczotkowice and Pierocice, leg. T. Tacik, 09.08.1957 (KRAM 086091, 014848); **35** - Górki Pierwsze, xer. gras., on gypsum, roadsides, on both sides of the road from the Szczerbaków–Strążyska dirt road (>10), 50°20'55,03" N, 20°44'01" N, 178 m, 50°20'54,72" N, 20°44'02" E, 179 m, leg. R. Piwowarczyk, 2006 (KTC) [13]; **41** - Gniazdowice, gypsum slope, leg. K. Towpasz, 11.07.1995, 29.06.1998 (KRA); **43** - Kazimierza Wielka, W part, between baulk and escarpment, leg. K. Towpasz, 11.07.1997 (KRA); **44** - Kaczkowice n. Kazimierza Wielka, leg. K. Towpasz, 21.06.1999 (KRA), vid. R. Piwowarczyk, 2008 [13]; **53** - Książnice Małe n. Koszyce, gras. on a baulk, leg. K. Towpasz, 26.06.1999 (KRA); **EE: 38** - N of Krzyżanowice, xer. gras. by a mid-field limestone excavation pit, leg. M. Nobis, 19.07.2000 (KRA, 0318683, 0318684) [13,108]; **46** - SSW of the Młodzawy estate in Skarżysko-Kamienna, between E side of the roadside and a forest complex (>10), 51°05'19,88" N, 20°52'33" E, 269 m, vid. Ł. Maślikowski, 2007 (phot.); **48** - 500 m N of the E part of Seredzice village [108]; **56** - Piaski n. Parszów, margin of a pine forest W of Piaski (2), 51°05'14,3" N, 20°55'09" E, 244 m, vid. M. Fałdziński, 2008 (phot.); **61** - NNW of Skałka Polska n. Ewelinów, xer. gras. and thermophilous scrubs on a S and SW-facing hill, 50°57'06" N, 20°10'03" E, 250 m, leg. R. Piwowarczyk, 09.07.2007 (KTC) [13]; **64** - Zachełmie n. Zagnańsk, xer. gras. on the edges of a closed quarry n. Chełmowa hill (<20), 50°58'10" N, 20°41'28" E, 354 m, leg. R. Piwowarczyk, 01.09.2005, leg. M. Nobis, A. Nobis, 2008 (KRA) [13]; **67** - Śniadka Druga, xer. gras. on the right slope of the Psarka river, so-called Wymysłowskie Doły, SW-facing (<50), 50°57'13,22" N, 21°00'45", 252 m, vid. A. T. Halamski, R. Piwowarczyk, 04.07.2006, leg. R. Piwowarczyk, 05.08.2008 (KTC) [13]; **74** - Kielce, a fallow behind the Institute of Chemistry, Jan Kochanowski University, Świętokrzyska street (3), 50°52'53" N, 20°39'31" E, 289 m, leg. B. Maciejczak, 2008 (KTC), vid. B. Piwowarski, 2009 [13]; **75** - Between Górno and Górno Zawada, xer. gras. on a hill S of a dirt road (>20), 50°50'38,3" N, 20°48'02,3" E, 279 m, leg. R. Piwowarczyk, 08.2009 (KTC) [13]; **84** - S of Podgórze n. Bilcza, fallows and xer. gras. NNW of an old warehouse, 50°46'57,4" N, 20°35'48" E, 271 m, vid. M. Fałdziński, 2008; Folwark n. Bilcza, gras., xer. scrub, fallows W and NW of a bus stop on the Kielce road, a few populations: W of the bus stop in Bilcza and of the road towards Kuby Młyny village (>20), 50°46'49,28" N, 20°36'59" E, 256 m, NW of the bus stop in Bilcza and S of Marmurowa street (>150), 50°46'58,91" N, 20°36'39" E, 253 m, vid. M. Fałdziński, 2008, leg. R. Piwowarczyk, 2008 (KTC) [13]; SSW of Kowala Mała, wastelands around a closed warehouse, NEE of Sukowianowa hill (<20), 50°47'04,20" N, 20°34'20" E, 277 m, 50°47'01" N, 20°34'35" E, 277 m, vid. M. Fałdziński, 23.07.2007; **87** - Łagów, Dule Gorge, xer. gras. on the Zbójecka Cave, S, SW-facing (>30), 50°45'55" N, 21°05'28" E, 295 m, leg. R. Piwowarczyk, 17.08.2006 (KTC), vid. B. Sępioł 2012, vid. R. Maniarski 2013 (phot.) [13]; Łagów, xer. gras. SE Zapłotnia street (>50), 50°46'16" N, 21°04'31" E, 296 m, vid. B. Sępioł, 06.08.2013 (phot.); NWW of Zagościniec n. Łagów, xer. gras. (>10), 50°47'15" N, 21°06'40" E, 309 m, leg. B. Piwowarski, 2010 (KTC); 150 m N of the Łagów–Nowy Staw road, xer. gras., S-facing (>20), 50°46'04,77" N, 21°06'43" E, 286 m , leg. R. Piwowarczyk, 17.08.2006 (KTC) [13]; **97** - Chańcza n. Raków, fallow NE of Chańcza (>5), 50°39'01,67" N, 21°03'34" E, 239 m , vid. M. Fałdziński, 2009; **FC: 97** - thermophilous gras., W of Drohiczyn, SE of Runice, (1), 52°23'57,4" N, 22°37'50,51" E, vid. W. Heise, P. Naks, 26.08.2014 (phot.); **FE: 24** - Iłki n. Celejów, Puławy distr., baulk, leg. research camp of a Łódź University student society, T. Załuski, 31.07.1971 (LOD, 13210); **54** - Sosnowa Wola n. Dzierzkowice, rendzina field, leg. B. Sałata, 11.07.1964 (LBL); **62** - Wesołówka, fallows, baulks and xer. gras. by the border of the Vistula v. ("Modrzywka"; >30), 50°58'41,9" N, 21°47'47,2" E, 170 m, vid. R. Piwowarczyk, 1999–2013; leg. R. Piwowarczyk, 29.06.2005, 01.08.2010 (KTC) [9,13,14,109]; **FE: 23** - Męćmierz, xer. gras. on the slopes of a vast ravine (one specimen), 51°18'07" N, 21°54'02" E, vid. R. Piwowarczyk & P. Chmielewski, 26.07.2009 [9,14]; **27** - Rury n. Lublin, in sown fields behind cemetery, leg. F. Karo, 21.07.1881 (WA, 018911 = 28451); **60** - Boria, cz. S, xer. gras. (<100), 50°58'15" N, 21°32'53" E, 167 m, vid. B. Sępioł, 27.06.2013, leg. R. Piwowarczyk, 29.06.2013, vid. R. Piwowarczyk, 07.2014 (with f. *citrina*, 2013–2014); **73** - Opoka Duża, fallow field bordering on the steep slopes of the Vistula v., overgrown with dense thermophilous scrub, ca. 1 km NW of the village (<5), 50°51'58" N, 21°51'23" E, vid. P. Chmielewski, 14.04.2009 [9]; **79** - Elizówka, midfield initial xer. gras. and scrub on hillsides, ca. 0.5 km of village buildings (two specimens), 50°51'37" N, 22°44'07" E, vid. P. Chmielewski, R. Piwowarczyk, 2008 [9]; **FF: 00** - cfr., Beszyce, xer. gras., leg. A. Pierścińska, 23.05.2009 (KRA); **90** - Kołaczyce, gras., leg. K. Towpasz, 06.08.1985 (KRA); **GD: 00** - in thermophilous gras. E of Osłów, 9 plants parasitic on *C. scabiosa* noted in 2011 ([110] and phot.); **GE: 13** - Tarnów in Chełm region, on baulks in clay fields, leg. K. Łapczyński, 18?? (WA, 018864 = 28390); **23** - cfr. vicinity of Chełm, on baulks under Góra Stawska hill, n. Staw village, leg. F. Karo, 28.06.1883 (WA, 018926 = 28449); **34** - Chełm, allotments, fields, leg. H. Koporska, 07.1932 (LBL); **45** - Wólka Leszczańska, calcareous slope, leg. D. Fijałkowski, 30.07.1949 (LBL); **55** - Turowiec, a vast complex of initial xer. gras. and juniper scrub, ca. 1.5 km N of the cemetery near Turowiec and Popówka villages (<5), 50°58'00" N, 23°38'12" E, vid. P. Chmielewski, 2007 [9]; **62** - Tarnogóra, xer. gras. of *Thalictro-Salvietum pratensis* association on a loess slope (>5), 50°53'42" N,

23°08'09" E, vid. A. Cwener, 2007 [9]; **64** - Wysokie n. Skierbieszów, sunny loes slope, leg. D. Fijałkowski, 15.07.1965 (LBL); **65** - cfr., Wola Uchańska, in a cultivation (rendzina), leg. M. Sowińska, 26.07.1961 (LBL); **74** - Wolica Uchańska, a xer. gras. and a fallow field on chalky rendzina on a W-, SW- and S-facing slope, by the Skierbieszów-Grabowiec road, immediately behind Wolica Uchańska (>30), 50°49'36" N, 23°29'29" E, leg. A. Cwener, 27.07.2006 (LBL), vid. P. Chmielewski, 2008 [9]; Kolonia Łaziska, a degraded xer. gras. on a S- and SW-facing slope, ca. 400 m E of the village (>20), 50°48'42" N, 23°24'05" E, leg. A. Cwener, 29.05.2005, vid. P. Chmielewski, 2007 [9]; Rogów, xer. gras., *Inuletum ensifoliae* association, in the "Rogów" res., on the calcareous slope of a water course, ca. 1 km SE of the village (>20), 50°47'52" N, 23°31'28" E, leg. A. Cwener, 02.07.2006 (LBL) [9]; Żuków, degraded xer. gras. on chalky slope (>20), 50°47'51" N, 23°28'09" E, leg. A. Cwener, 20.07.2006 (LBL) [9]; **55** - Maziarnia, roadsides on the margin of a forested xer. scrub complex, ca. 300 m S of Maziarnia n. Putnowice (>20), 50°58'32" N, 23°39'13" E, vid. P. Chmielewski, 25.04.2009 [9]; **71** - Gruszka Duża, a narrow belt of a xer. gras. by the roadside, ca. 1 km N of the village (>15), 50°49'10" N, 22°58'41" E, vid. P. Chmielewski, 2008 [9]; **77** - Gródek, xer. gras. on the slopes of the Bug v. stretching from the Kolonia Gródek village towards Czumów (>50), 50°47'46" N, 23°57'11" E; 50°46'57" N, 23°56'47" E, leg. P. Chmielewski, A. Cwener, 28.06.2008 [9]; Czumów n. Hrubieszów, loess slope on the Bug river v., leg. D. Fijałkowski, 18.07.1956, 26.07.1959 (LBL); **82** - Kąty II, initial xer. gras. near a dirt road from the Wymysłówka village to Wychody, on a midfield hill known as Wieprzecka Góra (<20), 50°40'46" N, 23°07'35" E; 50°40'35" N, 23°07'43" E, vid. R. Piwowarczyk, W. Michalczuk, P. Chmielewski, 2006 [9]; **83** - hills n. Zamość, (as *O. rapum-gensitae*), leg. F. Berdau, 06.18?? (WA, 018910=28424); **85** - Kazimierówka, the side of the road from the Kotlice village towards Perespa, ca. 300 m S of the Kotlice Forest. Over ten specimens of *O. elatior* were recorded in 2007; the species was not observed afterwards although the habitat did not changed; 50°40'35" N, 23°36'27" E, vid. P. Chmielewski, 2007 [9]; **94** - Sielec, Dzierążnia Kościelna, dry chalky slope, leg. M. Szozda, 16.07.1958 (LBL); **96** - Dobużek, Skarpa Dobużańska res., ca. 1 km NW of the village, a xer. gras. belonging to *Thalictro-Salvietum pratensis* association on the steep slope of the Huczwa river v. (>20), 50°34'50" N, 23°42'56" E, leg. P. Chmielewski, R. Piwowarczyk, 14.07.2006 (KTC) [9]; **GF: 04** - Tarnawatka-Tartak, fallow fields and the roadside of a field road over the hill from Tarnawatka S towards the Pańków village (>20), 50°31'28" N, 23°29'37" E, vid. P. Chmielewski, 27.07.2007 [9]; Huta Tarnawacka, an initial xer. gras. on the SW and S hillsides, ca. 0.5 km N of the village, by dirt road towards the Pauczno village (>50), 50°32'05" N, 23°29'37" E, vid. P. Chmielewski, 14.07.2006 [9]; **07** - Posadów, xer. gras., *Thalictro-Salvietum pratensis*, on chalky rendzina and chernozems, 0.5 km N of the village, on the edge of the Kamionka river v., on both sides of the road to Telatyn (>30), 50°30'27" N, 23°48'57" E; 50°30'14" N, 23°49'18" E, vid. P. Chmielewski, 2006 [9]; **25** - Machnów n. Uhnów, steppe slope, leg. J. Motyka, 28.08.1949 (LBL); Machnowska Góra res. (Roztocze), xer. gras., leg. K. Oklejewicz, 05.08.2009 (KRA); Machnów Stary, xer. gras. in the Machnowska Góra res., 50°22'10" N, 23°35'21" E; 50°22'17" N, 23°34'54" E, vid. P. Chmielewski, 2007 [9].

Localities difficult to locate in the field:
Pawłów, "Dąbrówka", dry hill, leg. prof. Polański, 13.07.1912 (KRAM, 037793); Ostrowiec, leg. W. Gajewski, 07.1930 (WA, 018879 = 28398).

List of undifferentiated and incorrect localities:
AB: 96 - Stargard Szczeciński [111]; **97** - Ulikowo [111]; **AD: 09** - Jezioro Paklicko Wielkie n. Nowy Dworek (Flußwerder im Packlitzsee) [112]; **BC: 56** - Goraj n. Czarnków (Czarnikau: Goray) [113]; **78** - Oborniki [113]; **89** - Murowana Goślina (Murowanna-Goślin) [114]; **BD: 08** - Poznań (Posen) [27]; **09** - Poznań Wschód (Posen-Ost) [113]; **70** - Nowe Żabno, Nowa Sól distr. [115]; **BE: 84** - Lubachów (Breitenhain) [116]; **CA: 87** - Kartuzy [117]; **99** - Pręgowo (Prangenau), among cereal crops n. Pręgowo [118–120]; **CD: 25** - Ciążeńskie Holendry (Brzeg, unpublished data, 1995 - ATPOL); **CE: 97** - Rąty (Ronty bei Karthaus), between potato cultivation and barley cultivation, specimen 61 cm [118]; **CF: 37** - Koźle (Dajdok et al. 1998 - ATPOL]; **DA: 80** - Gdańsk [117]; Westerplatte [118,121]; **87** - vicinty of Frombork, leg. Klinsmann 1844 [122]; leg. Hübner, 18?? [122], between Frombork and Stary Dwór [118]; **91** - Trutnowy [123]; **DB: 73** - Szynwałd, S slope of a mountain in scrub, specimen 68 cm [118,121]; **DC: 02** - in the forest between pine trees, n. Nielubie [118]; **DE: 61** - Opatów n. Kłobuck (Babczyńska-Sendek, unpublished data 1998 - US); **84** - Wrzosowa, Góra Kamionka hill, gras. on slopes (Babczyńska-Sendek, unpublished data 1998 - US); **DF: 02** - Babienica n. Psary, Skała, gras. (Babczyńska-Sendek, unpublished data 1998 - US); **03** - Woźniki, Coglowa Góra hill (Babczyńska-Sendek, unpublished data 1996 - US; mistaken with numerous *O. lutea*, not recorded by Babczyńska-Sendek); **04** - Rzeniszów-Świnica (Babczyńska-Sendek, unpublished data 1998 - US; mistaken with numerous *O. lutea*, not recorded by Babczyńska-Sendek); **22** - Niezdara (Babczyńska-Sendek, unpublished data 1988 - US); Nakło Śląskie (Babczyńska-Sendek, unpublished data 1995 - US); **23** - Najdziszów (Babczyńska-Sendek, unpublished data 1998 - US), Gołąsza Dolna (Babczyńska-Sendek, unpublished data 1997 - US); **25** - DG-Tucznawa (K. Rostański, unpublished data 1976, Jędrzejko, Żarnowiec unpublished data 1985 - all US; after misidentified specimen of *O. lutea*, leg. Sumara 1974, KTU); **28** - N of Sławniów n. Pilica [124]; **DF: 32** - Bobrowniki, border with Kamyce (Babczyńska-Sendek, unpublished data 1998 - US; only *O. lutea* found in 2011); **33** - between Rogoźnik and Wojkowice (Babczyńska-Sendek, unpublished data 1993 - US), Twardowice (Babczyńska-Sendek, unpublished data 1998 - US); **34** - DG, Ząbkowice-Wygiełzów (Babczyńska-Sendek, unpublished data 1996 - US; 2008–2013 only present numerous *O. lutea*); **35** - DG, Strzemieszyce Małe [103], **44** - Sosnowiec-Sielec [Sielce bei Bendzin; (cf. [71,125] - probably both after misidentified specimen of *O. lutea*, leg. G. Schneider 1876, B; still present in 2010, Ł. Krajewski, unpublished)]; **46** - Trzebinia-Czyżówka, Góra Wilcza hill (Babczyńska-Sendek, unpublished data 1995 - US; only found *O. lutea*); **51** - Ornontowice S, n.

Mikołów (Urbisz, unpublished data 1998 - US), **55** - Balin Duży (cf. [75,81] - mistaken with *O. lutea*, phot.); **DG: 01** - Leszna Góra, Cieszyn distr. [90]; **EA: 90** - Pieniężno [118]; **EB: 59** - Jezioro Mały, Piecki distr. [126]; **ED: 12** - Kampinos n. Warszawa [127]; **EE: 17** - Radom [127]; **EF: 37** - Szczuczyn [127]; **77** - Góra St. Marcin hill, S Tarnów [128]; **EG: 33** - Between Krościenko and Tylmanowa [128,129]; **FG: 13** - vicinity of Iwonicz [130].

Prefered habitats, plant communities and host

Orobanche elatior and *O. kochii* are clearly dinstinguished by their phytoceonoses. *Orobanche elatior* prefers sunny hillsides and slopes of hillocks, hills and valleys but also flat areas, infrequently also the colder horizon sector, at altitudes (40)200–400(700) m above sea level. Its highest sites are in the Góry Orlickie Mts (Central Sudetes Mts) and the Silesian Beskid Mts (Western Carpathians). It prefers alkaline soils, usually shallow rendzinas formed on chalk marl and limestone, dolomite, loess, gypsum, and chernozem. The abundance of specimens in populations is usually small and is 5–20 shoots but sometimes only a single inflorescence is noted. The most abundant populations occur in the Sudetes Mts (one site of a few hundreds shoots) [86] and in the central part of the Silesia-Cracow Upland (a few sites with over 100 shoots). Fluctuations in the population size are noticeable at localities observed over a few vegetative seasons.

Orobanche elatior prefers a broad phytocoenotic range. It comprises intermediate communities and is often difficult to classify. This reflects the transitional and mosaic type of habitats colonized by the species and their dynamic-succession condition. *Orobanche elatior* occurs in xerothermic grasslands of the alliance *Cirsio-Brachypodion pinnati* (*Festuco-Brometea* class) and communities of juniper scrub, e.g. in the Lublin Land. The species also often occurs in abandoned fields, fallows or wastelands, roadsides, along railways, in intermediate communities between xerothermic grasslands (*Festuco-Brometea, Festucetalia valesiacae*), thermophilous meadows (*Molinio-Arrhenetheretea, Arrhenatheretalia*), pioneer semiruderal xerothermic communities (*Agropyretea intermedio-repentis*) and ruderal communities (*Artemisietea vulgaris, Onopordetalia*, suballiance *Dauco-Melilotenion*, especially *Dauco-Picridetum hieracioidis*, and *Echio-Melilotetum* associations). It is fairly frequently recorded in fringe communities of the alliance *Geranion sanguinei* (*Trifolio-Geranietea* class). *Orobanche elatior* is also encountered within or on the margins of cultivations of, e.g. rape, cereal crops, or on the margins and baulks of calcareous fields, with a contribution of species of the alliance *Centauretalia cyani*, e.g. *Caucalidion lappulae* (*Stellarietea mediae* class) (Appendix S1). In the Sudetes Mts, the species was noted in thermophilous grasslands on forest margins of the alliance *Bromion erectii* [86].

Orobanche kochii prefers sunny sites, usually gentle S or SW-facing slopes but it is also found at N facing or flat sites; altitude about (40)200–400(430) m. These include xerothermic grasslands, thermophilous shrublands; located on slopes of hills, river valleys, gullies, margins of forests and fields, in former quarries, baulks, fallows and ruderal sites. It prefers alkaline soils, usually shallow rendzinas formed on chalk marl and limestone, dolomite, loess, and chernozem (black-earth soil). It is recorded most frequently in communities of the alliance *Cirsio-Brachypodion pinnati* (class *Festuco-Brometea*), especially on flowery or initial dry grasslands, i.e. *Adonido-Brachypodietum pinnati, Inuletum ensifoliae, Thalictro-Salvietum pratensis*, and on hills, in former gypsum excavation pits in the Nida Basin, also in *Seslerio-Scorzoneretum purpureae*, and in the community *Carex glauca-Tetragonolobus maritimus*. In the Silesia-Cracow Upland it occurs mostly on strongly xerothermic fallow rendzinas and on secondary calcareous grasslands formed close to old limestone quarries, usually on steep slopes, less frequently in sligtly thermophilous north-facing associations. In the Cracow-Częstochowa Upland, the species also sometimes grows in the ecotone of rocky grasslands at foothills of limestone mogotes (*Festucetum pallentis*) and thermophilous scrub (*Geranion sanguinei*), usually in the association *Origano-Brachypodietum pinnati*. As a result of natural plant succession in grasslands, the communities have a high contribution or a mosaic of species of thermophilous herbaceous forest edge communities of the alliance *Geranion sanguinei* (class *Trifolio-Geranietea sanguinei*) and shrub communities, particularly of the alliance *Berberidion* (class *Rhamno-Prunetea*), as well as xeric meadow species of the alliance *Arrhenatherion elatioris* (class *Molinio-Arrhenatheretea*). The species is also often encountered on former fallows and wastelands, especially in rendzina soil conditions, in communities of the class *Artemisietea vulgaris* with a high contribution of xerothermic and meadow species (Appendix S2). The abundance of populations varies greatly and ranges from year to year. Its populations are more abundant than those of *O. elatior* and comprise 15 shoots on average, rarely app. 100–250, e.g. 100 Gartatowice, 150 (Wojkowice) and 250 (between Wojsławice and Cynków). The abundance behaviour of *O. kochii* is similar to that of *O. picridis* [14]. It rapidly and densely colonizes abandoned areas after which its number drops suddenly as a result of succession changes. However, in the central part of the Silesia-Cracow Upland, the majority of its localities are less numerous (also less numerous than *O. elatior*), and sometimes only single specimens could be found. At these sites the species is difficult to find (in dry forms), especially because specimens are often very sparse in other *Orobanche*, mostly in the much more numerous *O. lutea*. All the communities mentioned above include a high contribution of host species of *C. scabiosa*, with the coverage being about (5)25(50)% (Appendix S1, Appendix S2).

Discussion

Orobanche kochii is more common and widely distributed in Poland, the Czech and Slovak Republics, and in much of Central Europe, with the distribution area shifting from central Europe towards Eastern Europe and Asia. *Orobanche elatior* is noticeably scarcer in Central Europe and its distribution is confined to Western and Central Europe. Their ranges overlap in Central Europe.

A revision of herbarium materials and field investigations into *O. elatior* and *O. kochii* in Poland was conducted between 1999 and 2014. Considerable differences in the distribution were found, especially in relation to the previous distribution map of *O. elatior* s.l. in Poland [70] and in Central Europe [21]. At present *O. elatior* occurs in Poland mainly in the Silesia-Cracow, Małopolska and Lublin Uplands, rarely in the Lower Vistula Valley, Central Sudetes Mts, Western Carpathian Mts and Carpathian Foothills. *Orobanche kochii*, which is definitely more common than *O. elatior*, has numerous localities especially in the Polish Uplands: Silesia-Cracow and Małopolska Uplands, and Lublin-Lviev, Volhynia Uplands, Roztocze and Polesie (Fig. 3, Fig. 4). Our research demonstrated that *O. kochii* and *O. lutea* are the most common representatives of this genus in Poland.

The abundance of populations of *O. elatior* and *O. kochii* varies greatly by region in Poland and fluctuates considerably in individual years. The both taxa are monophagous species and in Poland, like in other countries of Central Europe, they are parasitic on *C. scabiosa*.

Zázvorka [21] stated that *O. kochii* and *O. elatior* had never been found growing together at the same site, but they sometimes occur together in Poland (e.g. DG-Błędów, DG-Łosy, DG-Gołonóg, Podskale, Jaworzno-Ciężkowice, Lędziny, Imielin/Mysłowice).

The new form described as *O. kochii* f. *citrina*, and found in the populations of typical individuals, within the same habitat conditions, was previously known from Poland only, but may possibly be identified in a wider geographical range. We still need to confirm the cream color forms described as *O. kochii* and presented in photographs taken in the Crimea (Tepe-Oba Mts near Teodosia and Mys Fiolent), without any certain information on a host plant (http://www.plantarium.ru). However, herbarium materials at LE from the Crimea (Mys Fiolent) include individuals of *O. ritro*, parasitizing *Echinops ritro* [20], with a characteristic creamy or less common yellow color of corolla [27,62]. Many species of the genus *Orobanche*, e.g. *O. alba*, *O. alsatica* s.l., *O. caryophyllacea*, *O. hederae*, *O. laxissima* group, are often lighter in color than the typical, commonly occur form, but these forms almost never constitute the majority of a population. Such pigmentation is not typical and sometimes may also result from low insolation, soil pH, hosts, or incomplete plant pigmentation. Yellow and albinotic forms usually occur in shaded sites [10–12,68,131–133], but sometimes also in places exposed to strong sunlight.

The role of seed features in the identification of species *Orobanche* to relate such characters to the systematics of the group, have been presented in several studies (e.g. [11,40,68,132,134,135]). The use of micromorphological characters, i.e. seed and pollen ornamentation (SEM) or a cross-section of the exine of dust seeds (TEM), have revealed additional differences between related species of *O. alsatica* agg. (*O. alsatica*, *O. bartlingii*, *O. mayeri*) [132]. Concerning the seeds of the investigated species, the best taxonomic character is the diameter and the shape of wall perforations. In *O. elatior* the perforation diameter is much larger, 4–13 μm, and is usually elliptical, while in *O. kochii* the diameter is smaller, 2–4 μm, and subcircular. The size and shape of seeds is less important for species identification. The position of a very large number of seeds in the ovary may also determine their shape as well as that of their cells [19,136]. Palynological analysis also revealed auxiliary micromorphological subtle differences in the size and level of sculpture of pollen grains between both species [137]. Differences in morphology, preliminary molecular studies, seed and pollen sculpture, and some ecological preferences confirm the separation of the examined taxa at species level.

Orobanche elatior s.l. is a partially protected species in Poland (regulation 2014). It is not a common species and is included on the "Red list of vascular plants in Poland" as rare – R [138] and vulnerable (VU) in the Carpathian Mts [79]. It is also included on regional red lists and books: Gdańsk Pomerania, CR, Western Pomerania, E [139,140]; Western Pomerania and Wielkopolska, E [139]; Wielkopolska, CR [141]; Sudetes Mts, CR [142]; Opole province, CR [56], endangered, EN, in the Lower Silesia province [143], Kujawy-Pomerania region, I (indeterminate) [144], Małopolska Upland, VU [145]. The species is also vulnerable in neighboring countries, i.e. Germany [146,147], the Czech Republic [148] and Slovakia [149]. These threat categories require revision as they concern *O. elatior* s.l., which contained the rare *O. elatior* and considerably more common *O. kochii*. Both species are included together only in the vascular plant red list of the Silesia province (*O. elatior* – category VU, vulnerable and *O. kochii* – NT, near threatened) [150], coauthored by us.

The progressing process of secondary succession of tree- and shrub vegetation, the density of the herb layer and the cessation of former cultivation methods, especially the loss of grazing, are the main threats to both species. Their sites are often found near arable fields where they are exposed to chemical agents. *Orobanche elatior* is especially vulnerable as it often prefers habitats in the ecotone of arable fields, and tetralogic and deformed forms affected by weed-killing agents are often observed. Similarly due to the impact of EU subsidies on cultivation profitability, *O. kochii* localities on old fallows are reploughed. Land development is a direct threat in many cases as the species' habitats are often in scenic landscapes that are demanded for housing estates, especially in the central part of the Silesia-Cracow Upland, where numerous localities of both *O. kochii* and *O. elatior* are now on the margin of one of the largest European conurbations. Some localities are on the edge of operating, expanding quarries. A part of the population of *O. kochii* on the dolomite quarry in the DG-Ząbkowice has disappeared in this way. The arrival of invasive or expansive species also poses a considerable threat. *Heracleum sosnowskyi* has been moving worryingly close to its population. Mass self-sowing of *Solidago canadensis*, *S. gigantea* and *Calamagrostis epigejos* is observed at many sites. Only a few localities are protected: *O. elatior* – Bielinek res. (not confirmed here for many years), Gipsowa Góra hill, Machnowska Góra hill, Polana Polichno; *O. kochii* – Zimne Wody, Żmudź, Murawy Dobromierskie, Biała Góra hill, Skarpa Dobużańska, Rogów, Opalonki, Przęślin, Skorocice, Krzyżanowice, Skotniki Górne; also protected as ecological sites or within Natura 2000 sites. This does not ensure the preservation of the species as active protection is needed. Many localities are at disturbed sites or at operating excavation sites (limestone, gypsum). At least some of the localities should be protected as ecological sites and active protection measures should be implemented:

mowing or grazing of excessively developing field- and herbaceous vegetation, while preserving the unique mosaic systems and fringe communities, and nature monitoring. The abundance of the host species *C. scabiosa* should also be monitored as this is usually neglected or overlooked in the protection of parasitic plants. Special attention should be paid to the sites were *O. kochii* and *O. elatior* co-occur or have been recorded for a very long time (Góra Gipsowa Mt 1841–2010, Góra Tuł hill 1858–2008, between Pasieczki and Dziećkowice 1863–2010), or localities usually indicating exceptionally well-preserved patches of xerothermic grasslands, including other rare *Orobanche* species (*O. lutea*, *O. bartlingii*, *O. picridis*, *O. caryophyllacea*, *O. alsatica*).

Acknowledgments

The author thanks the curators of herbaria and other persons who kindly made their data on the species available. We are also grateful for the assistance during field studies. Special thanks are due to Jiri Zázvorka for invaluable comments and discussion and to Beata Babczyńska-Sendek for our field excursions and for providing access to the University of Silesia *Orobanche* database. This work was supported by the Polish Ministry of Science [grant No. NN303357733 (2008–2009), NN303551939 (2010–2013)] and partially Principal Research Project 612419 (2014–2015).

Authors' contributions

The following declarations about authors' contributions to the research have been made: concept of the study: RP; herbarium revision: RP; micromorphology of seeds, taxonomy and phytosociological analysis: RP; writing the manuscript and field research: RP, ŁK.

Competing interests

No competing interests have been declared.

Supplementary material

The following supplementary material for this article is available online at http://pbsociety.org.pl/journals/index.php/asbp/rt/suppFiles/asbp.2014.031/0:

1. Appendix S1: plant communities with *Orobanche elatior* in Poland.
2. Appendix S2: plant communities with *Orobanche kochii* in Poland.

References

1. Uhlich H, Pusch J, Barthel KJ. Die Sommerwurzarten Europa: Gattung Orobanche. Magdeburg: Westarp-Wiss.; 1995.
2. Pusch J, Günther KF. Orobanchaceae (Sommerwurzgewächse). In: Hegi G, editor. Illustrierte Flora von Mitteleuropa. Lieferung 1. Jena: Weissdorn-Verlag; 2009. p. 1–99. (vol 6/1A).
3. Joel DM, Gressel J, Musselman LJ (eds.). Parasitic Orobanchaceae: parasitic mechanisms and control strategies. Springer: Heidelberg; 2013.
4. Zázvorka J. Orobanchaceae Vent. Zárazovité. In: Goliášová K, editor. Flóra Slovenska. Bratislava: VEDA; 1997. p. 460–529. [vol 5(2)].
5. Zázvorka J. Orobanchaceae – zarazovite. In: Slavik B, editor. Kvetena Česke Republiky. Praha: Academia; 2000. p. 477–513. (vol 6).
6. Piwowarczyk R, Przemyski A. New locality of *Orobanche coerulescens* Stephan ex Willd. (Orobanchaceae) at the NW limit of its geographical range. Acta Soc Bot Pol. 2009;78(4):291–295.
7. Piwowarczyk R, Przemyski A. The distribution and habitat preferences of the declining species *Orobanche arenaria* (Orobanchaceae) at the northern limit of its geographical range. Acta Soc Bot Pol. 2010;79(1):43–50.
8. Piwowarczyk R, Chmielewski P, Gierczyk B, Piwowarski B, Stachyra P. *Orobanche pallidiflora* Wimm. & Grab. in Poland: distribution, habitat and host preferences. Acta Soc Bot Pol. 2010; 79(3):197–205.
9. Piwowarczyk R, Chmielewski P, Cwener A. The distribution and habitat requirements of the genus *Orobanche* L. (Orobanchaceae) in SE Poland. Acta Soc Bot Pol. 2011;80(1):37–48.
10. Piwowarczyk R. *Orobanche mayeri* (Suess. & Ronniger) Bertsch & F. Bertsch – the new species to Poland. Acta Soc Bot Pol. 2011;80(3):179–183.
11. Piwowarczyk R. *Orobanche alba* subsp. *alba* and subsp. *major* (Orobanchaceae) in Poland: current distribution, taxonomy, plant communities, hosts, and seed micromorphology. Biodiv Res Conserv. 2012;26:23–38.
12. Piwowarczyk R. Revised distribution and plant communities of *Orobanche alsatica* and notes on the Orobanchaceae series *Alsaticae* in Poland. Biodiv Res Conserv. 2012;26:39–51.
13. Piwowarczyk R. The genus *Orobanche* L. (Orobanchaceae) in the Małopolska Upland (S Poland): distribution, habitat, host preferences and taxonomic problems. Biodiv Res Conserv. 2012;26:3–22.
14. Piwowarczyk R. A revision of distribution and the ecological description of *Orobanche picridis* (Orobanchaceae) at the NE limit of its geographical range from Poland and Ukraine. Acta Agrobot. 2012;65(1):91–106.
15. Piwowarczyk R. *Orobanche bohemica* Čelak. (Orobanchaceae) at the eastern limit of its geographical range: new data on its distribution in Poland. Biodiv Res Conserv. 2012;26:53–59.
16. Piwowarczyk R. Revised distribution and phytosociological data of *Orobanche coerulescens* Stephan in Willd. (Orobanchaceae): Poland in relation to Central Europe. Biodiv Res Conserv. 2012;26:61–72.
17. Piwowarczyk R. *Orobanche purpurea* (Orobanchaceae) in Poland: current distribution, taxonomy, plant communities, and preferred hosts. Biodiv Res Conserv. 2012;26:73–81.
18. Piwowarczyk R. A revision of distribution and historical analysis of preferred hosts of *Orobanche ramosa* (Orobanchaceae) in Poland. Acta Agrobot. 2012;65(1):53–62.
19. Piwowarczyk R. Seed productivity in relation to other shoot features for endangered parasitic plant *Orobanche picridis* F.W. Schultz (Orobanchaceae). Pol J Ecol. 2013;61(1):55–64.
20. Nobis M, Nowak A, Nobis A, Paszko B, Piwowarczyk R, Nowak S, et al. Contribution to the flora of Asian and European countries: new national and regional vascular plants records. Acta Bot Gal. 2014;160(1):81–89.
21. Zázvorka J. *Orobanche kochii* and *O. elatior* (Orobanchaceae) in central Europe. Acta Musei Moraviae, Scientiae biologicae (Brno). 2010;95(2):77–119.
22. Hess HE, Landolt E, Hirzelr R. Flora der Schweiz. Stuttgart: Birkhäuser Verlag, Basel; 1972. (vol 3).
23. Frajman B, Krajšek SS, Dakskobler I. *Orobanche kochii* F. W. Schultz and *Orobanche elatior* Sutton (Orobanchaceae) – new species for the flora of Slovenia. Hladnikia. 2011;27:57–65.
24. Meusel H, Jäger E, Rauschert S, Weinert E. Vergleichende Chorologie der Zentraleuropäischen Flora. Jena: Gustav Fischer Verlag; 1978. (vol 2).
25. Hultén E, Fries M. Atlas of North European vascular plants north of the tropic of cancer. Königstein: Koeltz Scientific Books; 1986. (vol 1–3).
26. Beck G. Monographie der Gattung *Orobanche*. Bibl. Bot. 19, Cassel: Theodor Fischer; 1890.

27. Beck G. Orobanchaceae. In: Engler A, editor. Das Pflanzenreich IV. (261). Leipzig: Verlag von Wilhelm Engelmann; 1930. p. 1–348.

28. Novopokrovskij IV, Tzvelev NN. Sem. CXLVI. Zarazichovye – Orobachaceae Lindl. In: Flora SSSR. Moskva & Leningrad: Izdatel´stvo Akademii Nauk SSSR; 1958. p. 685–687. (vol 23).

29. Tzvelev NN. Sem. 146. Orobanchaceae Vent. – zarazichovye. In: Flora evropejskoj chasti SSSR. Leningrad: Nauka; 1981. p. 317–336. (vol 5).

30. Zhiyun Z, Tzvelev NN. Orobanchaceae. In: Zheng-Yi W, Raven PH, editors. Flora of China. Saint Louis, MO: Science Press, Beijing, and Missouri Botanical Garden Press; 1998. p. 229–243. (vol 18).

31. Mirek Z, Musiał L, Wójcicki JJ. Polish herbaria. Polish Bot Stud Guideb. 1997;18:1–110.

32. Thiers B. Index Herbariorum: a global directory of public herbaria and associated staff. New York Botanical Garden's Virtual Herbarium [Internet]. 2014 [cited 2014 Feb 20].

33. Braun-Blanquet J. Pflanzensoziologie, Grundzüge der Vegetationskunde. Vienna: Springer Verlag; 1964. (vol. 3).

34. Mirek Z, Piękoś-Mirkowa H, Zając A, Zając M. Flowering plants and pteridophytes of Poland – a checklist. In: Mirek Z, editor. Biodiversity of Poland. Kraków: W. Szafer Institute of Botany, Polish Academy of Sciences; 2002. (Biodiversity of Poland; vol 1).

35. Matuszkiewicz W. Przewodnik do oznaczania zbiorowisk roślinnych Polski. Vademecum Geobotanicum. Warszawa: Wydawnictwo Naukowe PWN; 2006.

36. Zając A. Założenia metodyczne "Atlasu rozmieszczenia roślin naczyniowych w Polsce". Wiad Bot. 1978;22(3):145–155.

37. Barthlott W. Epidermal and seed surface characters of plants: systematic applicability and some evolutionary aspects. Nord J Bot. 1981;1:345–355.

38. Barthlott W. Microstructural features of seed surfaces. In: Heywood VH, Moree DM, editors. Current concepts in plant taxonomy. London: Academic Press; 1984. p. 95–105.

39. Stearn WT. Botanical Latin. London: David & Charles; 1992.

40. Plaza L, Fernández I, Juan R, Pastor J, Pujadas A. Micromorphological studies on seeds of *Orobanche* species from the Iberian Peninsula and the Balearic Islands, and their systematic significance. Ann Bot. 2004;94:167–178.

41. Bojnansky V, Fargasová A. Atlas of seeds and fruits of Central and East-European flora. The Carpathian Mountains region. Dordrecht: Springer; 2007.

42. Black M, Bewley JD, Halmer P. The encyclopedia of seeds, science, technology and uses. Wallingford: CAB International; 2008.

43. Łomnicki A. Wprowadzenie do statystyki dla przyrodników. Warszawa: Wydawnictwo Naukowe PWN; 2000.

44. Sutton C. A description of five species of *Orobanche*. Trans Linn Soc Lond. 1798;4:173–181.

45. Wimmer F. Flora von Schlesien preussischen und österreichischen Anteils. Breslau: Verlag von Ferdinand Hirt; 1841.

46. Holandre JJJ. Nouvelle Flore de la Moselle. 2 ed. 1842. (vol 2).

47. Schultz F. *Orobanche kochii*, eine neue deutsche Art. Flora. 1847;30(5):65–67.

48. Turland N, Rumsey F. Proposals to reject the names *Orobanche major* and *O. laevis* (Orobanchacaea). Taxon. 1997;46:787–791.

49. Foley MJY. *Orobanche* L. In: Paiva J, Sales F, Hedge IC, Aedo C, Aldasoro JJ, Castroviejo S, et al., editors. Flora Iberica. Madrid: Real Jardín Botánico; 2001. p. 32-72. (vol 14).

50. Halacsy E, Braun H. Nachträge zur Flora von Nieder-Österreich, Wien: in Komm. W. Braumüller; 1882.

51. Beck G. Orobancheae novae. Feddes Repert. 1922;30(1):33–40.

52. Maly J. Die Orobanchen der Grätzer Legend. Flora. 1842;1:113–127.

53. Čelakowský L. Prodromus der Flora von Böhmen. Prague: Arch. Naturwiss. Landesdurchforsch. Böhm.; 1871. (vol 2).

54. Fiek E. Flora von Schlesien preussischen und österreichischen Antheils, enhaltend die wildwachsenden, verwilderten und angebauten Phanerogamen und Gefäss-Cryptogamen. Breslau: J.U. Kern's Verlag; 1881.

55. Schube T. Verbreitung der gefässpflanzen in Schlesien preussischen und österreichischen Anteils. Breslau: R. Nischowsky; 1903.

56. Nowak A, Nowak S. Zaraza wielka, *Orobanche elatior* Sutton. In: Nowak A, Spałek K, editors. Czerwona księga roślin województwa opolskiego. Rośliny naczyniowe wymarłe, zagrożone i rzadkie. Opole: OTPN; 2002. p. 141.

57. Teryokhin ES, Schibakova GV, Serafimovich NB, Kravcova TI. Opredelitel' zarazichovych flory SSSR (s atlasom plodov i semjan). Determinator of broomrapes of the USSR flora (with atlas of fruits and seeds). Sankt-Peterburg: Nauka; 1993.

58. Frajman B, Carlón L, Kosachev P, Sánchez Pedraja O, Schneeweiss GM, Schönswetter P. Phylogenetic position and taxonomy of the enigmatic *Orobanche krylowii* (Orobanchaceae), a predominatly Asian species newly found in Albania (SE Europe). Phytotaxa. 2013;137(1):1–14.

59. Grenier JCM, Godron DA. Flore de France 2. Paris: J.B. Baillière; 1853.

60. Chater AO, Webb DA. *Orobanche*. In: Tutin TG, Heywood VH, Burges NA, Moore DM, Valentine DH, Walters SM, et al., editors. Flora Europaea 3, Cambridge: Cambridge University Press; 1972.

61. Pignatti S. Flora d'Italia 2. Bologna: Edagricole; 1982.

62. Pujadas-Salvà AJ. Typification and characterization of *Orobanche ritro* Gren. & Godr. (Orobanchaceae) of Central and Southern Europe. Acta Bot Gal. 2012;159(1):67–72.

63. Carlón L, Laínz M, Moreno Moral G, Sánchez Pedraja Ó. A new species (*Orobanche loscosii*), a prioritary name for *O. icterica* (*O. leptantha*) and a new member of the Spanish flora (*O. elatior*). Flora Montiberica. 2011;48:89–101.

64. Pujadas-Salvà AJ. *Orobanche icterica* Pau y *Orobanche ritro* Gren. & Godr. (Orobanchaceae) en la Flora Ibérica. Acta Botanica Malacitiana. 2013;38:160–162.

65. Pujadas-Salvá AJ, Plaza-Arregui L, Triano-Muñoz E, Algarra-Ávila JA. *Orobanche icterica* (Orobanchaceae) revisited. Bot J Linn Soc. 2005;148:117–124.

66. Rumsey FJ, Jury SL. An account of *Orobanche* L. in Britain and Ireland. Watsonia. 1991;8:257–295.

67. Rumsey FJ. A reconsideration of *Orobanche maritima* Pugsley (Orobanchaceae) and related taxa in southern England and the Channel Islands. Watsonia. 2007;26:473–476.

68. Halamski AT, Piwowarczyk R. Graines d´Orobanches comme critère taxonomique: information sur les travaux en cours (Résumé de la communication du 9 mai 2007). Bull Mens Soc Linn Lyon. 2008;77(3–4):37–40.

69. Schneeweiss GM. Correlated evolution of life history and host range in the nonphotosynthetic parasitic flowering plants *Orobanche* and *Phelipanche* (Orobanchaceae). J Evol Biol. 2006;20(2):471–478.

70. Zając A, Zając M, editors. Atlas rozmieszczenia roślin naczyniowych w Polsce. Nakładem Pracowni Chorologii Komputerowej Instytutu Botaniki Uniwersytetu Jagiellońskiego, Kraków; 2001.

71. Uechtritz R. Resultate der Durchforschung der schlesischen Phanerogammenflora im Jahre 1878. Jahres-Bericht Schles Ges vater Cult., Breslau. 1879;56:154–176.

72. Kotula B. Spis roślin naczyniowych z okolicy Przemyśla. Spraw Komisji Fizjogr AU. 1881;15:1–90.

73. Fijałkowski D. Flora roślin naczyniowych Lubelszczyzny. Lublin: Lubelskie Towarzystwo Naukowe; 1994.

74. Kucharczyk M. Distribution atlas of vascular plants in the Middle Vistula River Valley. Lublin: UMCS Press; 2001.

75. Urbisz A. Konspekt flory roślin naczyniowych Wyżyny Krakowsko-Częstochowskiej, Prace Naukowe Uniwersytetu Śląskiego. Katowice: Wydawnictwo Uniwersytetu Śląskiego; 2004.

76. Urbisz A. Atlas rozmieszczenia roślin naczyniowych na Wyżynie Krakowsko-Częstochowskiej. Katowice: Centrum Dziedzictwa Przyrody Górnego Śląska; 2012.

77. Babczyńska-Sendek B. Problemy fitogeograficzne i syntaksonomiczne kserotermów Wyżyny Śląskiej. Prace Naukowe Uniwersytetu Śląskiego, Katowice; 2005.

78. Piękoś-Mirkowa H, Mirek Z. Flora Polski. Rośliny chronione. Warszawa: Multico Oficyna Wydawnicza; 2006.

79. Bartoszek W, Piwowarczyk R. Zaraza wielka. *Orobanche elatior* Sutton. In: Mirek Z, Piękoś-Mirkowa H, editors. Czerwona księga Karpat polskich. Rośliny naczyniowe. Kraków: Instytut Botaniki im.W. Szafera PAN i Instytut Ochrony Przyrody PAN; 2008.

80. Nowak-Dańda A, Dańda P. Nowe stanowisko *Orobanche elatior* (Orobanchaceae) na Wyżynie Krakowsko-Częstochowskiej. Fragm Flor Geobot Pol. 2008;15(1):137–139.

81. Dubiel E, Gawroński S, Grzegorzek P. Atlas roślin chronionych, rzadkich i ginących gminy Chrzanów. Chrzanów: Urząd Miejski; 2000.

82. Schulz R. Eine floristische und geologische Betrachtung des märkischen unteren Odertales. Verh Bot Ver Prov Brandenburg. 1916;58:76–105.

83. Schulz R. Zweiter Beitrag zur Flora des märkischen unteren Odertales. Bot Ver Prov Brandenburg. 1919;61:82–96.

84. Celiński F, Filipek M. Flora i zespoły roślinne leśnostepowego rezerwatu w Bielinku nad Odrą. Badania fizjograficzne nad Polską Zachodnią. 1958;4:5–198.

85. Rätzel S, Zimmermann F. Verbreitung der Arten der Gattung *Orobanche* L. in Brandenburg und Berlin. Verh Bot Ver Berlin Brandenburg. 1999;132:19–101.

86. Smoczyk M. Rzadkie i zagrożone rośliny naczyniowe polskiej części Pogórza Orlickiego (Sudety Środkowe) - część 2. Przyr Sudetów. 2012;15:3–16.

87. Gołąb Z. Inwentaryzacja stanowisk chronionych, rzadkich i zagrożonych gatunków roślin i grzybów na terenie gminy Lewin Kłodzki [manuscript]. In: Jankowski W, editor. Inwentaryzacja botaniczna gminy Lewin Kłodzki. Wrocław: "Fulica" Jankowski Wojciech na zlecenie Wojewódzkiego Konserwatora Przyrody we Wrocławiu; 2002.

88. Smoczyk M, Jakubska A. Rozmieszczenie storczykowatych Orchidaceae w polskiej części Gór Orlickich i Pogórza Orlickiego. Przyr Sudetów. 2004;7:41–54.

89. Smoczyk M. Rzadkie i zagrożone rośliny naczyniowe Pogórza Orlickiego (Sudety Środkowe) - część 1. Przyr Sudetów. 2010;13:53–70.

90. Schube T. Ergebnisse der Durchforschung der schlesischen Gefässpflanzenwelt im Jahre 1904. Jahres-Bericht Schles Ges vater Cult, Breslau. 1905;82:41–64.

91. Wimmer F. Flora von Schlesien. Breslau: Ferdinand Hirt; 1869.

92. Gierczyk B, Soboń J. Nowe stanowiska chronionych, zagrożonych i rzadko spotykanych gatunków roślin naczyniowych w Polsce. Przegl Przyr. 2008;19:3(4):19–31.

93. Schube T. Ergebnisse der Durchforschung der schlesischen Gefässpflanzenwelt im Jahre 1910. Jahres-Bericht Schles Ges vater Cult, Breslau. 1911;88:88–105.

94. Kolbenheyer C. Vorarbeiten zur Flora von Teschen und Bielitz. Ver zool-bot Ges in Wien. 1862;12:1185–1220.

95. Oborny A. Flora von Mähren und österr. Schlesien enthaltend die wildwachsenden, verwilderten und häufi gangebauten Gefässpflanzen. Brünn: C. Winiker; 1885.

96. Formánek E. Kvetena Moravy a rakouského Slezka. Brno: E. Formánek; 1887.

97. Pelc S. Rośliny naczyniowe Pogórza Cieszyńskiego. Rocz Nauk Dydak WSP Krak. 1967;28:109–208.

98. Pacyna A. Rośliny naczyniowe wschodniej części Pogórza Wielickiego i przylegającej części Beskidów (Karpaty Zachodnie). Prace Bot. 2004;38:1–367.

99. Jakubowska-Gabara J, Kucharski L, Zielińska K, Kołodziejek J, Witosławski P, Popkiewicz P. Atlas rozmieszczenia roślin naczyniowych w Polsce Środkowej. Gatunki chronione, rzadkie i narażone. Łódź: Wydawnictwo UŁ; 2011.

100. Karo F. Flora okolic Częstochowy. Pam Fizjogr. 1881;1:208–257.

101. Hereźniak J. Stosunki geobotaniczno-leśne północnej części Wyżyny Śląsko-Krakowskiej na tle zróżnicowania i przemian środowiska. Mon Bot. 1993;75:1–368.

102. Babczyńska-Sendek B, Błońska A, Hejdysz J. Characteristics of the flora of fallow lands on rendzina soils on the Twardowice Plateau (Silesian Upland). Acta Agrobot. 2012;65(4):75–90.

103. Sendek A. Rośliny naczyniowe Górnośląskiego Okregu Przemysłowego. Warszawa: OTPN, PWN; 1984.

104. Celiński F, Rostański K, Sendek A, Wika S, Cabała S. Nowe stanowiska rzadkich roślin naczyniowych na Górnym Śląsku i terenach przyległych. Cz. IV. Opolskie Tow Przyj Nauk Zesz Przyr. 1978/1979;18:3–18.

105. Uechtritz R. Zur Flora Schlesiens. Öst Bot Z XIII. 1863;4:116–119.

106. Fiek E. Ergebnisse der Durchforschung der schlesischer Phanerogamenflora im Jahre 1891. Jahr-Ber Schles Ges vaterl Kultur. 1892;69:155–180.

107. Nobis M, Nobis A. Nowe dane dotyczące występowania storczyka purpurowego *Orchis purpurea* Huds. (Orchidaceae) na Wyżynie Małopolskiej. Chrońmy Przyr Ojcz. 2010;66(2):149–155.

108. Nobis M. Rośliny naczyniowe zachodniej części Przedgórza Iłżeckiego (Wyżyna Małopolska). Prace Bot. 2007; 40:1–458.

109. Piwowarczyk R. Rośliny naczyniowe wschodniej części Przedgórza Iłżeckiego (Wyżyna Małopolska). Prace Bot. 2010;43:1–344.

110. Kalinowski P. Rzadkie rośliny naczyniowe Podlasia Nadbużańskiego – cz. 1. Gatunki siedlisk murawowych, łąkowych i szuwarowych. Fragm Flor Geobot Pol. 2012;19(2):361–377.

111. Schmidt WLE. Flora von Pommern und Rügen. 1. Aufl. 1840. In der Becker und Altendorff'schen Buchhandlung, Stettin; 2. Aufl . 1848. Vermehrt und verbessert von Dr. Baumgardt. Stettin: Verlag von Ferdinand Müller; 1848.

112. Decker P. Beiträge zur Flora der südlichen Neumark und der östlichen Niederlausitz. Verh Bot Ver Provinz Brandenburg. 1911;53:87–269.

113. Ascherson P, Graebner P. Flora des nordostdeutschen Flachlandes (ausser Ostpreussen). Berlin: Verlag von Gebrüder Borntraeger; 1898/99.

114. Ritschl G. Flora des Grossherzogthums Posen. Berlin; 1850.

115. Schube T. Ergebnisse der Durchforschung der schlesischen Gefässpflanzenwelt im Jahre 1928. Jahres-Bericht Schles Ges vater Cult, Breslau. 1929;101:88–96.

116. Schube T. Die Ergebnisse der Durchforschung der schlesischen Gefässpflanzenwelt im Jahre 1908. Jahres-Bericht der Schles Gesellsch vaterl Cultur. 1909;86:48–67.

117. Scholz JB. Die Pflanzengenossenschaften Westpreussens. Schriften der naturforschenden gesellschaft in danzig. Danzig: Kommissions Verlag von Wilhelm Engelman in Leipzig; 1905.

118. Abromeit J, Neuhoff W, Steffen H. Flora von Ost- und Westpreussen, 1898–1940, 1/1–25 (1898): 1–402, 2/26–43 (1903): 403–684, 3/44–49 (1926): 685–780. Berlin: Kommissionsverlag Gräfe und Unzer; 1940.

119. Klinsmann EF. Beiträge zur Flora von Danzig nebst einigen Mittheilungen aus andern Preußischen Gegenden in Preussische Provinzial-Blätter. Königsberg: D.W.L. Richter; 1836. (vol 15).

120. Wiess G. Die um Danzig wildwachsenden Pflanzen nach ihrem Geschlechtstheilen geordnet und beschrieben von Gottfried Reyger. Danzig: Anhuthsche Buchhandlung; 1825.

121. Preuss H. Die pontischen Pflanzenbestände im Weichselgebiet vom sandpunkt der Naturdenkmalpflege aus geschildert. Beiträge zur Naturdenkmalpflege. 1912;2:350–540.

122. Mądalski J. Atlas flory polskiej i ziem ościennych (Florae Polonica terrarumque adiacentium iconographia), Scrophulariaceae (Pars 4), Orobanchaceae, 17(1), Warszawa: Państwowe Wydawnictwo Naukowe PWN; 1973.

123. Preuss H. Das Herbarium Klinsmann unter besonderer Berücksichtigung der Danziger Adventivflora. Ber Westpreuss Bot-Zool Ver. 1928;50:201–230.

124. Babczyńska-Sendek B, Kloczkowska A, Błońska A. Xerothermic grasslands of Pilica surroundings – diversity, threats and directions of changes. Biodivers Res Conserv. 2011;23(1):59–70.

125. Raciborski M. Zmiany zaszłe we florze okolic Krakowa w ciągu ostatnich lat dwudziestu pięciu pod względem roślin dziko rosnących. Sprawozdania Komisji Fizjograficznej. 1884;18:99–126.

126. Polakowski B. Zespoły łąkowo-pastwiskowe Pojezierza Mazurskiego w świetle dotychczasowych badań. Zesz Nauk ART Olsztyn. 1979;28:113–121.

127. Rostafiński J. Florae Polonicae Prodromus, Uebersicht der bis jetzt im Königreiche Polen beobachteten. Phanerogamen. Verh Zool-Bot Ges. 1872; 81-202.

128. Knapp JA. Die bisher bekannten Pflanzen Galiziens und der Bukowina. Wien: Braumüller Verl.; 1872.

129. Herbich F. Botanischer Ausflug in die galizisch-karpatischen Alpen des Sandezer Kreises. Flora. 1834;17:561–587.

130. Knapp JA. Przyczynek do flory obwodów jasielskiego i sanockiego. Spraw Kom Fizjogr CK Tow Nauk Krak. 1869;3:74–109.

131. Halamski AT. *Orobanche hederae* Vaucher ex Duby (Orobanchaceae) – plasticité phénotypique et micro-morphologie des graines. Bull Mens Soc Linn Lyon. 2011;80(7-8):170–178.

132. Piwowarczyk R, Halamski AT, Durska E. Seed and pollen morphology in the *Orobanche alsatica* complex (Orobanchaceae) from central Europe and its taxonomic significance. Aust Syst Bot. 2014; 27(2):145–157.

133. Piwowarczyk R, Tatanov I. *Orobanche laxissima* Uhlich & Rätzel (Orobanchaceae) – a new species for Dagestan (Russia) and Azerbaijan. Biodivers Res Conserv. 2013;32:25–28.

134. Musselman LJ, Mann WF. A survey of surface characteristics of seeds of Scrophulariaceae and Orobanchaceae using scanning electron microscopy. Phytomorphology. 1976;26:370–378.

135. Abu Sbaih HA, Jury SL. Seed micromorphology and taxonomy in *Orobanche* (Orobanchaceae). Flora Mediterranea. 1994;4:41–48.

136. Piwowarczyk R, Jankowska-Blaszczuk M. Intra-specific diversity of seed productivity and morphological features in parasitic species *Orobanche bartlingii* Griseb. (Orobanchaceae). Pol J Ecol. 2014;62;723–738.

137. Piwowarczyk R, Madeja J, Nobis M. Pollen morphology of the Central European broomrapes (Orobanchaceae: *Orobanche*, *Phelipanche* and *Orobanchella*) and its taxonomical implications. Plant Syst Evol. 2015;301;795–808.

138. Zarzycki K, Szeląg Z. Red list of the vascular plants in Poland. In: Mirek Z, Zarzycki K, Wojewoda W, Szeląg Z., editors. Red list of plants and fungi in Poland. Kraków: W. Szafer Institute of Botany, Polish Academy of Sciences; 2006.

139. Żukowski W, Jackowiak B. List of endangered and threatened vascular plants in Western Pomerania and Wielkopolska (Great Poland). In: Żukowski W, Jackowiak B, editors. Endangered and threatened vascular plants of Western Pomerania and Wielkopolska. Poznań: Publications of the Departament of Plant Taxonomy of Adam Mickiewicz University in Poznań; 1995. (vol 3).

140. Markowski R, Buliński M. Ginące i zagrożone rośliny naczyniowe Pomorza Gdańskiego. Acta Bot Cassub Monogr. 2004;1:1–75.

141. Jackowiak B, Celka Z, Chmiel J, Latowski K, Żukowski W. Red list of vascular flora of Wielkopolska (Poland). Biodivers Res Conserv. 2007;15:95–127.

142. Fabiszewski J, Kwiatkowski P. Threatened vascular plants of the Sudeten Mountains. Acta Soc Bot Pol. 2002;71(4):339–350.

143. Kącki Z, Dajdok Z, Szczęśniak E. Czerwona lista roślin naczyniowych Dolnego Śląska. In: Kącki Z, editor. Zagrożone gatunki flory naczyniowej Dolnego Śląska. Wrocław: Instytut Biologii Roślin, Uniwersytet Wrocławski, Polskie Towarzystwo Przyjaciół Przyrody "Pro Natura"; 2003. p. 9–65.

144. Rutkowski L. Rośliny naczyniowe – Tracheophyta. In: Buszko J, Kasprzyk K, Pawlikowski T, Przystański A, Rutkowski L, editors. Czerwona lista roślin i zwierząt ginących i zagrożonych w regionie kujawsko-pomorskim. Acta Univ Nicolai Copernici. Biologia LIII, Suplement-Nauki Mat-Przyr Toruń. 1997;98:5–20.

145. Bróż E, Przemyski A. The red list of vascular plants in the Wyżyna Małopolska upland (S Poland). In: Mirek Z, Nikel A, editors. Rare, relict and endangered plants and fungi in Poland. Kraków: W. Szafer Intitute of Botany, Polish Academy of Sciences; 2009.

146. Korneck D, Schnittler M, Vollmer I. Rote liste der Farn- und Blütenpflanzen (Pteridophyta & Spermatophyta) Deutschlands. Schrift Vegetationsk. 1996;28:21–187.

147. Ludwig G, May R, Otto C. Verantwortlichkeit Deutschlands für die weltweite Erhaltung der Farn- und Blütenpflanzen – vorläufige Liste. BfN-Skripten 220; 2007.

148. Procházka F, editor. Červený a černý seznam cévnatých rostlin České republiky (stav v roce 2000). Praha: Příroda; 2001.

149. Feráková V, Maglocky Š, Marhold K. Červený zoznam papraďorastov a semenných rastlín Slovenska. In: Baláz D, Marhold K, Urban P, editors. Cerveny zoznam rastlín a zivocíchov Slovenska. Ochr Prír. 2001;20:44–76.

150. Parusel JB, Urbisz A, editors. Czerwona lista roślin naczyniowych województwa śląskiego. Raporty, Opinie. Katowice: CDPGŚ; 2012. (vol 6).

27

De novo sequencing and comparative transcriptome analysis of white petals and red labella in *Phalaenopsis* for discovery of genes related to flower color and floral differentation

Yuxia Yang[1], Jingjing Wang[1,2], Zhihu Ma[3], Guosheng Sun[3], Changwei Zhang[1]*

[1] College of Horticulture, Nanjing Agricultural University, 1# Weigang, Nanjing, Jiangsu Province, 210095, P.R. China

[2] Dongxin Department, Jiansu Provincial Agricultural Reclamation and Development Corporation, 26# Dongfang Middle Road, Lianyungang, Jiangsu Province, 222248, P.R. China

[3] Zhenjiang Institute of Agricultural Sciences in Hilly Area of Jiangsu Province, 112# Ninghang Road, Jurong, Jiangsu, 212400, P.R. China

Abstract

Phalaenopsis is one of the world's most popular and important epiphytic monopodial orchids. The extraordinary floral diversity of *Phalaenopsis* is a reflection of its evolutionary success. As a consequence of this diversity, and of the complexity of flower color development in *Phalaenopsis*, this species is a valuable research material for developmental biology studies. Nevertheless, research on the molecular mechanisms underlying flower color and floral organ formation in *Phalaenopsis* is still in the early phases. In this study, we generated large amounts of data from *Phalaenopsis* flowers by combining Illumina sequencing with differentially expressed gene (DEG) analysis. We obtained 37 723 and 34 020 unigenes from petals and labella, respectively. A total of 2736 DEGs were identified, and the functions of many DEGs were annotated by BLAST-searching against several public databases. We mapped 837 up-regulated DEGs (432 from petals and 405 from labella) to 102 Kyoto Encyclopedia of Genes and Genomes pathways. Almost all pathways were represented in both petals (102 pathways) and labella (99 pathways). DEGs involved in energy metabolism were significantly differentially distributed between labella and petals, and various DEGs related to flower color and floral differentiation were found in the two organs. Interestingly, we also identified genes encoding several key enzymes involved in carotenoid synthesis. These genes were differentially expressed between petals and labella, suggesting that carotenoids may influence *Phalaenopsis* flower color. We thus conclude that a combination of anthocyanins and/or carotenoids determine flower color formation in *Phalaenopsis*. These results broaden our understanding of the mechanisms controlling flower color and floral organ differentiation in *Phalaenopsis* and other orchids.

Keywords: *Phalaenopsis*; RNA-seq; transcriptome; flower color; diversity

Introduction

Phalaenopsis amabilis, an epiphytic monopodial orchid, is an important ornamental species with a huge global market, especially in western countries. Ornamental quality in *Phalaenopsis* is influenced by many factors, including flower color, fragrance and shape, cut-flower longevity, flowering control, and abiotic stress tolerance [1]. Among these, flower color and diversity are the two main factors visually impacting ornamental and commercial values of *Phalaenopsis*.

Phalaenopsis flowers are zygomorphic and include three outer tepals (T1–T3, also known as sepals) in the first whorl, two lateral inner tepals (t1 and t2, the petals), and a median inner tepal (t3, the lip or labella) [2,3]. Floral initiation and development are regulated by complex networks of exogenous and endogenous signals, including light stimuli, hormones, ligand–receptor interactions, signal transduction pathways, and transcription factor cascades [4]. A recent theory known as "the orchid code" proposes a reasonable model in which four different class B *AP3/DEF*-like MADS-box genes have played a vital role in the evolution of the orchid perianth by their combinatorial interaction. Orchid floral diversity is an evolutionary consequence of two duplication events and associated changes that occurred in the regulatory regions of the class B *AP3/DEF*-like MADS-box genes, which was followed by sub- and neo-functionalization. In orchids, class B *AP3/DEF*-like MADS-box genes are divided into four distinct clades: *PeMADS2*-like (clade 1), *OMADS3*-like (clade 2), *PeMADS3*-like (clade 3), and *PeMADS4*-like (clade 4), each having its own specific expression pattern. The combined expressions of clade 1 and clade 2 genes mediate the development of the three outer tepals, while the combination of clade 1, clade 2, and clade 3 genes leads to the development of the lateral inner tepals. Labella development is determined by a combination of genes

* Corresponding author. Email: changweizh@njau.edu.cn

Handling Editor: Przemysław Wojtaszek

from all four clades. Differential expression of clade 3 genes is obviously responsible for differences between inner and outer tepals, whereas differential expression of clade 4 genes differentiates the lateral inner tepals from the labella [2,3]. Orchid *PI/GLO*-like genes, found to be present in the *AP3/DEF*-like gene copies, are also necessary for current floral tissue development [4]. Despite this knowledge, however, the regulatory network controlling orchid floral development remains unclear.

Flower color is derived from the three major classes of plant pigments: anthocyanins, betalains, and carotenoids [5]. Of these, anthocyanins are the major contributors to flower color [6]. A class of water-soluble flavonoids, anthocyanins are synthesized in the cytosol and localized in vacuoles. Through the phenylpropanoid pathway, they provide a wide range of colors, ranging from orange-red to violet-blue in dark-colored flowers [5]. Despite their structural variety, anthocyanins are only categorized into six chromophore forms: pelargonidin, cyanidin, peonidin, delphinidin, petunidin, and malvidin [7,8]. The anthocyanin biosynthetic pathway has been well elaborated [9].

In orchids, the primary anthocyanin in red flowers is a cyanidin derivative that is typically modified by glycosylation and acylation [10]. The glycosylation-related gene *PeU-FGT3* plays a critical role in red color formation in *Phalaenopsis* [8]. Several important enzymes, such as chalcone synthase (CHS), chalcone isomerase (CHI), dihydroflavonol 4-reductase (DFR), and anthocyanidin synthase (ANS), are involved in the formation of colored anthocyanidins [5].

Compared with studies in other flowering plants, the molecular basis of floral color development has not been well characterized in orchids. A better understanding of the molecular mechanisms underlying orchid flower color and floral organ formation is thus needed. Furthermore, few transcriptomic-based investigations of the functions of genes related to flower color and floral differentiation have been reported for *Phalaenopsis*. To expand knowledge regarding flower color and diversity for *Phalaenopsis* breeding, in this study we analyzed differential gene expression between petals and labella using the Illumina RNA-Seq method. Our study generated a huge number of *Phalaenopsis* transcript sequences during floral formation that can be used to discover putative genes related to flower color and floral differentiation. By comparing relative gene expression levels between petals and labella, novel insights can be gleaned into orchid floral development. Our study therefore provides a foundation for future research on mechanisms underlying floral development in *Phalaenopsis* and other orchids.

Material and methods

Plant material and sample collection

Phalaenopsis plants with white petals and red labella (Fig. 1) were grown in greenhouses at Nanjing Agriculture University under natural light conditions and a controlled temperature of 22–27°C. Petals and labella were collected at the full-bloom stage. The two samples were immersed in liquid nitrogen and stored at −80°C until subjected to total RNA extraction.

Total RNA extraction, cDNA library construction, and Illumina deep sequencing

Total RNA was extracted from petal and labella samples using a Trizol kit (Takara, Japan). Total RNA quality and quantity were analyzed using a Nanodrop 2000 instrument (Thermo Scientific) and a ChipRNA 7500 Series II Bioanalyzer (Agilent). The two total RNA samples were delivered to Beijing Biomarker Biotechnology Co. (Beijing, China) for the construction of cDNA libraries using an mRNA-Seq Sample Preparation kit (Illumina) according to the manufacturer's instructions. The sequencing of the two samples was performed on an Illumina HiSeq 2000 system.

Sequence assembly and annotation

The raw image data produced from sequencing was transformed by base calling into raw reads. Transcriptome de novo assembly was carried out with the Trinity short-read assembly program, which generated in turn contigs, transcripts, and unigenes. To identify unigene putative functions, their sequences were aligned using BLASTX (*E*-value $\leq 10^{-5}$) against the following public protein databases: National Center for Biotechnology Information non-redundant (Nr) and nucleotide (Nt) databases and SwissProt, TrEMBL, Clusters of Orthologous Groups (COG), Gene Ontology (GO), and Kyoto Encyclopedia of Genes and Genomes (KEGG) databases. The Blast2GO software package was used to compare and determine unigene GO annotations. Finally, WEGO software was used to obtain GO functional classifications for all annotated unigenes.

Identification of differentially expressed genes (DEGs)

To identify DEGs between the two samples, the following formula was used to calculate significance (*P*) of differences in transcript accumulation for each gene:

$$P(y/x) = \left(\frac{N2}{N1}\right)^y \frac{(x+y)!}{x!\,(y)!\,\left(1+\frac{N2}{N1}\right)^{(x+y+1)}}$$

where *N1* and *N2* represent the total number of clean reads from petals and labella, respectively, and *x* and *y* represent the number of reads mapping to the given gene. We then

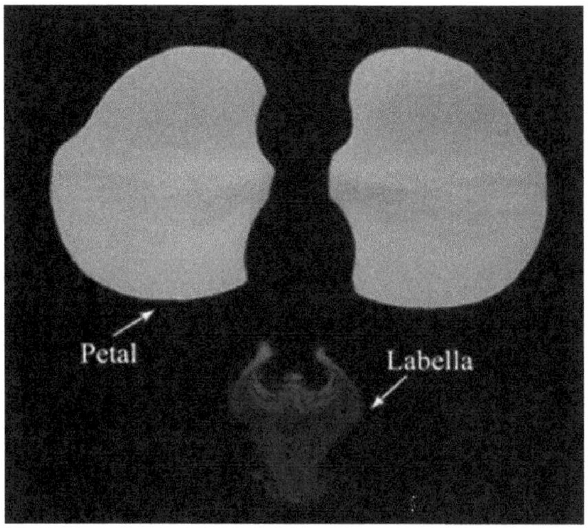

Fig. 1 Photographic image of selected flower materials

used false discovery rate (FDR) to determine the threshold of the *P*-value in multiple tests. To identify DEGs, we used a combination of two criteria: expression fold-change $|\log_2 \text{Ratio}| \geq 1$ and FDR-adjusted *P*-value < 0.001.

We performed GO and KEGG functional enrichment analysis to determine which DEGs were significantly enriched in GO terms and metabolic pathways ($P \leq 0.05$) compared with the selected transcriptome background. Significance was calculated according to the formula:

$$P = 1 - \sum_{i=0}^{m-1} \frac{\binom{M}{i}\binom{N-M}{n-i}}{\binom{N}{N}}$$

where *N* is the number of genes with GO/KEGG annotations, *n* is the number of DEGs in the set of *N* genes, *M* is the number of genes mapped to a certain GO/KEGG term, and m is the number of DEGs in the set of *M* genes.

Results

Short-read de novo sequencing and assembly

We pooled equal amounts of RNA from the two samples (petals and labella) to construct cDNA libraries for transcriptome sequencing and analysis on a Genome Analyzer IIx platform using Illumina sequencing technology. Using this sequencing approach, raw reads were generated from both ends of the cDNA fragments. After data filtering using stringent quality criteria (i.e., removal of sequences shorter than 65 bp or with CycleQ20 values less than 100%), we obtained 10 734 813 clean reads comprising 2 134 924 797 nucleotides (nt) from the petal library and 16 224 038 clean reads comprising 3 276 861 889 nt from the labella library. The clean reads were assembled de novo into contigs using the Trinity software package. The results of the sequencing assembly are presented in Tab. 1. A total of 1 057 077 and 2 105 819 contigs were assembled from petals and labella, respectively, with corresponding mean lengths of 75 bp and 67 bp and N50s of 99 bp and 65 bp. Of these, 16 999 (1.61%) contigs from petals and 17 334 (0.82%) contigs from labella were longer than 500 bp. The contigs from petals and labella were further assembled into 55 101 and 67 026 scaffolds with mean lengths of 811 bp and 948 bp, respectively. Scaffold N50 lengths were 1280 bp for petals and 1168 bp for labella. In total, 15 207 scaffolds assembled from petals and 24 853 scaffolds from labella coded for transcripts longer than 1 kb. Finally, 37 723 unigenes were obtained from the petals, with a final unigene N50 length of 1 125 bp and a total length of 26 656 163 nt; similarly, 34 020 unigenes were generated from the labella data, with a final unigene N50 length of 1398 bp and a total length of 27 636 110 nt. These unigenes were then organized into a transcriptome database for the identification of putative genes related to flower color and floral differentiation.

Tab. 1 Summary of sequencing and assembly results.

Assembled data	Statistics of data production	White petals	Red labella
Reads			
	No. of reads	10 734 813	16 224 038
	Total nucleotides (nt)	2 134 924 797	3 276 861 889
	GC percentage (%)	46.80	45.08
	Q20 percentage (%)	100.00	100.00
Contigs			
	No. of contigs	1 057 077	2 105 819
	Total nucleotides (nt) in contigs	79 618 808	141 300 083
	Length of N50 (bp)	99	65
	Mean length of contigs (bp)	75	67
	No. of contigs above 500 bp	16 999	17 334
Scaffolds			
	No. of scaffolds	55 101	67 026
	Total nucleotides (nt) in scaffolds	44 692 377	63 535 969
	Length of N50 (bp)	1280	1443
	Mean length of scaffolds (bp)	811	948
	No. of scaffolds above 500 bp	23 777	40 620
Unigenes			
	No. of unigenes	37 723	34 020
	Total nucleotides (nt) in unigenes	26 656 163	27 636 110
	Length of N50 (bp)	1125	1398
	Average length of unigenes (bp)	707	812
	No. of unigenes above 500 bp	15 903	15 984

DEG analysis between white petal and labella libraries

A primary goal of transcriptome sequencing is comparison of gene expression levels in different samples. In this study, a large number of DEGs were estimated by comparing gene expression between the petal and labella libraries. Unigene expression was calculated using the RPKM method [11]. DEGs detected with at least two-fold differences (FDR < 0.001 and $|\log_2 \text{Ratio}| \geq 1$) between petal and labella libraries are shown in Fig. 2. Using these criteria, we identified 2736 DEGs between petals and labella. Of these, 1277 were up-regulated and 1459 were down-regulated in petals. A greater number of genes were expressed only in the petals (243) than in the labella (62). This result suggests that petal and labella development may involve totally different processes.

DEG functional annotation

To validate and annotate the assembled DEGs, the 2736 DEGs were subjected to BLASTX comparisons (*E*-value ≤ 1×10^{-5}) against several public protein databases to identify putative functions of the unigene sequences. As a result, 2698, 2183, 2319, 2706, 2446, 837, and 1077 DEGs were found to have homologous sequences in Nr, Nt, SwissProt, TrEMBL, GO, KEGG, and COG databases, respectively.

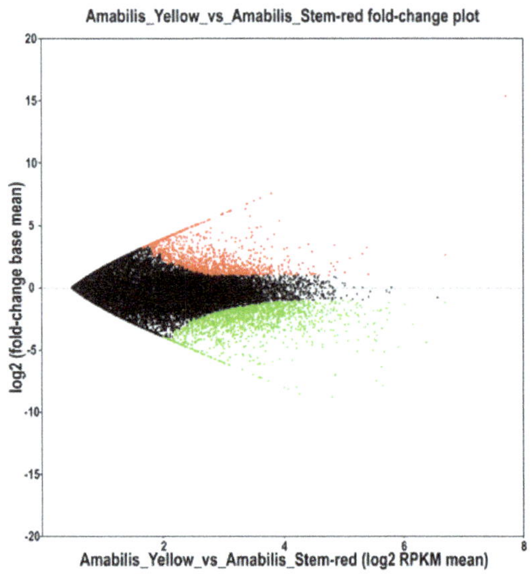

Fig. 2 Scatter plot of relative gene expression in petals vs. labella. Red dots above zero on the y-axis indicate genes having a higher expression level in petals, green dots below zero correspond to genes having a higher expression level in labella, and black dots represent genes that were similar in both libraries. False discovery rate < 0.001 and $|\log_2 \text{Ratio}| \geq 1$ were used as thresholds to judge the significance of gene expression differences.

COG ANALYSIS OF DEG PATTERNS. COG analysis is used to classify orthologous gene products, with all proteins in an orthologous gene cluster considered to have evolved from the same ancestral protein. The COG database, which contains putative protein sequences encoded by all genes from bacteria, archaea, and eukaryotes that have complete genome sequences, facilitates exploration via sequence analysis of evolutionary relationships among these groups [12].

The newly identified sequences were aligned against the known COG database to predict and classify their possible functions. In total, 1077 unigenes were assigned to at least one COG classification. Among the 25 COG categories (Fig. 3), the cluster for "General function prediction only" represented the largest group (262 unigenes; 24.3%), followed by "Posttranslational modification, protein turnover, chaperones" (146; 13.6%), "Transcription" (137; 12.7%) and "Translation, ribosomal structure and biogenesis" (118; 10.9%). The categories of "Cell motility" (3; 0.28%), "Nuclear structure" (1; 0.09%) and "Extracellular structures" (0; 0%) contained the smallest numbers of unigenes. This information should prove to be a valuable resource for studying specific processes and functions in *Phalaenopsis*.

GO FUNCTIONAL ENRICHMENT. GO is an international standardized gene functional classification system that describes properties of genes and their products in any organism [13]. Based on sequence homology, 2446 DEGs were categorized into three main categories (biological process, cellular component, and molecular function) and 62 subcategories (Fig. 4). Multiple terms were frequently assigned to the same transcript; thus, 15 159 GO term annotations were assigned to at least one GO category under biological

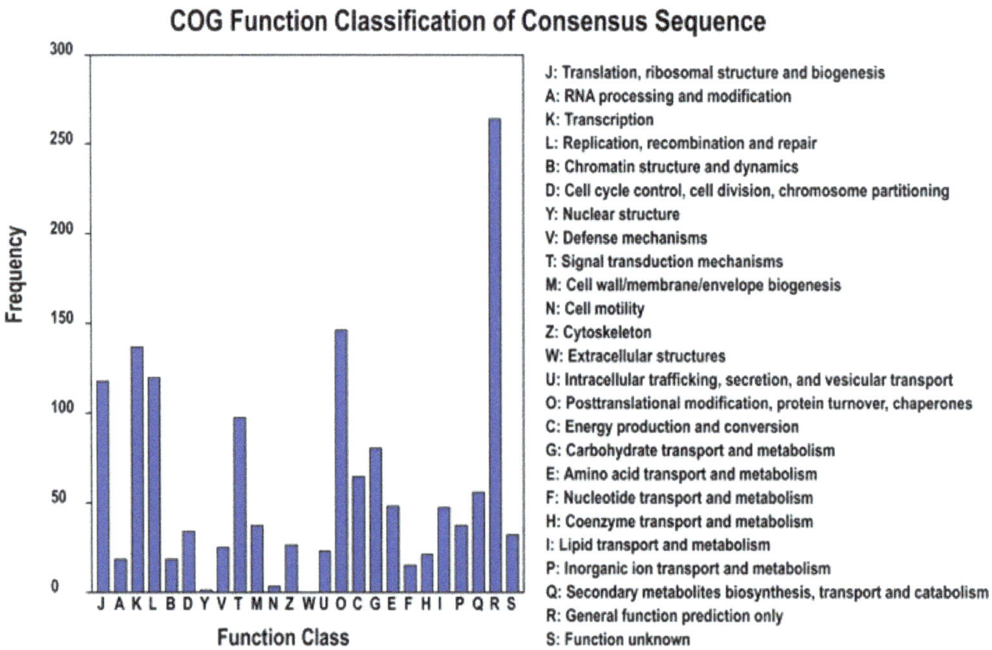

Fig. 3 Clusters of orthologous genes (COG) annotations of putative proteins. All putative proteins were aligned to the COG database and functionally classified into one or more of 25 molecular families. The capital letters on the x-axis correspond to the COG categories listed to the right of the histogram, and the y-axis indicates the numbers of DEGs assigned to the corresponding COG category.

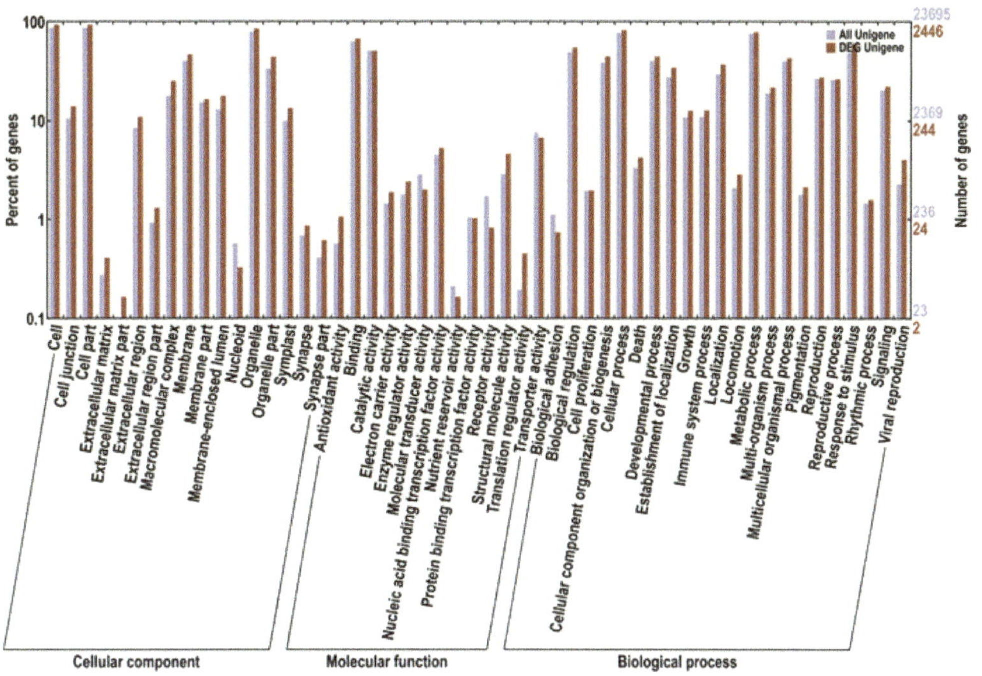

Fig. 4 Gene ontology (GO) classification of all annotated unigenes and DEGs. GO classification according to biological process, cellular component, and molecular function categories of all annotated 23 695 unigenes and 2446 differentially expressed unigenes. Subcategories are indicated on the x-axis. The right y-axis indicates the number of genes in each category, while the left y-axis represents the percentage of genes within each GO category falling into a specific subcategory. Gray bars correspond to all annotated unigenes and red bars indicate differentially expressed unigenes (DEGs).

processes, 11 346 under cellular components, and 3552 under molecular function. Within the molecular function category, the majority of DEGs were related to binding (1655; 67.6%). With respect to cellular components, most assignments were to cell (2283; 93.2%) and cell part (2283; 93.2%) subcategories. Among biological processes, the cellular subcategory (2016; 82.3%) was the most highly represented. These results suggest that the flowering stage in *Phalaenopsis* is predominantly controlled by genes related to cellular structure and molecular interactions. In contrast, almost no DEGs were assigned to virion (2; 0.8%) and virion part (2; 0.8%) under cellular components, cell killing (1; 0.4%) under biological processes, and chemoattractant activity (1; 0.4%) and receptor regulator activity (1; 0.4%) under molecular function.

KEGG PATHWAY ENRICHMENT ANALYSIS. Pathway enrichment analysis identifies significantly enriched metabolic pathways or signal transduction pathways in DEGs by comparing them with the whole genome background. Using pathway enrichment analysis, DEG-associated metabolic and signal transduction pathways can be identified. On the basis on sequence homology, 837 up-regulated DEGs (432 in petals and 405 in labella) were mapped to 102 KEGG pathways. This result suggests that the development and formation of petals and labella involve two totally different processes. Almost all KEGG pathways were found in the two floral parts: 102 pathways in petals and 99 in labella (Tab. 2).

Among DEGs, 441 could be mapped to 10 secondary metabolic sub-pathways. We identified many gene families that encode key enzymes involved in major secondary metabolic pathways, such as those associated with betalain, flavone and flavonol, phenylpropanoids, flavonoids, terpenoids, and carotenoids. A total of 2822 genes involved in the 10 sub-pathways were detected in our study. The broad coverage obtained for these secondary metabolic genes provides more abundant information for examining secondary metabolite biosynthesis in *Phalaenopsis*.

Expression levels of these genes exhibited great differences. The 432 DEGs up-regulated in petals mapped to 102 KEGG pathways distributed into the categories of metabolism (292 unigenes), genetic information processing (122 unigenes), organism systems (12 unigenes), cellular processes (4 unigenes),environmental information processing (9 unigenes), and human diseases (2 unigenes). The distribution of the 405 DEGs upregulated in labella into 94 KEGG pathways was as follows: genetic information processing (201 unigenes), metabolism (146 unigenes), cellular processes (38 unigenes), environmental information processing (19 unigenes), organism systems (15 unigenes), and human diseases (50 unigenes). The broad distribution of both petal and labella unigenes in these essential processes suggests that the associated biochemical pathway networks are complicated.

Tab. 2 Some unigenes mapped to Kyoto Encyclopedia of Genes and Genomes (KEGG) pathways.

KEGG pathways	Sub-pathways of KEGG pathway	Branches of sub-pathway	PN	PNU	PL	PLU
Metabolism	Energy metabolism	Photosynthesis-antenna proteins	11	0	11	11
		Photosynthesis	12	2	12	10
		Carbon fixation in photosynthetic organisms	14	0	14	14
		Oxidative phosphorylation	32	3	32	29
	Biosynthesis of other secondary metabolites	Tropane, piperidine and pyridine alkaloid biosynthesis	1	0	1	1
		Flavonoid biosynthesis	3	2	6	4
		Flavone and flavonol biosynthesis	1	0	2	2
		Stilbenoid, diarylheptanoid and gingerol biosynthesis	2	1	2	1
		Betalain biosynthesis	1	1	0	0
		Isoquinoline alkaloid biosynthesis	2	1	1	1
	Lipid metabolism	Glycerolipid metabolism	3	1	2	2
		Glycerophospholipid metabolism	4	1	3	3

PN – number of mapped unigenes from petals; PNU – number of mapped up-regulated unigenes from petals; PL – number of mapped unigenes from labella; PLU – number of mapped up-regulated unigenes from labella.

With respect to metabolic processes (Tab. 2), the distributions of petal and labella unigenes involved in energy metabolism were significantly different. As 67 unigenes from labella were participants in this pathway compared with 10 unigenes from petals, more DEGs in labella were accordingly related to this KEGG sub-pathway. These results suggest that the labella is associated with a more complex energy metabolism mechanism than are petals; the labella may thus not only be an energy sink, but may also serve as an energy source for its own function or supply energy to other floral organs.

Genes related to floral differentiation and color

Transcription factors control genomic expression and play vital roles in all aspects of higher plant life cycles. The distribution of transcription factors associated with floral differentiation – including MADS (16 DEGs), MYB, ARF, NAC, and TCP – was similar between petals and labella, but their expression levels were significantly different (Tab. 3).

Orchids produce flowers of a variety of colors ranging from greenish, white, and yellow to intense red, blue, and brown. Several anthocyanin biosynthetic DEGs, such as *CHI*, *CHS*, UDP-glucose: flavonoid-3-O-glucosyltransferase (*UFGT3*), *DFR*, and *ANS*, were uncovered in this study (Tab. 4).

Discussion

In this study, we obtained 37 723 and 34 020 unigenes from petals and labella, respectively. A total of 2736 DEGs were identified, many of which were functionally annotated by BLAST searches against several public databases. We mapped 837 DEGs (432 up-regulated in petals and 405 up-regulated in labella) to 102 KEGG pathways. Almost all of these pathways were found in both flower parts: 102 pathways in petals and 99 in labella. DEGs involved in energy metabolism were significantly differentially distributed between labella and petals. In addition, several DEGs related to flower color and floral differentiation were found in labella and petals.

Genes related to floral differentiation

MADS-box genes have long been noted for their roles in floral organ identity. In this study, we detected 16 genes differentially expressed between petals and labella that mapped to 29 MADS-like genes. All 16 of these DEGs were expressed in petals, with 9 found to be up-regulated. Only 12 of the 16 DEGs were expressed in labella, with 7 up-regulated in that floral organ. One of the latter seven DEGs was a *MADS4*-like unigene, which is specifically expressed in the labella and can be ectopically expressed in the labella-like petals of a peloric mutant [14]. Another identified unigene was homologous to AGL9-like genes, which may function at the top of a proposed regulatory hierarchy of MADS-box genes controlling floral development [15]. The TCP family is an ancient group of plant developmental transcription factors that regulate cell division in vegetative and reproductive structures and that are essential to the establishment of flower zygomorphy [16]. Three unigenes associated with TCP were identified as expressed in both petals and labella, although all three had much higher expression levels in labella than in petals. Studies have shown that TCP may have played a role in the diversification and radiation of Orchidaceae [17]. The elevated expression of these genes in labella imply that TCP has a more prominent role in patterning of labella than of petals. MYB transcription factors, which can affect overall petal morphology by influencing cell shape [18], were more numerous and more highly expressed in labella (9) than in petals (3). These genes may therefore also contribute to differences in differentiation between the labella and petals. We detected 10 unigenes that encode auxin response factor (ARF); although all of them were expressed in both labella and petals, more were up-regulated in labella (7 unigenes) than in petals (3 unigenes). Auxin signaling plays key roles in many plant growth and developmental processes from

Tab. 3 Some genes related to floral differentiation and their expressions in petals and labella.

Genes related to floral differentiation	Total No. of DEGs	No. of DEGs in petals	No. of up-regulated DEGs in petals	No. of DEGs in labella	No. of up-regulated DEGs in labella
MADS	16	16	9	12	7
MYB	12	9	3	12	9
ARF	10	10	3	10	7
NAC	6	6	1	6	5
bZIP	5	5	2	5	3
AP2	4	4	2	4	2
TCP	3	3	0	3	3
WRKY	3	3	1	3	2
YABBY	1	1	1	1	0
HD-ZIP	1	1	0	1	1
C3H	1	1	1	1	0

Tab. 4 Some genes related to flower color and their expressions in petals and labella.

Classification of flower color genes	Genes related to flower color	Total No. of DEGs	No. of DEGs in petals	No. of up-regulated DEGs in petals	No. of DEGs in labella	No. of up-regulated DEGs in labella
Genes related to anthocyanin synthesis	*CHI*	3	3	3	3	0
	CHS	1	0	0	1	1
	DFR	1	0	0	1	1
	UFGT3	1	1	0	1	1
	ANS(AS)	1	0	0	1	1
	F3'H	1	0	0	1	1
Genes related to carotenoid synthesis	*PSY*	1	1	1	1	0
	PAL	2	2	0	2	2
	FLS	1	1	0	1	1
	PDS	1	1	1	1	0
	DXS	1	1	1	1	0
	IPI	1	1	0	0	0
	GGPS	1	1	1	1	0
Some other related genes	*MYB*	12	9	3	12	9
	bZIP	5	5	2	5	3

embryogenesis to senescence [19]. Because more DEGs are up-regulated in labella, auxin may have a more complex role in labella patterning than in petals. We also found NAC transcription factors in both labella and petals, with a higher number of up-regulated DEGs identified in labella (5) than in petals (1). A previous study has shown that NAC transcription factors may be involved in pathogen infection responses in rice [20]. The much higher expression observed in labella compared with petals may serve as a defensive weapon in the labella, which functions as a landing platform and nectar guide for pollinators. WRKY transcription factors are key regulators of many plant processes, including biotic and abiotic stress response, senescence, seed dormancy, and seed germination [21]. We found WRKY transcription factors in both labella and petals, with a greater number of up-regulated DEGs identified in labella (2 DEGs) than in petals (1 DEGs). WRKY1 regulates sesquiterpene synthase gene expression and the accumulation of sesquiterpene induced by bacterial and fungal infection [22]; consequently, the observed coexpression of *WRKY* genes and monoterpene and sesquiterpene synthesis genes in labella and petals suggests that the function of these genes may be related to defense against biotic stress. The much higher expression in labella than in petals also suggests a defensive role for WRKY transcription factors in the labella, a landing platform for pollinators. We also identified genes that affect floral organ development [23–25], namely *YABBY*, basic region/leucine zipper motif (*bZIP*) transcription factor, and *C3H*-type zinc finger protein genes. These developmental genes were found to have different expression levels between labella and petals, suggesting that they are associated with labella and petal development but may have different functional mechanisms.

Genes related to flower color

A single *PeUFGT3* gene was detected in our study, and was found to be more highly expressed in red labella than in white petals. This observation is consistent with a report that *PeUFGT3* plays a critical role in red color formation in *Phalaenopsis* [8]. On the basis of these findings, we propose the following mechanism to explain red color formation in the labella (Fig. 5). First, CHS, the key enzyme in the flavonoid biosynthesis pathway, catalyzes the condensation of three acetate residues from malonyl-CoA with 4-coumaroyl-CoA to form tetrahydroxychalcone. Next, tetrahydroxychalcone is stereospecifically isomerized by CHI to colorless flavanone, which is catalyzed further to colorless dihydroquercetin by flavanone 3-hydroxylase (F3H) and flavonoid 3′-hydroxylase (F3′H). Dihydroquercetin is catalyzed in turn by DFR, ANS, and UFGT3 to form an anthocyanin (cyanidin-3-glucoside), thereby generating the red color of the labella [5,6]. In contrast to a previous study [8], however, we did not identify any DEGs encoding CHS, ANS, DFR, or F3′H in white petals. To explain this discrepancy, we suggest two possibilities. First, no method is perfect; we may thus have failed to detect the expression of these genes in white petals because of technical limitations. Alternatively, because different *Phalaenopsis* plants have inherited different characteristics, an anthocyanin synthesis pathway may not have existed in our chosen white petal samples (possibly the main reason for their white color) [26]. We also identified several DEGs that encode key enzymes regulating carotenoid synthesis in petals and labella, such as phytoene synthase (PSY), phenylalanine ammonialyase (PAL), and phytoene desaturase (PDS). In composites, floral coloration in the orange-red to red range is determined by a combination of anthocyanins and/or carotenoids [27]; the same observation has been made for orchids such as *Disa* hybrids [28]. The anthocyanin pathway may thus not be the only pathway controlling flower color formation in Phalaenopsis, especially red color. Pigment biosynthesis is regulated by a set of WD40, MYB, and bHLH transcription factors similar to those involved in trichome and cell shape development [9,29,30]. In this study, we detected 12 *MYB* genes in petals and labella, with more DEGs present and more strongly expressed in labella (9) than in petals (3). This differential expression may be responsible for the contrasting coloration of labella and petals.

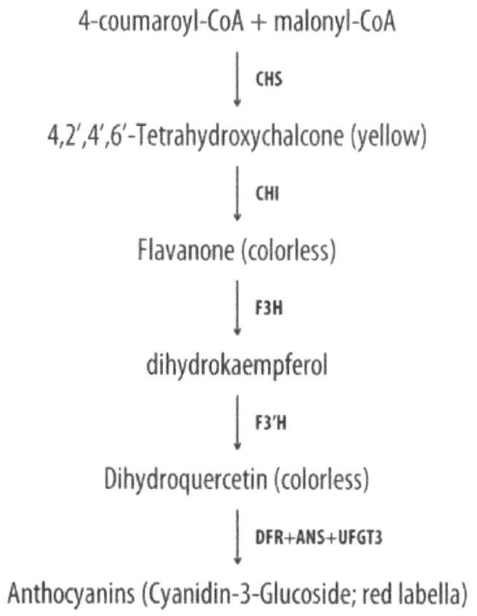

Fig. 5 Proposed pathway for synthesis of red coloration in labella.

The results of our study should serve as a valuable genetic resource for future investigations of the molecular basis of flower color and floral organ formation in *Phalaenopsis*, and inform *Phalaenopsis* flower breeding.

Acknowledgments

This work was supported by the National Natural Science Foundation of China (31101574), the Youth Science and Technology Innovation Fund of NJAU (KJ2011008), and the Agricultural Science and Technology Support Program of Zhenjiang (NY2011015).

Authors' contributions

The following declarations about authors' contributions to the research have been made: the experimental plan was conceived and designed by CZ; participated in sample collection and RNA preparation and performed the experiments: JW, YY; contributed to data analysis, bioinformatics analysis, and manuscript preparation: ZM, GS; drafted and revised the manuscript: JW, YY. JW and YY contributed equally to this work.

Supplementary material

The following supplementary material for this article is available online at http://pbsociety.org.pl/journals/index.php/asbp/rt/suppFiles/asbp.2014.023/0:

1. Tab S1: unigenes mapped to KEGG pathways.

References

1. Teixeira da Silva JA, Chin DP, Van PT, Mii M. Transgenic orchids. Sci Hortic. 2011;130(4):673–680.
2. Mondragón-Palomino M, Theißen G. MADS about the evolution of orchid flowers. Trends Plant Sci. 2008;13(2):51–59.
3. Mondragon-Palomino M, Theissen G. Why are orchid flowers so diverse? Reduction of evolutionary constraints by paralogues of class B floral homeotic genes. Ann Bot. 2009;104(3):583–594.
4. Salemme M, Sica M, Gaudio L, Aceto S. Expression pattern of two paralogs of the PI/GLO-like locus during *Orchis italica* (Orchidaceae, Orchidinae) flower development. Dev Genes Evol. 2011;221(4):241–246.
5. Tanaka Y, Sasaki N, Ohmiya A. Biosynthesis of plant pigments: anthocyanins, betalains and carotenoids. Plant J. 2008;54(4):733–749.
6. Tanaka Y, Brugliera F, Kalc G, Senior M, Dyson B, Nakamura N, et al. Flower color modification by engineering of the flavonoid biosynthetic pathway: practical perspectives. Biosci Biotechnol Biochem. 2010;74(9):1760–1769.
7. Fukuchi-Mizutani M. Biochemical and molecular characterization of a novel UDP-glucose:anthocyanin 3′-O-glucosyltransferase, a key enzyme for blue anthocyanin biosynthesis, from gentian. Plant Physiol. 2003;132(3):1652–1663.
8. Chen WH, Hsu CY, Cheng HY, Chang H, Chen HH, Ger MJ. Down-regulation of putative UDP-glucose: flavonoid 3-O-glucosyltransferase gene alters flower coloring in *Phalaenopsis*. Plant Cell Rep. 2011;30(6):1007–1017.
9. Koes R, Verweij W, Quattrocchio F. Flavonoids: a colorful model for the regulation and evolution of biochemical pathways. Trends Plant Sci. 2005;10(5):236–242.
10. Harborne JB, Williams CA. Anthocyanins and other flavonoids

(January 1998 to December 2000). Nat Prod Rep. 2001;18(3):310–333.

11. Mortazavi A, Williams BA, McCue K, Schaeffer L, Wold B. Mapping and quantifying mammalian transcriptomes by RNA-Seq. Nat Methods. 2008;5(7):621–628.
12. Xu P, Liu Z, Fan X, Gao J, Zhang X, Zhang X, et al. De novo transcriptome sequencing and comparative analysis of differentially expressed genes in *Gossypium aridum* under salt stress. Gene. 2013;525(1):26–34.
13. Hao QN, Zhou XA, Sha AH, Wang C, Zhou R, Chen SL. Identification of genes associated with nitrogen-use efficiency by genome-wide transcriptional analysis of two soybean genotypes. BMC Genomics. 2011;12(1):525.
14. Tsai WC, Hsiao YY, Pan ZJ, Hsu CC, Yang YP, Chen WH, et al. Molecular biology of orchid flowers. Adv Bot Res. 2008;47:99–145.
15. Yu H. Identification and characterization of three orchid MADS-box genes of the AP1/AGL9 subfamily during floral transition. Plant Physiol. 2000;123(4):1325–1336.
16. Mondragon-Palomino M, Trontin C. High time for a roll call: gene duplication and phylogenetic relationships of TCP-like genes in monocots. Ann Bot. 2011;107(9):1533–1544.
17. Sargent RD. Floral symmetry affects speciation rates in angiosperms. Proc R Soc Lond B Biol Sci. 2004;271(1539):603–608.
18. Baumann K, Perez-Rodriguez M, Bradley D, Venail J, Bailey P, Jin H, et al. Control of cell and petal morphogenesis by R2R3 MYB transcription factors. Development. 2007;134(9):1691–1701.
19. Guilfoyle TJ, Hagen G. Auxin response factors. Curr Opin Plant Biol. 2007;10(5):453–460.
20. Nakashima K, Tran LSP, van Nguyen D, Fujita M, Maruyama K, Todaka D, et al. Functional analysis of a NAC-type transcription factor OsNAC6 involved in abiotic and biotic stress-responsive gene expression in rice: rice OsNAC6 functions in stress responses. Plant J. 2007;51(4):617–630.
21. Rushton DL, Tripathi P, Rabara RC, Lin J, Ringler P, Boken AK, et al. WRKY transcription factors: key components in abscisic acid signalling. Plant Biotechnol J. 2012;10(1):2–11.
22. Rushton PJ, Bokowiec MT, Han S, Zhang H, Brannock JF, Chen X, et al. Tobacco transcription factors: novel insights into transcriptional regulation in the Solanaceae. Plant Physiol. 2008;147(1):280–295.
23. Golz JF, Hudson A. Plant development: YABBYs claw to the fore. Curr Biol. 1999;9(22):R861–R863.
24. Jakoby M, Weisshaar B, Dröge-Laser W, Vicente-Carbajosa J, Tiedemann J, Kroj T, et al. bZIP transcription factors in *Arabidopsis*. Trends Plant Sci. 2002;7(3):106–111.
25. Huang P, Ju HW, Min JH, Zhang X, Chung JS, Cheong HS, et al. Molecular and physiological characterization of the *Arabidopsis* thaliana oxidation-related zinc finger 2, a plasma membrane protein involved in ABA and salt stress response through the ABI2-mediated signaling pathway. Plant Cell Physiol. 2012;53(1):193–203.
26. van der Krol AR. Flavonoid genes in petunia: addition of a limited number of gene copies may lead to a suppression of gene expression. Plant Cell. 1990;2(4):291–299.
27. Kishimoto S, Sumitomo K, Yagi M, Nakayama M, Ohmiya A. Three routes to orange petal color via carotenoid components in 9 compositae species. J Jpn Soc Hortic Sci. 2007;76(3):250–257.
28. Matsui S. Varietal differences in flower colors of *Cattleya aurantiaca*. Res Bull Fac Agric Gifu Univ Jpn. 1992;57:181–185.
29. Tatsuzawa F, Ichihara K, Shinoda K, Miyoshi K. Flower colours and pigments in *Disa* hybrid (Orchidaceae). South Afr J Bot. 2010;76(1):49–53.
30. Grotewold E. The genetics and biochemistry of floral pigments. Ann Rev Plant Biol. 2006;57(1):761–780.

28

The effects of physical-chemical water parameters on the *Nymphaeion* alliance development in northwestern Serbia

Dejana Džigurski*, Branka Ljevnaić-Mašić, Ljiljana Nikolić
Faculty of Agriculture, University of Novi Sad, Trg. D. Obradovića 8, Novi Sad, Serbia

Abstract

Nymphaeion alliance vegetation is dominant floating-leaved vegetation in the Danube–Tisza–Danube hydrosystem in northwestern Serbia and comprises *Nymphaeetum albae*, *Nymphaeetum albo-luteae*, *Nymphoidetum peltatae* and *Trapetum natantis* associations. Comparative analysis of physical-chemical water parameters on localities where these - in most parts of Europe endangered and vulnerable stands - develop showed that most phytocenoses are associated with specific habitat conditions. Of the analyzed water properties, the factors that cause *Nymphaeion* alliance phytocenoses differentiation are primarily pH, alkalinity and COD-MnO_4. Formation of the *Nymphaeetum albae* stands is significantly associated with the highest values of pH, COD-MnO_4 and alkalinity, and the lowest nitrate, nitrite, dissolved and the total phosphorus content values, in comparison to the other studied associations. *Nymphoidetum peltatae* stands develop in waters characterized by the lowest pH and COD-MnO_4, low alkalinity, and the highest nitrate and nitrite values in relation to the other analyzed phytocenoses. *Trapetum natantis* stands, on the other hand, prefer the warmer sections of the canal network, neutral pH, and the highest values of BOD_5, dissolved and total phosphorus. Habitat conditions in which *Nymphaeetum albo-luteae* stands develop are of the widest range in comparison to other investigated phytocenoses.

Keywords: *Nymphaeion*; water properties; *Nymphaeetum albae*; *Nymphaeetum albo-luteae*; *Nymphoidetum peltatae*; *Trapetum natantis*; Serbia

Introduction

Floating-leaved plants are a common component of channels, rivers, lakes, ponds and other water basins. Floating-leaved phytocenoses of *Nymphaeion* Oberd. 1956 alliance play an important role in the shallow aquatic ecosystems [1]. *Nymphaeion* alliance (order *Potametalia* W. Koch 1926, class *Potametea* Tx. et Prsg. 1942, according to Soó [2]) vegetation is the dominant aquatic vegetation in the Hydrosystem Danube–Tisza–Danube (Hs DTD) in the northwestern Serbia. In the aforementioned sections, this alliance is represented by stands of the following associations: *Nymphaeetum albae* Vollmar 1947, *Nymphaeetum albo-luteae* Nowinski 1928, *Nymphoidetum peltatae* (Allorge 1922) Oberd. et Müller 1960 and *Trapetum natantis* Müller et Görs 1960.

Nymphaea alba, *Nuphar lutea* and *Nymphoides peltata* have least concern status on the "Red list of threatened species for Europe" [3], while *N. alba* and *N. lutea* have strictly protected national conservation status in Serbia [4]. *Nymphoidetum peltatae* stands in aquatic vegetation of Vojvodina are in the withdrawal phase [5]. *Trapa natans*, the edificator of *Trapetum natantis* association, is threatened in much of Europe and is subject of conservation efforts aimed at preserving and restoring its population (IUCN-LC). *T. natans* is located on the red lists of many European countries [3], and is included under "Appendix I" of the "Convention on the conservation of European wildlife and natural habitats" (Bern Convention) and IPA list (IPA criterion A, threatened species).

Factors affecting the development and distribution of *Nymphaeion* alliance phytocenoses were a subject of numerous studies, whereby physical-chemical water parameters [6–10], substrate composition [6,8,10], hydromorphological characteristics [7,11–19], water level fluctuations [18], presence of dams and weirs that disrupt the natural aquatic ecosystem regimen [13], waterflow velocity [20], etc. were investigated. According to Džigurski et al. [21,22], hydromorphological conditions in the studied hydromelioration facility (2–8 m canal depth, variable flow rate and navigation intensity, presence of more than 20 weirs, regulated water level changes, canal age 50–200 years) affect the development of macrophytes differently. Dawson and Szoszkiewicz [23] indicate that *Nymphaeion* alliance vegetation is typically found in habitats characterized by slow water flow. Findings of several extant studies suggest that, in the key phases of the *Nymphaea alba* lifecycle, dynamic changes and significant

* Corresponding author. Email: dejana@polj.uns.ac.rs
Handling Editor: Aleksandra Samecka-Cymerman

increases in the water level adversely affect the development of this species [24,25]. Optimum development of ass. *Nymphoidetum peltatae* stands is noted in slow-flowing waters [11,26], whereas, although *Nuphar luteum* favors more stagnant waters, it can tolerate smaller fluctuations in water flow and level [11]. On the other hand, *Trapetum natantis* is particularly tolerant of rapid water flow and fluctuations in water level [11,27].

The aim of this paper is to present floristic composition, distribution and physical-chemical water parameters of *Trapetum natantis, Nimphaeetum albo-luteae, Nimphaeetum albae* and *Nimphoidetum peltatae* stands in northwestern Serbia, which are valuable in the aquatic ecosystem biodiversity conservation efforts. The findings of this study, i.e., the defined optimal levels of the physical-chemical water parameters of these phytocenoses contribute to the implementation of the "European water framework directive" (WFD), which requires monitoring of the ecological status of surface waters [28–31].

Material and methods

Vojvodina, as the lowest part of the Pannonian Plain, due to the inflow of water masses formed in the Alps, Carpathians and Dinarides, is the largest hydro-confluence in Europe. Owing to this dense network of watercourses, regulation of their unfavorable water regimens was necessary, which was achieved via the construction of the Hs DTD canal network [32]. Hs DTD is a central part of the Rhine–Main–Danube navigational route and one of the largest hydrosystems in Europe (Fig. 1).

Hs DTD is located in the Province of Vojvodina, in northern Serbia (44°39′15.04″ and 46°10′11.32″ north latitude to the 18°50′15.00″ and 21°33′27.82″ east longitude). Bačka part of the Hs DTD comprises a network of nine canals, between the Danube River (in the west and south) and Tisza (in the east), 421 km in length. In terms of elevation, the network is partitioned into eight pools. In this context, a pool consists of one or more canals delimited by weirs, boat transits and pumping stations, with controlled water regimen. In the region, Hs DTD serves multiple purposes, providing drainage, irrigation, and navigation, and is used for fishing and recreation, as well as industrial and municipal wastewater acceptance.

Long-term studies of aquatic vegetation of Hs DTD in Bačka (Vojvodina Province, Serbia) were conducted in the vegetative period between 2004 and 2007, by applying the Braun-Blanquet method [33]. Plant species were determined according to Josifović [34], Tutin et al. [35,36] and Jávorka and Csapody [37]. All the relevés recorded in the investigated phytocoenoses were aggregated in one synoptic phytosociological table that shows the presence, abundance and coverage degree (from "+" to "5"). Only the stands that covered at least the minimum area recommended for these types of vegetation were recorded (>10 m^2).

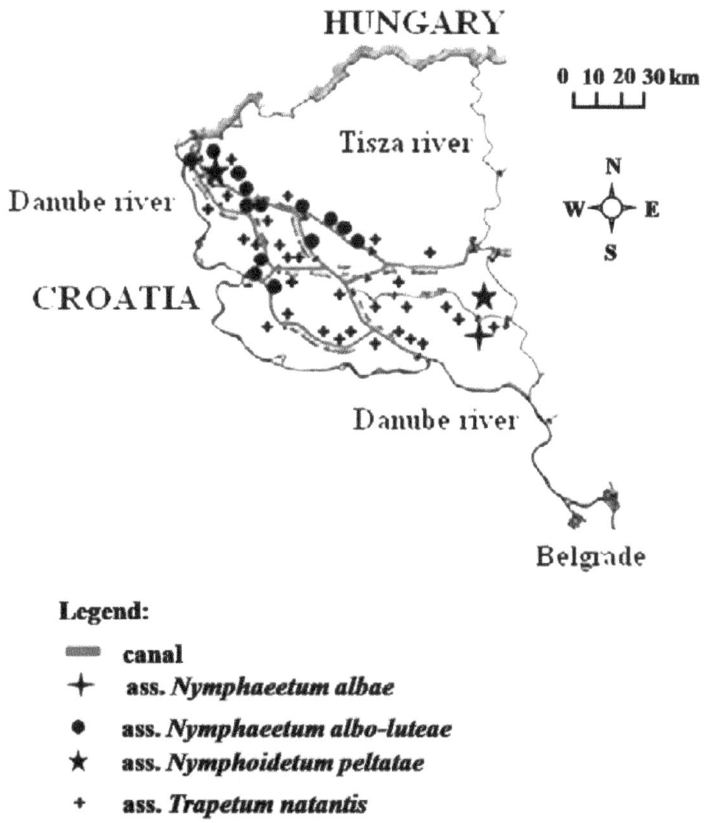

Fig. 1 Distribution of studied phytocoenoses of *Nymphaeion alliance* on Hs DTD in Bačka.

In the study period, the water samples were collected three times a year, during the vegetative period (May–August), 30 cm from water surface, on the canal sections where existence of *Nymphaeion* alliance phytocenoses were previously ascertained (Fig. 1). Water samples were taken at locations minimum 400 m apart. Basic physical and chemical water properties were determined according to standard methods [38]. The following parameters were monitored: temperature, pH value (with pH-meter), alkalinity, chemical (COD-MnO_4) and biological (BOD_5) oxygen demand, and nutrient content (NH_4^+, NO_3^-, NO_2^-, PO_4^{3-} and total phosphorus).

The water chemical analysis data pertaining to the aforementioned parameters was processed using STATISTICA 7.0 software [39]. The statistical significance of the differences among the studied phytocenoses, in terms of the analyzed water parameters, was established by applying ANOVA and Fisher LSD-test.

Results

Species composition of the studied plant associations

Floating-leaved vegetation *Nymphaeion* of Hs DTD is represented by stands of associations *Nymphaeetum albae*, *Nymphaeetum albo-luteae*, *Nymphoidetum peltatae* and *Trapetum natantis*.

Ass. *Nymphaeetum albae* stands develop at water depths of 0.5–2 m, on the section that is a natural depression and is used as a fishpond (Fig. 1). This section is characterized by a wide waterbed, slow water flow and water level fluctuations of significantly lower intensity compared to the other canal network sections. The studied stands consist of 12 plant species (Tab. 1) – a floristic poverty characteristic of this phytocenosis, on average comprising 2–15 taxa [40]. In addition to floating-leaved species *Nymphaea alba*, this distinctive collection also includes submerged species *Ceratophyllum demersum* and *Myriophyllum spicatum*; thus, this association is significantly bilayer. In the ecological series in the study site, these stands build on the stands of *Ceratophylletum demersi* (Soó 27) Hild. 1956 and ass. *Nymphoidetum peltatae*, and towards the shore, emerged stands of *Scirpo-Phragmitetum* W. Koch 1926 and ass. *Typhetum angustifoliae* Pign. 1953.

At a significant number of sites within the canal network, ass. *Nymphaeetum albo-luteae* forms a narrow fringe towards emerged vegetation, only occasionally forming broad bands that extend to the middle of the canal. These stands achieve optimum development in water depths of 1–2.5 m. The floristic composition consists of 18 hydrophytes, whereby the distinctive collection, in addition to the characteristic association species, comprises *C. demersum*, *M. spicatum*, *Vallisneria spiralis* and *Salvinia natans* (Tab. 1). Ass. *Nymphaeetum albo-luteae* stands exhibit a typical bilayer structure. In the ecological series, towards deeper water, abundant populations of *Myriophyllum spicatum* and stands of *Ceratophylletum demersi*, *Ceratophyllo demersi-Vallisnerietum spiralis* Lazić 2006, *Potametum nodosi* Soó (1928) 1960, Segal 1964 and *Trapetum natantis* are found, and towards the shore, associations *Hydrocharidetum morsus-ranae* van Langendonck 1935, *Salvinia-Spirodeletum polyrrhizae* Slavnić 1956, *Scirpo-Phragmitetum*, *Typhetum angustifoliae* and *Typhetum latifoliae* G. Lang 1973.

Ass. *Nymphoidetum peltatae* was established on only two, mutually distant sites, at water depths of 0.5–2.5 m. The analyzed stands comprise 15 macrophyte species (Tab. 1). In addition to the edificator, whose abundance and wide coverage provides a unique character to these stands, species of the characteristic group comprising *T. natans*, *Hydrocharis morsus-ranae*, *C. demersum*, *V. spiralis*, *Lemna minor*, *Spirodela polyrrhiza* and *S. natans* contribute to the heterogeneous structure of these stands. Low values of abundance and coverage of the submerged plants in these stands attest to the significant influence of ass. *Nymphoidetum peltatae* stands on the reduced growth of submersed macrophytes [41]. At study sites, towards the middle of the canal, ass. *Nymphoidetum peltatae* stands build on the stands of ass. *Ceratophylletum demersi* and ass. *Nymphaeetum albae*, or *Sagittaria sagittifolia* L. and *Potamogeton perfoliatus* L. populations. Towards the shore, they are in contact with the stands of ass. *Scirpo-Phragmitetum* and ass. *Typhetum angustifoliae*.

Tab. 1 Synoptic table of the *Nymphaeion* alliance in Hs DTD in Bačka.

Phytocoenoses	NA	NL	NP	TN
Number of relevés	15	32	20	44
Number of species	12	18	15	20
Nymphaeion				
Nymphaea alba L.	V$^{3-5}$	III$^{1-3}$	II^{+}	I^{+}
Nuphar lutea (L.) Sm.		V$^{2-5}$	III^{+-1}	I^{+}
Nymphoides peltata Hill.	I^{1}	I^{+}	V$^{2-5}$	
Trapa natans L.	I^{+}	III^{+-1}	IV^{+}	V$^{3-5}$
Hydrocharis morsus-ranae L.	I^{+}	III^{+-1}	IV^{+-2}	IV^{+-2}
Polygonum amphibium L.	I^{+}			I^{+}
Potametalia* and *Potametea				
Myriophyllum spicatum L.	IV^{+-2}	IV^{+-1}	III^{+-2}	IV^{+-1}
Ceratophyllum demersum L.	IV^{+-3}	V^{+-3}	V^{+-2}	IV^{+-2}
Potamogeton perfoliatus L.	I^{+}	I^{+}	II^{+}	I^{+-1}
Najas marina L.	I^{+}	I^{+}		I^{+}
Vallisneria spiralis L.		IV^{+-1}	IV^{+-1}	II^{+-1}
Elodea canadensis Rich		I^{+}		I^{+}
Potamogeton crispus L.		I^{1}		I^{+}
Potamogeton pectinatus L.			I^{+}	II^{+-1}
Others				
Lemna minor L.	I^{+}	III^{+}	IV^{+}	III^{+}
Spirodela polyrrhiza (L). Schleid.	I^{+}	II^{+-3}	IV^{+}	IV^{+-1}
Utricularia vulgaris L.	I$^{1-3}$	I^{+}		I^{+}
Salvinia natans (L.) Allioni		IV^{+-2}	V^{+-1}	II^{+-1}
Lemna trisulca L.		III^{+}	III^{+}	I^{+}
Sagittaria sagittifolia L.		I^{1}	II^{+}	I^{+}
Lemna gibba L.				I^{+}

NA – *Nymphaeetum albae*; NL – *Nymphaeetum albo-luteae*; NP – *Nymphoidetum peltatae*; TN – *Trapetum natantis*.

Ass. *Trapetum natantis* is the dominant floating-leaved vegetation in the Hs DTD canal network (Fig. 1) [21]. These stands achieve optimal development in the central, deeper parts of the canal (about 3 m), characterized by frequent water fluctuations, as well as in the calm and shallower sections (about 0.5 m). They have developed in the form of narrow or wider bands, while covering the entire canal width in some places. The floristic composition consists of 20 species (Tab. 1). In addition to the edificator, the characteristic group includes free-floating *H. morsus ranae* and *S. polyrrhiza*, and submerged *C. demersum* and *M. spicatum*. In deeper waters, in the absence of competition from other floating plants, these stands independently form a wide belt. At lower depths, they form a belt of floating vegetation, together with the stands of communities *Nymphaeetum albo-luteae, Hydrocharidetum morsus-ranae, Salvinio-Spirodeletum polyrrhizae* and *Lemno-Spirodeletum*. Apart from the floating community stands, in the ecological series, stands of ass. *Trapetum natantis* build on the submerged stands of associations *Ceratophylletum demersi, Ceratophyllo demersi-Vallisnerietum spiralis* and *Myriophyllo-Potametum* Soó 1934 and the emerged stands of associations *Scirpo-Phragmitetum, Typhetum angustifoliae* and *Typhetum latifoliae*.

Physical-chemical water parameters of the analyzed phytocenoses habitat

The physical-chemical parameter water analysis revealed that pH, alkalinity and COD-MnO_4 are the key factors that cause the phytocenoses differentiation in *Nymphaeetum albae, Nymphaeetum albo-luteae, Nymphoidetum peltatae* and *Trapetum natantis* (Fig. 2, Tab. 2, Tab. 3, Fig. 3).

The water temperature analysis revealed that the existing differences are not statistically significant. Mean values ranged from 23.86°C on the localities where the stands of ass. *Nymphaeetum albo-luteae* are formed, through 24.35°C in the stands of ass. *Nymphoidetum peltatae* and 24.5°C in the stands of ass. *Nymphaeetum albae*, to 24.65°C in the stands ass. *Trapetum natantis* (Fig. 2). However, in the ass. *Trapetum natantis* stand habitats, the extreme temperature of 29°C was noted.

The pH value proved to be a very important factor that caused the phytocenoses differentiation within the *Nymphaeion alliance* (Tab. 2, Fig. 3). The highest average value (8.172) was noted at sites where stands of ass. *Nymphaeetum albae* develop, while in all other phytocenoses, it was around 7.6 (Fig. 2). Very broad ecological valence with respect to pH was observed at the habitats of the stands *Nymphaeetum albo-luteae* (min = 7.1, max = 8). Statistically highly significant differences between the water pH values in the habitat of ass. *Nymphaeetum albae* and ass. *Nymphoidetum peltatae* ($P = 0.0023$) were noted, between those measured in the habitat of ass. *Nymphaeetum albae* and ass. *Trapetum natantis* ($P = 0.001$), and between ass. *Nymphaeetum albae* and ass. *Nymphaeetum albo-luteae* ($P = 0.002$), Tab. 3.

Alkalinity significantly differentiated the analyzed associations (Tab. 2, Fig. 3). The highest mean value was measured for the sites populated by the stands of ass. *Nymphaeetum albae* (5.310), which are also the most tolerant to alkalinity variations (min = 2.745, max = 6.893). These are followed by the bicarbonate content of water in which the stands of ass. *Trapetum natantis* (3.507) and ass. *Nymphoidetum peltatae* (3.114) develop, and finally by the values associated with water dominated by stands of ass. *Nymphaeetum albo-luteae* (2.844), Fig. 2. Statistically highly significant differences were observed between the habitat alkalinity values of the ass. *Nymphaeetum albae* and *Nymphaeetum albo-luteae* stands ($P = 0.001$), and between ass. *Nymphaeetum albae* and ass. *Nymphoidetum peltatae* ($P = 0.009$), as well as statistically significant differences between water alkalinity in habitats ass. *Nymphaeetum albae* and ass. *Trapetum natantis* ($P = 0.013$), Tab. 3.

The highest mean chemical oxygen demand value COD-MnO_4 (11.7 mg/l) and its greatest variation was noted in stands of *Nymphaeetum albae* (Fig. 2). In the water where stands of ass. *Nymphaeetum albo-luteae* develop, the mean value of 7.927 was obtained, whereas for ass. *Trapetum natantis* and ass. *Nymphoidetum peltatae* habitats, the mean COD-MnO_4 values of 7.272 and 6.927 were noted, respectively. Statistically highly significant differences in terms of COD-MnO_4 were found between localities populated by ass. *Nymphaeetum albae* and ass. *Trapetum natantis* stands ($P = 0.001$), as well as between the stands of *Nymphaeetum albae* and *Nymphaeetum albo-luteae* ($P = 0.003$), and between the *Nymphaeetum albae* and *Nymphoidetum peltatae* ($P = 0.001$), Tab. 2, Tab. 3.

The obtained mean values of BOD_5 – slightly above 2 mg O_2/l in water where the stands of ass. *Nymphaeetum albae* (2.1775) and ass. *Nymphaeetum albo-luteae* (2.488) develop – indicate moderate water pollution. On the other hand, higher mean BOD_5 values obtained for water populated by ass. *Trapetum natantis* stands (3.808), with the extremes exceeding 7 mg/l, are indicative of polluted water. At the border between a relatively clean and moderately polluted waters are the sites where ass. *Nymphoidetum peltatae* stands develop (mean value 1.70). Overall, the results indicate a "good ecological status" (BOD_5 below 6 mg/l) in the category of artificial water bodies, for the majority of sites where the analyzed stands develop, with the exception of a few sites at which stands of ass. *Trapetum natantis* are found (Fig. 2).

The differences in the ammonium ion content of the investigated stands were not statistically significant. The lowest values were noted at localities populated by ass. *Nymphaeetum albo-luteae* stands (mean value 0.245), while the highest were obtained for those characterized by presence of ass. *Nymphaeetum albae* stands (mean value 0.437; Fig. 2). Of particular significance is the ammonium ion content on some localities populated by stands of ass. *Trapetum natantis*, reaching max = 0.62, and thus indicating that the quality of water is poor (Fig. 2).

The water nitrate content analysis revealed that development of ass. *Nymphoidetum peltatae* stands favors water richer in these compounds (mean value 1.217), compared to the other stands (Fig. 2). Stands of *Trapetum natantis* (mean value 0.874), *Nymphaeetum albo-luteae* (mean value 0.827) and *Nymphaeetum albae* (mean value 0.68) were also characterized by relatively high nitrate content. No statistically significant differences were noted in the water nitrate content where the analyzed stands develop (Tab. 2).

In the studied aquatic ecosystem, nitrite content had no significant effect on the differentiation of *Nymphaeion*

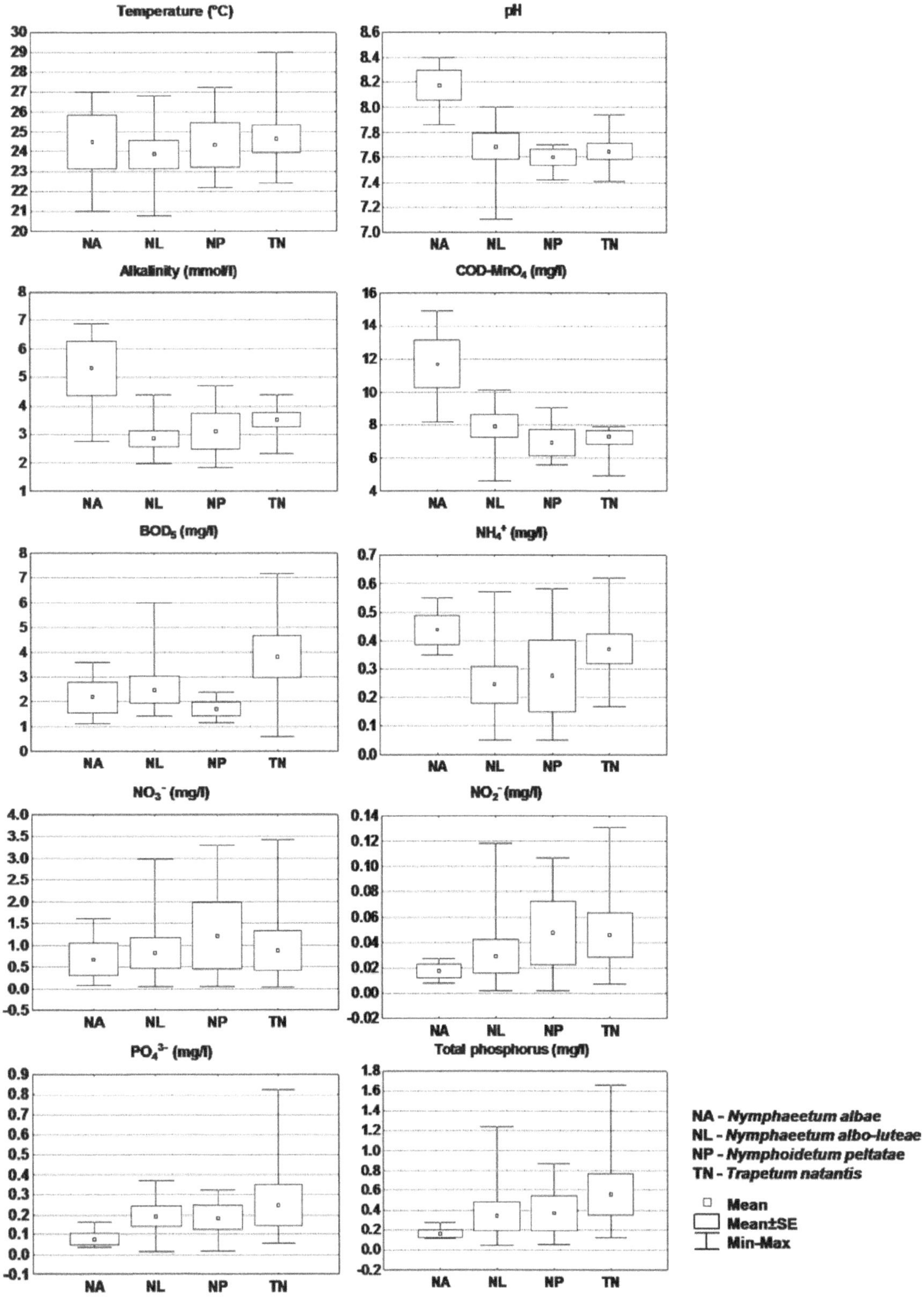

Fig. 2 The main physical-chemical water parameters of the *Nymphaeion* alliance phytocenoses habitat.

Tab. 2 Significance of differences between studied phytocoenoses compared to the analyzed water properties (ANOVA).

Variable	*SS* effect	*df* effect	*MS* effect	*SS* error	*df* error	*MS* error	*F*-ratio	*P*-value
Temperature	2.677	3	0.892	93.169	20	4.658	0.191	0.901
pH	0.916	3	0.305	1.067	20	0.053	5.725	0.005**
Alkalinity	17.134	3	5.711	23.439	20	1.172	4.873	0.010*
COD-MnO_4	62.933	3	20.978	68.225	20	3.411	6.149	0.004**
BOD_5	15.172	3	5.057	61.969	20	3.098	1.632	0.214
NH_4^+	0.129	3	0.043	0.608	20	0.030	1.414	0.268
NO_3^-	0.638	3	0.212	27.512	20	1.375	0.154	0.925
NO_2^-	0.003	3	0.001	0.035	20	0.002	0.585	0.632
PO_4^{3-}	0.078	3	0.026	0.808	20	0.040	0.641	0.597
Total phosphorus	0.445	3	0.148	3.999	20	0.200	0.743	0.539

SS – sum of squares; *df* – degrees of freedom; *MS* – mean squares. * $P < 0.05$; ** $P < 0.01$.

alliance phytocenoses. Mean values ranged from 0.017 in *Nymphaeetum albae* stands to 0.047 in *Nymphoidetum peltatae* stands (Fig. 2).

The water-dissolved phosphorus content analysis revealed that the stands of ass. *Trapetum natantis* formed in water with a higher dissolved phosphorus content (mean value 0.246). The dissolved phosphorus mean content in water with ass. *Nymphaeetum albo-luteae* stands was 0.191, whereas 0.185 was measured for water with ass. *Nymphoidetum peltatae* (Fig. 2). Stands of ass. *Nymphaeetum albae* developed in water with a much lower dissolved phosphorus value in comparison to the other stands (mean value 0.08), where minimum variation in the measured values was also noted.

Although notable, the differences in the total phosphorus content for different localities are not statistically significant (Tab. 2). Stands of *Trapetum natantis* developed in water relatively richer in phosphates (mean value 0.556), compared to the other analyzed phytocenoses, in which a wide range of values was noted (min = 0.123, max = 1.662). These are followed by habitats in which ass. *Nymphoidetum peltatae* (mean value 0.371), ass. *Nymphaeetum albo-luteae* (mean value 0.350) and ass. *Nymphaeetum albae* (mean value 0.164) develop.

Tab. 3 Significant of differences in water properties between studied phytocoenoses (Fisher LSD-test).

Water properties	NA-NL	NA-NP	NA-TN
pH	0.002	0.002	0.001
Alkalinity (mmol/l)	0.001	0.009	0.013
COD-MnO_4 (mg/l)	0.003	0.001	0.001

NA – *Nymphaeetum albae*; NL – *Nymphaeetum albo-luteae*; NP – *Nymphoidetum peltatae*; TN – *Trapetum natantis*. The differences between NL-NP, NL-TN and NP-TN are insignificant.

Discussion

Among the water properties analyzed as a part of this study, pH, alkalinity and COD-MnO_4 are the most significant, and are the main contributors to the differentiation of phytocenoses *Nymphaeetum albae, Nymphaeetum albo-luteae, Nymphoidetum peltatae* and *Trapetum natantis* (Tab. 2, Fig. 3). Dawson and Szoszkiewicz [23] also reported that, in British rivers, pH and alkalinity, in addition to total phosphorus, are the key environmental parameters that significantly affect the differentiation of aquatic phytocenoses. In many European river systems, the role of alkalinity and pH is crucial in the differentiation and development of macrophytes, i.e., these parameters are directly correlated with the macrophyte distribution and abundance [42].

In the stands of *Nymphaeetum albae*, the highest mean pH value in relation to other *Nympheion alliance* phytocenoses, as well as the measured extreme values, are in accordance with the results reported by Szańkowski and Kłosowski [6] for aquatic ecosystems in Poland. However, all measured values are below the limits that define a "good ecological status" for artificial water bodies. High water alkalinity levels, as well as wide ecological amplitude for this parameter, were noted by Smits et al. [43] and Dimopoulos et al. [24] in the ass. *Nymphaeetum albae* stands. The highest average COD-MnO_4 value and ammonium ion content, relative to other phytocenoses, indicates that these stands are periodically exposed to water laden with organic matter (attributed to the close proximity of the food industry factories). The lowest nitrate, nitrite, dissolved and total phosphorus content compared to other analyzed stands characterizes the conditions in which these stands develop as mesotrophic [44], which is in accordance with the previously reported results [24,45,46]. Lukács et al. [45] found that this association prefers water with markedly poor organic matters, and the stands of this phytocenosis are vulnerable in waters rich in nutrients. Mesotrophic status is confirmed by the significant

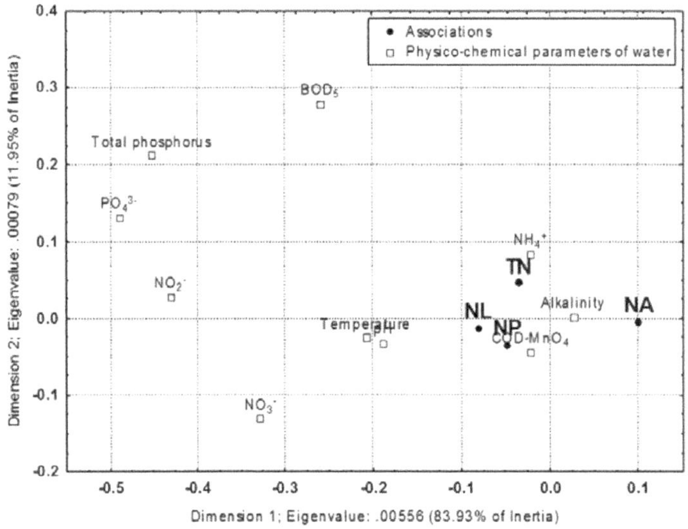

Fig. 3 Correspondence analysis between the identified *Nymphaeion* alliance associations and physical-chemical water quality parameters.

presence of *M. spicatum* in the stands of this community (Tab. 1). As *M. spicatum* growth decreases with further increases in the trophic level, it appears to act as its growth modulator [47,48]. This premise is supported by the highest abundance and coverage values for this macrophyte in the ass. *Nymphaeetum albae* stands, in relation to the other analyzed phytocenoses developing in waters characterized by higher trophic levels (Tab. 1, Fig. 2).

Our results, as well as the data reported in pertinent literature [6,49], confirm that the habitat conditions in which ass. *Nymphaeetum albo-luteae* stands develop are subject to a wide range of variation. Slightly lower pH value and wide ecological valence of the *Nuphar lutea* edificator is confirmed by Kočić et al. [7] and Jabłońska and Kłosowski [8]. In the studied stands, the lowest alkalinity was noted, in comparison with the water in which stands of other phytocenoses develop. BOD_5 value is in the range of "good ecological status" (below 6 mg/l) [50]. Lower nitrate content in the water in which the stands of *Nuphar lutea* develop was also found in the rivers of neighboring Croatia [7]. The dissolved and total phosphorus values define these habitat conditions as moderately eutrophic [44], as confirmed by the data reported in pertinent literature [9,11,51,52]. Moreover, the species of the *C. demersum* characteristic group, which forms thick "underwater meadows" in the analyzed stands, is an indicator of eutrophic habitat conditions, in line with the results reported for other studies [9,11,12,24,26,45,49,53,54].

In the stands of ass. *Nymphoidetum peltatae*, the lowest pH and the narrowest range of variation of this parameter were recorded. Szańkowski and Kłosowski [6] reported similar pH value for water flows in Poland. Bicarbonate values (min = 1.830, max = 4.734) are in the range of variation noted by Smits et al. [43] most likely to allow development of edificator *Nymphoides peltata*. Significantly, the lowest COD-MnO_4 and BOD_5 values, as well as very narrow range of variation in these parameters, was noted in comparison to other analyzed phytocenoses. The highest nitrate and nitrite content in ass. *Nymphoidetum peltatae* habitats, in comparison to other analyzed phytocenoses, is in line with the findings reported by other authors [6,46,53]. The dissolved and total phosphorus values are in the range of eutrophic status, as indicated by the results of previous studies [6]. In addition to being an edificator, *H. morsus-ranae*, which was highly abundant in these stands, is also an indicator of eutrophic status [11,55]. On the other hand, according to Sýkora [51], development of this association is unaffected by water quality.

The most widespread floating-leaved vegetation in the canal network, ass. *Trapetum natantis*, is one of the most thermophilic aquatic phytocenoses [21,56,57]. The study findings indicate that the analyzed phytocenosis stands prefer neutral water, which is consistent with the results reported by other authors [6,27]. In contrast to the high COD-MnO_4 values recorded by Szańkowski and Kłosowski [6], and low values determined by Kočić et al. [7], our results indicate that the habitats on which the analyzed stands form are moderately loaded with organic substances. Periodically measured, BOD_5 values significantly higher than the "good ecological status" limit [50] are typically due to organic pollution caused by the discharge of untreated sewage, industrial effluents and leaching from agricultural land. These conditions are confirmed by the presence of *Potamogeton pectinatus*, which often develops in polluted waters, rich in phosphates [7,9,23,58]. Of particular significance is the ammonium ion content that, in certain localities, reaches max = 0.62, indicating that the water in which the stands developed is of impaired quality. *Trapetum natantis* stands developed in waters laden with phosphates, characterizing these conditions as eutrophic [44]. The highest dissolved and total phosphorus content, and very wide ecological amplitude of this important parameter, as an indicator of eutrophication, is characteristic of stands in other regions [6,11,45]. In addition to the edificator, hydrophytes *S. polyrrhiza* [9,12,21,51,54,59], *H. morsus-ranae, C. demersum* and *L. minor* [21,51,59] are also well adapted to the eutrophic levels and are widely represented. Within *Nymphaeion* alliance phytocenoses, the mean value of phosphorus significantly exceeds the limit of "good ecological status" only in habitats of *Trapetum natantis* stands [50]. Moreover, the ability of *T. natans* to tolerate significant water level fluctuations in the studied hydromelioration facility, in conjunction with the physical-chemical water parameters, is driving its expansion [21].

Comparative analysis of the physical-chemical water parameters at the localities where the ass. *Nymphaeion* stands develop revealed that most phytocenoses are closely linked to specific habitat conditions. Water parameter values measured at most localities are the consequence of the anthropogenic eutrophication process. Owing to these adverse effects, it is likely that the biodiversity and survival of ass. *Nympheion* phytocenoses will be jeopardized [12]. Given that the phytocenoses that are endangered and vulnerable in most of the Europe have developed in this anthropogenic hydroecosystem, it is of paramount importance that its ecological equilibrium is maintained, in order to achieve "good ecological status" and preserve the stands of these immensely valuable phytocenoses.

Acknowledgments

This study was supported by the public water management company Vode Vojvodine, Novi Sad (Serbia). The authors would like to acknowledge two anonymous reviewers for constructive and helpful comments.

Authors' contributions

The following declarations about authors' contributions to the research have been made: wrote the paper: DD; field research: DD, BLM, LN; phytosociological analyses: DD; chemical analysis: DD, BLM, LN; preparation of distribution map: BLM; statistical analysis: BLM.

References

1. Kolada A. The use of aquatic vegetation in lake assessment: testing the sensitivity of macrophyte metrics to anthropogenic pressures and water quality. Hydrobiologia. 2010;656(1):133–147.
2. Soó R. A Magyar Flòra ès Vegetàciò Rendszertani-növènyföldrajzi Kèzikönyve I–VI. Budapest: Akadèmiai kiadò; 1964.
3. IUCN red list of threatened species [Internet]. 2013 [cited 2013 Nov 29].
4. Radulovic S, Boon PJ, Laketic D, Simonovic P, Puzovic S, Zivkovic M, et al. Preliminary checklists for applying SERCON (system for evaluating rivers for conservation) to rivers in Serbia. Arch Biol Sci. 2012;64(3):1037–1056.
5. Butorac B. Review of aquatic vegetation of the regional park "Stari Begej". Tiscia. 1995;29:27–32.
6. Szankowski M, Kłosowski S. Habitat conditions of nymphaeid associations in Poland. Hydrobiologia. 1999;415:177–185.
7. Kočić A, Hengl T, Horvatić J. Water nutrient concentrations in channels in relation to occurrence of aquatic plants: a case study in eastern Croatia. Hydrobiologia. 2008;603(1):253–266.
8. Jabłońska E, Kłosowski S. Ecology of rare water plant communities in lakes of north-eastern Poland. Acta Soc Bot Pol. 2012;81(1):3–9.
9. Balevičienė J, Balevičius A. Qualitative and quantitative parameters of phytocenoses in Lithuanian lakes of different trophic state. Ekologija. 2006;2:34–43.
10. Pełchaty M. Does nymphaeid distribution reflect the spatial heterogeneity of abiotic conditions in a shallow lake? Belg J Bot. 2007;140(1):73–82.
11. Schneider E. Aquatic macrophytes in the Danube Delta – indicators for water quality and habitat parameters. Stud Univ Babes Bolyai Biol. 2009;1:21–31.
12. Steffen K, Becker T, Herr W, Leuschner C. Diversity loss in the macrophyte vegetation of northwest German streams and rivers between the 1950s and 2010. Hydrobiologia. 2013;713(1):1–17.
13. Bernez I, Daniel H, Haury J, Ferreira MT. Combined effects of environmental factors and regulation on macrophyte vegetation along three rivers in western France. River Res Appl. 2004;20(1):43–59.
14. O'Hare MT, Baattrup-Pedersen A, Nijboer R, Szoszkiewicz K, Ferreira T. Macrophyte communities of European streams with altered physical habitat. Hydrobiologia. 2006;566(1):197–210.
15. Hrivnak R, Otahelova H, Valachovic M. The relationship between macrophyte vegetation and habitat factors along a middle-size European river. Pol J Ecol. 2007;55(4):717–729.
16. Breugnot E, Dutartre A, Laplace-Treyture C, Haury J. Local distribution of macrophytes and consequences for sampling methods in large rivers. Hydrobiologia. 2008;610(1):13–23.
17. Grinberga L. Environmental factors influencing the species diversity of macrophytes in middle-sized streams in Latvia. Hydrobiologia. 2010;656(1):233–241.
18. Hrivnák R, Ot'aheľová H, Valachovič M, Paľove-Balang P, Kubinská A. Effect of environmental variables on the aquatic macrophyte composition pattern in streams: a case study from Slovakia. Fundam Appl Limnol. 2010;177(2):115–124.
19. Radulovic S, Laketic D, Vukov D. A riverside tale: assessment of altered habitat effects on macrophyte assemblage on the river Tamiš, Serbia. Arch Biol Sci. 2010;62(4):1163–1174.
20. Madsen JD, Chambers PA, James WF, Koch EW, Westlake DF. The interaction between water movement, sediment dynamics and submersed macrophytes. Hydrobiologia. 2001;444(1–3):71–84.
21. Džigurski D, Ljevnaić-Mašić B, Nikolić L. *Trapetum natantis* Müller et Görs 1960 in hydromeliorative facilities in Serbia. Acta Soc Bot Pol. 2013;82(2):125–133.
22. Džigurski D, Knežević A, Stojanović S, Nikolić L, Ljevnaić-Mašić B. The vegetation of canal Novi Sad-Savino Selo. Thaiszia. 2010;20:137–145.
23. Szoszkiewicz K, Dawson FH. Relationships of some ecological factors with the associations of vegetation in British rivers. In: Caffrey J, Barrett PRF, Ferreira MT, Moreira IS, Murphy KJ, Wade PM, editors. Biology, ecology and management of aquatic plants. Dordrecht: Springer Netherlands; 1999. p. 117–122.
24. Dimopoulos P, Sykora KV, Gilissen C, Wiecherink D, Georgiadis T. Vegetation ecology of Kalodiki Fen (NW Greece). Biologia (Bratisl). 2005;60(1):69–82.
25. Paillisson JM, Marion L. Can small water level fluctuations affect the biomass of *Nymphaea alba* in large lakes? Aquat Bot. 2006;84(3):259–266.
26. Nikolić L, Pajević S, Ljevnaić B. Primary production dynamics of dominant hydrophytes in Lake Provala (Serbia). Cent Eur J Biol. 2009;4(2):250–257.
27. Hrivnák R. Aquatic plant communities in the catchment area of the Ipeľ river in Slovakai and Hungary. Part II. Class Potametea. Thaiszia. 2002;12:137–160.
28. European Commission. Directive 2000/60/EC of the European Parliament and of the Council of 23rd October 2000: establishing a framework for Community action in the field of water policy. Off J Eur Communities. 2000;327:1–72.

29. Hrivnák R, Oťaheľová H, Jarolímek I. Diversity of aquatic macrophytes in relation to environmental factors in the Slatina river (Slovakia). Biologia (Bratisl). 2006;61(4):413–419.

30. Ferreira MT, Aguiar FC. Riparian and aquatic vegetation in Mediterranean-type streams (western Iberia). Limnetica. 25(1–2):411–424.

31. Fabris M, Ghetti PF. Application and development of river quality bioindication methods based on macrophytes. In: 4th ECRR conference on river restoration Italy, Venice S. Servolo Island 16–21 June 2008. Venice: ECRR; 2008.

32. Milošev Ž. Hidrotehnički radovi u Banatu i Bačkoj pre izgradnje hidrosistema Dunav–Tisa–Dunav. Hidrosistem Dunav–Tisa–Dunav, 25 godina kasnije. Novi Sad: JVP Vode Vojvodine; 2002.

33. Braun-Blanquet J. Pflanzensoziologie. 3rd ed. Wien: Springer; 1964.

34. Josifović M, editor. Flora of SR Serbia. Beograd: SANU; 1970. (vol 1–9).

35. Tutin TG, Heywood VH, Burges NA, Valentine DH, Walters SM, Webb DA, editors. Flora Europaea. Cambridge: Cambridge University Press; 1964. (vol 1).

36. Tutin TG, Heywood VH, Burges NA, Moore DM, Valentine DH, Walters SM, et al., editors. Flora Europaea. Cambridge: Cambridge University Press; 1968. (vol 2–5).

37. Jávorka S, Csapody V. Icanographie der Flora des Südostlichen Mitteleuropa. Budapest: Akademiai Kiado; 1975.

38. APHA, AWWA, WPCF. Standard methods for examination of water and wastewater. 17th ed. Washington, DC: APHA; 1995.

39. STATISTICA 7.0. StatSoft; 2004.

40. Spałek K. *Nymphaeetum albae* Vollmar 1947 em. Oberd. in Oberd. et al. 1967, a plant association new to Poland. Acta Soc Bot Pol. 2007;76(1):81–84.

41. Larson D. Growth of three submerged plants below different densities of *Nymphoides peltata* (S.G. Gmel.) Kuntze. Aquat Bot. 2007;86(3):280–284.

42. Lewin I, Szoszkiewicz K. Drivers of macrophyte development in rivers in an agricultural area: indicative species reactions. Cent Eur J Biol. 2012;7(4):731–740.

43. Smits AJM, de Lyon MJH, van der Velde G, Steentjes PLM, Roelofs JGM. Distribution of three nymphaeid macrophytes [*Nymphaea alba* L., *Nuphar lutea* (L.) Sm. and *Nymphoides peltata* (Gmel.) O. Kuntze] in relation to alkalinity and uptake of inorganic carbon. Aquat Bot. 1988;32(1–2):45–62.

44. OECD. Eutrophication of waters: monitoring, assessment and control. Paris: Organisation for Economic Co-operation and Development; 1982.

45. Lukács BA, Dévai G, Tóthmérész B. Aquatic macrophytes as bioindicators of water chemistry in nutrient rich backwaters along the Upper-Tisza river (in Hungary). Phytocoenologia. 2009;39(3):287–293.

46. Papastergiadou E, Babalonas D. The relationships between hydrochemical environmental factors and the aquatic macrophytic vegetation in stagnant and slow flowing waters. I. Water quality and distribution of aquatic associations. Arch Für Hydrobiol Suppl. 1993;90(4): 475–491.

47. Smith DH, Madsen JD, Dickson KL, Beitinger TL. Nutrient effects on autofragmentation of *Myriophyllum spicatum*. Aquat Bot. 2002;74(1):1–17.

48. Chase JM, Knight TM. Effects of eutrophication and snails on Eurasian watermilfoil (*Myriophyllum spicatum*) invasion. Biol Invasions. 2006;8(8):1643–1649.

49. Lacoul P, Freedman B. Environmental influences on aquatic plants in freshwater ecosystems. Env Rev. 2006;14(2):89–136.

50. Uredba o graničnim vrednostima zagadujućih materija u površinskim i podzemnim vodama i sedimentu i rokovima za njihovo dostizanje. Sl Glas RS. 2012;(50).

51. Sýkora KV. Field guide Dutch plant communities. Species composition and ecology. 2006.

52. Szoszkiewicz K, Kayzer D, Staniszewski R, Dawson H. Measures of central tendency of aquatic habitat parameters: application to river macrophyte communities. Pol J Ecol. 2010;58(4):693–706.

53. Nikolić L, Čobanović K, Lazić D. *Nymphoides peltata* (Gmel.) Kuntze, *Myriophyllum spicatum* L. and *Ceratophyllum demersum* L. biomass dynamics in Lake Provala (the Vojvodina Province, Serbia). Cent Eur J Biol. 2007;2(1):156–168.

54. Grīnberga L, Spriņge G. Potential impact of climate change on aquatic vegetation of River Salaca, Latvia. Proc Latv Acad Sci Sect B Nat Exact Appl Sci. 2008;62(1–2):34–39.

55. Staniszewski R. Estimation of river trophy in the Kujawskie Lakeland using mean trophic rank and chemical index of trophy. Rocz AR Pozn Bot. 2001;4:165–173.

56. Coldea G, Sanda V, Popescu A, Ştefan N, editors. Les associations végétales de Roumanie. Cluj: Presses Universitaires de Cluj; 1997. (vol 1).

57. Olszewski T, Markowski T. *Trapa natans* L. s.l. as extinct in the 19th century component of Gdańsk Pomerania flora. In: Olszewski T, Afranowicz R, Bociąg K, editors. Contemporary trends of botanical research – on Professor Hanna Piotrowska 80th birthday anniversary. 2007. p. 153–156. (Acta Bot Cassub; vol 6).

58. Janauer G, Dokulil M. Macrophytes and algae in running waters. In: Ziglio G, Siligardi M, Flaim G, editors. Water quality measurements. Chichester: John Wiley & Sons; 2006. p. 89–109.

59. Kufel L, Strzałek M, Wysokińska U, Biardzka E, Oknińska S, Ryś K. Growth rate of duckweeds (Lemnaceae) in relation to the internal and ambient nutrient concentrations – testing the Droop and Monod models. Pol J Ecol. 2012;60(2):241–249.

Permissions

All chapters in this book were first published in ASBP, by Polish Botanical Society; hereby published with permission under the Creative Commons Attribution License or equivalent. Every chapter published in this book has been scrutinized by our experts. Their significance has been extensively debated. The topics covered herein carry significant findings which will fuel the growth of the discipline. They may even be implemented as practical applications or may be referred to as a beginning point for another development.

The contributors of this book come from diverse backgrounds, making this book a truly international effort. This book will bring forth new frontiers with its revolutionizing research information and detailed analysis of the nascent developments around the world.

We would like to thank all the contributing authors for lending their expertise to make the book truly unique. They have played a crucial role in the development of this book. Without their invaluable contributions this book wouldn't have been possible. They have made vital efforts to compile up to date information on the varied aspects of this subject to make this book a valuable addition to the collection of many professionals and students.

This book was conceptualized with the vision of imparting up-to-date information and advanced data in this field. To ensure the same, a matchless editorial board was set up. Every individual on the board went through rigorous rounds of assessment to prove their worth. After which they invested a large part of their time researching and compiling the most relevant data for our readers.

The editorial board has been involved in producing this book since its inception. They have spent rigorous hours researching and exploring the diverse topics which have resulted in the successful publishing of this book. They have passed on their knowledge of decades through this book. To expedite this challenging task, the publisher supported the team at every step. A small team of assistant editors was also appointed to further simplify the editing procedure and attain best results for the readers.

Apart from the editorial board, the designing team has also invested a significant amount of their time in understanding the subject and creating the most relevant covers. They scrutinized every image to scout for the most suitable representation of the subject and create an appropriate cover for the book.

The publishing team has been an ardent support to the editorial, designing and production team. Their endless efforts to recruit the best for this project, has resulted in the accomplishment of this book. They are a veteran in the field of academics and their pool of knowledge is as vast as their experience in printing. Their expertise and guidance has proved useful at every step. Their uncompromising quality standards have made this book an exceptional effort. Their encouragement from time to time has been an inspiration for everyone.

The publisher and the editorial board hope that this book will prove to be a valuable piece of knowledge for researchers, students, practitioners and scholars across the globe.

List of Contributors

Ewa Sucharzewska
Department of Mycology, University Warmia and Mazury in Olsztyn, Oczapowskiego 1a, 10-719 Olsztyn-Kortowo, Poland

Maria Dynowska
Department of Mycology, University Warmia and Mazury in Olsztyn, Oczapowskiego 1a, 10-719 Olsztyn-Kortowo, Poland

Dariusz Kubiak
Department of Mycology, University Warmia and Mazury in Olsztyn, Oczapowskiego 1a, 10-719 Olsztyn-Kortowo, Poland

Elżbieta Ejdys
Department of Mycology, University Warmia and Mazury in Olsztyn, Oczapowskiego 1a, 10-719 Olsztyn-Kortowo, Poland

Anna Biedunkiewicz
Department of Mycology, University Warmia and Mazury in Olsztyn, Oczapowskiego 1a, 10-719 Olsztyn-Kortowo, Poland

Yunus Dogan
Buca Faculty of Education, Dokuz Eylul University, 35150 Buca-Izmir, Turkey

Ingvar Svanberg
Uppsala Centre for Russian and Eurasian Studies, Uppsala University, Box 514, 751 20 Uppsala, Sweden

Sigurður Ægisson
Hvanneyrarbraut 45, 580 Siglufjörður, Iceland

Raivo Kalle
Institute of Veterinary Medicine and Animal Sciences, Estonian University of Life Sciences, Kreutzwaldi 62, 51014 Tartu, Estonia

Renata Sõukand
Estonian Literary Museum, Vanemuise 42, 51003 Tartu, Estonia

Danuta Wiss
Institute of Experimental Biology, Wrocław University, Kanonia 6/8, 50-328 Wrocław, Poland

Beata Zagórska-Marek
Institute of Experimental Biology, Wrocław University, Kanonia 6/8, 50-328 Wrocław, Poland

Yujing Liu
College of Life and Environmental Sciences, Minzu University of China, Beijing 100081, China

Wujisguleng Wujisguleng
College of Life and Environmental Sciences, Minzu University of China, Beijing 100081, China

Mongolian Medicine College, Inner Mongolia Medical University, Hohhot 010110, China

Chunlin Long
College of Life and Environmental Sciences, Minzu University of China, Beijing 100081, China

Ingvar Svanberg
Uppsala Centre for Russian and Eurasian Studies, Uppsala University, Box 514, 751 20 Uppsala, Sweden

Andrea Dénes
Natural History Department, Janus Pannonius Museum, Box 158, 7601 Pécs, Hungary

Nóra Papp
Department of Pharmacognosy, University of Pécs, Rókus 2, 7624 Pécs, Hungary

Dániel Babai
Centre for Humanities, Hungarian Academy of Sciences, Országház 30, 1014 Budapest, Hungary

Bálint Czúcz
Centre for Ecology, Hungarian Academy of Sciences, Alkotmány 2-4, 2163 Vácrátót, Hungary

Zsolt Molnár
Centre for Ecology, Hungarian Academy of Sciences, Alkotmány 2-4, 2163 Vácrátót, Hungary

Yongxiang Kang
College of Forestry, Northwest A&F University, Yangling 712100, China

Łukasz Łuczaj
Department of Botany and Biotechnology of Economic Plants, University of Rzeszów, Werynia 502, 36-100 Kolbuszowa, Poland

Sebastian Ye
Skargi 11/70, 39-300 Mielec, Poland

Shijiao Zhang
College of Forestry, Northwest A&F University, Yangling 712100, China

Jin Kang
College of Forestry, Northwest A&F University, Yangling 712100, China

Will C. McClatchey
Botanical Research Institute of Texas, 1700 University Drive, Fort Worth, Texas 76107, USA
Department of Botany, University of Hawai'i at Manoa, Honolulu, Hawai'i 96822, USA
Environmental Studies Program, Texas Christian University, Fort Worth, Texas 76107, USA

Marta Kolanowska
Department of Plant Taxonomy and Nature Conservation, University of Gdańsk, Wita Stwosza 59, 80-308 Gdańsk-Oliwa, Poland

Dariusz L. Szlachetko
Department of Plant Taxonomy and Nature Conservation, University of Gdańsk, Wita Stwosza 59, 80-308 Gdańsk-Oliwa, Poland

Emilia Ossowska
Department of Plant Taxonomy and Nature Conservation, University of Gdańsk, Wita Stwosza 59, 80-308 Gdańsk, Poland

Rafał Szymczyk
EKOPROJEKT Environmental Survey Laboratory, Nowica 24, 14-405 Wilczęta, Poland

Adam Bohdan
Society Workshop for All Beings, Świętojańska 22/1, 15-082 Białystok, Poland

Martin Kukwa
Department of Plant Taxonomy and Nature Conservation, University of Gdańsk, Wita Stwosza 59, 80-308 Gdańsk, Poland

Esther Katz
Institut de Recherche pour le Développement, UMR 208 IRD/MNHN, Cuvier 57, 75005 Paris, France

Claudia Leonor López
Museu Paraense Emílio Goeldi, Belém, Avenida Perimetral 1901, Box 399, 66077-530, Belém, Pará, Brazil

Marie Fleury
Muséum National d'Histoire Naturelle, UMR 208 IRD/ MNHN, Cayenne, French Guiana

Robert P. Miller
United Nations Development Program, Project GATI, Brasilia, DF, Brazil

Valeria Payê
Institute of Social Sciences, University of Brasilia, Campus Darcy Ribeiro, Asa Norte, 70910-900 Brasilia, DF, Brazil

Terezinha Dias
Empresa Brasileira de Pesquisa Agropecuária, Recursos genéticos e biotecnologia, Parque Estação Biológica - W 5 Norte Final Asa Norte, Box 02372, 70700-900 Brasília, DF, Brazil

Franklin Silva
Center of Sustainable Development, University of Brasilia, Campus Darcy Ribeiro, Asa Norte, 70904-970 Brasilia, DF, Brazil

Zelandes Oliveira
Center of Sustainable Development, University of Brasilia, Campus Darcy Ribeiro, Asa Norte, 70904-970 Brasilia, DF, Brazil

Elaine Moreira
Department of Anthropology, Federal University of Roraima, Avenida Capitao Ene Garcez, 64000-000 Boa Vista, Roraima, Brazil

Katarzyna Szczepańska
Department of Botany and Plant Ecology, Wrocław University of Environmental and Life Sciences, pl. Grunwaldzki 24a, 50-363 Wrocław, Poland

Maria Kossowska
Institute of Environmental Biology, University of Wrocław, Kanonia 6/8, 50-328 Wrocław, Poland

Łukasz Łuczaj
Department of Botany and Biotechnology of Economic Plants, University of Rzeszów, Werynia 502, 36-100 Kolbuszowa, Poland

Eva Holá
Department of Botany, University of South Bohemia, Branišovská 31, 370 05 České Budějovice, Czech Republic

Nature Conservation Agency of the Czech Republic, Kaplanova 1931/1, 148 00 Prague 11, Czech Republic

Jan Vrba
Nature Conservation Agency of the Czech Republic, Kaplanova 1931/1, 148 00 Prague 11, Czech Republic

Renata Linhartová
Nature Conservation Agency of the Czech Republic, Kaplanova 1931/1, 148 00 Prague 11, Czech Republic

Eva Novozámská
Prague Botanic Garden, Nádvorní 134, 171 00 Prague 7, Czech Republic

Magda Zmrhalová
Regional Museum in Šumperk, Hlavní tř. 22, 787 31 Šumperk, Czech Republic

Vítězslav Plášek
Department of Biology and Ecology, University of Ostrava, Chittussiho 10, 710 00 Ostrava, Czech Republic

Jan Kučera
Department of Botany, University of South Bohemia, Branišovská 31, 370 05 České Budějovice, Czech Republic

Joanna Świerczyńska
Department of Plant Cytology and Embryology, University of Gdańsk, Wita Stwosza 59, 80-308 Gdańsk, Poland

Jerzy Bohdanowicz
Department of Plant Cytology and Embryology, University of Gdańsk, Wita Stwosza 59, 80-308 Gdańsk, Poland

Sylwia Swigonska
Department of Biochemistry, Faculty of Biology and Biotechnology, University of Warmia and Mazury in Olsztyn, M. Oczapowskiego 1A, 10-957 Olsztyn-Kortowo, Poland

Ryszard Amarowicz
Division of Food Science, Institute of Animal Reproduction and Food Research of Polish Academy of Sciences, Tuwima 10, Box 55, 10-718 Olsztyn 5, Poland

Angelika Król
Department of Biochemistry, Faculty of Biology and Biotechnology, University of Warmia and Mazury in Olsztyn, M. Oczapowskiego 1A, 10-957 Olsztyn-Kortowo, Poland

Agnieszka Mostek
Department of Biochemistry, Faculty of Biology and Biotechnology, University of Warmia and Mazury in Olsztyn, M. Oczapowskiego 1A, 10-957 Olsztyn-Kortowo, Poland

Anna Badowiec
Department of Biochemistry, Faculty of Biology and Biotechnology, University of Warmia and Mazury in Olsztyn, M. Oczapowskiego 1A, 10-957 Olsztyn-Kortowo, Poland

Stanisław Weidner
Department of Biochemistry, Faculty of Biology and Biotechnology, University of Warmia and Mazury in Olsztyn, M. Oczapowskiego 1A, 10-957 Olsztyn-Kortowo, Poland

Natalia Urbańska
Department of Pharmaceutical Biology and Medicinal Plants Biotechnology, Medical University of Warsaw, Banacha 1, 02-097 Warsaw, Poland

Joanna Giebułtowicz
Department of Bioanalysis and Drug Analysis, Medical University of Warsaw, Banacha 1, 02-097 Warsaw, Poland

Olga Olszowska
Department of Pharmaceutical Biology and Medicinal Plants Biotechnology, Medical University of Warsaw, Banacha 1, 02-097 Warsaw, Poland

Wojciech J. Szypuła
Department of Pharmaceutical Biology and Medicinal Plants Biotechnology, Medical University of Warsaw, Banacha 1, 02-097 Warsaw, Poland

Bożenna Czarnecka
Department of Ecology, Faculty of Biology and Biotechnology, Maria Curie-Skłodowska University, Akademicka 19, 20-033 Lublin, Poland

Anna Rysiak
Department of Ecology, Faculty of Biology and Biotechnology, Maria Curie-Skłodowska University, Akademicka 19, 20-033 Lublin, Poland

Łukasz Chabudziński
Geoinformation Laboratory, Faculty of Earth Sciences and Spatial Management, Maria Curie-Skłodowska University, Kraśnicka 2c,d, 20-718 Lublin, Poland

Krzysztof Spalik
Department of Plants Systematics and Geography, Faculty of Biology, University of Warsaw, Aleje Ujazdowskie 4, 00-478 Warsaw, Poland

Rosália Rubel
Bioprocess Engineering and Biotechnology Division, Federal University of Paraná (UFPR), Jardim das Américas, CEP 81531-990, Curitiba, Paraná (PR), Brazil

Leandro Freire dos Santos
Bioprocess Engineering and Biotechnology Division, Federal University of Paraná (UFPR), Jardim das Américas, CEP 81531-990, Curitiba, Paraná (PR), Brazil

Herta Stutz Dalla Santa
Bioprocess Engineering and Biotechnology Division, Federal University of Paraná (UFPR), Jardim das Américas, CEP 81531-990, Curitiba, Paraná (PR), Brazil

Luciana Porto de Souza Vandenberghe
Bioprocess Engineering and Biotechnology Division, Federal University of Paraná (UFPR), Jardim das Américas, CEP 81531-990, Curitiba, Paraná (PR), Brazil

Adenise Lorenci Woiciechowski
Bioprocess Engineering and Biotechnology Division, Federal University of Paraná (UFPR), Jardim das Américas, CEP 81531-990, Curitiba, Paraná (PR), Brazil

Júlio Cesar de Carvalho
Bioprocess Engineering and Biotechnology Division, Federal University of Paraná (UFPR), Jardim das Américas, CEP 81531-990, Curitiba, Paraná (PR), Brazil

Sascha Habu
Bioprocess Engineering and Biotechnology Division, Federal University of Paraná (UFPR), Jardim das Américas, CEP 81531-990, Curitiba, Paraná (PR), Brazil

Francisco Menino Destéfanis Vítola
Bioprocess Engineering and Biotechnology Division, Federal University of Paraná (UFPR), Jardim das Américas, CEP 81531-990, Curitiba, Paraná (PR), Brazil

Christopher Augur
Institute of Research for Development, Unite Biotrans - IMEP, Boite 441, Paul Cézanne University, Av. Escadrille Normandie-Niemen, F-13391 Marseille Cedex 20, France

Carlos Ricardo Soccol
Bioprocess Engineering and Biotechnology Division, Federal University of Paraná (UFPR), Jardim das Américas, CEP 81531-990, Curitiba, Paraná (PR), Brazil

Joanna Zalewska-Gałosz
Institute of Botany, Jagiellonian University, Kopernika 31, 31-501 Cracow, Poland

Dagmara Kwolek
Institute of Botany, Jagiellonian University, Kopernika 31, 31-501 Cracow, Poland

Katija Dolina
Institute for Marine and Coastal Research, Botanical Garden on the island Lokrum, University of Dubrovnik, Kneza Damjana Jude 12, P.O. Box 83, 20000 Dubrovnik, Croatia

Łukasz Łuczaj
Department of Botany, Institute of Applied Biotechnology and Basic Sciences, University of Rzeszów, Werynia 502, 36-100 Kolbuszowa, Poland

Renata Piwowarczyk
Department of Botany, Institute of Biology, Jan Kochanowski University, Świętokrzyska 15, 25-406 Kielce, Poland

Łukasz Krajewski
Department of Nature Protection and Rural Landscape, Institute of Technology and Life Sciences, Hrabska 3, Falenty, 05-090 Raszyn, Poland

Yuxia Yang
College of Horticulture, Nanjing Agricultural University, 1# Weigang, Nanjing, Jiangsu Province, 210095, P.R. China

Jingjing Wang
College of Horticulture, Nanjing Agricultural University, 1# Weigang, Nanjing, Jiangsu Province, 210095, P.R. China

Dongxin Department, Jiansu Provincial Agricultural Reclamation and Development Corporation, 26# Dongfang Middle Road, Lianyungang, Jiangsu Province, 222248, P.R. China

Zhihu Ma
Zhenjiang Institute of Agricultural Sciences in Hilly Area of Jiangsu Province, 112# Ninghang Road, Jurong, Jiangsu, 212400, P.R. China

Guosheng Sun
Zhenjiang Institute of Agricultural Sciences in Hilly Area of Jiangsu Province, 112# Ninghang Road, Jurong, Jiangsu, 212400, P.R. China

Changwei Zhang
College of Horticulture, Nanjing Agricultural University, 1# Weigang, Nanjing, Jiangsu Province, 210095, P.R. China

Dejana Džigurski
Faculty of Agriculture, University of Novi Sad, Trg. D. Obradovića 8, Novi Sad, Serbia

Branka Ljevnaić-Mašić
Faculty of Agriculture, University of Novi Sad, Trg. D. Obradovića 8, Novi Sad, Serbia

Ljiljana Nikolić
Faculty of Agriculture, University of Novi Sad, Trg. D. Obradovića 8, Novi Sad, Serbia